ALGEBRAIC ANALYSIS, GEOMETRY, AND NUMBER THEORY

ALGEBRAIC ANALYSIS, GEOMETRY, AND NUMBER THEORY

Proceedings of the JAMI Inaugural Conference

Edited by Jun-Ichi Igusa
Supplement to the
American Journal of Mathematics

The Johns Hopkins University Press
Baltimore and London

The Johns Hopkins University Press, 701 West 40th Street, Baltimore, Maryland 21211
The Johns Hopkins University Press, Ltd., London

The paper used in the publication meets the minimum requirements of American National Standard for Information Sciences—Permanence of Paper for Printed Library Materials, ANSI Z39.48-1984.

Library of Congress Cataloging-in-Publication Data

JAMI Inaugural Conference (1988: Johns Hopkins University)
Algebraic analysis, geometry, and number theory: proceedings of the JAMI Inaugural Conference / edited by Jun-ichi Igusa.
p. cm.
"Supplement to the American journal of mathematics."
ISBN 0-8018-3841-X (alk. paper)
1. Mathematical analysis—Congresses. 2. Algebra—Congresses. 3. Geometry—Congresses. 4. Numbers, Theory of—Congresses. I. Igusa, Jun-ichi, 1924- II. Japan-U.S. Mathematics Institute. III. American journal of mathematics. Supplement. IV. Title.
QA300.J36 1988
515—dc20 89-8103
CIP

TABLE OF CONTENTS

REMARKS BY DR. STEVEN MULLER
PRESIDENT OF THE JOHNS HOPKINS UNIVERSITY
AT THE INAUGURAL CONFERENCE FOR THE JAPAN-U.S. MATHEMATICS INSTITUTE
MAY 16, 1988

Good morning, ladies and gentlemen. It is my great pleasure to welcome you to The Johns Hopkins University and to the Inaugural Conference of the Japan-U.S. Mathematics Institute. The Institute represents both the long-standing ties between Japan and Johns Hopkins and the international world in which we are now living.

As you may know, The Johns Hopkins University has had an international character since its founding in 1876. The University owes its very existence to the German statesman and educator, Wilhelm von Humboldt, who was the founding spirit of the modern German university. Modeled explicitly on von Humboldt's precepts, Johns Hopkins was America's first true modern research university, granting the doctorate, committed to freedom of teaching and research, and dedicated to the unity of research and teaching.

It was this emphasis on the freedom of inquiry that attracted an extraordinary graduate student from Japan, who studied at Johns Hopkins from 1884 to 1887 in the same group of students as Woodrow Wilson, who was earning his Hopkins doctorate at that time. This student was Dr. Inazo Nitobe, who wrote his well-known book on Japanese-American relations during his Johns Hopkins years. Dr. Nitobe, that remarkable intellectual, educator and statesman, who was the leader of the Liberal Movement in Japan in the early years of this century, made it his life's aim to be a bridge across the Pacific. In 1984, on the centennial of Dr. Nitobe's matriculation at Johns Hopkins, it was my special pleasure to travel to Morioka in Iwate Prefecture and to plant two Maryland dogwood trees at Dr. Nitobe's birthplace. That year, the Japanese government also recognized Dr. Nitobe's extraordinary achievements and placed his portrait on the newly issued 5000 yen note.

I referred earlier to Woodrow Wilson whose life's work was "in the nation's service." His words must now be modified to embrace the world. The Johns Hopkins University has always had strong ties to Europe, and we are increasingly reaching out to the whole world but especially to East Asia and to China and Japan. In 1986, we dedicated The Johns Hopkins

University-Nanjing University Center for Chinese and American Studies in the People's Republic of China.

Since Dr. Nitobe's time at Hopkins, the University has developed many links to Japan. The School of Hygiene and Public Health and the School of Medicine have trained numerous Japanese practitioners and researchers. The Edwin O. Reischauer Center at the School of Advanced International Studies was established in 1984 as the primary graduate institution in Washington for training future leaders to work in U.S.-East Asian affairs. The SAIS Japan Forum sponsors research, publications and speakers on economic, political and security issues affecting our two nations. Last fall, the Reischauer Center was graced by a visit from Crown Prince Akihito who planted a Japanese cherry tree in front of the Rome Building where the Center is located. It is no surprise then that former Prime Minister Yasuhiro Nakasone returned to SAIS last week to deliver a major address on U.S.-Japanese economic relations. Nor is it any surprise that Sohei Nakayama, the Founding Chairman of the Board of the International University of Japan in Niigata, will visit Homewood next week to receive an honorary degree at the University's Commencement Exercises.

There is one other link between Hopkins and Japan—lacrosse. For the past two years, members of the University's championship lacrosse team have exchanged visits with athletes from Japanese universities where they are learning to play lacrosse.

Johns Hopkins thus has always been an international university, attracting scholars, students—and athletes—from around the world. Today, however, a profoundly interdependent global society requires us to do yet more. Again and again we must reach out to learn in and from other lands in new ways, both sending and receiving knowledge and the scholars who bear it. The bridge of Nitobe's vision must carry even greater traffic, and in both directions.

The Japan-U.S. Mathematics Institute now becomes one of the vehicles that travels Nitobe's bridge. The Institute fosters improved relationships between Japan and the United States as well as scientific research in the mathematical sciences. Under the direction of Professor Jun-Ichi Igusa, the Institute provides opportunity for scholars from our two countries to work together and to benefit from the stimulation of scholarly exchange as they seek to solve the mysteries of mathematics.

The Institute is blessed by the interest and support of Professor Hironaka who will present the opening lecture of the conference this morning. We are also grateful to the Hironaka Foundation and the National

Science Foundation for their generous support of the Conference. These meetings represent an important first step for an Institute which holds great promise for Mathematical Sciences in general and for the exchange of knowledge between our two countries in particular. I am pleased to participate in the inauguration of the Japan–U.S. Mathematics Institute, and I hope that you enjoy your stay at Johns Hopkins.

PREFACE

During the week of May 16–19, 1988 an Inaugural Conference for the Japan-U.S. Mathematics Institute, abbreviated as JAMI, was held at the Johns Hopkins University. The idea of JAMI was conceived in 1986 by Professors J. Morava and W. S. Wilson. Through the devoted efforts of Professor J.-P. Meyer, the chairman of the Department, and with the support of the President of the University, Dr. S. Muller, and Dean Armstrong, it has taken a concrete form. The organizing committee of the conference consisted of the above-mentioned professors, Professor T. Ono and myself. The conference was supported by generous grants from the National Science Foundation and the Japan Association for Mathematical Sciences.

The conference was opened on the morning of May 16 with a welcoming address by Dr. Muller and the various sessions were presided by Professors S. S. Abhyankar, A. Borel, G. D. Mostow and J. Tate. The seventeen invited addresses in algebraic analysis, algebraic geometry and number theory were delivered by Professors J. Arthur, G. Faltings, D. Goldfeld, H. Hida, H. Hironaka, Y. Ihara, K. Kato, M. Kashiwara, G. Kempf, R. McPherson, J. Morava, S. Mori, T. Oshima, M. Sato, J. Shalika, G. Shimura and S. Zucker. The conference was attended by about one hundred forty mathematicians. We would like to express our sincere appreciation to, above all, the speakers and to all the participants who made the Inaugural Conference a memorable and important occasion.

We have received the text of President Muller's address and the manuscripts of all invited speakers. The publication of these Proceedings has been made with the cooperation of the Johns Hopkins Press, its director, Mr. J. Goellner and Professor J. H. Sampson, another Hopkins editor of the *American Journal of Mathematics.*

Jun-Ichi Igusa
Director of JAMI

ALGEBRAIC ANALYSIS, GEOMETRY, AND NUMBER THEORY

TOWARDS A LOCAL TRACE FORMULA

By James Arthur*

---

1. Suppose that G is a connected, reductive algebraic group over a local field F. We assume that F is of characteristic 0. We can form the Hilbert space $L^2(G(F))$ of functions on $G(F)$ which are square integrable with respect to the Haar measure. The regular representation

$$(R(y_1, y_2)\phi)(x) = \phi(y_1^{-1}xy_2), \qquad \phi \in L^2(G(F)), \qquad x, y_1, y_2 \in G(F),$$

is then a unitary representation of $G(F) \times G(F)$ on $L^2(G(F))$. Kazhdan has suggested that there should be a local trace formula attached to R which is analogous to the global trace formula for automorphic forms. The purpose of this note is to discuss how one might go about proving such an identity, and to describe the ultimate form the identity is likely to take.

To see the analogy with automorphic forms more clearly, consider the diagonal embedding of F into the ring

$$A_F = F \oplus F.$$

The group $G(A_F)$ of A_F-valued points in G is just $G(F) \times G(F)$. The group $G(F)$ embeds into $G(A_F)$ as the diagonal subgroup. Observe that we can map $L^2(G(F))$ isomorphically onto $L^2(G(F)\backslash G(A_F))$ by sending any $\phi \in L^2(G(F))$ to the function

$$(g_1, g_2) \to \phi(g_1^{-1}g_2), \qquad (g_1, g_2) \in G(F)\backslash G(A_F).$$

In this way, the representation R becomes equivalent to the regular representation of $G(A_F)$ on $L^2(G(F)\backslash G(A_F))$.

As is well known, R may be interpreted as a representation of the convolution algebra

---

Manuscript received Nov. 30, 1988.

*Supported in part by NSERC Operating Grant A3483.

$$C_c^\infty(G(A_F)) = C_c^\infty(G(F) \times G(F)).$$

Consider a function in this algebra of the form

$$f(y_1, y_2) = f_1(y_1)f_2(y_2), \qquad y_1, y_2 \in G(F),$$

for functions f_1 and f_2 in $C_c^\infty(G(F))$. Then $R(f)$ is an operator on $L^2(G(F))$ which maps a function ϕ to the function

$$\begin{aligned}(R(f)\phi)(x) &= \int_{G(A_F)} f(g)(R(g)\phi)(x)dg \\ &= \int_{G(F)}\int_{G(F)} f_1(u)f_2(y)\phi(u^{-1}xy)dudy \\ &= \int_{G(F)}\left(\int_{G(F)} f_1(xu)f_2(uy)du\right)\phi(y)dy.\end{aligned}$$

Thus, $R(f)$ is an integral operator with kernel

$$(1.1) \qquad K(x, y) = \int_{G(F)} f_1(xu)f_2(uy)du, \qquad x, y \in G(F).$$

In the 1970's, Harish-Chandra studied the values of this kernel on the diagonal. He introduced a certain truncation of the resulting function, which he used in the case of p-adic F to show that the restriction of $R(f)$ to the space of cusp forms has finite rank [3(b)]. Harish-Chandra's truncation remains somewhat mysterious, and it is not clear what role it might play in the local trace formula.

Implicit in Kazhdan's suggestion is that one should play off (1.1) with the formula for $K(x, y)$ given by Eisenstein integrals. The identity obtained by equating these two formulas for $K(x, x)$ could then be taken as the starting point. The function $K(x, x)$ will not be integrable unless $G(F)$ is compact. However, one can always multiply each expression for $K(x, x)$ by the characteristic function of a large compact set. This is the most naive form of truncation, but, perhaps surprisingly, it appears that it will be possible to compute the resulting integrals. One obtains the weighted orbital integrals and weighted characters which are the local terms in the global trace formula [1(h)]. It has always seemed strange that these local objects should have occurred only in a global context. It now appears that they do have a genuine local interpretation, as objects which arise naturally from a problem in local harmonic analysis.

2. As an example, consider the case that G is anisotropic over F. Then $G(F)$ is a compact group. The function $K(x, x)$ is of course smooth, and is therefore integrable. Changing variables, and applying the Weyl integration formula, we obtain

$$\int_{G(F)} K(x,x)dx = \int_{G(F)} \int_{G(F)} f_1(xu)f_2(ux)dudx$$

$$= \int_{G(F)} \int_{G(F)} f_1(u)f_2(x^{-1}ux)dudx$$

$$= \int_{G(F)} \left(\int_{(G(F))} \int_{G(F)} |D(\gamma)| f_1(x_1^{-1}\gamma x_1) f_2(x^{-1}x_1^{-1}\gamma x_1 x) dx_1 d\gamma \right) dx,$$

where $(G(F))$ stands for the set of regular conjugacy classes in $G(F)$, $d\gamma$ is the measure on $(G(F))$ induced from appropriate Haar measures on maximal tori of $G(F)$, and $D(\gamma)$ is the Weyl discriminant. Thus

$$D(\gamma) = det(1 - Ad(\gamma))_{\mathfrak{g}/\mathfrak{g}_\gamma},$$

where $\mathfrak{g}$ is the Lie algebra of G, and $\mathfrak{g}_\gamma$ is the Lie algebra of the centralizer of γ in G. Since $G(F)$ is compact, we can take the integral over x inside the integrals over x_1 and γ. Changing variables from x to $x_2 = x_1 x$, we obtain

$$\int_{G(F)} K(x,x)dx = \int_{(G(F))} J_G(\gamma, f)d\gamma,$$

where

$$J_G(\gamma, f) = |D(\gamma)| \int_{G(F)} f_1(x_1^{-1}\gamma x_1)dx_1 \int_{G(F)} f_2(x_2^{-1}\gamma x_2)dx_2.$$

We have expressed the integral of $K(x, x)$ in terms of products of orbital integrals.

On the other hand, the integral of $K(x, x)$ equals the trace of $R(f)$. This operator decomposes into a direct sum over the irreducible constituents of R. As motivation for the noncompact case, let us describe the decomposition formally in terms of the function $K(x, x)$.

If (σ, V_σ) is an irreducible representation of $G(F)$, let $HS(V_\sigma)$ be the Hilbert space of Hilbert-Schmidt operators on V_σ. (This of course is just a finite dimensional matrix algebra in the present case, since $G(F)$ is compact.) The L^2-direct sum

$$\bigoplus_\sigma HS(V_\sigma),$$

taken over the set of equivalence classes of σ, and relative to the inner product

$$\Big(\bigoplus_\sigma S_\sigma, \bigoplus_\sigma S'_\sigma\Big) = \sum_\sigma (S_\sigma, S'_\sigma)\deg(\sigma) = \sum_\sigma \operatorname{tr}(S_\sigma(S'_\sigma)^*)\deg(\sigma),$$

is a Hilbert space which supports a unitary representation

$$\bigoplus_\sigma S_\sigma \to \bigoplus_\sigma(\sigma(y_2)S_\sigma\sigma(y_1)^{-1}), \qquad y_1, y_2 \in G(F),$$

of $G(F) \times G(F)$. The map

$$\bigoplus_\sigma S_\sigma \to \sum_\sigma \operatorname{tr}(\sigma(x)S_\sigma)\deg(\sigma), \qquad x \in G(F), \tag{2.1}$$

is then an isometric isomorphism from $\oplus_\sigma HS(V_\sigma)$ onto $L^2(G(F))$ which intertwines the two representations of $G(F) \times G(F)$. For each σ, let $\mathcal{B}_\sigma$ be an orthonormal basis of $HS(V_\sigma)$. By pulling back the operator $R(f)$ to $\oplus_\sigma$ HS(V_σ), we obtain a second formula

$$\sum_\sigma \sum_{S\in\mathcal{B}_\sigma} \operatorname{tr}(\sigma(x)\sigma(f_2)S\sigma(\tilde{f}_1))\overline{\operatorname{tr}(\sigma(y)S)}\deg(\sigma)$$

for the kernel $K(x, y)$. We are writing $\tilde{f}_1$ here for the function $x_1 \to f_1(x_1^{-1})$. In particular, $\operatorname{tr}(\sigma(\tilde{f}_1))$ equals $\operatorname{tr}(\tilde{\sigma}(f_1))$, where $\tilde{\sigma}$ stands for the contragradient of σ. Since f is smooth, the terms in the series are rapidly decreasing. We obtain

$$\int_{G(F)} K(x,x)dx = \sum_{\sigma} \sum_{S \in \mathcal{B}_\sigma} \int_{G(F)} \mathrm{tr}(\sigma(x)\sigma(f_2)S\sigma(\tilde{f}_1))\overline{\mathrm{tr}(\sigma(x)S)}dx \deg(\sigma)$$

$$= \sum_{\sigma} \sum_{S \in \mathcal{B}_\sigma} \mathrm{tr}(\sigma(f_2)S\sigma(\tilde{f}_1)S^*)$$

$$= \sum_{\sigma} \mathrm{tr}(\sigma(\tilde{f}_1))\mathrm{tr}(\sigma(f_2))$$

$$= \sum_{\pi=(\tilde{\sigma},\sigma)} J_G(\pi, f),$$

where

$$J_G(\pi, f) = \mathrm{tr}(\tilde{\sigma}(f_1))\mathrm{tr}(\sigma(f_2)).$$

We have expressed the integral of $K(x, x)$ in a second way, in terms of products of characters.

The local trace formula for compact groups is then the identity

$$\int_{(G(F))} J_G(\gamma, f)d\gamma = \sum_{\pi=(\tilde{\sigma},\sigma)} J_G(\pi, f), \tag{2.2}$$

which is thus a simple consequence of the Weyl integration formula and the Peter-Weyl theorem.

3. Now, suppose that G is a general reductive group. Again there are two parallel formulas for $K(x, x)$. One is a geometric expansion in terms of regular semisimple conjugacy classes of $G(F)$. The other is a spectral expansion in terms of irreducible tempered representations of $G(F)$. We shall describe each of these in turn.

The geometric expansion is again a consequence of the Weyl integration formula. Rather than writing this simply in terms of conjugacy classes in $G(F)$, we prefer to keep track of the elliptic conjugacy classes in Levi subgroups of G. Let M_0 be a fixed minimal Levi subgroup of G over F, and let $\mathcal{L}$ stand for the (finite) set of Levi subgroups of G which contain M_0. For any $M \in \mathcal{L}$, we have the split component A_M of the center of M, and the (restricted) Weyl group W_0^M of (M, A_{M_0}). We shall write $M(F)_{ell}$ for the set of elements in $M(F)$ whose centralizer in G is a torus which is F-anisotropic modulo A_M. Then $(M(F)_{ell})$ will denote the set of $M(F)$-conjugacy classes in $M(F)_{ell}$. Any strongly G-regular element in $G(F)$ is $G(F)$-conjugate to a class in one of the sets $(M(F)_{ell})$. This class is unique modulo conjugation by the Weyl group W_0^G of G. Applying the Weyl integration formula to the right hand side of the identity

$$K(x, x) = \int_{G(F)} f_1(u) f_2(x^{-1}ux)\,du$$

obtained from (1.1), we arrive at the expression

$$(3.1)\quad \sum_{M \in \mathcal{L}} |W_0^M|\,|W_0^G|^{-1} \int_{(M(F)_{ell})} |D(\gamma)| \cdot \int_{A_M(F)\backslash G(F)} f_1(x_1^{-1}\gamma x_1) f_2(x^{-1}x_1^{-1}\gamma x_1 x)\,dx_1\,d\gamma$$

for $K(x, x)$. As before, $D(\gamma)$ is the Weyl discriminant, and $d\gamma$ stands for the measure on $(M(F)_{ell})$ induced from Haar measures on maximal tori of $M(F)$. This is the geometric expansion of $K(x, x)$.

The spectral expansion is a consequence of Harish-Chandra's Plancherel formula [3(a)], [3(b)]. Suppose that $M \in \mathcal{L}$. Let $M(F)^1$ be the kernel of the usual homomorphism H_M from $M(F)$ to

$$\mathfrak{a}_M = \mathrm{Hom}(X(M)_F, \mathbf{R}).$$

We shall write $\Pi_2(M(F)^1)$ for the set of equivalence classes of irreducible square integrable representations of $M(F)^1$. Observe that a representation $\sigma \in \Pi_2(M(F)^1)$ may be identified with an orbit

$$\sigma_\lambda(m) = \sigma_0(m) e^{\lambda(H_M(m))}, \qquad \lambda \in i\mathfrak{a}_M^*, \qquad m \in M(F),$$

of irreducible representations of $M(F)$ under the action of $i\mathfrak{a}_M^*$. If P is a fixed parabolic subgroup with Levi component M, we can then form the induced representation $\mathcal{I}_P(\sigma_\lambda)$ of $G(F)$. For each such σ, we fix a suitable orthonormal basis $\mathcal{B}_\sigma$ of the space of Hilbert-Schmidt operators acting on the underlying space of $\mathcal{I}_P(\sigma_0)$. The Plancherel formula provides an analogue of the map (2.1). This in turn leads to a second formula

$$(3.2)\quad \sum_{M\in\mathcal{L}} |W_0^M||W_0^G|^{-1} \sum_{\sigma\in\Pi_2(M(F)^1)} \sum_{S\in\mathcal{B}_\sigma} \int_{i\mathfrak{a}_M^*/i\mathfrak{a}_M^\vee} \operatorname{tr}(\mathcal{I}_P(\sigma_\lambda, x)\mathcal{I}_P(\sigma_\lambda, f_2) \cdot S\mathcal{I}_P(\sigma_\lambda, \tilde{f}_1)) \cdot \overline{\operatorname{tr}(\mathcal{I}_P(\sigma_\lambda, x)S)}m(\sigma_\lambda)d\lambda$$

for the kernel $K(x, x)$. Here

$$m(\sigma_\lambda) = d_\sigma \mu(\sigma_\lambda)$$

is the Plancherel density, given by the product of the formal degree of σ with Harish-Chandra's μ-function, while

$$\mathfrak{a}_M^\vee = \operatorname{Hom}(H_M(M(F)), 2\pi\mathbf{Z}).$$

(The subgroup $\mathfrak{a}_M^\vee \subseteq \mathfrak{a}_M^*$ is a lattice if F is a p-adic field, and is trivial if F is Archimedean.) This is the spectral expansion of $K(x, x)$.

The question becomes how one might integrate (3.1) and (3.2) over x in $A_G(F)\backslash G(F)$. It can be shown that the terms in (3.1) and (3.2) corresponding to $M = G$ are both integrable. However, none of the other terms turn out to be integrable. This can be seen most clearly in (3.1). For suppose that for some proper M,

$$\int_{A_G(F)\backslash G(F)} \int_{(M(F)_{ell})} \int_{A_M(F)\backslash G(F)} |D(\gamma)| f_1(x_1^{-1}\gamma x_1) f_2(x^{-1}x_1^{-1}\gamma x_1 x)dx_1 d\gamma dx$$

converged (as a triple integral). Taking the integral over x inside the other integrals, and making a change of variables, we would obtain

$$\int_{(M(F)_{ell})} |D(\gamma)| \int_{A_M(F)\backslash G(F)} f_1(x_1^{-1}\gamma x_1)dx_1 \int_{A_G(F)\backslash G(F)} f_2(x_2^{-1}\gamma x_2)dx_2.$$

Since the integrand in x_2 is left-invariant under $A_M(F)$, and $A_G(F)\backslash A_M(F)$ has infinite volume, we reach a contradiction. It will therefore be necessary to truncate (3.1) and (3.2) in some fashion before the integration can be attempted.

4. Let K be a fixed maximal compact subgroup of $G(F)$. We assume that K is in good position relative to M_0, and that K corresponds to a special vertex if F is p-adic. Then

$$G(F) = KA_{M_0}(F)K.$$

Let T be a point in $\mathfrak{a}_{M_0}$ which is highly regular, in the sense that the infimum

$$d(T) = \inf_{\alpha} |\alpha(T)|,$$

taken over the roots α of (G, A_{M_0}), is large. We then define $u(x, T)$ to be the characteristic function of the set of points

$$x = k_1hk_2, \qquad k_1, k_2 \in K, \qquad h \in A_G(F)\backslash A_{M_0}(F),$$

in $A_G(F)\backslash G(F)$ such that $H_{M_0}(h)$ lies in the convex hull

$$H_{cx}(\{sT: s \in W_0^G\}/\mathfrak{a}_G).$$

(For each M, there is a canonical decomposition $\mathfrak{a}_M = \mathfrak{a}_M^G \oplus \mathfrak{a}_G$, so in particular, $\mathfrak{a}_G$ can be regarded as a subspace of $\mathfrak{a}_{M_0}$. We are writing $H_{cx}(S/\mathfrak{a}_G)$ here for the convex hull of the projection onto $\mathfrak{a}_{M_0}/\mathfrak{a}_G$ of a subset S of $\mathfrak{a}_{M_0}$.) The function $u(x, T)$ is $A_G(F)$-invariant. It can be regarded as the characteristic function of a large compact subset of $A_G(F)\backslash G(F)$.

The product of $u(x, T)$ with each of the expressions (3.1) and (3.2) is integrable over $A_G(F)\backslash G(F)$. The problem is to make sense of the integrals. There are several questions here, but we can see that the essential computational step would be as follows. For fixed elements $\gamma \in (M(F)_{ell})$ and $\sigma \in \Pi_2(M(F)^1)$, find asymptotic formulas (as $d(T)$ becomes large) for the integrals

$$\int_{A_G(F)\backslash G(F)} \int_{A_M(F)\backslash G(F)} f_1(x_1^{-1}\gamma x_1)f_2(x^{-1}x_1^{-1}\gamma x_1 x)u(x, T)dx_1dx, \tag{4.1}$$

and

$$\int_{A_G(F)\backslash G(F)} \operatorname{tr}(\mathcal{I}_P(\sigma_\lambda, x)S)\overline{\operatorname{tr}(\mathcal{I}_P(\sigma_{\lambda'}, x)S')}u(x, T)dx. \tag{4.2}$$

In (4.2), S and S' are K-finite Hilbert-Schmidt operators on the space of $\mathcal{I}_P(\sigma_0)$, and λ and λ' are points in $i\mathfrak{a}_M^*$ whose projections onto $i\mathfrak{a}_G^*$ are equal.

5. We shall look at (4.1) and (4.2) separately. The discussion will be slightly simpler if we do not have to deal with lattices in the spaces $\mathfrak{a}_M$. Let us therefore assume until further notice that the field F is Archimedean. The maps H_M are then surjective, and the subgroups $\mathfrak{a}_M^\vee$ are trivial.

Consider first the expression (4.1). The integrals over x and x_1 are both over compact sets, so we may interchange their order. The expression becomes

$$\int_{A_M(F)\backslash G(F)} \int_{A_G(F)\backslash G(F)} f_1(x_1^{-1}\gamma x_1) f_2(x^{-1}x_1^{-1}\gamma x_1 x) u(x, T)\, dx\, dx_1$$

$$= \int_{A_M(F)\backslash G(F)} \int_{A_G(F)\backslash G(F)} f_1(x_1^{-1}\gamma x_1) f_2(x_2^{-1}\gamma x_2) u(x_1^{-1}x_2, T)\, dx_2\, dx_1$$

$$= \int_{A_M(F)\backslash G(F)} \int_{A_M(F)\backslash G(F)} f_1(x_1^{-1}\gamma x_1) f_2(x_2^{-1}\gamma x_2) u(x_1, x_2, T)\, dx_2\, dx_1$$

where

$$u(x_1, x_2, T) = \int_{A_G(F)\backslash A_M(F)} u(x_1^{-1}ax_2, T)\, da.$$

Since the centralizer of γ in $G(F)$ is compact modulo $A_M(F)$, the integrals over x_1 and x_2 in the last expression may be taken over compact sets. In particular, T may be taken to be highly regular in a sense which is uniform in x_1 and x_2.

The next lemma is the main point. Let $\mathcal{P}(M)$ be the (finite) set of parabolic subgroups $P = MN_P$ of G with Levi component M. For any such P, and any point

$$x = nmk, \qquad n \in N_P(F), \qquad m \in M(F), \qquad k \in K,$$

in $G(F)$, we set

$$H_P(x) = H_M(m),$$

as usual. We also write T_P for the projection onto $\mathfrak{a}_M$ of any Weyl translate

$$sT, \qquad s \in W_0^G,$$

such that sT lies in the positive chamber of some minimal parabolic subgroup $P_0 \in \mathcal{P}(M_0)$, with $P_0 \subset P$.

LEMMA 5.1. *Assume that $M \neq G$. Then there is a subset $S(x_1, x_2, T)$ of $A_G(F)\backslash A_M(F)$ with the following properties.*

(i) *$vol(S(x_1, x_2, T)) \leq C(x_1, x_2)e^{-\epsilon d(T)}$,*
for a locally bounded function $C(x_1, x_2)$ and a positive constant ϵ.

(ii) *If a lies in the complement of $S(x_1, x_2, T)$, then $u(x_1^{-1}ax_2, T)$ equals 1 if and only if $H_M(a)$ lies in*

$$H_{cx}(\{T_P + H_P(x_1) - H_{\overline{P}}(x_2) : P \in \mathcal{P}(M)\}/\mathfrak{a}_G). \tag{5.1}$$

This lemma is a generalization of [1(g), Lemma 3]. In the p-adic case treated in [1(g)], $S(x_1, x_2, T)$ is actually empty for $d(T)$ large.

The lemma allows us to relate the weight factor $u(x_1, x_2, T)$ to the volume of a convex hull. Let $\overline{v}_M(x_1, x_2, T)$ be the volume in $\mathfrak{a}_M/\mathfrak{a}_G$ of the set (5.1). Since H_M defines a proper map of $A_M(F)$ onto $\mathfrak{a}_M$, we can choose the Haar measures on these groups so that $u(x_1, x_2, T)$ is asymptotic to $\overline{v}_M(x_1, x_2, T)$. The original expression (4.1) will then equal the integral

$$\int_{A_M(F)\backslash G(F)} \int_{A_M(F)\backslash G(F)} f_1(x_1^{-1}\gamma x_1)f_2(x_2^{-1}\gamma x_2)\overline{v}_M(x_1, x_2, T)dx_2dx_1, \tag{5.2}$$

modulo a function of T which is $O(e^{-\epsilon d(T)})$. The function $\overline{v}_M(x_1, x_2, T)$ is, incidentally, a polynomial in T. Its constant term equals

$$\overline{v}_M(x_1, x_2) = (-1)^{\dim(A_M/A_G)}v_M(x_1, x_2),$$

where

$$v_M(x_1, x_2) = \mathrm{vol}(H_{cx}(\{H_{\overline{P}}(x_1) - H_P(x_2) : P \in \mathcal{P}(M)\}/\underline{\mathfrak{a}}_G)). \tag{5.3}$$

6. Now consider the second expression (4.2). It is convenient to write

$$\mathrm{tr}(\mathcal{I}_P(\sigma_\lambda, k_1xk_2)S) = E(x, \psi_S, \lambda)_{(k_1,k_2)}, \qquad k_1, k_2 \in K,$$

in Harish-Chandra's notation for Eisenstein integrals. (See [3(a), Section 7] or [1(c), Section I.3–I.4].) Here, ψ_S is a $(K \cap M(F))$-spherical function from $M(F)$ to a K-finite space V_K of functions on $K \times K$, and $E(x, \psi_S, \lambda)$ is the Eisenstein integral with values in V_K. Then (4.2) becomes an integral

$$\int_{A_G(F)\backslash G(F)} (E(x, \psi_S, \lambda), E(x, \psi_{S'}, \lambda'))u(x, T)dx \tag{6.1}$$

of inner products in V_K.

In the special case of K-bi-invariant functions on $GL(n, F)$ (and with F a p-adic field), Waldspurger has found an exact formula for (6.1) in [5(a)]. It is valid whenever $d(T)$ is large, and is given in terms of Harish-Chandra's c-functions. Remarkably, it is the exact analogue of Langlands' formula ([4], [1(d)]) for the inner product of truncated cuspidal Eisenstein series. It is likely that Waldspurger's techniques will carry over to the general case of (6.1), yielding an asymptotic formula analogous to the asymptotic formula [1(d)] for arbitrary Eisenstein series. Alternatively, recent ideas of Casselman on inner products of distributions apply to (6.1), and will perhaps lead to a slick proof of the same inner product formula. In any case, the result will be an expression

$$\sum_{P_1} \sum_{s,s' \in W(\mathfrak{a}_M, \mathfrak{a}_{M_1})} (c(s, \lambda)\psi_S, c(s', \lambda')\psi_{S'})e^{(s\lambda - s'\lambda')(T_{P_1})}\theta_{P_1}(s\lambda - s'\lambda')^{-1} \tag{6.2}$$

a reader familiar with Eisenstein series will recognize. The outer sum is over parabolic subgroups $P_1 = M_1 N_{P_1}$ which contain a fixed minimal parabolic subgroup P_0, which is in turn contained in the original group $P = MN_P$. The inner sum is over the set of isomorphisms from $\mathfrak{a}_M$ onto $\mathfrak{a}_{M_1}$ which are the restrictions of elements in W_0^G. The function $c(s, \lambda)$ is of course Harish-Chandra's c-function. Finally,

$$\theta_{P_1}(s\lambda - s'\lambda') = \mathrm{vol}(\mathfrak{a}_{M_1}^G/\mathbf{Z}(\Delta_{P_1}^\vee))^{-1} \prod_{a \in \Delta_{P_1}} (s\lambda - s'\lambda')(\alpha^\vee),$$

where Δ_{P_1} is the set of simple roots of (P_1, A_{M_1}). In the case of Archimedean F that we are considering, the relation between (6.2) and (4.2) will be asymptotic. The difference between the two expressions should be bounded by

$$\rho(\lambda, \lambda')\|S\| \, \|S'\|e^{-\epsilon d(T)},$$

where $\rho(\lambda, \lambda')$ is a locally bounded function on $i\mathfrak{a}_M^* \times i\mathfrak{a}_M^*$, and ϵ is a positive constant.

7. The source of the trace formula is to be the identity between the geometric and spectral expansions (3.1) and (3.2). We are proposing to truncate these expressions simply by multiplying them by the characteristic function $u(x, T)$. The resulting integrals over x in $A_G(F)\backslash G(F)$ will of course be equal. We would like to obtain explicit formulas for the integrals by substituting the expressions (5.2) and (6.2).

On the geometric side there is an immediate question of uniformity. The asymptotic approximation (5.2) of (4.1) is only valid for fixed γ. However, γ is to be integrated over all regular elements in (3.1). I have not investigated whether there is an estimate which will take care of the elements γ in (3.1) which approach the singular set. If this is not possible, we may require a second kind of truncation, the sole purpose of which is to handle such questions of uniformity. This was the case for the global trace formula.

On the spectral side, we must compute the contribution of (6.2) to (3.2). This entails changing (6.2) by replacing S and S' with $\mathcal{I}_P(\sigma_\lambda, f_2)S\mathcal{I}_P(\sigma_\lambda, \tilde{f}_1)$ and S, respectively. We would then take the limit as λ' approached λ, and finally integrate the product of the resulting expression with $m(\sigma_\lambda)$ over $\lambda \in i\mathfrak{a}_M^*$. The combinatorics of this procedure are similar to the case of Eisenstein series [1(f)], and have been carried out by Waldspurger [5(a)], at least in a special case. Again, I do not know whether there will be a serious problem in general concerning the uniformity in λ of the asymptotic approximation (6.2). However, the analogous problem has been solved for Eisenstein series [1(e)], where it is presumably more difficult.

The end result would be an explicit trace formula which we can now describe. On the geometric side will be the distributions

$$(7.1)\quad J_M(\gamma, f)$$

$$= |D(\gamma)| \int_{A_M(F)\backslash G(F)} \int_{A_M(F)\backslash G(F)} f_1(x_1^{-1}\gamma x_1) f_2(x_2^{-1}\gamma x_2) v_M(x_1, x_2)\, dx_1 dx_2,$$

where γ belongs to $M(F)_{ell}$ and $v_M(x_1, x_2)$ is the volume (5.3). The terms on the spectral side require a little more description.

Consider the subgroup

$$G(A_F)^1 = \{(y_1, y_2) \in G(F) \times G(F): H_G(y_1) = H_G(y_2)\}$$

of $G(A_F) = G(F) \times G(F)$. We shall write $\Pi(G)$ for the set of (equivalence classes of) representations of this subgroup obtained by restricting irreducible constituents $\tilde{\pi}_1 \otimes \pi_2$ of the induced representations

$$\mathcal{I}_P(\tilde{\sigma}_{-\lambda} \otimes \sigma_\lambda) = \mathcal{I}_{\bar{P}}(\tilde{\sigma}_{-\lambda}) \otimes \mathcal{I}_P(\sigma_\lambda), \qquad M \in \mathcal{L}, \quad \sigma \in \Pi_2(M(F)^1), \quad \lambda \in i\mathfrak{a}_M^*,$$

of $G(A_F)$. (This notation is motivated by (5.3), which leads us to identify P with the parabolic subgroup $\bar{P} \times P$ of $G \times G$, instead of $P \times P$.) Associated to these induced representations, we have normalized intertwining operators

$$R(w, \tilde{\sigma}_{-\lambda} \otimes \sigma_\lambda) = R(w, \tilde{\sigma}_{-\lambda}) \otimes R(w, \sigma_\lambda), \qquad w \in W(\mathfrak{a}_M, \mathfrak{a}_M),$$

from $\mathcal{I}_P(\tilde{\sigma}_{-\lambda} \otimes \sigma_\lambda)$ to $\mathcal{I}_P(w(\tilde{\sigma}_{-\lambda}) \otimes w(\sigma_\lambda))$. These are independent of how w is represented in the normalizer of $M(F)$. It is really not $\Pi(G)$ that we want, however, but the subset $\Pi_{disc}(G)$ of such representations in which $w(\sigma_\lambda) = \sigma_\lambda$ for some element w in

$$W(\mathfrak{a}_M)_{reg} = \{w \in W(\mathfrak{a}_M, \mathfrak{a}_M): \det(w - 1)_{\mathfrak{a}_M^G} \neq 0\}.$$

Now, in analogy with automorphic forms, we write

$$I_{disc}(f) = I_{disc}^G(f)$$

for the expression obtained by taking the sum over $M \in \mathcal{L}$, $w \in W(\mathfrak{a}_M)_{reg}$, $\sigma \in \Pi_2(M(F)^1)$, and $\lambda \in i\mathfrak{a}_M^*/i\mathfrak{a}_G^*$, of the product of

$$|W_0^M||W_0^G|^{-1}\,|\det(w - 1)_{\mathfrak{a}_M^G}|^{-1}\epsilon_{\sigma_\lambda}(w)$$

with

$$\text{(7.2)} \qquad \int_{i\mathfrak{a}_G^*} \operatorname{tr}(R(w, \tilde{\sigma}_{-(\lambda+\mu)} \otimes \sigma_{\lambda+\mu})\mathcal{I}_P(\tilde{\sigma}_{-(\lambda+\mu)} \otimes \sigma_{\lambda+\mu}, f))d\mu.$$

Here $\epsilon_{\sigma_\lambda}(w)$ is a certain sign character which is peculiar to the local setting, and is defined in terms of the zeros of the Plancherel density. Observe that the factor (7.2) depends only on the restriction f^1 of f to $G(A_F)^1$. This factor actually vanishes unless $w(\sigma_\lambda) = \sigma_\lambda$, a condition which can be satisfied for only finitely many λ. (In the present case of Archimedean F, the condition can be satisfied for at most one λ. We have written the formula as sum over λ so that the p-adic analogue will be more transparent.) We can therefore write

$$I_{disc}(f) = \sum_{\pi \in \Pi_{disc}(G)} a^G_{disc}(\pi) \operatorname{tr} \pi(f^1), \tag{7.3}$$

a linear combination of irreducible characters in $\Pi_{disc}(G)$. The complex numbers

$$a^M_{disc}(\pi), \qquad M \in \mathcal{L}, \qquad \pi \in \Pi_{disc}(M),$$

can be defined in this way for all Levi subgroups, and will appear as coefficients on the spectral side.

Suppose that $M \in \mathcal{L}$. We shall write $\Pi_{temp}(M(A_F)^1)$ for the set of equivalence classes of irreducible tempered representations of $M(A_F)^1$. Each representation $\pi \in \Pi_{temp}(M(A_F)^1)$ can be identified with an orbit

$$\pi_\lambda = \tilde{\pi}_{1,-\lambda} \otimes \pi_{2,\lambda}, \qquad \lambda \in i\mathfrak{a}^*_M,$$

of irreducible representations of $M(A_F) = M(F) \times M(F)$ under the action of $i\mathfrak{a}^*_M$. For any such π, and any $P \in \mathcal{P}(M)$, we can form the induced representations

$$\mathcal{I}_P(\pi_\lambda, f) = \mathcal{I}_{\bar{P}}(\tilde{\pi}_{1,-\lambda}, f_1) \otimes \mathcal{I}_P(\pi_{2,\lambda}, f_2), \qquad \lambda \in i\mathfrak{a}^*_M.$$

We also have the standard unnormalized intertwining operators

$$J_{P'|P}(\pi_\lambda) = J_{\bar{P}'|\bar{P}}(\tilde{\pi}_{1,-\lambda}) \otimes J_{P'|P}(\pi_{2,\lambda}), \qquad P' \in \mathcal{P}(M),$$

from $\mathcal{I}_P(\pi_\lambda)$ to $\mathcal{I}_{P'}(\pi_\lambda)$, each of which can be written as a product of a rational function

$$r_{P'|P}(\pi_\lambda) = r_{\bar{P}'|\bar{P}}(\tilde{\pi}_{1,-\lambda}) r_{P'|P}(\pi_{2,\lambda})$$

with a normalized intertwining operator

$$R_{P'|P}(\pi_\lambda) = R_{\bar{P}'|\bar{P}}(\tilde{\pi}_{1,-\lambda}) \otimes R_{P'|P}(\pi_{2,\lambda})$$

[1(j), Theorem 2.1]. It is easy to show that the limit

$$\mathcal{J}_M(\pi_\lambda, P) = \lim_{\nu\to 0} \sum_{P'\in\mathcal{P}(M)} (J_{P'|P}(\pi_\lambda)^{-1} J_{P'|P}(\pi_{\lambda+\nu}))\theta_{P'}(\nu)^{-1}$$

exists [1(b), Lemma 6.2]. In general, the function

$$J_M(\pi_\lambda, f) = \mathrm{tr}(\mathcal{J}_M(\pi_\lambda, P)\mathcal{I}_P(\pi_\lambda, f))$$

will have singularities in λ. However, it can be shown that if π belongs to $\Pi_{disc}(M)$, then $J_M(\pi_\lambda, f)$ is a Schwartz function $\lambda \in i\mathfrak{a}_M^*$. The distributions

$$j_M(\pi, f) = \int_{i\mathfrak{a}_M^*} J_M(\pi_\lambda, f)d\lambda, \qquad \pi \in \Pi_{disc}(M), \tag{7.4}$$

will be the remaining terms on the spectral side.

The local trace formula will be the identity of distributions

$$\sum_{M\in\mathcal{L}} |W_0^M||W_0^G|^{-1}(-1)^{dim(A_M/A_G)} \int_{(M(F)_{ell})} J_M(\gamma, f)d\gamma \tag{7.5}$$

and

$$\sum_{M\in\mathcal{L}} |W_0^M||W_0^G|^{-1}(-1)^{dim(A_M/A_G)} \sum_{\pi\in\Pi_{disc}(M)} a_{disc}^M(\pi)j_M(\pi, f). \tag{7.6}$$

We have described it for the function $f = f_1 \times f_2$. However, the distributions $J_M(\gamma, f)$ and $j_M(\pi, f)$ make sense for any function f in $C_c^\infty(G(A_F))$, and the identity would hold in this generality. Of course there are still some analytic questions to be answered, so the identity must remain conjectural for the present. (Incidentally, the notation in (7.6) is slightly at odds with that used in connection with automorphic forms. In the papers [1(h)–1(k)] we defined $J_M(\pi_\lambda, f)$ in terms of the normalized intertwining operators $R_{P'|P}(\pi_\lambda)$ instead of the unnormalized operators $J_{P'|P}(\pi_\lambda)$ used here. Moreover, we denoted the corresponding integral (7.4) simply by $J_M(\pi, f)$.)

8. A distribution I on $G(A_F)$ is said to be *invariant* if it is left unchanged under conjugation. That is,

$$I(f^g) = I(f), \qquad f \in C_c^\infty(G(A_F)), \qquad g \in G(A_F),$$

where

$$f^g(g_1) = f(gg_1g^{-1}), \qquad g, g_1 \in G(A_F).$$

The distributions $J_M(\gamma)$ and $j_M(\pi)$, defined by (7.1) and (7.4), are not invariant if $M \neq G$. Following the methods of the global trace formula [1(h)], [1(i)], we shall sketch how the local identity we have just described could be converted into an *invariant* local trace formula.

It is best to restrict our attention to the Hecke algebra $\mathcal{H}(G(A_F))$ of functions in $C_c^\infty(G(A_F))$ which are left and right finite under the maximal compact subgroup $K \times K$. The results of Clozel and Delorme [2] allow one to characterize the topological vector space $\mathcal{I}(G(A_F)^1)$ of functions on $\Pi_{temp}(G(A_F)^1)$ of the form

$$f_G^1(\pi) = \mathrm{tr}(\pi(f^1)), \qquad \pi \in \Pi_{temp}(G(A_F)^1), \qquad f \in \mathcal{H}(G(A_F)).$$

Suppose that θ is a continuous linear map from $\mathcal{H}(G(A_F))$ to a topological vector space $\mathcal{V}$. We can assume that $\theta(f)$ depends only on f^1. Then θ is said to be *supported on characters* if $\theta(f)$ depends only on the function f_G^1. When θ has this property, there is a unique continuous map

$$\hat{\theta}\colon \mathcal{I}(G(A_F)^1) \to \mathcal{V}$$

such that $\hat{\theta}(f_G^1) = \theta(f)$ for all f.

Suppose that $M \in \mathcal{L}$. Given a representation $\pi \in \Pi_{temp}(M(A_F)^1)$, we can form the limit

$$\mathcal{R}_M(\pi_\lambda, P) = \lim_{\nu \to 0} \sum_{P' \in \mathcal{P}(M)} R_{P'|P}(\pi_\lambda)^{-1} R_{P'|P}(\pi_{\lambda+\nu})\theta_{P'}(\nu)^{-1}.$$

Then for any $f \in \mathcal{H}(G(A_F))$, we define $\phi_M(f)$ to be the function on $\Pi_{temp}(M(A_F)^1)$ whose value at π equals

$$\phi_M(f, \pi) = \int_{i\mathfrak{a}_M^*} \mathrm{tr}(\mathcal{R}_M(\pi_\lambda, P)\mathcal{I}_P(\pi_\lambda, f))\,d\lambda.$$

Thus, ϕ_M is a transform from functions on $G(A_F)$ to functions on $\Pi_{temp}(M(A_F)^1)$. Indeed, one can show that ϕ_M is a continuous linear mapping from $\mathcal{H}(G(A_F))$ into $\mathcal{I}(M(A_F)^1)$. (See [1(j), Theorem 12.1].)

Consider first an element $\gamma \in M(F)_{ell}$. It is not hard to show that the variance of ϕ_M under conjugation is the same as that of the distribution $J_M(\gamma, f)$. The extent to which these objects differ can therefore be measured by invariant distributions. For each $\gamma \in M(F)_{ell}$, one can define an invariant distribution

$$I_M(\gamma, f) = I_M^G(\gamma, f), \qquad f \in \mathcal{H}(G(A_F)),$$

which is supported on characters, and which satisfies the inductive formula

$$J_M(\gamma, f) = \sum_{L \in \mathcal{L}(M)} \hat{I}_M^L(\gamma, \phi_L(f)). \tag{8.1}$$

Here $\mathcal{L}(M)$ denotes the set of Levi subgroups which contain M. (See [1(h), Section 2].)

Next suppose that π is a representation in $\Pi_{disc}(M)$. The variance of ϕ_M under conjugation also matches that of $j_M(\pi, f)$. It follows without difficulty that there is an invariant distribution

$$i_M(\pi, f) = i_M^G(\pi, f), \qquad f \in G(A_F),$$

which is supported on characters, and satisfies the inductive formula

$$j_M(\pi, f) = \sum_{L \in \mathcal{L}(M)} \hat{i}_M^L(\pi, \phi_L(f)). \tag{8.2}$$

This distribution can also be defined directly. The limit

$$r_M(\pi_\lambda, P) = \lim_{\nu \to 0} \sum_{P' \in \mathcal{P}(M)} r_{P'|P}(\pi_\lambda)^{-1} r_{P'|P}(\pi_{\lambda+\nu}) \theta_{P'}(\nu)^{-1} \tag{8.3}$$

provides a rational function of λ whose poles do not meet $i\mathfrak{a}_M^*$. It is then an easy consequence of [1(b), Corollary 6.5] that

$$i_M(\pi, f) = \int_{i\mathfrak{a}_M^*} r_M(\pi_\lambda, P) \operatorname{tr}(\mathcal{I}_P(\pi_\lambda, f)) d\lambda. \tag{8.4}$$

The following proposition gives the final invariant local trace formula. As in the global case, it is a formal consequence of the definitions (8.1) and (8.2).

PROPOSITION 8.1. *The identity of the noninvariant expressions* (7.5) *and* (7.6) *implies that the invariant expressions*

$$\sum_{M\in\mathcal{L}} |W_0^M||W_0^G|^{-1}(-1)^{\dim(A_M/A_G)} \int_{(M(F)_{ell})} I_M(\gamma, f)d\gamma, \tag{8.5}$$

and

$$\sum_{M\in\mathcal{L}} |W_0^M||W_0^G|^{-1}(-1)^{\dim(A_M/A_G)} \sum_{\pi\in\Pi_{disc}(M)} a_{disc}^M(\pi)i_M(\pi, f) \tag{8.6}$$

are also equal.

Proof. Write $J(f)$ for the two equal quantities (7.5) and (7.6). We shall set $I(f) = I^G(f)$ equal to the expression (8.5). Substituting (8.1) into (7.5), and then applying the definition of I^L, we obtain

$$\begin{aligned} J(f) &= \sum_{M\in\mathcal{L}} \sum_{L\in\mathcal{L}(M)} |W_0^M||W_0^G|^{-1}(-1)^{\dim(A_M/A_G)} \int_{(M(F)_{ell})} \hat{I}_M^L(\gamma, \phi_L(f))d\gamma \\ &= \sum_{L\in\mathcal{L}} |W_0^L||W_0^G|^{-1}(-1)^{\dim(A_L/A_G)}\hat{I}^L(\phi_L(f)). \end{aligned}$$

Similarly, if $i(f) = i^G(f)$ denotes the expression (8.6), we can write

$$J(f) = \sum_{L\in\mathcal{L}} |W_0^L||W_0^G|^{-1}(-1)^{\dim(A_L/A_G)}\hat{i}^L(\phi_L(f)),$$

for (7.6). We are trying to show that I^G equals i^G. We may assume inductively that

$$\hat{I}^L(\phi_L(f)) = \hat{i}^L(\phi_L(f)),$$

for any $L \subsetneq G$. The corresponding terms in the two expansions of $J(f)$ therefore cancel. All that remains is the required equality of the two distributions

$$\hat{I}^G(\phi_G(f)) = \hat{I}^G(f_G^1) = I^G(f)$$

and

$$\hat{i}^G(\phi_G(f)) = \hat{i}^G(f_G^1) = i^G(f). \qquad \square$$

9. There are two special cases of our putative local trace formula which have already been established. Before discussing these, we shall first say a word about p-adic groups. From Section 5 through Section 8, we were assuming that F was Archimedean. Now, take F to be a p-adic field. Then $H_M(A_M(F))$ is only a lattice in $\mathfrak{a}_M$. The volume $\bar{v}_M(x_1, x_2, T)$ in (5.2) must be replaced by the number of lattice points in a convex hull. There will be an analogous change in the asymptotic formula (6.2) for (4.2). However, these difficulties are not serious, and may be handled by the methods of [1(g), Sections 4–5]. We can expect that the changes caused by replacing $\mathfrak{a}_M$ by a lattice will run parallel on the geometric and spectral sides. The discrepancies should cancel, leaving intact the identity of (7.5) and (7.6). The invariant identity of (8.5) and (8.6) would continue to hold, and the definitions (7.1), (7.4), (8.1) and (8.2) would remain the same, except with the domain of integration in (7.4) changed to $i\mathfrak{a}_M^*/i\mathfrak{a}_M^\vee$.

The first special case comes from the noninvariant identity of (7.5) with (7.6). Take G to be the general linear group $GL(n)$, and F to be a p-adic field. Choose f_2 to be supported on $G(F)_{ell}$. Then the terms with $M \neq G$ in (7.5) vanish, and the expression reduces to an average

(9.1)

$$\int_{(G(F)_{ell})} \left(|D(\gamma)| \int_{A_G(F)\backslash G(F)} f_1(x_1^{-1}\gamma x_1)dx_1 \cdot \int_{A_G(F)\backslash G(F)} f_2(x_2^{-1}\gamma x_2)dx_2 \right) d\gamma$$

of products of elliptic orbital integrals. The spectral expression (7.6) also simplifies. Suppose that $\pi \in \Pi_{disc}(M)$ is the restriction of $\tilde{\pi}_1 \otimes \pi_2$ to $M(A_F)^1$. Applying a general splitting property [1(h), Corollary 7.4] to the operators $\mathcal{J}_M(\pi_\lambda, P)$, and using the fact that f_2 is supported on the elliptic set, we can show that

$$j_M(\pi, f) = \int_{i\mathfrak{a}_M^*/i\mathfrak{a}_M^\vee} \tilde{f}_{1,M}(\pi_{1,\lambda}) \operatorname{tr}(\mathcal{R}_M(\pi_{2,\lambda}, P)\mathcal{J}_P(\pi_{2,\lambda}, f_2))d\lambda,$$

where

$$\tilde{f}_{1,M}(\pi_{1,\lambda}) = \mathrm{tr}(\mathcal{I}_{\overline{P}}(\tilde{\pi}_{1,-\lambda}, f_1)).$$

We are in the case of $GL(n)$, in which induced tempered representations are all irreducible. Therefore, π_1 equals π_2, and is induced from a discrete series. Applying a general descent property [1(h), Corollary 7.2] to the operator $\mathcal{R}_M(\pi_{2,\lambda}, P)$, we conclude that $j_M(\pi, f)$ vanishes unless π_1 is actually a discrete series. Now, specialize f_2 to a K-bi-invariant function on $G(F)$. Then $\tilde{f}_{1,M}(\pi_{1,\lambda})$ will vanish unless π_1 is unramified. In other words, M equals M_0, and the restriction of $\pi_1 = \pi_2$ to $M_0(F)^1$ is the trivial representation τ. Since

$$|W_0^{M_0}|\,|W_0^G|^{-1}(-1)^{dim(A_{M_0}/A_G)} = (-1)^{n-1}(n!)^{-1},$$

the expression (7.6) becomes

$$(9.2) \qquad (-1)^{n-1}(n!)^{-1}\int_{i\mathfrak{a}^*_{M_0}/i\mathfrak{a}^\vee_{M_0}} \tilde{f}_{1,M_0}(\tau_\lambda)\mathrm{tr}(\mathcal{R}_{M_0}(\tau_\lambda, P_0)\mathcal{I}_{P_0}(\tau_\lambda, f_2))d\lambda.$$

The identity of (9.1) with (9.2) is due to Waldspurger, and is the main result of [5(a)]. In another paper [5(b)], Waldspurger uses this identity in a remarkable way to establish some cases of Rogawski's conjecture on Shalika germs for p-adic orbital integrals.

For the second special case, we take F to be either real or p-adic. We allow G to be any connected group, except we assume for simplicity that the split component A_G is trivial. Take f_1 to be a pseudo-coefficient of a discrete series representation $\pi_1 \in \Pi_2(G(F))$. That is, if π_1' is any irreducible tempered representation of $G(F)$, $\mathrm{tr}(\tilde{\pi}_1'(f_1))$ equals 1 or 0, according to whether π_1' is equivalent to π_1 or not. Then the expression (8.6) equals $\mathrm{tr}(\pi_1(f_2))$. By the splitting and descent formulas [1(h), Proposition 9.1 and Corollary 8.3], the expression (8.5) equals

$$\sum_{M\in\mathcal{L}} |W_0^M|\,|W_0^G|^{-1}(-1)^{dim(A_M/A_G)}\int_{(M(F)_{ell})} I_M(\gamma, f_1)I_G(\gamma, f_2)d\gamma.$$

The function f_2 is supposed to belong to the Hecke algebra on $G(F)$. However, it is clear by density that this simpler form of the invariant trace formula holds for any $f_2 \in C_c^\infty(G(F))$. Fix a group M and an element $\gamma_1 \in$

$M(F)_{ell}$, and let f_2 approach the Dirac delta measure on $G(F)$ at γ_1. Then $\mathrm{tr}(\pi_1(f_2))$ approaches $\Theta_{\pi_1}(\gamma_1)$, the value of the character of π_1 at γ_1. By the Weyl integration formula, the function

$$|D(\gamma)|^{1/2}I_G(\gamma, f_2) = |D(\gamma)| \int_{A_M(F)\backslash G(F)} f_2(x_2^{-1}\gamma x_2)\,dx_2, \qquad \gamma \in (M(F)_{ell}),$$

approaches the Dirac measure on $(M(F)_{ell})$ at the conjugacy class of γ_1 in $M(F)$. Taking into account the different W_0^G-orbits of γ_1 which occur in (8.5), we obtain

$$I_M(\gamma_1, f_1) = (-1)^{dim(A_M)}|D(\gamma_1)|^{1/2}\Theta_\pi(\gamma_1). \tag{9.3}$$

For real F, this is essentially Theorem 6.4 of [1(k)].

Suppose that f_1 is actually a matrix coefficient of π_1. If π_1 is not supercuspidal, this presents the technical problem of extending the distributions $I_M(\gamma)$ to the Schwartz space. Leaving this question aside, we see that the function $\phi_L(f_1)$ will vanish for any $L \neq G$. This implies that

$$I_M(\gamma_1, f_1) = |D(\gamma_1)|^{1/2} \int_{A_M(F)\backslash G(F)} f_1(x_1^{-1}\gamma_1 x_1)v_M(x_1)\,dx_1,$$

where $v_M(x_1) = v_M(x_1, 1)$. The formula (9.3) becomes

$$\int_{A_M(F)\backslash G(F)} f_1(x_1^{-1}\gamma_1 x_1)v_M(x_1)\,dx_1 = (-1)^{dim(A_M)}\Theta_{\pi_1}(\gamma_1). \tag{9.4}$$

In this form the identity is the main result of [1(a)], when F is real, and of [1(g)], when F is p-adic and π_1 is supercuspidal. (The author of these papers seems to have had some trouble distinguishing between a representation and its contragradient. In [1(a)], $\Theta_\omega(h)$ should be replaced by $\Theta_{\tilde{\omega}}(h)$, while $\Theta_\pi(\gamma)$ should be replaced by $\Theta_{\tilde{\pi}}(\gamma)$ in [1(g)].) If F is p-adic and π_1 is special, the formulas (9.3) and (9.4) have not been established. The local trace formula would be a natural way to prove them.

10. We shall conclude with some brief general remarks. As in the global case, the local trace formula should be a special case of a local twisted trace formula. For this, we would allow G to be any connected component of a (nonconnected) reductive group over F. Let G^0 be the identity component of the group generated by G. We would then define

$$(R(y_1, y_2)\phi)(x) = \phi(y_1^{-1}xy_2), \qquad x \in G^0(F), \qquad y_1, y_2 \in G(F),$$

for any function $\phi \in L^2(G^0(F))$. This provides a canonical extension of the regular representation of $G^0(A_F)$ to the group generated by $G(A_F)$. In this setting, the definitions (7.1), (7.4), (8.1) and (8.2), the identity between (7.5) and (7.6), and the identity between (8.5) and (8.6), should all remain valid.

As we mentioned in Section 1, there are strong similarities between the local and global trace formulas. The reader can compare the invariant local formula with the invariant global formula (3) in the introduction of [1(h)]. The local formula is actually less complicated. One reason for this is that the geometric terms are parametrized only by semisimple elements. In the global formula, there are also terms on the geometric side parametrized by unipotent classes in the discrete subgroup. These account for the coefficients $a^M(\gamma) = a^M(S, \gamma)$ in [1(h), (3)]. The spectral analogue of a semisimple class is a tempered representation. The only spectral terms in the local formula come from tempered representations. In the global formula there are also terms coming from nontempered representations in the discrete spectrum. These are responsible for the extra local terms $I_M(\pi, f)$ which occur on the spectral side of the global formula.

Nevertheless, the basic local ingredients of the global formula are the invariant distributions $I_M(\gamma, f)$. In the global setting, f stands for a function on an adèle group, whereas in the definition (8.1), f is a function on the product of $G(F)$ with itself. However, the splitting formula [1(h), Proposition 9.1] can be applied in both situations. In each case, it allows one to write the distributions on the product of groups in terms of similar distributions on a single group $G(F)$. Now, there are believed to be relations between the values these distributions assume on different groups (that is, on groups related by endoscopy). Such identities would in turn provide a stable (global) trace formula, and would lead to reciprocity laws between automorphic forms on different groups. A natural question, which was perhaps behind Kazhdan's original suggestion, is whether one can use the local trace formula to establish these identities.

REFERENCES

1. Arthur, J.,
   (a) *The characters of discrete series as orbital integrals*, Invent. Math. **32** (1976), 205-261.
   (b) *The trace formula in invariant form*, Ann. of Math. **114** (1981), 1-74.
   (c) *A Paley-Wiener theorem for real reductive groups*, Acta Math. **150** (1983), 1-89.
   (d) *On the inner product of truncated Eisenstein series*, Duke Math. J. **49** (1982), 35-70.
   (e) *On a family of distributions obtained from Eisenstein series I: Applications of the Paley-Wiener theorem*, Amer. J. Math. **104** (1982), 1243-1288.
   (f) *On a family of distributions obtained from Eisenstein series II: Explicit formulas*, Amer. J. Math. **104** (1982), 1289-1336.
   (g) *The characters of supercuspidal representations as weighted orbital integrals*, Proc. Indian Acad. Sci. **97** (1987), 3-19.
   (h) *The invariant trace formula I. Local theory*, J. Amer. Math. Soc. **1** (1988), 323-383.
   (i) *The invariant trace formula II. Global theory*, J. Amer. Math. Soc. **1** (1988), 501-554.
   (j) *Intertwining operators and residues I. Weighted characters*, to appear in J. Funct. Anal.
   (k) *Intertwining operators and residues II. Invariant distributions*, to appear in Comp. Math.
2. Clozel, L. and Delorme, P., *Le théorème de Paley-Wiener invariant pour les groupes de Lie réductifs II*, preprint.
3. Harish-Chandra,
   (a) *Harmonic analysis on real reductive groups III. The Maas-Selberg relations and the Plancherel formula*, Ann. of Math. **104** (1976), 117-201; also in Collected Papers, Vol. IV, Springer-Verlag, 259-343.
   (b) *The Plancherel formula for reductive p-adic groups*, Collected Papers, Vol. IV, Springer-Verlag, 353-367.
4. Langlands, R. P., *Eisenstein series*, in *Algebraic Groups and Discontinuous Subgroups*, Proc. Sympos. Pure Math., Vol. 9, Amer. Math. Soc., Providence, R.I., 1966, 235-252.
5. Waldspurger, J-L.,
   (a) *A propos des intégrales orbitales pour GL(N)*, preprint.
   (b) *Un exemple de germes de Shalika pour GL(n)*, preprint.

UNIVERSITY OF TORONTO

CRYSTALLINE COHOMOLOGY AND P-ADIC GALOIS-REPRESENTATIONS

By Gerd Faltings

I. Introduction. The paper which follows deals with the relation between crystalline and étale cohomology of algebraic varieties. This had been first observed by A. Grothendieck (the "mysterious functor), and a precise conjecture was made by J. M. Fontaine (conjectures C_{cris}, C'_{cris} and $C_{\text{cris}}{}^{\text{pot}}$ in [Fo2]). Recently this conjecture has been shown by him and by W. Messing ([MF]), but no detailed proof has appeared so far. Their approach uses the syntomic topology, while the author has found a different way, the method of almost étale extensions. This method goes back to J. Tate and had been enhanced for applications to p-adic Hodge-theory. If it is combined with Fontaine's ideas it works quite well in the present context, even in some cases where the other approaches do not succeed.

The main results can be described as follows: Suppose we are given a smooth algebraic variety X over a p-adic field K, which is supposed to have good reduction. Associated to X we have the crystalline cohomology $H^*_{\text{cr}}(X)$, and the p-adic étale cohomology of the geometric fibre $H^*_{\text{et}}(X)$. $H^*_{\text{cr}}(X)$ is a vector-space over a subfield $K_0 \subset K$, the maximal unramified extension of $\mathbf{Q}_p$. As additional structure it has a semilinear Frobenius-automorphism, and after base-extension to K it acquires a Hodge-filtration. Let us call this structure a filtered F-crystal. $H^*_{\text{et}}(X)$ is a $\mathbf{Q}_p$-vector-space on which the Galois group $\text{Gal}(\bar{K}/K)$ acts. The theory of Fontaine predicts that each one of those cohomologies determines the other, and that under good circumstances ($K = K_0$, $\dim(X) < p - 1$) there is even a fully faithful functor from filtered F-crystals to p-adic Galois representations which converts crystalline into étale cohomology. Moreover this theory then also covers p-torsion.

The underlying philosophy might be explained as follows: If we consider l-adic étale cohomology, l a prime different from p, then the hypothe-

Manuscript received 26 June 1988.

ses "good reduction" implies that the Galois representation is unramified, that is trivial on the inertia group. In case $l = p$ the picture is more complicated: There is a notion of crystalline representation, and instead of being trivial on the inertia the Galois-representation is crystalline (which is a strong restriction). Often this can be used to gain relevant information. Historically this has been known for a long time for H^1, where "crystalline" means "generic fibre of a finite flat group scheme." As an example let us cite K. Ribet's proof of the converse to Herbrand's theorem, where one has to construct unramified extensions of cyclotomic fields. These are given by the action of Galois on p-adic étale cohomology. It is wellknown that this action tends to be unramified outside p, but for the prime p we need more local information which can be provided by our theory. Now his methods can be applied also to cohomology in degree bigger than one.

The paper proceeds as follows: We first review relevant facts from commutative algebra, étale and crystalline cohomology. These are mostly somehow known, or at least dwell on well known ideas, but they cannot be found in the literature in such a form that we can use them directly. We also use the occasion to generalise many results. Some of the main new features are:

—We extend the comparison-theory in [FL] to families of F-crystals.

—We construct logarithmic versions of crystalline cohomology, for non proper varieties with a good compactification (hidden behind this there is a whole theory of "logarithmic commutative algebra," which however we choose not to develop in detail).

—We make precise the relation between étale cohomology and Galois-cohomology.

—We allow nontrivial systems of coefficients.

After that we prove the comparison results, first in the absolute case, and then more general for direct images under "log-smooth" maps. We conclude by an application to the theory of finite flat group schemes, giving a complete description in terms of "semilinear algebra." Finally we show that our method also allow to settle the "de Rham conjecture."

It should be clear that I am very much indebted to J. M. Fontaine and W. Messing. They proved first most of the relevant results, and they supplied a framework in which the author could make good use of the techniques developed in [Fa]. As usual reproving and generalising a known result is much easier than showing it in the first place. Hopefully a more detailed overview about their methods will appear soon. I would also like to mention the work of P. Berthelot ([B2]) and A. Ogus ([O]) on rigid cohomology respectively convergent isocrystals, which provides just the right context for crystalline cohomology with $\mathbf{Q}_p$-coefficients. Research on this paper was partially supported by the National Science Foundation, grant DMS-8502316.

II. Commutative algebra. a) In this chapter we introduce notations and give some basic facts from commutative algebra. The setup will be as follows, as in [Fa]: V denotes a complete discrete valuation-ring, with fraction-field K of characteristic 0, and a perfect residue field k of positive characteristic p. Then the normalisation of V in any finite extension is also such a discrete valuation-ring, and we normalise all valuations by the rule that the prime p has valuation $\mathrm{v}(p) = 1$. In general R denotes a smooth V-algebra, of relative dimension d. Usually R is of finite type over V, but we also allow localisations of such rings, or even henselisations or strict henselisations. Sometimes we assume that R is small, which means that there exists an étale map $V[T_1^{\pm 1}, \ldots, T_d^{\pm 1}] \to R$.

By $\bar{R}$ we denote the maximal extension of R which is étale in characteristic zero. That is if R is geometrically integral we take the maximal field extension of its fraction-field such that the normalisation of $R[1/p]$ in this field is unramified over $R[1/p]$. Then $\bar{R}$ is the normalisation of R in this field. In general $R \otimes_V \bar{K}$ is a product of integral domains, and $\bar{R}$ is the product of the corresponding normalisations. Let us now assume for the moment that $R \otimes_V \bar{K}$ is an integral domain, and that R is small. Then we obtain a well understood subextension R_∞ of $\bar{R}$ by first extending V to its integral closure $\bar{V}$ in the algebraic closure $\bar{K}$ of K, and then adjoining all p-power roots of the T_i (which occur in the definition of small). One of the main results in [Fa] states that $\bar{R}$ is almost étale over R_∞, which means roughly that for many purposes we may replace the big ring $\bar{R}$ by R_∞. If R is not geometrically integral we can first replace V by an unramified extension, so that R becomes a product of geometrically integral domains. We then apply these considerations to each factor.

Unfortunately, the result in [Fa] is not stated in quite the necessary generality: We also need it for the p-adic completion $\hat{R}$ of R. However an inspection of the proof shows that it also works in this case, with no changes necessary except that one has to insert these p-adic completions in the beginning.

b) The next point is the construction of a ring $B^+(R)$, following [Fo2] and [Fo3] (Here and in the following, we generally follows Fontaine's notation, but usually dispense with some indices, as we already have too many of those): Consider the ring $S = \text{proj.lim}(\bar{R}/p\bar{R})$. Here the limit is over a projective system of rings indexed by numbers $n \geqslant 0$, with transition-maps given by Frobenius. So elements of S are sequences $\{r_n, n \geqslant 0\}$ of elements of $\bar{R}/p\bar{R}$, such that $r_n = r_{n+1}{}^p$. S is a ring of characteristic p, and the Frobenius is bijective on S.

We now form the Witt-vectors $W(S)$. Elements of $W(S)$ are given by sequences $[s_0, s_1, \ldots]$ with $s_n \in S$, with addition and multiplication given by the usual universal formulas. Let us only remark that $[s, 0, 0, \ldots] \cdot [t, 0, 0, \ldots] = [s \cdot t, 0, 0, \ldots]$, and that $p \cdot [s_0, s_1, \ldots] = [0, s_0{}^p, s_1{}^p, \ldots]$. Furthermore we obtain a homomorphism from $W(S)$ to the p-adic completion $\bar{R}^\wedge$ of $\bar{R}$, as follows: If $[s_0, s_1, s_2, \ldots]$ is an element of $W(S)$, write each s_n as a sequence $\{s_{nm}, m \geqslant 0\}$ of elements of $\bar{R}/p\bar{R}$. Lift the s_{nm} to elements r_{nm} of $\bar{R}$, and consider $r_{0m}{}^{p^m} + p \cdot r_{1m}{}^{p^{m-1}} + p^2 \cdot r_{2m}{}^{p^{m-2}} + \cdots p^m \cdot r_{mm} \in \bar{R}/p^{m+1}R$. For varying m these form a compatible projective system, and we obtain the desired ring-homomorphism $W(S) \to \bar{R}^\wedge$. If the Frobenius is surjective on $\bar{R}/p\bar{R}$, (for example if R is small) this is a surjection. Denote by I its kernel, and construct the divided power hull $D_I(W(S))$. Note that we could also have used divided powers for the ideal $I + pW(S)$, as $pW(S)$ already admits divided powers. $D_I(W(S))$ has a topology defined by the divided powers $(I + pW(S))^{[n]}$, and $B^+(R)$ is its completion. It is an algebra over $B^+(V)$, which in turn is an algebra over $W(k)$, the Witt-vectors of the residue field k of V. The Galois-group $\text{Gal}(\bar{R}/R)$ operates continuously on $B^+(R)$.

Let us describe things in more detail: Choose a sequence of p-power roots of p in $\bar{V} \subset \bar{R}$, that is a sequence of elements v_n with $v_0 = p$, $v_n = v_{n+1}{}^p$. Denote by $\underline{p} \in S$ the element defined by reducing the v_n modulo p. Then $\underline{p}$ is a nonzero-divisor in S, and an element $s = (s_0, s_1, \ldots)$ of S is divisible by $\underline{p}$ if and only if s_0 vanishes. We also define an element $\underline{-1}$ of S by taking p-power roots of -1 (If $p \neq 2$ we may choose $\underline{-1} = -1$). Let $\xi = [\underline{p}, (\underline{-1})^p, 0, 0, \ldots] \in W(S)$. Then one checks easily ([Fo2], Proposition 2.4) that ξ generates I, and that ξ and p form a regular sequence in $W(S)$. Especially it follows that $D_I(W(S))$ is obtained by adjoining divided powers $\xi^n/n!$ to $W(S)$.

Also the Frobenius-automorphism on S induces an automorphism Φ of $W(S)$, which respects the ideal generated by p and I (or by p and ξ). It follows that Φ extends to an automorphism of $D_I(W(S))$. We define a decreasing filtration $F^{\cdot}$ on $D_I(W(S)$ by defining F^n to consists of those elements in the n-th divided power of I whose image under Φ is divisible by p^n.

It follows that for $n < p$ F^n coincides with the n-th divided power of I, and up to p-torsion this holds even for all n. Furthermore the F^n are closed in the topology of $D_I(W(S))$, and so $B^+(R) = \text{proj.lim } B^+(R)/F^n(B^+(R))$. One also derives that $B^+(R)$ becomes a filtered ring, with $gr^0{}_F(B^+(R)) = \bar{R}^\wedge$ (if Frobenius is surjective on $\bar{R}/p\bar{R}$, which it will be in applications), and that the map (induced from $V \to R$) $gr^{\cdot}{}_F(B^+(V)) \otimes_{\bar{V}^\wedge} \bar{R}^\wedge \to gr^{\cdot}{}_F(B^+(R))$ is an isomorphism. It can be shown that $B^+(R)/p^m B^+(R)$ is equal to the crystalline cohomology of $(\bar{R}/p^m\bar{R})/(V_0/p^m V)$ (compare [Fo3], Theorem 1: The key point is that there exists a unique map from $B^+(R)$ into any nilpotent PD-thickening of $\bar{R}/p^m\bar{R}$).

If R is an integral domain the Galois-group $\text{Gal}(\bar{R}/R)$ obviously acts continuously on $B^+(R)$, respecting all structures. In general everything factors as a product over the integral factors of R. Also the whole construction can be made with the integral closure of $\hat{R}$ instead of R. We then call the result $B^+(\hat{R})$.

There exists a canonical map $\alpha : \mathbf{Q}_p(1) \to B^+(R)^*$ (multiplicative group), defined as follows: $\mathbf{Q}_p(1)$ is isomorphic to the group of maps from $\mathbf{Q}_p$ into μ_{p^∞}, the p-power roots, or equivalently to sequences of p-power roots $\{\zeta_n, n \geqslant 0\}$, with $\zeta_n = \zeta_{n+1}{}^p$. Reducing such a sequence modulo p defines an element $\underline{\zeta}$ of S, and hence we define its image under α as $[\underline{\zeta}, 0, 0, \ldots] \in W(S)$, which is multiplicative in $\underline{\zeta}$. α is Galois linear, and $\Phi(\alpha(x)) = \alpha(x)^p$. Also $\alpha(\mathbf{Z}_p(1)) \subset 1 + I$, so taking logarithms defines an additive map $\beta : \mathbf{Z}_p(1) \to F^1$. Again β is Galois linear, and $\Phi(\beta(x)) = p\beta(x)$. If x is a generator of $\mathbf{Z}_p(1)$ we can form $\beta(x)/\xi \in B^+(R)$. Under the map into $\bar{R}^\wedge$ this goes to an element of the form $(\text{unit}) \cdot p^{1/(p-1)}$. It follows that the obvious map (constructed from β) $\bigoplus_{n \geqslant 0} gr^0{}_F(B^+(R))(n) \to gr^{\cdot}{}_F(B^+(R))$ is an isomorphism modulo p-torsion.

Finally we define $B(R) = B^+(R)[p^{-1}, \beta(x)^{-1}]$, where x denotes any generator of $\mathbf{Z}_p(1)$. It has an induced filtration $F^{\cdot}$ (with $\beta(x)^{-1}$ of degree -1), as well as Frobenius and Galois-action. Its associated graded is isomorphic to $\bigoplus_{n \in \mathbf{Z}} gr^0{}_F(B^+(R))(n)$.

If $K_0 \subset K$ denotes the fraction-field of the Witt vectors $W(k) \subset V$, $B(R)$ is a K_0-algebra, and one obtains a new filtration of $B(R) \otimes_{K_0} K$ as follows: Let $J \subset B^+(R) \otimes_{W(k)} V$ denote the kernel of the map $B^+(R) \otimes_{W(k)} V \to \bar{R}^\wedge \otimes_{W(k)} V \to \bar{R}'$. One checks that the natural map $gr_I(B^+(R)) \to gr_J(B^+(R) \otimes_{W(k)} V)$ is an isomorphism up to p-torsion, as $\bar{V}^\wedge \otimes_{W(k)} V$ is up to p-torsion isomorphic to a product of copies of $\bar{V}^\wedge$. It follows that just as before the J-adic filtration on $B^+(R) \otimes_{W(k)} K$ extends to a filtration $F^{\cdot}$ on $B(R) \otimes_{K_0} K$ ($\beta(x)^{-1}$ has degree -1), such that $gr^{\cdot}{}_F(B(R)) \cong gr^{\cdot}{}_F(B(R)$

$\otimes_{K_0} K$). Despite that fact the filtration is separated, which is only possible because $B(R) \otimes_{K_0} K$ is not topologically complete (see [Fo2], Proposition 4.7, for a similar statement. The essential point is that the kernel of the multiplication-map $V \otimes_{V_0} V \to V$ is nilpotent modulo p).

For many purposes the complicated object $B^+(R)$ (or more precisely $B^+(R)/pB^+(R)$) can be replaced by a simpler one, as follows: Assume that Frobenius is surjective on $\bar{R}/p\bar{R}$. Consider the map $W(S)/pW(S) \cong S \to \bar{R}/p\bar{R}$, where the second arrow is projection onto the second (= degree 1) component in $S = \text{proj.lim}(\bar{R}/p\bar{R})$. It is easy to see that it induces a map from $B^+(R)/pB^+(R)$ onto $\bar{R}/p\bar{R}$, whose kernel contains $F^p(B^+(R)/pB^+(R))$ and is contained in F^{p-1} (here $F^{\cdot}$ denotes the quotient-filtration on $B^+(R)/pB^+(R)$). More precisely $\xi = [\underline{p}, (\underline{-1})^p, 0, 0, \ldots]$ is mapped to $p^{1/p}$, and one derives that the induced filtration on $\bar{R}/p\bar{R}$ is given by $F^i(\bar{R}/p\bar{R}) = p^{i/p}\bar{R}/p\bar{R}$. Also the map $\varphi^i = \Phi/p^i$ from $F^i(B^+(R))$ into $B^+(R)$ induces φ^i: $F^i(\bar{R}/p\bar{R}) \to \bar{R}/p\bar{R}$, which sends x to $x^p/(-p)^i$ (This depends only on x modulo p).

c) We have to define certain categories $\mathfrak{MF}(R)$, $\mathfrak{MF}_{\text{big}}(R)$, $\mathfrak{MF}^\nabla(R)$ and $\mathfrak{MF}^\nabla{}_{\text{big}}(R)$, following [ML]. For this we assume $V = V_0 = W(k)$ is the ring of Witt vectors over a perfect field k of characteristic 0, that R is a smooth V-algebra, and that we are given a semilinear endomorphism Φ : $\hat{R} \to \hat{R}$ of the p-adic completion $\hat{R}$ of R which lifts the Frobenius on R/pR. Φ induces a map $d\Phi_*: \Omega_{R/V} \otimes_{R\ \Phi} \hat{R} \to \Omega_{R/V} \otimes_R \hat{R}$, which is divisible by p as in characteristic p the differential of Frobenius vanishes. An object of $\mathfrak{MF}_{\text{big}}(R)$ consists of a p-torsion R-module M, a sequence of p-torsion R-modules $F^i(M)$, and sequences of R-linear maps $F^i(M) \to F^{i-1}(M)$, $F^i(M) \to M$, and $\varphi^i : F^i(M) \otimes_{R\ \Phi} R \to M$, subject to the following conditions:

i) The composition $F^i(M) \to F^{i-1}(M) \to M$ is the map $F^{i-1}(M) \to M$.

ii) The map $F^i(M) \to M$ is an isomorphism for $i \ll 0$.

iii) The composition of φ^{i-1} with $F^i(M) \to F^{i-1}(M)$ is $p\varphi^i$.

Maps between such objects are just compatible R-linear maps between M's and $F^i(M)$'s, and this makes $\mathfrak{MF}_{\text{big}}(R)$ an abelian category. Axiom iii) can be formulated slightly different, by introducing the R-module $\tilde{M}$ which is the colimit of the following diagram:

$$\cdots \to F^{i+1}(M) \leftarrow F^{i+1}(M) \to F^i(M) \leftarrow F^i(M) \to F^{i-1}(M) \leftarrow \cdots$$

The arrows "$\rightarrow$" are the ones used in the definition of $\mathfrak{MF}_{\text{big}}(R)$, and the other arrows "$\leftarrow$" are multiplication by p. Now axiom iii) is equivalent to the fact that the φ^i induces a Φ-linear map $\tilde{M} \rightarrow M$, or an R-linear map φ: $\tilde{M} \otimes_{R\ \Phi} R \rightarrow M$.

$\mathfrak{MF}(R)$ is the full subcategory of $\mathfrak{MF}_{\text{big}}(R)$ whose objects consist of tuples $\{M, F^i(M), \varphi\}$ such that M as well as all $F^i(M)$ are finitely generated p-torsion R-modules, that $F^i(M) = (0)$ for $i \gg 0$, and such that φ induces an isomorphism $\tilde{M} \otimes_{R\ \Phi} R \rightarrow M$. It should be noted that $\mathfrak{MF}(R)$ and $\mathfrak{MF}_{\text{big}}(R)$ depend on the choice of the Frobenius-lift Φ.

It turns out the $\mathfrak{MF}(R)$ has a variety of nice properties which are not totally obvious:

THEOREM 2.1.

i) *Suppose* $\{M, F^i(M), \varphi\}$ *is an object of* $\mathfrak{MF}(R)$. *Then all maps* $F^{i+1}(M) \rightarrow F^i(M) \rightarrow M$ *are injections onto direct summands (as R-modules). So we may consider the* $F^i(M)$ *as submodules of* M, *giving a finite filtration.*

ii) *Locally in the Zariski-topology on* Spec(R) M *is isomorphic to a direct sum of R-modules* R/p^eR.

iii) *If* M, N *are objects in* $\mathfrak{MF}(R)$, *any map* $f: M \rightarrow N$ *is strict for the filtrations. That is* $f(M) \cap F^i(N) = f(F^i(M))$.

Proof. We allow R to be a localisation of a V-algebra of finite type, so that we can localise in prime ideals and use induction over $\dim(R/pR)$. We may thus assume that R is a local ring with maximal ideal $\mathfrak{m}$, and that all assertions are already true up to $\mathfrak{m}$-torsion. If $\dim(R/pR) = 1$ R is a discrete valuation ring. $\mathfrak{m} = pR$, and all assertions (except "direct summand" in i)) are already shown in [FL], Proposition 1.8. The main trick there is to use that the length of $\tilde{M}$ is always $\geqslant$ the length of M, with equality only if all maps $F^{i+1}(M) \rightarrow F^i(M)$ are injections. The assertion about direct summands in i) follows by applying this remark to quotients M/p^iM, which are again in $\mathfrak{MF}(R)$.

In general we assume first that $pM = (0)$, so that we are dealing with finite modules over R/pR. For any such module L we define the i-th Fitting ideal $\mathfrak{F}_i(L)$ as follows: Choose a presentation $(R/pR)^a \rightarrow (R/pR)^b \rightarrow M \rightarrow 0$. The first map is given by an $a \times b$ matrix A, and $\mathfrak{F}_i(L)$ is the ideal generated by the minors of A of size $(b - i) \times (b - i)$. One checks that this is independent of all choices involved, commutes with base-extension, and that for an exact sequence $L_1 \to L_2 \to L_3 \to 0$ one has $\mathfrak{F}_n(L_2) \supset \Sigma_{i+j=n} \mathfrak{F}_i(L_1) \cdot \mathfrak{F}_j(L_3)$. Equality holds if L_2 is the direct sum of L_1 and L_3. Also, if R is a local ring L is free if and only if all $\mathfrak{F}_i(L)$ are either zero or R/pR, and if this is true the rank of L is the smallest index i for which the second alternative holds. We apply this as follows: By definition $\tilde{M}$ is the cokernel of a map $\bigoplus_{h<i<\infty} F^i(M) \to \bigoplus_{h+1<i<\infty} F^i(M)$, $h \ll 0$. If r_i denotes the rank of $F_i(M)$ in the generic prime of R/pR (which is normal, hence integral), r_i is the smallest index r for which $\mathfrak{F}_r(F^i(M)) \neq (0)$. Call these Fitting-ideals $\mathfrak{a}_i$. They are $\mathfrak{m}$-primary, and from above we infer that for $r = \text{rank}(M) = \text{rank}(\tilde{M})$, $\Pi_{h<i<\infty}\, \mathfrak{a}_i \supset \mathfrak{F}_r(\tilde{M}) \cdot \Pi_{h+1<i<\infty}\, \mathfrak{a}_i$. On the other hand base-extension by the Frobenius Φ on R/pR sends $\tilde{M}$ to M, so $\mathfrak{a}_h = \mathfrak{F}_r(M) = \Phi_*(\mathfrak{F}_r(\tilde{M})) =$ ideal generated by $\Phi(\mathfrak{F}_r(\tilde{M}))$. As all ideals involved are non-zero, one derives easily that $\mathfrak{F}_r(\tilde{M}) = R/pR$: For example consider for each ideal $\mathfrak{a}_i$ the maximal power of $\mathfrak{m}$ which contains it, and use that $gr_{\mathfrak{m}}(R)$ has no zero divisors.

It follows that $\tilde{M}$ is a free R/pR-module. On the other hand it is the direct sum of the cokernels of the maps $F^{i+1}(M) \to F^i(M)$, so these cokernels are free as well, and the images are direct summands. Finally we already know that these maps are injective up to $\mathfrak{m}$-torsion, and by decreasing induction over i we deduce i) and ii), using: $F^{i+1}(M)$ free $\Rightarrow$ it has no $\mathfrak{m}$-torsion $\Rightarrow F^{i+1}(M) \to F^i(M)$ is injective $\Rightarrow F^i(M)$ is free.

For iii) we show that if p annihilates both M and N, then the image and the kernel of f are both filtered direct summands. We may assume that R/pR is local, and we choose basis for M and N such that all $F^i(M)$ respectively $F^I(N)$ are generated by subsets of these bases. The map f is then given by a matrix A which has supertriangle form. We denote its rank by r, and by α the ideal generated by its $r \times r$-minors. The induced map $\tilde{f} : \tilde{M} \to \tilde{N}$ is described by the superdiagonal in A, so the ideal generated by its $r \times r$-minors is contained in $\mathfrak{a}$. On the other hand after base-extension via Frobenius it coincides with $\mathfrak{a}$, and the same reasoning as above gives that both ideals are equal to R/pR. This easily implies iii), and so the theorem is proved for modules which are annihilated by p. For the general case we make induction over the smallest integer e such that p^e annihilates all modules involved. Choose M as in i) and ii). As $\tilde{M}$ is a right exact functor of M we already know that the assertions hold for M/pM, especially that the maps $F^i(M)/pF^i(M) \to M/pM$ are injections. If we let $N \subset M$ denote the subobject pM, it follows easily that the map $\tilde{N} \to \tilde{M}$ is injective, identifying $\tilde{N}$ with $p\tilde{M}$. As Frobenius is flat we derive that $\varphi : \tilde{N} \otimes_{R\ \Phi} R \to N$ is an isomorphism, so that N is in $\mathfrak{M}\mathfrak{F}(R)$ and we can apply induction to it. So i) and ii) already hold for N. Finally one derives them for M itself, using (for example) that as all $F^i(M)/pF^i(M)$ are free, for any R-module L maps $F^i(N) = pF^i(M) \to pL$ extend to $F^i(M) \to L$.

The same technique of devissage works for iii): We already know the assertion for the induced map $M/pM - N/pN$. It follows easily that we may replace N by $f(M) + pN$. Repeating this several times we arrive at the case where f is surjective. Dually we already know the assertion for ${}_pM \to {}_pN$, the induced map on the kernels by p-multiplication (which are in $\mathfrak{MF}(R)$ by i) and ii)). Replacing M by $M/(\mathrm{Ker}(f) \cap {}_pM)$ and repeating we reduce to the case where f is also injective. So finally f is an isomorphism, and iii) obviously holds. This concludes the proof of Theorem 1.

COROLLARY. *$\mathfrak{MF}(R)$ is an abelian subcategory of $\mathfrak{MF}_{\mathrm{big}}(R)$.*

Finally we need a technical fact:

PROPOSITION 2.2. *Let M and N denote elements of $\mathfrak{MF}(R)$ respectively $\mathfrak{MF}_{\mathrm{big}}(R)$. Assume that $pN = (0)$, that all $F^i(N)$ are finitely generated R-modules and submodules of N (under $F^i(N) \to N$), and that $gr_F(N)$ is torsion-free (i.e., has no associated prime-ideal which is not a divisor of p). Then any map $f : M \to N$ is strict, and $f(M)$ is in $\mathfrak{MF}(R)$.*

Proof. By induction over $\dim(R/pR)$ we may assume that R is local, with maximal ideal $\mathfrak{m}$, and that the assertion already holds up to $\mathfrak{m}$-torsion. If $\dim(R/pR) = 0$ the proposition follows easily from [FL], Proposition 1.6. In general ($\dim(R/pR) > 0$) may replace N by the saturation of $f(M)$, that is its largest subobject which coincides with $f(M)$ after localising in the prime ideal pR. In other words, we replace $F^i(N)$ by the preimage of $\mathfrak{m}$-torsion in $F^i(N)/f(F^i(M))$. Then $N/f(M)$ is torsion, and the map $\varphi : \tilde{N} \otimes_R {}_\Phi R \to N$ is injective (this holds after localisation in pR). It follows that $(N/f(M))^\sim \otimes_R {}_\Phi R \to N/f(M)$ is also injective. As both sides are of finite length over R, as $\mathrm{length}((N/f(M))^\sim) \geqslant \mathrm{length}(N/f(M))$, and as base-extension by Frobenius multiplies length by $p^{\dim(R/pR)} > 1$, it follows that $N = f(M)$ is in $\mathfrak{MF}(R)$. The rest follows from Theorem 2.1.

d) Now we introduce integrable connections: An object of $\mathfrak{MF}^\nabla(R)$ is a tuple $\{M, F^i(M), \varphi\}$ in $\mathfrak{MF}(R)$ together with an integrable connection $\nabla : M \to M \otimes_R \Omega_{R/V}$. These should be satisfy Griffiths-transversality ($\nabla(F^i(M)) \subset F^{i-1}(M) \otimes_R \Omega_{R/V}$), and the maps $\varphi^i : F^i(M) \otimes_R {}_\Phi R \to M$ should be parallel. The latter condition means that $\nabla \circ \varphi^i = (\varphi^{i-1} \otimes_R d\Phi_*/p) \circ \nabla : F^i(M) \otimes_R {}_\Phi R \to F^{i-1}(M) \otimes_R \Omega_{R/V}$. Here $d\Phi_*/p : \Omega_{R/V} \otimes_R {}_\Phi R \to \Omega_{R/V}$ denotes the differential of Frobenius divided by p. Another way to look at this is to note that $\tilde{M} \otimes_R {}_\Phi R$ gets an integrable connection, as follows: $\tilde{M}$ is the quotient of the direct sum of all $F^i(M)$, under the equivalence relation which identifies $m \in F^i(M)$ with $p \cdot m \in F^{i-1}(M)$. The connection maps $m \otimes 1$ ($m \in F^i(M)$) to $(1 \otimes d\Phi_*/p)(\nabla(m)) \in F^{i-1}(M) \otimes_R {}_\Phi\Omega_{R/V}$. One checks that this indeed defines an integrable connection on $\tilde{M} \otimes_F {}_\Phi R$. Finally the condition on the φ^i means that $\varphi : \tilde{M} \otimes_R {}_\Phi R \to M$ is parallel.

It also follows that the connection ∇ on M is nilpotent: We may assume that $pM = (0)$. Then the connection on $\tilde{M} \otimes_R {}_\Phi R \cong gr_F(M) \otimes_R {}_\Phi R$ is manufactured from the endomorphism of the first factor given by the curvature of ∇, which has degree one and is therefore nilpotent. One can also define $\mathfrak{MF}^\nabla{}_{\text{big}}(R)$ in an obvious way, and it is an abelian category which contains $\mathfrak{MF}^\nabla(R)$ as full subcategory. However we do not need it for the moment. Again we note that the definitions depend on the choice of Φ. However sometimes we can get rid of this difficulty: For integers $a \leqslant b$ we define $\mathfrak{MF}_{[a,b]}(R)$ is the full subcategory of $\mathfrak{MF}(R)$ consisting of objects with $F^a(M) = M$, $F^{b+1}(M) = (0)$. Similar for $\mathfrak{MF}_{[a,b]}{}^\nabla(R)$ and the corresponding "big" categories. We claim that for $b - a < p$ the category $\mathfrak{MF}^\nabla{}_{[a,b]}(R)$ is independent of the choice of Φ:

THEOREM 2.3. *Assume* $0 \leqslant b - a \leqslant p - 1$, *and* $p > 2$. *Then for any two choices of* Φ *there is an equivalence between the corresponding categories* $\mathfrak{MF}_{[a,b]}{}^\nabla(R)$. *These equivalences satisfy the obvious cocycle condition, so that up to canonical isomorphism* $\mathfrak{MF}_{[a,b]}{}^\nabla(R)$ *is independent of the choice of* Φ.

Proof. Suppose we have two Frobenius lifts Φ and Ψ. In general the proof works for any pair (Φ, Ψ) of maps from R to a ring S without p-torsion, such that Φ and Ψ coincide module p. Define a map $\alpha : \tilde{M} \otimes_R {}_\Phi S \to \tilde{M} \otimes_R {}_\Psi S$ by the following rule: It is enough to define α locally in Spec(R). Also we may assume that $a = 0$, $b = p - 1$: Otherwise shift the numbering. Choose local coordinates $t_1, \ldots, t_d$, i.e. an étale morphism $V[t_1, \ldots, t_d] \to R$. Let $\partial_i = \partial/\partial t_i$ denote the dual base of R-derivations. Via ∇ these operate on M, and for any multi-index $I = (i_1, \ldots, i_d)$ we get an endomorphism $\nabla(\partial)^I$ of M. Also $(\Phi(t) - \Psi(t))^I$ denotes the monomial $\Pi\, (\Phi(t_j) - \Psi(t_j))^{i_j}$. It is divisible by $p^{|I|}$, $|I| = i_1 + \cdots + i_d$ the order of I. Finally $I! = i_1! \cdot \cdots \cdot i_d!$.

Now choose $m \in F^i(M)$, which defines an element of $\tilde{M}$. Its image under α is given by the formula

$$\alpha(m \otimes 1) = \sum_I \nabla(\partial)^I(m) \otimes (\Phi(t) - \Psi(t))^I/(I! \cdot p^{\min\{|I|,i\}}).$$

Here $\nabla(\partial)^I(m)$ is considered as an element of $F^{\max(0,i-|I|)}(M)$, and the sum is over all multiindices I. One checks that the second terms $(\Phi(t) - \Psi(t))^I/(I! \cdot p^{\min\{|I|,i\}})$ are elements of S and converge to zero in the p-adic topology (here we need that $p > 2$), so that the sum is finite.

Finally we have to verify the necessary formal properties:

—α as above gives indeed a well defined map from $\tilde{M} \otimes_{R\,\Phi} S$ to $\tilde{M} \otimes_{R\,\Psi} S$: Use that $\Phi(r) = \Sigma_I \Psi(\nabla(\partial)^I(r)) \otimes (\Phi(t) - \Psi(t))^I/I!$ (Taylor's formula).

—For three different Frobenius-lifts Φ_1, Φ_2, Φ_3 the α's satisfy transitivity: Use the binomial formula. It follows (by applying this remark to the triple Φ, Ψ, Φ) that they are isomorphisms.

—α is independent of the choice of local coordinates $t_1, \ldots, t_d$: There is a coordinate free expression for α: Let $J \subset R \otimes_V R$ denote the kernel of multiplication $R \otimes_V R \to R$, $R_1 = D_J(R \otimes_V R)^\wedge$ is the completed divided power hull. The pair (Φ, Ψ) defines a homomorphism from R_1 into $\hat{S}$ (p-adic completion), and the two modules $M \otimes_V R$ and $R \otimes_V M$ become isomorphic over R_1, by definition of an integrable connection. Call this module M_1. The isomorphism is given by Taylor's formula. It respects the filtration on M_1 which is the product of the filtrations given by $F^i(M)$ on M and the divided powers $J^{[n]}$ on R_1. There exists a map from $F^n(M_1) = \Sigma_{i+j=n} F^i(M) \cdot J^{[j]}$ to $\tilde{M} \otimes_{R\,\Psi} S$, sending $\Sigma\, x_i \cdot y_j$ ($x_i \in F^i(M)$, $y_j \in J^{[j]}$) to $\Sigma_{i+j=n}\, x_i \otimes (\Phi, \Psi)(y_i)/p^j$. That this is well defined follows from a computation in local coordinates. Now we obtain α by first applying the isomorphism $M \otimes_V R \cong R \otimes_V M$ (over R_1), and then the map above.

—α is parallel for the connections: Straightforward.

It follows that $\tilde{M} \otimes_{R\,\Phi} R$, together with its connection, is up to canonical isomorphism independent of the specific choice of Φ. As this is the only way in which Φ enters into the definition of $\mathfrak{M}\mathfrak{F}^\nabla(R)$ we are through. As an application we can define $\mathfrak{M}\mathfrak{F}^\nabla(X)$ for any smooth V-scheme X, by glueing the data obtained from local Frobenius lifts Φ.

e) Now we define a functor from $\mathfrak{M}\mathfrak{F}^\nabla(R)$ to Galois-representations, as in [FL]. We suppose that there exists an étale map $V[T_1^{\pm 1}, \ldots, T_d^{\pm 1}] \to R$, so especially Frobenius is surjective on R/pR. It also follows that there exists a Frobenius-lift Φ on $\hat{R}$, for example the one which raises the T_i to their pth power. This one has the additional advantage that it is finite, flat, and étale in characteristic 0. Let us recall the definition of $B^+(\hat{R})$, which has been constructed from the integral closure S of the p-adic completion $\hat{R}$. If $D = B^+(\hat{R})[1/p]/B^+(\hat{R})$, this has a "filtration" (not necessarily by subobjects) $F^i(D)$ and maps φ^i = Frobenius/p^i from $F^i(D)$ to D, but it is not in $\mathfrak{MF}_{\text{big}}(R)$ because there is no natural R-module structure. However as R is naturally contained in $B^+(\hat{R})/F^1(B^+(\hat{R}))$, and as R is smooth over V, we can lift the inclusion above to a V-linear $R \to B^+(\hat{R})$. Furthermore as $F^1(B^+(\hat{R}))$ admits divided powers for any R-module M with an integrable connection the tensor product $M \otimes_R B^+(\hat{R})$ is up to canonical isomorphism independent of the choice of the lift. So if we are given an object M in $\mathfrak{MF}^\nabla(R)$ we see that its base-extension to $B^+(\hat{R})$ as well as that of $\tilde{M} \otimes_R {}_\Phi R$ are both independent of choices. If $M \in \mathfrak{MF}_{[a,b]}{}^\nabla(R)$ with $b - a < p$ the latter is also canonically isomorphic to $(M \otimes_R B^+(\hat{R}))^\sim \otimes_{B^+(\hat{R})}\, {}_\Phi B^+(\hat{R})$, as $(M \otimes_R B^+(\hat{R}))^\sim \cong \tilde{M} \otimes_R \tilde{B}^+(\hat{R})$: Use that as filtered module $M \cong gr_F(M))$ and as the two ways to compose the morphism from $\hat{R}$ to $B^+(\hat{R})$ with Frobenius-lifts Φ on $\hat{R}$ respectively $B^+(\hat{R})$ give the same result modulo p (as required in the proof of Theorem 3). It follows that $\varphi : \tilde{M} \otimes_R {}_\Phi R \to M$ induces a map $(M \otimes_R B^+(\hat{R}))^\sim \to M \otimes_R B^+(\hat{R})$ which up to canonical isomorphism is independent of all choices. So we can define $\mathbf{D}(M) = \mathrm{Hom}(M \otimes_R B^+(\hat{R}), D)$. The homomorphisms should be $B^+(\hat{R})$-linear, and respect filtrations and the φ's.

If Φ happens to be étale in characteristic 0 we may find a lift $\hat{R} \to B^+(\hat{R})$ which respects Frobenius-lifts, see below. It follows that using this lift $\mathbf{D}(M)$ is equal to the set of homomorphisms in $\mathfrak{MF}_{\text{big}}(R)$ from M to D (which now has become an $\hat{R}$-module).

Now to the construction of such a good lift: Form $R_\infty = \text{dir.lim}\, \hat{R}$. The direct limit is over copies of R indexed by the natural numbers, and the maps are given by Φ mapping from one copy to the next. So $R \subset R_\infty$, and Φ extends to an automorphism of R_∞ which lifts Frobenius. As R_∞/pR_∞ has bijective Frobenius the p-adic completion of R_∞ is isomorphic (respecting Frobenius) to the Witt vectors over R_∞/pR_∞. Finally by assumption R_∞ is contained in the maximal extension of $\hat{R}$ which is étale in characteristic 0, so by the construction of $B^+(\hat{R})$ the Witt-vectors over R_∞/pR_∞ map to it, and we are done.

In the following we fix such good lifts. However everything we do will be independent of the choices involved, especially also of that of a Frobenius lift Φ on $\hat{R}$.

The Galois-group Γ of $\hat{R}^-$ over $\hat{R}$ acts on $M \otimes_R B^+(\hat{R})$. However the action involves the connection and differs from the natural action on the second factor. If we chose local parameters $\{T_1, \ldots, T_d\}$ (as usual) and denote by R_∞ the subextension generated by adjoining p-power roots of the T_i, then the subgroup $\Gamma_\infty \subset \Gamma$ which fixes R_∞ acts via the second factor. To describe the action of all of Γ it now suffices to specify how the geometric fundamental group Δ (the stabiliser of $\bar{V}$) acts, as Γ is generated by Γ_∞ and Δ. To do this note that adjoining the p-power roots of the T_i defines a homomorphism $\Delta \to \mathbf{Z}_p(1)^d$ (with kernel $\Gamma_\infty \cap \Delta$). For $\sigma \in \Delta$ we denote the components of its image by $\sigma_i \in \mathbf{Z}_p(1)$, and note that we may apply the homomorphism $\beta : \mathbf{Z}_p(1) \to F^1(B^+(\hat{R}))$ to them. On the other hand let ∂_i denote the derivations dual to $d\log(t_i) = dT_i/T_i$. Then $\sigma \in \Delta$ acts as $\exp(\Sigma\, \nabla(\partial_i) \otimes \beta(\sigma_i))$, that is it sends $m \otimes 1$ to $\Sigma_I\, \nabla(\partial^I)(m) \otimes \beta(\sigma)^I/I!$. This Galois-action also defines a Galois-action on $\mathbf{D}(M)$.

We say that an object $M \in \mathfrak{MF}(R)$ and an abelian group L have the same type if locally in $\mathrm{Spec}(R)$ $M \cong L \otimes R$. If $\mathrm{Spec}(R/pR)$ is connected (that is R/pR is integral) this defines the type of any $M \in \mathfrak{MF}(R)$.

THEOREM 2.4. *Suppose $M \in \mathfrak{MF}_{[0,p-2]}(R)$, and R/pR integral. Then* $\mathrm{Hom}(M, D)$ *is a finite abelian group of the same type as M, and* $\mathrm{Ext}^1(M, D) = (0)$. *The* Hom*'s and* Ext*'s are computed in the abelian category $\mathfrak{MF}_{\mathrm{big}}(R)$.*

Proof. We may assume that $pM = (0)$. By devissage the general result follows easily from this. As D is p-divisible it suffices to compute the order of $\mathrm{Hom}(M, B^+(\hat{R})/pB^+(\hat{R}))$, and that any extension annihilated by p of M by $B^+(R)/pB^+(R)$ splits. Let $S = \hat{R}^-$ (maximal extension of $\hat{R}$ étale in characteristic 0), $E = S/pS$, which becomes an (R/pR)-module via $R/pR \to R_\infty/pR_\infty \to R_\infty/pR_\infty \to S/pS$ (the second map is the inverse of Φ on R_∞), and is filtered by $F^i(E) = p^{i/p}S/pS$. By defining, for $x \in F^i(E)$, $\varphi^i(x) = x^p/(-p)^i$ E becomes an object of $\mathfrak{MF}_{\mathrm{big}}(R)$. Furthermore we have defined a surjection $B^+(\hat{R})/pB^+(\hat{R}) \to E$ whose kernel X is equal to $F^{p-1}(X)$, i.e. the maps $F^{p-1}(X) \to F^i(X)$ are surjective for $i < p$. It follows that $\mathrm{Hom}(M, X) = \mathrm{Ext}^1(M, X) = (0)$, by the restraints on the filtration on M, and so the long exact Ext-sequence allows us to replace $B^+(R)/pB^+(R)$ by E.

We know that M is a free R/pR-module. Choose a basis $m_1, \ldots, m_h$ of M and integers e_i, such that each $F^j(M)$ is generated by the m_i with $e_i \geqslant j$. By assumption the e_i lie between 0 and $p-2$. The object $M \in \mathfrak{MF}(R)$ is now completely described by an invertible (φ bijective) $h \times h$ matrix $A = (a_{ij})$, such that $\varphi^{e_i}(m_i) = \Sigma_j\, a_{ij} \cdot m_j$. Choose elements b_{ij} in S such that $b_{ij}{}^p \equiv a_{ij}$ modulo p. These form an invertible matrix B. Then the set $\mathrm{Hom}(M, E)$ can be described as the set of solutions $x_i \in S$ of the congruences $x_i{}^p \equiv (-p)^{e_i} \cdot \Sigma_j\, b_{ij} \cdot x_j$ modulo p^{e_i+1}, $1 \leqslant i \leqslant h$ (Two solutions are considered equal if the x_i coincide modulo p). To investigate the set of solutions let us define for any $\sigma \geqslant 0$ a solution modulo p^σ as an h-tupel $x_i \in S$

such that the congruences above hold modulo $p^{e_i+\sigma}$. If $\sigma \geqslant (p-2)/(p-1)$ one checks that this depends only on x_i modulo $p^\sigma \cdot S$ (x_i is divisible by $p^{e_i/p}$), so we identify two solutions if they coincide modulo p^σ. An application of Newton's method gives that for $\sigma \geqslant 1$ any solution modulo p^σ lifts uniquely to a solution module $p^{\sigma+2/p}$. It follows that solutions modulo p (in fact modulo p^σ with $\sigma > (p-2)/(p-1)$ suffices) correspond bijectively to solutions modulo p^σ, any $\sigma \geqslant 1$, or to solutions of the corresponding equations in the p-adic completion of S. In fact these solutions lie already in S, as the whole lifting can be done in a sufficiently big finite $\hat{R}$-subalgebra of S.

So consider the S-algebra $\mathfrak{A} = S[X_1, \ldots, X_h]/(\{X_i^p - (-p)^{e_i} \cdot \Sigma_j b_{ij} \cdot X_j\})$: It is finite and flat of degree p^h over S. Its discriminant is the $\mathfrak{A}/S$-norm of the determinant of the matrix

$$C = p \cdot \operatorname{diag}(X_i^{p-1}) - \operatorname{diag}((-p)^{e_i}) \cdot B = -\operatorname{diag}((-p)^{e_i}) \cdot B(\mathbf{1} - Z),$$

where $Z = B^{-1} \cdot \operatorname{diag}(\pm X_i^{p-1}/p^{e_i-1})$). As in the algebra $\mathfrak{A}$ any monomial X^I ($I = (i_1, \ldots, i_h)$) is divisible by p to the power $(1 \cdot \underline{e})/p - c$, a constant independent of 1, it follows that Z^n is divisible by $p^{2n/p-c}$. So Z is topologically nilpotent, the discriminant is a unit up to p-powers, and $\mathfrak{A}[1/p]$ is étale over $S[1/p]$. As S is closed under such extensions $\operatorname{Hom}_S(\mathfrak{A}, S)$ has order p^h, and this can be identified with the set of solutions, or with $\operatorname{Hom}(M, E)$. So we have shown the first part of the theorem.

For extensions we use a similar argument: Any such extension $0 \to E \to\ ? \to M \to 0$ splits as an extension of filtered R/pR-modules. To get a splitting which respects the φ^i we need to solve congruences $x_i^p \equiv (-p)^{e_i} \cdot (y_i + \Sigma_j b_{ij} \cdot x_j)$ modulo p^{e_i+1}, $1 \leqslant i \leqslant h$, with $y_i \in S$. As before we define solutions modulo p^σ, which lift uniquely to solutions in S. The solutions in S are given by a finite flat S-algebra which is étale over $S[1/p]$, so there are p^h of them. Thus Theorem 4 has been shown.

f) We denote by Γ the Galois group of S over $\hat{R}$, and by $\Delta \subset \Gamma$ the kernel of the map $\Gamma \to \operatorname{Gal}(\bar{K}/K)$. Γ acts continuously on $B^+(R)$ and D, and so also on $\mathbf{D}(M)$ for any $M \in \mathfrak{MF}^\nabla(R)$. If $pM = (0)$ and $M \in \mathfrak{MF}_{[0,p-2]}{}^\nabla(R)$ this action cannot be seen directly in the description above as $\mathbf{D}(M) = \operatorname{Hom}_S(M, S \otimes \mathbf{Q}_p/\mathbf{Z}_p)$, as the choices involved are not Γ-invariant. We also have to use the connection ∇, as already described above: The ideal $F^i(S/pS)$ of $S/pS = D$ has p-th power zero, so it acquires divided powers by the rule $x^{[n]} = 0$ for $n \geqslant p$. Furthermore Frobenius induces an isomorphism $(S/pS)/F^1 \cong S/pS$. Choose any lifting of the resulting homomorphism $R \to S/pS \cong (S/pS)/F^1$ to $R \to S/pS$. Then for any R/pR-module M with integrable connection ∇ $M \otimes_R (S/pS)$ is up to canonical isomorphism independent of all choices, and Γ acts on it. If $M \in \mathcal{MF}_{[0,p-2]}{}^{\nabla}(R)$ this induces the Γ-action on $\mathbf{D}(M) \subset \mathrm{Hom}_S(M \otimes_R (S/pS), S/pS)$.

Now assume that Γ acts continuously on a finite $\mathbf{F}_p$-vectorspace L, and that we are given a Γ-linear map $L \to \mathbf{D}(M)$, $M \in \mathcal{MF}_{[0,p-2]}{}^{\nabla}(R)$ annihilated by p. This amounts to giving a pairing $L \times M \to S/pS$, or a map α : $M \to \mathrm{Hom}(L, S/pS)$, satisfying certain requirements which will be explained now.

First let us introduce some notation. Choose an étale map $V[T_1^{\pm 1}, \ldots, T_d^{\pm 1}] \to R$, and let $R_1 \supset R$ denote the extension generated by adjoining the p-th roots $T_i{}^{1/p}$. Furthermore $\Gamma_1 \subset G$ is the stabiliser of R_1, and Δ_1 its intersection with Δ. Δ_1 is normal in Δ, with the quotient isomorphic to $\mu_p{}^d$, operating in the obvious manner on the $T_i{}^{1/p}$. Also Δ and Γ_1 generate Γ. As Frobenius on S/pS induces an isomorphism from R_1/pR_1 onto R/pR, we may use the inverse of this isomorphism to make S/pS into an R-algebra. This is Γ_1-invariant, so α will be Γ_1-linear. The action of Γ on $M \otimes_R (S/pS)$ is then determined by that of $\mathbf{D}$, and a check of the definitions reveals the following: The truncated power-series for logarithm and exponential define isomorphisms between $F^1(S/pS)$ and $1 + F^1(S/pS)$. An element $\sigma \in \Delta$ sends $T_i{}^{1/p}$ to $\zeta_i \cdot T_i{}^{1/p}$, ζ_i a p'th root of unity. Let $\lambda_i = \log(\zeta_i)$. Then σ sends $m \otimes 1$ to $\Sigma_I \nabla(\partial^I)(m) \otimes \lambda^I/I!$.

The sum is over all multi-indices $I = (i_1, \ldots, i_d)$ with $0 \leqslant i_j < p$, $\nabla(\partial^I)$ denotes the action of the derivations $T_i \cdot \partial/\partial T_i$, and λ^I denotes the corresponding monomial in the elements λ_i. It follows that the conditions on α are:

i) α respects filtrations and the φ_i.
ii) $\alpha(M)$ is contained in the Γ_1-invariants of $\mathrm{Hom}(L, S/pS)$.
iii) For $\sigma \in \Delta$ $\sigma(\alpha(m)) = \Sigma_I \alpha(\nabla(\partial^I)(m)) \cdot \lambda^I/I!$.

As $\mathrm{Hom}(L, S/pS)$ is a filtering union of objects in $\mathcal{MF}_{\mathrm{big}}(R)$ which satisfy the conditions of Proposition 2.2 it follows that α is strict, and that its kernel N is also in $\mathcal{MF}(R)$. From condition iii) we derive that N is ∇-stable: Suppose that for some $m \in N$ and some i $\nabla(\partial_i)(m)$ does not map to zero in $M/N \cong \alpha(M) \subset \mathrm{Hom}(L, S/pS)$. Choose a σ such that $\zeta_i = \zeta$ is a primitive p'th root of unity, and all other ζ_j vanish. It follows that

$$0 = \sigma(\alpha(m)) = \sum_{1 \leqslant n \leqslant p-2} \alpha(\nabla(\partial_i{}^n)(m)) \cdot \lambda^n/n!. \qquad (\lambda = \log(\zeta))$$

We want to conclude that all the individual terms in the sum vanish. For this we consider the action of Γ_1 on the right side. Except for the ζ's everything is invariant, and Γ_1 acts on ζ via a surjection $\Gamma_1 \to \mathbf{F}_p^*$. It follows that the $\{\lambda^n | 0 \leqslant n \leqslant p-2)\}$ are Γ_1-eigenvectors to different characters, hence the assertion.

We derive that $\alpha(\nabla(\partial_i)(m)) \in p^{1-1/(p-1)} \cdot \mathrm{Hom}(L, S/pS) \subset F^{p-2}(\mathrm{Hom}(L, S/pS))$, hence (as α is strict) that $\nabla(\partial_i(m)) \in N + F^{p-2}(M)$. Finally the associated degree one map (= curvature of ∇} $\tilde{M} = gr_F(M) \to gr_F(M) \otimes_R \Omega_{R/V}$ respects $gr_F(N) \subset gr_F(M)$, and as $\varphi : \tilde{M} \otimes_{R\,\Phi} R \to M$ is parallel the assertion follows. We conclude that $L \to \mathbf{D}(M)$ factors over $\mathbf{D}(M/N) \subset \mathbf{D}(M)$. Assume now that the map $L \to \mathbf{D}(M)$ does not factor over $\mathbf{D}(M/N)$, for any nontrivial subobject $N \neq (0)$ of M in $\mathfrak{MF}_{[0,p-2]}{}^{\nabla}(R)$, which implies that α is injective. Then we may apply the whole theory to $M_1 = M \otimes_R R_1$, and conclude that the kernel of the map $\alpha_1 : M_1 \to \mathrm{Hom}(L, S/pS)$ lies in $\mathfrak{MF}^{\nabla}(R_1)$, hence is generated by its ∇-parallel elements. As these elements are in M it follows that α_1 is injective as well. Going on by induction we derive that $\alpha_\infty : M_\infty = M \otimes_R R_\infty \to \mathrm{Hom}(L, S/pS)$ is injective, where R_∞ is the extension of R obtained by adjoining all p-power roots of the T_i.

Now localise R_∞ at the prime-ideal pR_∞ to get a discrete valuation ring W with perfect residue field. $M \otimes_R W$ is an element of $\mathfrak{MF}_{[0,p-2]}(W)$, and $\mathbf{D}(M \otimes_R W) = \mathbf{D}(M)$ (there is an injection, and both sides have the same order). By [FL] (using the analysis of simple objects in Chapter 4 there) the image of the map $L \to \mathbf{D}(M \otimes_R W)$ is a crystalline representation of $\mathrm{Gal}(\bar{W}/W)$, hence of the form $\mathbf{D}(N)$ for a quotient N of $\mathbf{M} \otimes_R W$. As $M \otimes_R W$ injects into $\mathrm{Hom}(L, S/pS)$ N must coincide with M, so the dimension of the image is equal to the rank of M. We derive:

LEMMA 2.5. *Suppose $M \in \mathfrak{MF}_{[0,p-2]}{}^{\nabla}(R)$ with $pM = (0)$, L is an $\mathbf{F}_p - \Gamma$-modul, and we have a map $L \to \mathbf{D}(M)$ which does not factor over $\mathbf{D}(M/N)$ for any nontrivial quotient M/N of M in $\mathfrak{MF}_{[0,p-2]}{}^{\nabla}(R)$. Then $\mathrm{rank}_{R/pR}(M) \leqslant \dim_{\mathbf{F}_p}(L)$, and equality implies $L \cong \mathbf{D}(M)$.*

The same holds for arbitrary $M \in \mathfrak{MF}_{[0,p-2]}{}^{\nabla}(R)$ and finite $\mathbf{Z}_p - \Gamma$-modules L, provided we replace $\dim_{\mathbf{F}_p}$ by $\mathrm{length}_{\mathbf{Z}_p}$ and rank by "length after localisation in the prime-ideal pR."

Proof. Except for the last sentence everything has been shown. The rest follows by induction over the smallest e with $p^e \cdot L = (0)$: If $e = 1$, the hypothesis implies that $pM = (0)$, and the assertion has been shown. In general consider the induced map ${}_pL \to \mathbf{D}(M/pM)$, where ${}_pL \subset L$ denotes the elements annihilated by p. Let $M_1 \subset M$ denote the maximal subobject in $\mathcal{MF}_{[0,p-2]}{}^{\nabla}(R)$ such that this map factors over $\mathbf{D}(M/M_1)$, so that the length of M/M_1 is less or equal to that of ${}_pL$. Now apply induction to the induced map $L/{}_pL \to \mathbf{D}(M_1)$.

g) We define an adjoint $\mathbf{E}$ of $\mathbf{D}$ by the rule $\mathbf{E}(L) = \text{dir.lim}\{M|L \to \mathbf{D}(M)\}$. The direct limit is over all pairs consisting of $M \in \mathcal{MF}_{[0,p-2]}{}^{\nabla}(R)$ and a map $L \to \mathbf{D}(M)$, the morphisms being the obvious ones. It is enough to consider pairs such that the map does not factor over $\mathbf{D}(M/N)$ for a nontrivial quotient M/N of M, and these form an ordered set. By Lemma 5 this ordered set has a maximal element which we call $\mathbf{E}(L)$, and we know that the length of it is less than that of L: $l(\mathbf{E}(L) \leqslant l(L)$, with equality only if $L \cong \mathbf{D}(\mathbf{E}(L))$. Quite formally it follows that $\text{Hom}(M, \mathbf{E}(L)) \cong \text{Hom}(L, \mathbf{D}(M))$, where the left side denotes homomorphisms in $\mathcal{MF}_{[0,p-2]}{}^{\nabla}(R)$, and on the right we have Galois linear maps. Also $\mathbf{E}$ is left exact, and $\mathbf{D}$ faithful implies that the map from M into $\mathbf{E}(\mathbf{D}(M))$ is always injective. By the inequality about length we derive that is an isomorphism, and so $\mathbf{D}$ is fully faithful. Finally the image of $\mathbf{D}$ consists of those L for which $l(\mathbf{E}(L)) = l(L)$. As we have always inequality, and as $\mathbf{E}$ is left exact, and follows that the set of those L is closed under subobjects and quotients. Call them dual-crystalline representations. Hence:

THEOREM 2.6. $\mathbf{D}$ *induces an equivalence of* $\mathcal{MF}_{[0,p-2]}{}^{\nabla}(R)$ *with the full subcategory of finite* $\mathbf{Z}_p - \Gamma$*-modules whose objects are dual-crystalline representations. This subcategory is closed under subobjects and quotients.*

Remark. As $\mathcal{MF}_{[0,p-2]}{}^{\nabla}(R)$ as well as the category of étale sheaves both satisfy étale descent, it follows that a representation is already dual-crystalline if it becomes so after passage to an étale covering. Also the proof of the fact that the set of dual crystalline representations is closed under subobjects used reduction to the case of a discrete valuation ring, and then the classification of simple objects in $\mathcal{MF}$ over such a ring if the residue field is algebraically closed. There seems to be no more direct way, except that in our case this classification ([FL], Chapter 4) can be a little bit simplified.

In the proof we have assumed that $\hat{R}$ is integral, and that there exists an étale map from $V[T_1^{\pm 1}, \ldots, T_d^{\pm 1}]$ to R. The first condition is harmless: In general $\hat{R}$ is a product of integral domains, and our reasoning applies to each factor. We just have to replace Galois-representations by locally constant p-torsion sheaves on $\text{Spec}(\hat{R}[1/p])$. For the second assumption we can do the following: Assume X is a proper smooth V-scheme. Cover X by open affines $\mathrm{Spec}(R)$ with R satisfying the condition above. The categories $\mathcal{MF}_{[0,p-2]}{}^{\nabla}(R)$ glue to give a category $\mathcal{MF}_{[0,p-2]}{}^{\nabla}(X)$, and the functors $\mathbf{D}$ associate to any object of this category compatible system of étale sheaves on $\mathrm{Spec}(\hat{R}[1/p])$. These can be expressed in terms of certain finite étale coverings. If we extend these by normalisation to $\mathrm{Spec}(\hat{R})$ the results glue to a finite covering of the formal V-scheme $\mathfrak{X}$ associated by X. By EGA3 this covering is algebraic, and we obtain a locally constant sheaf on $X \otimes_V K$. It follows:

THEOREM 2.6*. *Suppose X is a proper smooth V-scheme. There exists a fully faithful contravariant functor* $\mathbf{D}$ *from* $\mathcal{MF}_{[0,p-2]}{}^{\nabla}(X)$ *to the category of locally constant étale p-torsion sheaves on $X \otimes_V K$. The image (called dual-crystalline sheaves) is closed under subobjects and quotients. Locally* $\mathbf{D}$ *is given by the procedure above.*

Remark. For $0 \leqslant i \leqslant p-2$ the Galois modules $\mathbf{Z}/p^r\mathbf{Z}(i)$ (Tate-twist) are dual-crystalline, as they correspond to the following object $R/p^rR\{i\}$ in $\mathcal{MF}^{\nabla}(R)$: The underlying R-module with connection is just R/p^rR, with filtration given by $F^i(R/p^rR\{i\}) = R/p^rR\{i\}$, $F^{i+1}(R/p^rR\{i\}) = (0)$, and $\varphi^i(1) = 1$. That indeed $\mathbf{D}(R/p^rR\{i\}) = \mathbf{Z}/p^r\mathbf{Z}(i)$ follows easily from the existence of the map $\beta^{\otimes i} : \mathbf{Z}_p(i) \to \mathbf{F}^i(B^+(R))$, as $\varphi_i \circ \beta^{\otimes i} = \beta^{\otimes i}$. Tensoring with $\hat{R}\{i\}$ is the $\mathcal{MF}$-analogue of Tate-twist and will be denoted by "$\{i\}$".

h) The functor $\mathbf{D}$ is given by homomorphisms into the universal object D. Sometimes we need a description using tensor products instead. The purpose of this section is to provide the necessary facts. Fix an integer a between 0 and $p-2$. In the following we consider objects M of $\mathcal{MF}_{[0,a]}{}^{\nabla}(R)$. For such an M consider $M^t = \mathrm{Hom}_R(M, R[1/p]/R\{a\})$. By this we mean the internal home in $\mathcal{MF}^{\nabla}(R)$, that is the usual R-homomorphisms filtered by the rule that $F^i(M^t) = \mathrm{Hom}_R(M/F^{d-i+1}(M), R[1/p]/R\{a\})$, and the appropriate choice of φ derived from the fact that the map $\tilde{M} \times \tilde{M}^t \to R\{a\}[1/p]/R\{a\}$ is a perfect duality. M is canonically isomorphic to M^{tt}, so we have defined an involution on $\mathcal{MF}_{[0,a]}{}^{\nabla}(R)$. Also $\mathbf{D}(M^t) =$ Frobenius-invariants in F^0 of $M \otimes_R B^+(R)\{-a\}$, so any element of $\mathbf{D}(M)$ induces a transformation from $\mathbf{D}(M^t)$ to $\mathrm{Hom}(B^+(R)\{a\}, D)$. This defines a pairing $\mathbf{D}(M) \times \mathbf{D}(M^t) \to \mathbf{Q}_p/\mathbf{Z}_p(a)$, so if we define the corresponding duality on finite $\mathbf{Z}_p - \Gamma$-modules by $L^t = \mathrm{Hom}(L, \mathbf{Q}_p/\mathbf{Z}_p(a))$ we get a natural transformation $\mathbf{D}(M^t) \to \mathbf{D}(M)^t$. This transformation is always an isomorphism: Extend R to R_∞, then localise to reduce to the case of a discrete valuation ring. After an unramified extension we may assume that the discrete valuation ring has alalgebraically closed residue field. It is easy to check the assertion for simple objects in $\mathcal{MF}_{[0,d]}$, using their description in [FL], Chapter 4, and the rest follows by devissage.

Also let $L^* = \mathrm{Hom}(L, \mathbf{Q}_p/\mathbf{Z}_p)$, and similar for M^*. The definition of $\mathbf{D}(M)$ gives a natural transformation $M \otimes_R B^+(R) \to \mathbf{D}(M)^* \otimes B^+(R)$, respecting Γ-actions, filtrations and φ's. Its transpose is a map $M^* \otimes_R B^+(R) \leftarrow \mathbf{D}(M) \otimes B^+(R)$. Replacing M by M^t and shifting filtrations by a (that is tensoring with $\hat{R}\{a\}$) we get $M \otimes_R B^+(R) \leftarrow \mathbf{D}(M)^t \otimes B^+(R) \otimes_R \hat{R}\{a\}$. It is easily seen that this way we obtain an inverse up to $\beta^{\otimes a}$ of $M \otimes_R B^+(R) \to \mathbf{D}(M)^* \otimes B^+(R)$, that is a transformation the other way such that the composition either way is induced by $\beta^{\otimes a} : \hat{R}\{a\}(a) \to B^+(R)$.

Now suppose that we have a finitely generated $\hat{R}$-module M with an integrable connection, a filtration $F^i(M)$ and maps $\varphi^i : F^i(M) \otimes_{R\,\Phi} R \to M$, such that for each r the quotient $M/p^r M$ lies in $\mathcal{MF}_{[0,a]}{}^{\nabla}(R)$. If $L^* = \text{proj.lim}\, \mathbf{D}(M/p^r M)^*$, it follows that there is a natural transformation $M \otimes_R B^+(R) \to L^* \otimes B^+(R)$, which has an inverse up to $\beta^{\otimes a}$. As in $B(R)$ β becomes inverted we have $M \otimes_R B(R) \cong L^* \otimes B(R)$. This shows that M is associated to L^* in the sense of [Fo1], Chapter 3.

i) Finally let us mention that everything extends to the logarithmic context: Now we consider rings R with an étale map $V[T_1, \ldots, T_d] \to R$, and étale covers of $R[1/(p \cdot T_1 \cdot \cdots \cdot T_d)]$. $\bar{R}$ denotes the normalisation of R in the maximal covering, $\Gamma \supset \Delta$ the corresponding Galois groups. The connections have to be logarithmic, i.e. given by maps $\nabla : M \to M \otimes_R \Omega_{R/V}(d \log \infty)$, and the Frobenius-lifts have to respect the divisor at infinity in the sense that $\Phi(T_i) = \text{unit} \cdot T_i{}^p$. After that all theorems as well as their proofs carry over verbatim. Globally we obtain categories $\mathcal{MF}^{\nabla}(X^\circ)$, for X a smooth scheme over V, and $X^\circ = X - D$ the complement of a relative normal crossings divisor. This notation is slightly misleading, as everything depends on the chosen compactification of X°, but we hope that this will not confuse the reader too much.

Remark. The monodromy along ∞ of any dual-crystalline sheaves is of p-power order. This follows (locally) by base extension $R \to R_n = R[T_i{}^{p^{-n}}]$, for if $M \in \mathcal{MF}^{\nabla}(R)$ is annihilated by p^n the connection on $M_n = M \otimes_R R_n$ has no poles at ∞, so that the monodromy operates trivially on the associated Galois representation. As $R \to R_n$ is ramified of p-power order the assertion follows.

III. Etale cohomology. a) Let us review some facts about étale cohomology: We consider schemes which are smooth over an algebraically closed field K of characteristic 0, and locally constant sheaves with finite stalks on their étale site. If $(X, \mathbf{L})$ is such a pair $H^*(X, \mathbf{L}) = H^*_e(X, \mathbf{L})$ denotes its étale cohomology. It satisfies the usual axioms (cup-product, functoriality, Künneth-formula), and if X is proper and smooth of pure dimension d we have Poincaré-duality: There is a transformation $\mathrm{tr}_X : H^{2d}(X, \underline{\mathbf{Z}/n\mathbf{Z}}) \to \mathbf{Z}/n\mathbf{Z}(-d)$ such that for any $\mathbf{L}$ annihilated by n we obtain a perfect duality $H^*(X, \mathbf{L}) \times H^{2d-*}(X, \mathbf{L}^*) \to H^{2d}(X, \underline{\mathbf{Z}/n\mathbf{Z}}) \to \mathbf{Z}/n\mathbf{Z}(-d)$. The Tate-twists refer to Galois-action in case X and $\mathbf{L}$ are already defined over a subfield of K, and we have made use of the fact that $\mathbf{Z}/n\mathbf{Z}$ is injective as a module over itself. It follows that for any map $f : X \to Y$ between such schemes the functorial $f^* : H^*(Y, \mathbf{L}^*) \to H^*(X, f^*(\mathbf{L}^*))$ has an adjoint $f_* : H^*(X, f^*(\mathbf{L})) \to H^{*+2e}(Y, \mathbf{L})(e)$, with $e = \dim(Y) - \dim(X)$. Especially for any smooth $Z \subset X$ of pure codimension e we obtain a characteristic class $c_Z = i_*(1) \in H^{2e}(X, \mathbf{Z}/n\mathbf{Z})(e)$ (i is the inclusion map). Also i_* can be described (using purity for local cohomology) as $H^*(Z, i_*(\mathbf{L})) \cong H^{*+2e}{}_Z(X, \mathbf{L})(e) \to H^{*+2e}(X, \mathbf{L})(e)$. Finally Poincaré-duality is to a large extent described by the class $c_\Delta \in H^{2d}(X \times X, pr_1{}^*(\mathbf{L}) \otimes pr_2{}^*(\mathbf{L}^*))(d)$ which under Künneth corresponds to the duality isomorphism. c_Δ is the image of the identity-map in $H^0(X, \mathbf{L} \otimes \mathbf{L}^*)$ under the diagonal inclusion $X = \Delta \subset X \times X$.

b) Everything has a variant for open varieties. Still assume that X is proper and smooth over K, of pure dimension d, and that $D \subset X$ is a divisor with simple normal crossings, that is $D = \bigcup_{i \in I} D_i$ is the union of smooth divisors D_i such that for any subset $M \subset I$ the intersection $X_M = \bigcap_{i \in M} D_i$ is smooth of codimension $|M|$ in X. If $\mathbf{L}$ is a locally constant torsion-sheaf on $X - D$ we may consider the cohomology $H^*(X - D, \mathbf{L}) = H^*(X, \mathbf{R}j_*(\mathbf{L}))$ as well as cohomology with compact support $H^*_!(X - D, \mathbf{L}) = H^*(X, j_!(\mathbf{L}))$. More general (as in [DI] 4.2) for any disjoint decomposition $I = A \cup B$ of the components of D we construct cohomology with compact support along B as $H^*(X - D_A, j_!(\mathbf{L}))$, $D_A = \bigcup_{i \in A} D_i$. This is equal to the hypercohomology of the object in the derived category of X obtained by successively applying to $\mathbf{L}$ the operations $\mathbf{R}j_{a^*}$ for $a \in A$ and $j_{b!}$ for $b \in B$ ($j_i : X - D_i \to X$ the inclusions). The result is independent of the order in which we apply these operations, and it commutes with restriction to any smooth subscheme $Y \subset X$ which meets each X_M transversally, as well as with taking local cohomology with support in such a Y: These assertions can be checked locally, that is along the strict henselisation of X in a closed point $x \in X(K)$. There the fundamental group of the complement of D is the free abelian profinite group generated by "loops around the D_i," so the possible structures of $\mathbf{L}$ are known. Especially the simple objects (others are treated by devissage) are given by characters of the fundamental group, so for them everything looks like a product of one-dimensional situations, and by Künneth the assertion follows. Furthermore cohomology with compact support along D_B is functorial for maps $f: X_1 \to X_2$ with $f^{-1}(D_{A2}) \subset D_{A1}$ and $f^{-1}(D_{B2}) \supset D_{B1}$. Poincaré-duality gives a perfect pairing between cohomology of $\mathbf{L}$ with compact support along D_B, and cohomology of $\mathbf{L}^*$ with compact support along D_A. So we obtain a diagonal-class in the cohomology over $X \times X$ of $pr_1^*(\mathbf{L}) \otimes pr_2^*(\mathbf{L}^*)$ with compact support along $D_B \times X \cup X \times D_A$.

b) We wish to describe this class via local cohomology with support in the diagonal. We have the difficulty that the diagonal is not transverse to the stratification on $X \times X$. To resolve this blow up $X \times X$ along all its smooth subschemes $D_i \times D_i$, for $i \in I$. The result $\pi : Y \to X \times X$ is independent of the order in which we do these blow-ups, and it has a divisor with simple normal crossings D_Y (the divisor at infinity) whose components are the proper transforms $D_i^{(1)}$ and $D_i^{(2)}$ of $D_i \times X$ and $X \times D_i$, and the exceptional divisors E_i. Furthermore the proper transform of the diagonal is isomorphic to it, so we get a closed immersion $X \to Y$. It is transverse to D_Y. More precisely X does not meet any $D_i^{(1)}$ or $D_i^{(2)}$, and it intersects the E_i along D_i.

We denote by $\mathbf{M}$ the object in the derived category of $X \times X$ which is the exterior product of $\mathbf{R}j_*(\mathbf{L})$ and $j_!(\mathbf{L}^*)$, that is the extension of $pr_1^*(\mathbf{L}) \otimes pr_2^*(\mathbf{L}^*)$ by direct image with compact support along $X \times D$. On Y we define two objects $\mathbf{M}_1$ and $\mathbf{M}_2$, by extending $pr_1^*(\mathbf{L}) \otimes pr_2^*(\mathbf{L}^*)$ either with compact support along all $D_i^{(2)}$'s and E_i;s, or just along the $D_i^{(2)}$'s. There exist maps $\mathbf{M}_1 \to \mathbf{M}_2 \leftarrow \mathbf{L}\pi^*(\mathbf{M})$.

LEMMA 3.1. *These maps induce isomorphisms* $H^*(Y, \mathbf{M}_1) \cong H^*(Y, \mathbf{M}_2) \cong H^*(X \times X, \mathbf{M})$.

Proof. We even show the lemma for sheaves $pr_1^*(\mathbf{L}_1) \otimes pr_2^*(\mathbf{L}_2)$, $\mathbf{L}_1$ and $\mathbf{L}_2$ locally constant but not necessarily related. We claim that $\mathbf{R}\pi_*(\mathbf{M}_1) \cong \mathbf{R}\pi_*(\mathbf{M}_2) \cong \mathbf{M}$. This assertion is local in $X \times X$, so we can reduce to strict henselisations. By devissage it suffices to treat simple $\mathbf{L}$'s, and for them everything behaves as a product of situations where $X = \mathbf{P}^1$ is the projective line, $D = \{\infty\} \subset X$. We may pass to some ramified covering

$t \to t^m$ of $\mathbf{P}^1$ (m prime to n) and assume that in addition $\mathbf{L}_1$ and $\mathbf{L}_2$ are constant and equal to $\mathbf{Z}/n\mathbf{Z}$.

Now Y is the blow-up of $\mathbf{P}^1 \times \mathbf{P}^1$ in $\{\infty \times \infty\}$. As the assertion is already obvious away from this point, any discrepancy would show up in the global cohomologies $H^*(X \times X, \mathbf{M})$ and $H^*(Y, \mathbf{M}_{1/2})$. The cohomology of $\mathbf{M}$ is $H^*(\mathbf{A}^1, \underline{\mathbf{Z}/n\mathbf{Z}}) \otimes H^*_!(\mathbf{A}^1, \underline{\mathbf{Z}/n\mathbf{Z}}) = \mathbf{Z}/n\mathbf{Z}(-1)$ concentrated in degree 2 (generated by the class of the diagonal), and one easily shows that the same result holds for $H^*(Y, \mathbf{M}_{1/2})$. Thus the lemma is shown.

As an application we give a geometric construction of the class $c_\Delta \in H^{2d}(X \times X, \mathbf{M})(d)$ given by Poincaré-duality: It is the image of the identity under $H^0(X, \mathbf{R}j_*(\mathbf{L} \otimes \mathbf{L}^*) \cong H^{2d}{}_X(Y, \mathbf{M}_2)(d) \to H^{2d}(Y, \mathbf{M}_2)(d) \cong H^{2d}(X \times X, \mathbf{M})(d)$. A proof goes as follows. Define $\mathbf{N}$, $\mathbf{N}_1$ and $\mathbf{N}_2$ as the pullbacks of $\mathbf{M}$, $\mathbf{M}_1$ and $\mathbf{M}_2$ under exchange of the two factors in $X \times X$. That is we replace $\mathbf{L}$ by $\mathbf{L}^*$, and exchange the divisors for which we require compact support. There exist natural pairings $\langle\ ,\ \rangle : \mathbf{M} \times \mathbf{N} \to \underline{\mathbf{Z}/n\mathbf{Z}}$, $\mathbf{M}_2 \times \mathbf{N}_1 \to \underline{\mathbf{Z}/n\mathbf{Z}}$, and we have to check that for the class c defined above for any $x \in H^{2d}(X \times X, \mathbf{N})$ $\mathrm{tr}_{X \times X}(\langle c, x\rangle) = \mathrm{tr}_X(\langle 1, x|X\rangle)$. However this is fairly obvious if we use Lemma 3.1 to lift x to a class in $H^{2d}(Y, \mathbf{N}_1)$.

c) Etale cohomology is a generalisation of Galois cohomology. In our case this can be made more explicit. From now on we denote by X a scheme smooth and proper over Spec(V), where V is a complete discrete valuation ring with fraction field K of characteristic 0, and perfect residue-field k of characteristic $p > 0$. Furthermore $D \subset X$ is a divisor with normal crossings over V, and $X^\circ = X - D$ denotes its complement. If the geometric generic fibre of X is irreducible, and if $\mathbf{L}$ is a locally constant étale sheaf on $X^\circ \otimes_V \bar{K}$ we may compute its cohomology as follows (see [Fa]): For each open affine $U \subset X$ let $\Delta(U) = \pi_1(U^\circ \otimes_V \bar{K})$, and form the canonical complex $C^*(\Delta(U), \mathbf{L})$ which computes the cohomology of the $\Delta(U)$-module associated with $\mathbf{L}$. For varying U these complexes form a presheaf, and so for each hypercovering of X we obtain a double complex, combining the differentials in C^* with the generalised Cech-differentials. The inductive limit (over all hypercoverings) of these complexes computes the étale cohomology $H^*(X^\circ \otimes_V \bar{K}, \mathbf{L})$. We may replace the Zariski topology on X by the étale topology, provided we use pointed étale maps $U \to X$ (pointed means that we choose a lifting of the geometric generic point of X, which is automatic for open immersions). We will usually do this as the local rings become strictly henselian, which simplifies many arguments.

There is an essentially equivalent description which also works if $X \otimes_V \bar{K}$ is not irreducible. For it we only assume that X is either of finite type and smooth over V, or a localisation or strict henselisation of such. Consider the topos $\mathfrak{X}^\circ$ whose objects are the following: For each étale $U \to X$ we have given a locally constant étale sheaf $\mathbf{L}(U)$ on $U^\circ \otimes_V \bar{K}$, and for each X-map $U_1 \to U_2$ a transformation $\mathbf{L}(U_2)|U_1 \to \mathbf{L}(U_1)$, satisfying the usual compatibility. Furthermore for any étale covering $U = \cup U_i$ $\mathbf{L}(U)$ should be the maximal locally constant subsheaf of the kernel of (j denotes suitable étale maps) $\Pi j_*(\mathbf{L}(U_i)) \to \Pi j_*(\mathbf{L}(U_i \times_X U_j))$. The maps are the obvious ones. Also the construction can be done with the Zariski topology on X instead of the étale topology.

There exists a map ρ from the étale topos of $X^\circ \otimes_V K$ to $\mathfrak{X}^\circ$: The direct image of a sheaf $\mathbf{F}$ on $X^\circ \otimes_V K$ associates to U the locally constant subsheaf associated to the global sections of $\mathbf{F}$ on the universal covering of $U^\circ \otimes_V K$. The inverse image of a collection $\mathbf{L}(U)$ is the direct limit, over all hypercoverings $U. \to X$, of $j_*(\mathbf{L}(U.) = \ker(j_*(\mathbf{L}(U_0)) \to j_*(\mathbf{L}(U_1)))$. Here $j_*(\mathbf{L}(U))$ stands for a suitable product of direct images under inclusions $j : U^\circ \otimes_V K \to X^\circ \otimes_V K$. The cohomology $H^*(\mathfrak{X}^\circ, \mathbf{L})$ is given by an inductive limit of group cohomologies, as above, and the inverse image induces an isomorphism $H^*(\mathfrak{X}^\circ, \mathbf{L}) \cong H^*(X^\circ \otimes_V K, \rho^*(\mathbf{L}))$: We may replace X by $\operatorname{Spec}(R)$ for R a local ring in X. Then for purposes of cohomology $\mathfrak{X}^\circ$ is equivalent to the topos of locally constant sheaves on $X^\circ \otimes_V K$, and its cohomology is the cohomology of the fundamental group, so everything follows from the fact ([Fa], Chapter II, Lemma 2.3) that $X^\circ \otimes_V K$ is a $K(\pi, 1)$. Similar $H^*(X^\circ \otimes_V \bar{K}, \rho^*(\mathbf{L}))$ can be computed as the stalk (at the unique geometric point) of the direct image under $\mathfrak{X}^\circ \to \{$étale topos of $K\}$. We denote this stalk by $H^*(\mathfrak{X}^\circ \otimes \bar{K}, \mathbf{L})$.

d) We need the corresponding description for cohomology with compact support. For this we have to describe sheaves $\mathbf{F}$ on the étale site of $X \otimes_V K$. The divisor D at infinity defines a stratification of X. If all irreducible components $D_1, \ldots, D_n$ of D are smooth and their intersections are irreducible the strata can be indexed by subsets $M \subset \{1, \ldots, n\}$, corresponding to $X(M) = \cap_{i \in M} D_i$ which has pure codimension $|M|$. In general we still denote the strata by $X(M)$, with an index M, and define $|M|$ as the codimension of $X(M)$. Also $X(M)^\circ \subset X(M)$ is the complement of all substrata of smaller dimension. Sometimes we assume that all strata are irreducible and defined over K. This holds if the residue field k of V is algebraically closed.

An étale sheaf $\mathbf{F}$ on $X \otimes_V K$ can be described by its restrictions $\mathbf{F}(M)$ to $X(M)^\circ \otimes_V K$, together with the specialisation maps $\mathbf{F}(M) \to i^*j_*(\mathbf{F}(N))$ for inclusions $i : X(M) \otimes_V K \subset X(N) \otimes_V K, j : X(N)^\circ \otimes_V K \subset X(N) \otimes_V K$.

This is known in the case we have only two strata (glueing X from a closed subset and its complement), and in general we use induction, adding one stratum after the other. So let us define a topos $\mathfrak{X}$ as follows: An object of $\mathfrak{X}$ consists of a collection $\mathbf{L}(M) = \{\mathbf{L}(M, U)\}$ of objects of $\mathfrak{X}(M)^\circ$, together with compatible and transitive maps (for $i : X(M) \subset X(N)$, $j : X(N)^\circ \subset X(N)$, $U \to X(N)$ étale) $\mathbf{L}(M, U \cap X(M)) \to i^*j_*(\mathbf{L}(N, U))$. As before we obtain a homomorphism ρ from the étale topos of $X \otimes_V K$ to $\mathfrak{X}$, by glueing the previously defined maps for each $X(M)$, and ρ induces isomorphisms on cohomology: Replace X by $\mathrm{Spec}(R)$, R a strictly local ring in X. Now $\mathfrak{X}$ is equivalent to the topos of étale sheaves on $X \otimes_V K$ which are locally constant on each open stratum. For example for any locally constant $\mathbf{L}(M)$ on some $X(M)^\circ$ we obtain a complex of such sheaves with represents $\mathbf{R}j_{M*}(\mathbf{L}(M))$ ($j_M : X(M)^\circ \to X$ the inclusion). For such a complex ρ induces an isomorphism on cohomology, by the previous result applied to $X(M)$. The general case follows by devissage. Again $H^*(X \otimes_V \bar{K}, \rho^*(\mathbf{L}))$ is computed using the stalk $H^*(\mathfrak{X} \otimes \bar{K}, \ldots)$ of the direct image under $\mathfrak{X} \to \{$étale topos of $K\}$.

d) The inclusion $j : X^\circ \to X$ induces a map of topoi $j : \mathfrak{X}^\circ \to \mathfrak{X}$, commuting with ρ. For a sheaf of abelian groups $\mathbf{L}$ on $\mathfrak{X}^\circ$ we define an extension $j_!(\mathbf{L})$ in $\mathfrak{X}$ by the rule $j_!(\mathbf{L})(M) = (0)$ for $X(M) \neq X$, so that $\rho^*(j_!(\mathbf{L})) = j_!(\rho^*(\mathbf{L}))$. If $X = \mathrm{Spec}(R)$ with R strictly local (and V also strictly local) we can describe $H^*(\mathfrak{X}, j_!(\mathbf{L}))$ via Galois cohomology: Let $\Gamma = \pi_1(X^\circ \otimes_V K)$ and $\Gamma(M) = \pi_1(X(M)^\circ \otimes_V K)$ denote the fundamental groups, $I(M) \subset D(M) \subset \Gamma$ the inertia and decomposition group of $X(M)$, so that $I(M)$ is isomorphic to a product of $|M|$ copies of $\hat{\mathbf{Z}}$. We may assume that for $X(M) \subset X(N)$ we have inclusions $I(N) \subset I(M) \subset D(M) \subset D(N)$. Furthermore there exist isomorphisms (the inclusion $X(N) \subset X$ has a section) $\Delta(N) \to D(N)/I(N)$, which map the inertia of $X(M) \subset X(N)$ isomorphic onto $I(M)/I(N)$.

Objects of $\mathfrak{X}$ are given by systems of $\Gamma(M)$-modules $\mathbf{L}(M)$, and $\Gamma(M)$-maps $\mathbf{L}(M) \to \mathbf{L}(N)$ if $X(M) \subset X(N)$. For example for any Γ-module $\mathbf{L}$ the derived direct image $\mathbf{R}j_*(\mathbf{L})$ can be represented by the complexes which associate to $X(M)$ $D^*(\Gamma, \mathbf{L})^{I(M)}$, the $I(M)$-invariants in the canonical acyclic resolution of $\mathbf{L}$ by $D^n(\Gamma, \mathbf{L}) =$ continuous maps $\Gamma^{n+1} \to \mathbf{L}$, etc., and its cohomology is represented by $C^*(\Gamma, \mathbf{L}) = \Gamma$-invariants of $D^*(\Gamma, \mathbf{L})$. Similar for each stratum $X(N)$ we define a complex $\mathbf{R}j_{N*}(D^*(\Gamma, \mathbf{L})^{I(N)})$ by associating to M either (0) (if $M \not\subset N$) or $D^*(\Gamma, \mathbf{L})^{I(M)}$ (if $M \subset N$). It corresponds to the derived direct image of the restriction of $\mathbf{R}j_*(\mathbf{L})$ to $X(N)^\circ$. There is a resolution of $j_!(\mathbf{L})$ by the total complex of a double complex made from these, by putting in degree n the direct sum of the $\mathbf{R}j_{N*}(D^*(\Gamma, \mathbf{L})^{I(N)})$ for $|N| = n$. It follows that $H^*(\mathfrak{X}, j_!(\mathbf{L}))$ can be represented by the corresponding complex made out of the $D^*(\Gamma, \mathbf{L})^{D(N)}$, with maps given by the identity on $D^*(\Gamma, \mathbf{L})$ (as $D(M) \subset D(N)$ if $X(M) \subset X(N)$). Finally we may replace each $D^*(\Gamma, \mathbf{L})^{D(N)}$ by the quasiisomorphic complex $D^*(D(N), \mathbf{L})^{D(N)} = C^*(D(N), \mathbf{L})$. Similarly for geometric cohomology, using the geometric fundamental groups $\Delta \subset \Gamma$ and $\Delta(M) \subset \Gamma(M)$.

Essentially the same construction can be performed for any punctured irreducible affine $U = \operatorname{Spec}(R)$ (punctured means that we prescribe an embedding of R into the algebraic closure of the function-field $K(X)$ of X), provided all strata $X(M)$ are geometrically irreducible: Let $\Gamma = \pi_1(U^\circ \otimes_V K)$, $\bar{R}$ the integral closure of R in the covering defined by Γ. For each stratum $U(M) = \operatorname{Spec}(R(M))$ chose a prime ideal $\mathfrak{q}(M) \subset \bar{R}$ which lies over the prime $\mathfrak{p}(M)$ of R which defines $U(M)$. This defines decomposition groups $D(M) \subset \Gamma$. If $U(M) \subset U(N)$ there exists $\sigma \in \Gamma$ with $\mathfrak{q}(M) \subset \sigma(\mathfrak{q}(N))$, and σ is unique up to right multiplication by $D(N)$. The action of σ defines a well defined map from $D^*(\Gamma, \mathbf{L}(U))^{D(N)}$ to $D^*(\Gamma, \mathbf{L}(U))^{D(M)}$, and we may form a double complex from the $D^*(\Gamma, \mathbf{L}(U))^{D(N)}$. Up to canonical isomorphism this double complex is independent of the choice of the $\mathfrak{q}(M)$'s, as two such choices differ by conjugation by a σ unique up to $D(M)$. Furthermore this double complex behaves functorial in U: For a map $U_1 \to U_2$, corresponding to $R_2 \subset R_1$, consider $\bar{R}_1$ and $\bar{R}_2$ as subrings of the algebraic closure of the function field of X. Then $\bar{R}_2 \subset \bar{R}_1$, and we get $\Gamma_1 \to \Gamma_2$ and $D^*(\Gamma_2, \mathbf{L}(U_2)) \to D^*(\Gamma_1, \mathbf{L}(U_1))$, which induce a map of double complexes. It follows that for any hypercovering of X by pointed U's we obtain a triple complex. Form the inductive limit, over all hypercoverings, of the associated simple complexes. The complex computes $H^*(\mathfrak{X}, j_!(\mathbf{L}))$: There is a map one way, as the complexes $D^*(\Gamma, \mathbf{L}(U))^{I(M)}$ define a (not necessarily acyclic) resolution of $j_!(\mathbf{L})$ related to the direct image of the complex on X° given by $D^*(\Gamma, \mathbf{L}(U))$. Now reduce to the case of a strictly local X, and there the result has been checked before.

A similar description holds for the extension by zero along part of the boundary, as in the beginning of this chapter: Only use $D^*(\Gamma, \mathbf{L})^{D(N)}$ or $C^*(D(N), \mathbf{L})$ related to strata $X(N)$ appearing in this part of the divisor at ∞.

Again for geometric cohomology we have to use geometric fundamental groups Δ, and then everything works as well.

e) Define a sheaf of rings $\bar{R}$ on $\mathfrak{X}^\circ$ by associating to each étale $U \to X$, with $U = \operatorname{Spec}(R)$ affine, small and irreducible, the locally constant sheaf on $U^\circ \otimes_V K$ given by $\bar{R}$, the maximal extension of R which is étale over $U^\circ \otimes_V K$. Similar for $\bar{R}/p \cdot \bar{R}$, and the ideal $\bar{J} \subset \bar{R}$ "defining the boundary" (compare [Fa], Chapter I, f)). If $\mathbf{L}$ denotes any sheaf of p-torsion-modules on $\mathfrak{X}$ we can form the cohomologies $\mathcal{H}^*(X^\circ, \mathbf{L}) = H^*(\mathfrak{X}^\circ, \mathbf{L} \otimes \bar{R})$ and $\mathcal{H}^*_!(X^\circ, \mathbf{L}) = H^*(\mathfrak{X}^\circ, \mathbf{L} \otimes \bar{J})$. They take values in the category of $\bar{V}$-modules modulo almost isomorphism. There are obvious maps $\mathcal{H}^*_!(X^\circ, \mathbf{L}) \to \mathcal{H}^*(X^\circ, \mathbf{L}) \leftarrow H^*(\mathfrak{X}^\circ, \mathbf{L})$. We also construct a map $H^*_!(\mathfrak{X}^\circ, \mathbf{L}) \to \mathcal{H}^*_!(X^\circ, \mathbf{L})$, at least in the case where all strata are geometrically irreducible: For any affine punctured $U = \operatorname{Spec}(R) \to X$ choose prime ideals $\mathfrak{q}(M) \subset \bar{R}$ lying above the defining ideals of the strata $U(M)$. As before we construct a double complex from the complexes $D^*(\Gamma, \mathbf{L}(U) \otimes (\bar{R}/\mathfrak{q}(M)))^{D(M)}$, and by letting U vary in a hypercovering we get a triple complex. We claim that the inductive limit (over all hypercoverings) of these complexes has cohomology almost equal to $\mathcal{H}^*_!(X^\circ, \mathbf{L})$: There is a map from $\mathcal{H}^*_!(X^\circ, \mathbf{L})$ into this, induced from $\bar{J} \subset \bar{R}$. The rest is a local statement, that is we may assume that $X = \operatorname{Spec}(R)$ with R strictly local. Also we may assume that p annihilates $\mathbf{L}$. Choose an étale map $V[T_1, \ldots, T_d] \to R$ such that the divisor at ∞ has equation $\{T_1 \cdot \cdots \cdot T_d = 0\}$. By the theory of almost unramified extensions we may replace $\bar{R}$ by R_∞, the extension of R obtained by enlarging V to $\bar{V}$ and adjoining all roots of the T_i, and $\mathbf{L} \otimes J$ by its $\operatorname{Gal}(\bar{R}/R_\infty)$-invariants (which is almost flat over R_∞/pR_∞). However now all $D(M)$ are equal to Γ, and J has a resolution which in degree m consists of the direct sum of all $R_\infty/\mathfrak{q}(M)$ with $|M| = m$. As these are torsion-free the assertion follows easily. Again there is a variant for the geometric case.

Of course for any decomposition $D = D_A \cup D_B$ of the divisor at ∞ we obtain a version of $\mathcal{H}^*$ with compact support along D_B. These are functorial for maps $f: X_1 \to X_2$ with $f^{-1}(D_{A2}) \subset D_{A1}, f^{-1}(D_{B2}) \supset D_{B1}$. Finally from the description of $H^*_!(\mathfrak{X}^\circ, \mathbf{L})$ via Galois cohomology we get functorial transformations $H^*_!(\mathfrak{X}^\circ, \mathbf{L}) \to \mathcal{H}^*_!(X^\circ, \mathbf{L})$, induced by $\mathbf{L} \to \mathbf{L} \otimes (\bar{R}/\mathfrak{q}(M))$. This also works for compact support along only a part of the boundary. All these maps respect cup-products like $\mathcal{H}^* \times \mathcal{H}^* \to \mathcal{H}^*$ and $\mathcal{H}^* \times \mathcal{H}^*_! \to \mathcal{H}^*_!$. We intend to show that they induce almost isomorphisms from $H^*(X^\circ \otimes_V \bar{K}, \mathbf{L}) \otimes \bar{V}$ to $\mathcal{H}^*(X^\circ \otimes \bar{K}, \mathbf{L})$, but for this we first have to derive some properties of $\mathcal{H}^*$. Let us remark that by [Fa], Chapter II this is already known for the case of constant sheaves $\mathbf{L} = \mathbf{Z}/n\mathbf{Z}$.

f) The most important fact which we need is Poincaré-duality: Let $\mathbf{L}^* = \underline{\operatorname{Hom}}(\mathbf{L}, \underline{\mathbf{Z}/n\mathbf{Z}})$, assume that X is proper and smooth over V of pure relative dimension d, and that $\mathbf{L}$ is a locally constant finite étale sheaf on

$X^\circ \otimes_V \bar{K}$. The cup product induces a pairing $\mathcal{H}^*(X^\circ \otimes \bar{K}, \mathbf{L}) \times \mathcal{H}^{2d-*}{}_!(X^\circ \otimes \bar{K}, \mathbf{L}^*) \to \mathcal{H}^{2d}{}_!(X^\circ \otimes \bar{K}, \underline{\mathbf{Z}/n\mathbf{Z}}) \to \bar{V}/n\bar{V}(-d)$, which is an almost perfect duality (note that $\bar{V}/n\bar{V}$ is almost injective as a module over itself): As in [Fa] we use a combination of Serre-duality and local duality. Also we may (by devissage) assume that $n = p$, and that the residue field k of V is algebraically closed. For the local part we replace X by Spec(R), R strictly henselian. Choose an étale map $V[T_1, \ldots, T_d] \to R$ with the usual conditions about strata, and let $R_\infty \subset \bar{R}$ denote the extension of R obtained by adjoining all roots of the T_i, and by extending V to $\bar{V}$. Then $\mathcal{H}^*(X^\circ \otimes \bar{K}, \mathbf{L})$ and $\mathcal{H}^*{}_!(X^\circ \otimes \bar{K}, \mathbf{L}^*)$ are given by Galois cohomology of $\mathbf{L} \otimes \bar{R}$ and $\mathbf{L}^* \otimes \bar{J}$. We claim that the cup product pairs them almost perfectly into $\Omega_{R/V} \otimes (R \otimes_V \bar{V})\hat{}(-d)$, as in [Fa]. For this we may replace $\mathbf{L} \otimes \bar{R}$ and $\mathbf{L}^* \otimes \bar{J}$ by their invariants L_∞ and $J_\infty \cdot L_\infty$ under Gal$(\bar{R}/R_\infty)$ (which are almost projective over R_∞/pR_∞), and the group Δ by $\hat{\mathbf{Z}}(1)^d$. Now the complexes computing the cohomology are already in duality, and the local assertion follows. We also note that the cohomology satisfies certain finiteness conditions: Write R_∞ as the increasing union of subrings R_n which are of finite type over R. If $L_n \subset L_\infty$ denotes suitable finitely generated submodules of the invariants under Gal(R_∞/R_n), the maps $L_n \otimes_{R_n} R_\infty \to L_\infty$ are isomorphisms up to p^ϵ, provided n is big enough (depending on $\epsilon > 0$). It follows (using the Hochschild spectral sequence, and the known result about cohomology of Gal(R_∞/R_n) with values in R_∞/pR_∞ and J/pJ) that for any positive $\epsilon > 0$ the cohomologies $\mathcal{H}^*(X^\circ \otimes \bar{K}, \mathbf{L})$ and $\mathcal{H}^*{}_!(X^\circ \otimes \bar{K}, \mathbf{L}^*)$ are finitely presented up to p^ϵ, that is there exist maps from finitely presented $R \otimes_V \bar{V}$-modules into them with kernel and cokernel annihilated by p^ϵ.

Having made these arguments we note that there is no need to assume that R is strictly local. It suffices that R is small, that is there is an étale map $V[T_1, \ldots, T_d] \to R$ as above. It also follows that the corresponding cohomology $\mathcal{H}^*(X^\circ \otimes \bar{K}, \mathbf{L})$ and $\mathcal{H}^*{}_!(X^\circ \otimes \bar{K}, \mathbf{L}^*)$ behaves well under étale localisation: For any étale $R_1 \to R_2$ the cohomology of Spec(R_2) is obtained (up to $\mathfrak{m}$-torsion) from that of Spec(R_1) by tensoring with R_2. So if we choose any hypercovering of X by small $U =$ Spec(R), and form the direct image of the complexes above (constructed from the $D^*(\Delta, \mathbf{L}(U) \otimes (\bar{R}/q(M)))^{D(M)}$) which compute $\mathcal{H}^*(X^\circ \otimes \bar{K}, \mathbf{L})$ and $\mathcal{H}^*{}_!(X^\circ \otimes \bar{K}, \mathbf{L}^*)$, we obtain complexes of quasicoherent sheaves on the Zariski topos of X which compute $\mathcal{H}^*(X^\circ \otimes \bar{K}, \mathbf{L})$ and $\mathcal{H}^*{}_!(X^\circ \otimes \bar{K}, \mathbf{L}^*)$. Refinement of hypercoverings induces almost quasiisomorphisms. Also the cohomologies are finitely presented up to p^ϵ, for any $\epsilon > 0$. Finally cup-product gives a pairing into the complex which computes $\mathcal{H}^*_!(X^\circ \otimes \bar{K}, \underline{\mathbf{Z}/n\mathbf{Z}}^*)$, and into the differentials. As Serre duality now applies (easily checked) we derive a perfect pairing on cohomology, hence Poincaré-duality holds:

THEOREM 3.2. *Assume X is proper and smooth over V, of pure relative dimension d, and that $\mathbf{L}$ is a locally constant étale sheaf of finite $\mathbf{Z}/n\mathbf{Z}$-modules on $X^\circ \otimes_V \bar{K}$. Then cup-product induces an almost perfect pairing $\mathcal{H}^*(X^\circ \otimes \bar{K}, \mathbf{L}) \times \mathcal{H}^*_!(X^\circ \otimes \bar{K}, \mathbf{L}^*) \to \bar{V}/n\bar{V}(-d)$.*

These pairings are compatible with the map from étale cohomology into $\mathcal{H}^*$. Also there are versions involving cohomology with compact support along part of the boundary.

We also need the analogue of Lemma 3.1. For this assume that we blow up X along the intersection of two geometrically irreducible components D_1 and D_2 of D. The blow-up $\tilde{X}$ has as divisor at infinity the proper transforms of the components of D, as well as the exceptional divisor E as a new component. For $\mathbf{L}$ on X as before, we consider the cohomology of its pullback on $\tilde{X}$. The most interesting case occurs if use cohomology with compact support along one of the D_i's, say D_1, and along some other components different from D_2. On $\tilde{X}$ we then chose proper support along the proper transform of D_1 and these other components, and along E we have a choice: Denote by $\mathcal{H}^*(\tilde{X}^\circ \otimes \bar{K}, \mathbf{L}_1)$ the result of requiring compact support along E, by $\mathcal{H}^*(\tilde{X}^\circ \otimes \bar{K}, \mathbf{L}_2)$ the other choice. Then the maps $\mathcal{H}^*(\tilde{X}^\circ \otimes \bar{K}, \mathbf{L}_1) \to \mathcal{H}^*(\tilde{X}^\circ \otimes \bar{K}, \mathbf{L}_2) \leftarrow \mathcal{H}^*(\tilde{X}^\circ \otimes \bar{K}, \mathbf{L})$ are almost isomorphisms: Reduce to the case of $X = \mathrm{Spec}(R)$, R strictly local, and choose an étale $V[T_1, \ldots, T_d] \to R$, as usual. We may assume that D_i is defined by $\{T_i = 0\}$. The blow-up $\tilde{X}$ is covered by two small affines $\tilde{X} = U_1 \cup U_2$, with intersection U_{12}. Here $U_i = \mathrm{Spec}(R_i)$, $R_1 = R[T_2/T_1]$, $R_2 = R[T_1/T_2]$, $R_{12} = R[T_1/T_2, T_2/T_1]$. The group Δ is the same for R as well as all R_i, and the sequence $0 \to \bar{R} \to \bar{R}_1 \oplus \bar{R}_2 \to \bar{R}_{12} \to 0$ is almost exact (use the usual R_∞'s), so $\bar{R}$ is almost quasi-isomorphic to the complex $\bar{R}_1 \oplus \bar{R}_2 \to \bar{R}_{12}$. The corresponding results holds for $R(M)$'s, and also for the ideals J defining the boundary conditions. Finally tensoring with $\mathbf{L}$ and applying $H^*(\Delta,)$ gives the required isomorphisms. It follows that Lemma 3.1 carries over.

g) As $\mathcal{H}^*$ satisfies Poincaré-duality we obtain maps $f_* : \mathcal{H}^*(Y^\circ \otimes \bar{K}, f^*(\mathbf{L})) \to \mathcal{H}^*(X^\circ \otimes \bar{K}, \mathbf{L})$ adjoint to f^*, for any proper map of smooth schemes $f: Y \to X$ with $f^{-1}(D_X) \supset D_Y$. The same for proper (if $f^{-1}(D_X) \subset D_Y$) and also for mixed support. We want to give a geometric description of f_* provided f is a closed immersion of a smooth divisor, $f^{-1}(D_X) = D_Y$, and Y meets all strata in X transversally. In étale cohomology f_* is represented by the connecting homomorphism associated to the exact sequence $0 \to \mathbf{L} \to \mathbf{R}j_*(\mathbf{L}) \to i_*(\mathbf{L})(-1)[-1] \to 0$ ($j : X - Y \to X$ and $i : Y \to X$ the inclusions). We want a similar description for the theory $\mathcal{H}^*$, at least in the case that the residue-field k of V is algebraically closed. Let us first start locally (compare [Fa], Chapter I, Theorem 4.5):

Assume $X = \mathrm{Spec}(R)$ is affine, and that there exists an étale map $V[T_0, T_1, \ldots, T_d] \to R$ such that Y is defined by $\{T_0 = 0\}$, and that the divisor at infinity has equation $\{T_1 \cdot \cdots \cdot T_d = 0\}$. Furthermore we assume that all strata in X and Y are small, that is there exist étale maps into products of copies of the multiplicative group $\mathcal{G}_m$. We denote by Δ the fundamental group of $(X - Y)^\circ \otimes_V \bar{K}$, by $\tilde{\Delta} = \Delta/I$ the fundamental group of $X^\circ \otimes_V \bar{K}$, and by $\bar{R}$ and $\tilde{R}$ the corresponding maximal extensions of R. If $\mathbf{L}$ is any finite Δ-module the cohomology $H^i(I, \mathbf{L} \otimes \bar{R})$ almost vanishes for $i \neq 0, 1$: The standard-trick allows us to replace $\bar{R}$ by R_∞, $\mathbf{L} \otimes \bar{R}$ by its invariants under $\mathrm{Gal}(\bar{R}/R_\infty)$, and I by the corresponding inertia which now has cohomological dimension one. Also $H^0(I, \mathbf{L} \otimes \bar{R})$ contains $\mathbf{L} \otimes \tilde{R}$. Furthermore if we choose a prime $\mathfrak{q} \subset \bar{R}$ lying above the ideal $T_0 \cdot R$ defining Y we obtain $H^1(I, \mathbf{L} \otimes \bar{R}) \to H^1(\hat{\mathbf{Z}}(1), \mathbf{L} \otimes \bar{R}/\mathfrak{q}) = \mathbf{L} \otimes (\bar{R}/\mathfrak{q})(-1)$. So finally we have an exact sequence of complexes $0 \to C^*(\tilde{\Delta}, \mathbf{L} \otimes \tilde{R}) \to C^*(\Delta, \mathbf{L} \otimes \bar{R}) \to ? \to 0$, where we do not need to know very much about ? except that it is functorial, admits a cup product multiplication by $C^*(\tilde{\Delta}, \mathbf{L} \otimes \tilde{R})$, and a map (in the sense of derived categories) $? \to C^*(\Delta_Y, \mathbf{L} \otimes \bar{R}_Y(-1))[-1]$ (Δ_Y the fundamental group of $Y^\circ \otimes_V \bar{K}$, etc.). As all these are functorial in R we can globalise: Let $\mathrm{Spec}(R)$ vary in a hypercovering of X to get double complexes. Then form the inductive limit over all hypercoverings to obtain a long exact sequence $\cdots \to \mathcal{H}^i(X^\circ \otimes \bar{K}, \mathbf{L}) \to \mathcal{H}^i((X - Y)^\circ \otimes \bar{K}, \mathbf{L}) \to ?^i \to \mathcal{H}^{i+1}(X^\circ \otimes \bar{K}, \mathbf{L}) \to \cdots$, with maps $?^i \to \mathcal{H}^{i-1}(X^\circ \otimes \bar{K}, \mathbf{L})(-1)$ which are almost isomorphisms if $\mathbf{L}$ is constant ([Fa], Chapter II, Theorem 1.2, v)). The same holds for cohomology with compact support, or compact support along only part of the boundary (Use the description of $\mathcal{H}^*$ using triple complexes with coefficients $\mathbf{L} \otimes \bar{R}(M)$, instead of double complexes with $\mathbf{L} \otimes J$). We claim that the connection map $?^i \to \mathcal{H}^{i+1}(X^\circ \otimes \bar{K}, \mathbf{L})$ is the composition of $?^i \to \mathcal{H}^{i-1}(Y^\circ \otimes \bar{K}, \mathbf{L})(-1)$ with the direct image (under the inclusion $i : Y \to X$) $i_* : \mathcal{H}^{i-1}(Y^\circ \otimes \bar{K}, \mathbf{L})(-1) \to \mathcal{H}^{i+1}(X^\circ \otimes \bar{K}, \mathbf{L})$. Taking cup-products with elements from $\mathcal{H}^*_!(X^\circ \otimes \bar{K}, \mathbf{L}^*)$ we reduce to the case of a constant $\mathbf{L} = \underline{\mathbf{Z}/n\mathbf{Z}}$, where the result is known by [Fa], Chapter II, proof of Theorem 2.4. As a corollary we derive that the transformation from étale cohomology $H^*(X^\circ$

$\otimes_V \bar{K}, \mathbf{L}) \to \mathcal{H}^*(X^\circ \otimes \bar{K}, \mathbf{L})$ commutes with i_*: In the construction above replace the continuous coefficients $\mathbf{L} \otimes \bar{R}$ by $\mathbf{L}$, to get a geometric description of i_* in the étale topology. The inclusion of $\mathbf{L}$ into $\mathbf{L} \otimes \bar{R}$ induces a map on all complexes, and the result follows. It also holds for cohomology with compact support, or the mixed cases.

We derive that the transformation $\alpha : H^*(X^\circ \otimes_V \bar{K}, \mathbf{L}) \to \mathcal{H}^*(X^\circ \otimes \bar{K}, \mathbf{L})$ respects direct images i_* for closed immersions of smooth divisors (transversal to infinity). We now show that this holds for any closed immersion $i : Y \subset X$ of a smooth subscheme of pure codimension $\mathbf{e}$: Denote by $\tilde{X}$ the blow-up of X along Y, and by $\tilde{i} : E \subset \tilde{X}$ the inclusion of the exceptional divisor. The projections $\tilde{X} \to X$ and $E \to Y$ are both denoted by π. $E = \mathbf{P}(\mathcal{E})$ is the projective bundle over Y associated to the conormal bundle $\mathcal{E}$ of Y in X, so that on E there is a canonical quotient $\mathcal{O}(1) = \mathcal{E}/\mathcal{N}$ of $\pi^*(\mathcal{E})$. Let $c \in H^{2e-2}(E \otimes_V \bar{K}, \underline{\mathbf{Z}/n\mathbf{Z}})(e-1)$ denote the highest Chern-class $(-1)^{e-1} \cdot c_{e-1}(\mathcal{N})$, and similar for its image in $\mathcal{H}^*(E \otimes \bar{K}, \underline{\mathbf{Z}/n\mathbf{Z}})(e-1)$. In both (almost isomorphic) theories $\pi_*(c) = 1$. It is well-known that in étale cohomology for any $x \in H^*(Y^\circ \otimes_V \bar{K}, \mathbf{L})$ we have $\pi^* i_*(x) = \tilde{i}_*(\pi^*(x) \cdot c))$. Applying α (which commutes with $\tilde{i}_*$) we obtain $\pi^*\alpha(i_*(x)) = \tilde{i}_*(\pi^*(\alpha(x)) \cdot c)$. Finally as $\pi_*\pi^* = id$ on $\mathcal{H}^*(X^\circ, \mathbf{L})$ we apply π^* to obtain

$$\alpha(i_*(x)) = \pi_*\tilde{i}_*(\pi^*(\alpha(x)) \cdot c)$$

$$= i_*\pi_*(\pi^*(\alpha(x)) \cdot c) = i_*(\alpha(x) \cdot \pi_*(c)) = i_*(\alpha(x)).$$

This argument also applies to cohomology with compact support, or with compact support along part of the boundary.

h) We intend to apply all this to the diagonal embedding $X \subset X \times X$, X proper and smooth over V of pure relative dimension d. As this is not transversal to the stratification we first blow up $X \times X$ along all $D_i \times D_i$, D_i a component of the divisor at infinity. Call the result Y, and denote by $\pi : Y \to X \times X$ the projection. If $\mathbf{M}$ denotes the exterior product of $\mathbf{R}j_*(\mathbf{L})$ and $j_!(\mathbf{L}^*)$, the identity of $H^*(X^\circ \otimes_V \bar{K}, \mathbf{L})$ corresponds by Poincaré-duality to the class of the diagonal in $H^{2d}((X \times X) \otimes_V \bar{K}, \mathbf{M})(d)$. Identifying this with $H^{2d}(Y \otimes_V \bar{K}, \mathbf{M}_2)(d)$ (Lemma 3.1) this class is the direct image of the identity in $H^0(X, \mathbf{M}|X)$ under the diagonal inclusion $X \to Y$. The same holds for $\mathcal{H}^*(X^\circ \otimes \bar{K}, \mathbf{L})$, and it follows that the transformation $H^*(X^\circ \otimes_V \bar{K}, \mathbf{L}) \otimes_V \bar{V} \to \mathcal{H}^*(X^\circ \otimes \bar{K}, \mathbf{L})$ respects the characteristic class of the diagonal. Equivalently its adjoint is an isometry for Poincaré-duality (as it is itself), so both maps are isomorphisms. As usual the same argument applies to cohomology with compact support, or compact support along only part of the boundary. So far we have assumed that the residue-field k of V is algebraically closed, so that all strata are geometrically irreducible and we can use Galois-cohomology. However this is not necessary, as we may always replace V by its maximal unramified extension. We have shown:

THEOREM 3.3. *Suppose* $\mathbf{L}$ *is a locally constant étale sheaf of finite abelian groups on* $X \otimes_V \bar{K}$. *Then the transformation* $H^*(X^\circ \otimes_V \bar{K}, \mathbf{L}) \otimes_V \bar{V} \to \mathcal{H}^*(X^\circ \otimes \bar{K}, \mathbf{L})$ *is an almost isomorphism. The same holds for cohomologies with proper support.*

i) In the definition of $\mathcal{H}^*$ we use étale coverings of X. Instead we could have worked with étale coverings of the formal scheme $\hat{X}$ obtained by completing X along its closed fibre. In the description via Galois-cohomology this would amount to replace the extensions $\bar{R}$ by $\hat{R}^-$, the maximal extensions of the p-adic completion $\hat{R}$ unramified in characteristic 0. As for small R's the corresponding Galois-cohomologies of $\mathbf{L} \otimes \bar{R}$ and $\mathbf{L} \otimes \hat{R}^-$ almost coincide (by the standard trick) the final result would be the same. So in the following we usually work with the formal scheme $\hat{X}$, which allows us to use the results of Chapter I.

IV. Crystalline cohomology. a) As usual V denotes a complete discrete valuation-ring of mixed characteristics, with perfect residue field k. During most of this chapter we assume that $V = W(k)$ is the ring of Witt vectors over k, but occasionally if this is not true we denote by $V_0 \subset V$ this subring, and $K_0 \subset K$ the corresponding extension of fraction-fields. V_0 and K_0 admit the canonical Frobenius automorphism. If X is a smooth V-scheme of relative dimension d the de Rham cohomology of X over V can be identified with the crystalline cohomology of X over V, or even that of $X_0 = X \otimes_V V/pV$ (see [B1]). This also holds with coefficients: Good coefficients are given by crystals of vector bundles, that is by bundles on X with an integrable connection. Also usually we consider filtered bundles, with the connection satisfying Griffiths transversality. Then the de Rham cohomology $H^*_{\text{crys}}(X/V, \mathcal{E}) = H^*(X, \mathcal{E} \otimes \Omega^*_{X/V})$ admits a filtration by the images of $H^*_{\text{crys}}(X/V, F^q(\mathcal{E})) = H^*(X, F^q(\mathcal{E} \otimes_V \Omega^*_{X/V}))$.

In general the crystalline cohomology of a crystal $\mathcal{E}$ on X_0/V can be computed as follows: Choose any open affine covering $X_0 = \bigcup_{i \in I} U_{i0}$, and embed the U_{i0} into smooth V-schemes U_i. For any finite subset $J \subset I$ we let U_{J0} denote the intersection of the U_{i0} for $i \in J$, by U_J the product $\Pi_{i\in J} U_i$, and by W_J the divided power-hull of the diagonally embedded subscheme $U_{J0} \subset U_J$. The W_J form a simplicial scheme (in degree n we use the disjoint union of all W_J where J has cardinality $n + 1$), and so from the global sections of the de Rham complexes $\Gamma(W_J, \mathcal{E} \otimes \Omega^*_{U_J/V})$ we form a double complex whose associated total complex computes the crystalline cohomology of $\mathcal{E}$. The result is independent of the choices involved, and the result also holds in the filtered context. Finally in this description it is quite easy to see how endomorphisms of X_0 operate on the cohomology.

Now suppose that $V = V_0$. The Frobenius on Spec(V) extends to the Frobenius endomorphism of $X \otimes_V k$, and we want to study the cohomology of crystals with Frobenius-action, mostly that of objects in $\mathcal{MF}^\nabla(X)$. The Frobenius operates on this cohomology, and we intend to show that under some technical restrictions the cohomology will be in $\mathcal{MF}(V)$. So suppose $(\mathcal{E}, F^i(\mathcal{E}), \varphi, \nabla)$ is an object of $\mathcal{MF}^\nabla_{[0,a]}(X)$, for some integer a between 0 and $p - 2$. Choose an open affine covering $\{U_{i0}\}$ of X_0, as before, embed U_{i0} into U_i which is smooth over V, and choose Frobenius lifts Φ_i on the p-adic completion $\hat{U}_i$. By definition of $\mathcal{MF}^\nabla$ and the theory in Chapter I we obtain compatible systems of parallel maps $\varphi : \Phi_i^*(\tilde{\mathcal{E}}) \to \mathcal{E}$, which induce transformations on the de Rham complex $\mathcal{E} \otimes \Omega^*_{U_J/V}$ as well as their global sections $\Gamma(W_J, \mathcal{E} \otimes \Omega^*_{U_J/V})$. If $I \subset \mathcal{O}_{W_J}$ denotes the ideal defining U_{J0} we know that for $i \leqslant p - 1$ $\Phi^*(I^{[i]})$ is divisible by p^i, and we obtain for such i transformations (induced by Φ^*/p^i) $H^*_{\text{crys}}(X/V, F^i(\mathcal{E})) \otimes_{V\ \Phi} V \to H^*_{\text{crys}}(X/V, \mathcal{E})$ which define objects $H^*(X, \mathcal{E})$ in $\mathcal{MF}_{\text{big}}(V)$. Note that $H^b_{\text{crys}}(X/V, F^i(\mathcal{E}))$ vanishes if $i > a + \min(b, d)$.

b) We claim that for $a + \min(b, d) \leqslant p - 2$ $H^a(X, \mathcal{E})$ is actually in $\mathcal{MF}(V)$. For this let us assume first that $\mathcal{E}$ is annihilated by p, so that $\tilde{\mathcal{E}} = gr_F(\mathcal{E})$. We shall show the result by a local computation, so first assume that $X = \text{Spec}(R)$ is affine, and that we have chosen a Frobenius lift Φ on $\hat{R}$. Now $\mathcal{E}$ is given by an $M \in \mathcal{MF}^\nabla(R)$ and $H^*_{\text{crys}}(X/V, \mathcal{E})$ is represented by the de Rham cohomology $M \otimes \Omega^*_{R/V}$. As the differential $d\Phi_*$ is divisible by p the maps $d\Phi_*/p^q$ on $\Omega^q_{R/V}$ define a transformation $gr_F(\Omega^*_{R/V}) \otimes_{R\ \Phi} R/pR \to \Omega^*_{R/V} \otimes R/pR$, which induce the Φ-linear Cartier-isomorphisms $\Omega^q_{R/V} \otimes_R R/pR \cong H^q(\Omega^*_{R/V} \otimes R/pR)$. The same holds after tensoring with M, that is $\Omega^q_{R/V} \otimes_R \tilde{M} \cong H^q(\Omega^*_{R/V} \otimes M)$: $\tilde{M} = gr_F(M)$ is a graded R/pR-module with a map $\tilde{M} \to \tilde{M} \otimes_R \Omega_{R/V}$ of degree one induced from ∇. These data determine the connection on M and thus $M \otimes_R \Omega_{R/V}$, as $\varphi : \tilde{M} \otimes_{R\ \Phi} R \cong M$ is parallel. It follows that we can make a devissage on M, by filtering $\tilde{M}$ by its submodules generated by homogeneous elements of degree $\geqslant i$, and thus reduce to $M = R/pR$ with the trivial connection. However there the result is known.

Now globalising we get that φ induces for $a + \min(b, d) \leqslant p - 2$ Φ-linear isomorphisms $H^b{}_{\mathrm{crys}}(X/V, gr_F(\mathcal{E})) \cong H^b{}_{\mathrm{crys}}(X/V, \mathcal{E})$. The restriction on b comes in because globally the transformations $\varphi_i : H^*{}_{\mathrm{crys}}(X/V, gr_F{}^i(\mathcal{E})) \to H^*{}_{\mathrm{crys}}(X/V, \mathcal{E})$ are well-defined only for $i \leqslant p - 2$, so we need that $H^b{}_{\mathrm{crys}}(X/V, gr_F{}^i(\mathcal{E}))$ vanishes for other i's. For $i = p - 1$ we still get a semilinear transformation $H^*{}_{\mathrm{crys}}(X/V, F^{p-1}(\mathcal{E})) \to H^*{}_{\mathrm{crys}}(X/V, \mathcal{E})$ which for $a + \min(b, d) = p - 1$ induces an injection $\bigoplus_{0 \leqslant i \leqslant p-2} H^b{}_{\mathrm{crys}}(X/V, gr_F{}^i(\mathcal{E})) \otimes_{V\,\Phi} V \oplus H^b{}_{\mathrm{crys}}(X/V, F^{p-1}(\mathcal{E})) \to H^*{}_{\mathrm{crys}}(X/V, \mathcal{E})$.

If in addition X is proper over V all cohomology groups have finite length, and a simple argument shows that the spectral sequence associated to the filtration $F^*(\mathcal{E})$ of $\mathcal{E}$ degenerates in small degrees, that is all differentials affecting $H^b{}_{\mathrm{crys}}(X/V, \mathcal{E})$ for $a + \min(b, d) \leqslant p - 2$ vanish. It also follows that $H^b(X, \mathcal{E})$ is in $\mathcal{MF}(V)$ for such b's. If $\min(b, d) = p - 1 - a$ this is no longer true, but at least the maps φ give an injection of $H^b{}_{\mathrm{crys}}(X/V, \mathcal{E})^{\sim}$ into $H^b{}_{\mathrm{crys}}(X/V, \mathcal{E})$. So far we have assumed that p annihilates $\mathcal{E}$, but by devissage these results hold for general $\mathcal{E}$ as well. We derive:

THEOREM 4.1. *Suppose X is proper and smooth over $V = V_0$ of relative dimension d, $\mathcal{E} \in \mathcal{MF}^{\nabla}{}_{[0,a]}(X)$. For $a + \min(b, d) \leqslant p - 2$ $H^a(X, \mathcal{E})$ lies in $\mathcal{MF}(V)$. Furthermore in the spectral sequence $E_1{}^{mn} = H^{m+n}{}_{\mathrm{crys}}(X/V, gr_F{}^m(\mathcal{E})) \Rightarrow H^{m+n}{}_{\mathrm{crys}}(X/V, \mathcal{E})$, the differentials $d_r{}^{mn}$ vanish for $a + m + n \leqslant p - 2$.*

c) Crystalline cohomology also works in the logarithmic context, i.e. using differentials with logarithmic poles instead of ordinary differentials. For this we assume that X is smooth over V, and $D \subset X$ a divisor with simple normal crossings relative V. That is $D = \bigcup_{a \in A} D_a$ is the union of smooth V-schemes D_a meeting transversally. A PD-thickening of X consists of a PD-immersion $U \subset \mathcal{U}$, $U \subset X$ open, and line bundles $\{\mathcal{L}_a | a \in A\}$ on $\mathcal{U}$ together with global sections $l_a \in \Gamma(\mathcal{U}, \mathcal{L}_a)$, lifting the line bundles $\mathcal{O}(D_a)$ and their trivial sections 1 on U. A morphism between such thickenings is a map of PD-immersions, together with an isomorphism of the $\mathcal{L}_a$'s which respects the l_a. With these definitions the usual theory of crystalline cohomology works: The product of two immersions $U_1 \subset \mathcal{U}_1$ and $U_2 \subset \mathcal{U}_2$ almost exists, as in the usual theory: Locally if $\mathcal{U} = \mathrm{Spec}(B_1)$ and $\mathcal{U}_2 = \mathrm{Spec}(B_2)$ are both affine and if the $\mathcal{L}_a$'s are trivial, we use the ind-object obtained by forming $B =$ divided power hull of $(B_1 \otimes_V B_2)[l_{a1}/l_{a2}, l_{a2}/l_{a1}]$, and using the diagonal immersions into $\mathrm{Spec}(B/I^{[n]})$,

$I \subset B$ is the ideal defining the diagonal. We can define logarithmic crystals, which are equivalent to modules $\mathcal{E}$ on X together with an integrable connection $\nabla : \mathcal{E} \to \mathcal{E} \otimes \Omega_{X/V}(d \log \infty)$, and their cohomology $H^*_{\text{crys}}(X^\circ, \mathcal{E})$ can be computed by the logarithmic de Rham complex. As usual we abuse notation, as this cohomology may depend on the compactification X of X°. For example the line bundles $\mathcal{L}_a$ give the crystal defined by $\mathcal{O}(D_a)$, with its natural logarithmic connection. More generally we have variants of cohomology with compact support. If for any $a \in A$ we give an integer n_a, we define these as the cohomology of $\mathcal{E}(\Sigma_{a \in A} n_a \cdot D_a)$, the usual cohomology with compact support corresponding to the choice $n_a = -1$ for all $a \in A$.

Finally we have to worry about functoriality: Assume we are given a map $f : X \to Y$, divisors $D = \bigcup_{a \in A} D_a \subset X$ and $E = \bigcup_{b \in B} E_b \subset Y$, such that $f^*(E_b) = \Sigma\, n_{ab} \cdot D_a$, with integers n_{ab}. For PD-immersions $X \supset U \subset \mathcal{U}$ and $Y \supset W \subset \mathcal{W}$, together with line bundles $\mathcal{L}_a$, $\mathcal{M}_b$, and global sections l_a and m_b, consider extensions $f : \mathcal{U} \to \mathcal{W}$ and isomorphisms $f^*(m_b) \cong \Pi\, \mathcal{L}_a{}^{n_{ab}}$ which send $f^*(m_b)$ to $\Pi\, l_a{}^{n_{ab}}$. These allow us to define the pull-back of a logarithmic crystal on Y, and also the map $f^* : H^*_{\text{crys}}(Y^\circ, \mathcal{E}) \to H^*_{\text{crys}}(X^\circ, f^*(\mathcal{E}))$ in crystalline cohomology.

Furthermore we can define "log-smooth" maps $f : X \to Y$: We could mimic the usual definition using lifting over nilpotents. However in our case of simple normal crossings divisors in smooth schemes (which means log-smooth over V) this comes down to requiring that $df : f^*(\Omega_{Y/V}(d \log \infty) \to \Omega_{X/V}(d \log \infty)$ is locally a direct summand. Its cokernel deserves to be called $\Omega_{X/V}(d \log \infty)$, and the direct image $\mathbf{R}f_{*\text{crys}}(\mathcal{E})$ of a logarithmic crystal $\mathcal{E}$ on X can be represented by the relative de Rham compledx $\mathbf{R}f_*(\mathcal{E} \otimes \Omega^*_{X/Y}(d \log \infty))$. However in general this is only a crystal in the sense of derived categories, as its cohomology groups may not be locally free and may not commute with base change.

Of course there is also the usual notion of a filtered crystal (corresponding to connections satisfying Griffiths transversality), and one can also define crystals on X_0/V ($X_0 = X \otimes_V V/pV$), which however correspond one to one to crystals on X/V. If $V = V_0$, so that the Frobenius on V/pV lifts to V, we can also define Frobenius crystals as crystals $\mathcal{E}$ on X_0/V together with a transformation of crystals $\Phi^*(\mathcal{E}) \to \mathcal{E}$, Φ the Frobenius endomorphism on X_0 (which has multiplicities $n_{ab} = p\delta_{ab}$ at infinity). This induces a semilinear Frobenius endomorphism on $H^*_{\text{crys}}(X^\circ/V, \mathcal{E})$. This also works for cohomology with compact support, or compact support along part of the boundary. With these definitions the whole theory about $\mathfrak{M}\mathfrak{F}$ goes through, the key being a local calculation, and we obtain:

THEOREM 4.1*. *Suppose X is proper and smooth over V of relative dimension d, $D \subset X$ a relative divisor with simple normal crossings, $\mathcal{E} \in \mathfrak{M}\mathfrak{F}^{\nabla}_{[0,a]}(X^{\circ})$. For $a + \min(b, d) \leqslant p - 2$ $H^a(X^{\circ}, \mathcal{E})$ lies in $\mathfrak{M}\mathfrak{F}(V)$. Furthermore in the spectral sequence $E_1^{mn} = H^{m+n}{}_{\text{crys}}(X^{\circ}/V, gr_F{}^m(\mathcal{E})) \Rightarrow H^{m+n}{}_{\text{crys}}(X^{\circ}/V, \mathcal{E})$, the differentials d_r^{mn} vanish for $a + m + n \leqslant p - 2$.*

The same holds for cohomology with compact support, or compact support along part of the boundary.

d) We also need Poincaré-duality: Assume that X is proper and smooth over V, of pure relative dimension d. Choose a p-power n which annihilates all $\mathcal{E}$'s in sight, let $(\mathcal{O}_X/n\mathcal{O}_X)\{a\}$ denote the object in $\mathfrak{M}\mathfrak{F}^{\nabla}(X)$ which has underlying crystal $\mathcal{O}_X/n\mathcal{O}_X$ with the trivial connection, whose filtration lives in degree a, and for which $\varphi_a(1) = 1$. For $\mathcal{E} \in \mathfrak{M}\mathfrak{F}^{\nabla}_{[0,a]}(X)$ define $\mathcal{E}^t = \underline{\text{Hom}}(\mathcal{E}, \mathcal{O}_X/n\mathcal{O}_X\{a\})$. This gives an involution on $\mathfrak{M}\mathfrak{F}^{\nabla}_{[0,a]}(X)$, and cup product and trace map define a pairing $H^*(X/V, \mathcal{E}) \times H^{2d-*}(X/V, \mathcal{E}^t) \to H^{2d}(X/V, \mathcal{O}_X/n\mathcal{O}_X\{a\}) \to V/nV\{a + d\}$, which is a perfect duality (use Serre-duality). It follows that we can define direct images f_*, and if f is the inclusion of a smooth divisor $Y \subset X$ the direct image can be represented by the connecting homomorphism in the exact sequence $0 \to \mathcal{E} \otimes \Omega^*{}_{X/V} \to \mathcal{E} \otimes \Omega^*{}_{X/V}(d \log Y) \to \mathcal{E} \otimes \Omega^*{}_{Y/V}[-1] \to 0$.

More generally in the logarithmic context we have to pair ordinary cohomology with cohomology with compact support, and the description above of f_* works for immersions transverse to the divisor at infinity. For immersions of higher codimension we have to use blow-ups. Finally the constructions in Chapter III (modification of $X \times X$ by blowing up along all $D_i \times D_i$, and Lemma 3.1) work in this case and give a geometric construction of the characteristic class of the diagonal.

e) What do we do if $V \neq V_0$, or if $a + d > p - 2$? We use $\mathbf{Q}_p$-coefficients and apply the theory of convergent F-isocrystals developed by A. Ogus in [O]. Unfortunately this reference does not cover the definition of crystalline cohomology for such crystals, so we have to introduce this here. We do not try to do things in full generality, but rather confine ourselves to a special case which suffices for our applications. A general theory has been announced by P. Berthelot ([B2]), which however uses a slightly different definition of cohomology with compact support (he does not assume a simple normal crossings situation).

Let V denote a discrete valuation ring as before, and X a scheme over V. An affine enlargement of X is given by a flat V-algebra A, I-adically complete for some ideal $I \subset A$ which is nilpotent modulo p, and a map (over V) $\mathrm{Spec}(A/I) \to X$. A map of enlargements $(A, I) \to (B, J)$ is a V-homomorphism sending I into J, and respecting the maps into X. This category has a coproduct, which for A and B as above is given by a suitable completion of $A \otimes_V B$. A p-adic enlargement is an enlargement with $I = pA$. A convergent isocrystal of $\mathcal{O}_X$-modules $\mathcal{E}$ on X/V associates to any enlargement A an $A \otimes \mathbf{Q}_p = A \otimes_V K$-module $\mathcal{E}(A)$, and for any homomorphism of enlargements $A \to B$ an isomorphism $\mathcal{E}(B) \cong \mathcal{E}(A) \otimes_A B$, with the usual compatabilities. Similar for p-convergent isocrystals, by using p-adic enlargements. We say that $\mathcal{E}$ is finitely generated if $\mathcal{E}(A)$ is a finite $A \otimes_V K$-module, for any enlargement A. In this case $\mathcal{E}(A)$ has a natural p-adic topology. In applications all isocrystals will be finitely generated.

For any V-map $X \to Y$ we define the pullback of convergent isocrystals on Y in the obvious way. If X is smooth over V a convergent isocrystal on X corresponds to an $\hat{\mathcal{O}}_X \otimes_V K$-module with integrable connection on the formal completion $\hat{X}$, such that for each local section m the Taylor-series $\Sigma\, \nabla(\partial)^I(m) \cdot T^I/I!$ converges in the open unit disk $\|T\| < 1$ ($\partial_i = \partial/\partial t_i$ for local coordinates t_i). Also for X smooth over k convergent isocrystals of finite type are locally free, that is $\mathcal{E}(A)$ is projective of finite rank over $A \otimes \mathbf{Q}$, for any enlargement A: It suffices to consider A's arising from smooth local lifts of X. Then the Fitting-ideals of $\mathcal{E}(A)$ are stable under derivations, hence either equal to (0) or to $A \otimes \mathbf{Q}$.

If $V = V_0$ and X is a scheme over $k = V/pV$, a convergent F-isocrystal $\mathcal{E}$ is a convergent isocrystal with an isomorphism $\Phi^*(\mathcal{E}) \cong \mathcal{E}$, Φ the Frobenius on X. By [O], Proposition 2.18 any p-convergent F-isocrystal extends to a convergent F-isocrystal. As usual convergent isocrystals $\mathcal{E}$ can be given by covering X by affines, embedding into smooth V-schemes, and then evaluating $\mathcal{E}$ on some universal enlargements (subject to some glueing). Let us try to define the crystalline cohomology of a convergent F-isocrystal $\mathcal{E}$ on X/V, at least if X is smooth over the residue-field k of V. First assume that X is affine, and embed it into a affine smooth formal V-scheme: $X = \mathrm{Spec}(R/I) \subset \mathrm{Spec}(R)$. If we define $R_n = R[I^n/p]$ ($\subset R \otimes_V K$) one checks that for n big enough the projection from R to R/I extends to R_n, mapping I^n/p to 0: For any formal V-lift $\tilde{R}$ of X the I-adic completion $\hat{R}$ is isomorphic to $\tilde{R}$^$[[T_1, \ldots, T_d]]$, with I generated by the maximal ideal of V and the T_i. The assertion now follows by an explicit calculation. It follows that we can evaluate $\mathcal{E}$ on the I-adic completion $\hat{R}_n$. Furthermore (using as usual the diagonal embeddings) one checks that $\mathcal{E}(\hat{R}_n)$ has an integrable connection $\nabla : \mathcal{E}(\hat{R}_n) \rightarrow \mathcal{E}(\hat{R}_n) \otimes_R \Omega_{R/V}$, and we may form the de Rham complex $\mathcal{E}(\hat{R}_n) \otimes_R \Omega^*_{R/V}$. We define $H^*_{\text{cris}}(X/V, \mathcal{E})$ as the cohomology of the projective limit (over n) of these complexes. This is independent of the choice of the smooth embedding: It suffices to check that any surjection $R \rightarrow R'$ induces an isomorphism. But then $\hat{R} \cong \hat{R}'[[T_i]]$ (as before), the inclusion $R' \subset R$ induces parallel isomorphisms $\mathcal{E}(\hat{R}_n) \cong \mathcal{E}(\hat{R}'_n) \otimes_{\hat{R}'_n} R_n$, $\hat{R}_n \cong \hat{R}'_n[[p^{-1/n} \cdot T]]$, and the de Rham complex for R is the topological tensor-product of the de Rham complex for R' and the de Rham complex for the open unit ball (power-series in $K[[T_i]]$ which converge in the open ball). However the latter is acyclic, as integrations works (if $\Sigma\, a_n \cdot T^n$ converges for $\|T\| < 1$, so does $\Sigma\, a_n/(n + 1) \cdot T^{n+1}$). In fact this argument even shows that the map on de Rham complexes induced by $R \rightarrow R'$ is a topological homotopy equivalence (if $\mathcal{E}$ is finitely generated), by chosing the projective limit topology on all complexes. It also follows that for any map $f : X \rightarrow Y$ we get an induced f^* : $H^*_{\text{cris}}(Y/V, \mathcal{E}) \rightarrow H^*_{\text{cris}}(X/V, f^*(\mathcal{E}))$, by lifting f to the $\hat{R}$'s (By standard methods the map on cohomology is independent of the liftings).

In the general (global) case we first cover X by affines U_i and embed these into smooth $\text{Spec}(R_i)$. Then $U_{ij} = U_i \cap U_j$ embeds into $R_{ij} = R_i \otimes_V R_j$, etc., and from the de Rham complexes we form a double complex which computes $H^*_{\text{cris}}(X, \mathcal{E})$. This is independent of all choices, and for V-maps $f : X \rightarrow Y$ we still get a well-defined $f^* : H^*_{\text{cris}}(Y/V, \mathcal{E}) \rightarrow H^*_{\text{cris}}(X/V, f^*(\mathcal{E}))$. We intend to show that $H^*_{\text{cris}}(X/V, \mathcal{E})$ is a finite-dimensional K-vector-space, in case X is smooth and proper over k, and $\mathcal{E}$ is finitely generated convergent F-isocrystal. This is done as follows: Choose local liftings $\text{Spec}(R_i)$ of affines $U_i \subset X$, and lattices $M_i \subset \mathcal{E}(\hat{R}_i)$ (the M_i are finitely generated over $\hat{R}_i$, and span $\mathcal{E}(\hat{R}_i)$ over $\hat{R}_i \otimes \mathbf{Q}$). Replacing the M_i by transforms under a sufficiently high power of Frobenius we may assume that the connection ∇ maps M_i to $M_i \otimes_{R_i} \Omega_{R_i/V}$. It then can be easily checked that the M_i can be modified so that they glue and define a finitely generated crystal (in the usual sense of crystalline cohomology) on X/V. The (usual) crystalline cohomology of this crystal is finitely generated over V (it suffices to check modulo p, and there we use de Rham cohomology on X), and tensoring with K gives $H^*_{\text{cris}}(X/V, \mathcal{E})$. If X is smooth and proper of pure dimension d over k there exists a trace-map $H^{2d}_{\text{cris}}(X, \mathcal{O}_X) \rightarrow K$ ([O], Theorem 3.12), and Poincaré-duality holds (as for the finiteness reduce to de Rham cohomology on the special fibre). Furthermore if X lifts to a smooth formal V-scheme $\mathfrak{X}$ (which is an enlargement of X) $H^*_{\text{cris}}(X, \mathcal{E})$ is isomorphic to the de Rham cohomology of $\mathcal{E}(\mathfrak{X})$.

Of course the whole theory works in the logarithmic context, using differentials with logarithmic poles, liftings of the divisors at infinity, etc. It can be translated into a theory of bundles with an integrable logarithmic connection satisfying some (complicated) convergence property. However in the logarithmic context convergent isocrystals of finite type need not be locally free, as the Fitting-ideals could have components in the boundary. However this cannot happen for convergent F-isocrystals, as requiring invariance under Frobenius now again forces these ideals to be either equal to (0) or to $A \otimes \mathbf{Q}$.

V. Comparison. a) We now can start to compare étale and crystalline cohomology. Assume as usual that X is proper and smooth over V, $D \subset X$ a simple normal crossings divisor which is the union of smooth components. Define sheaves $B^+/(I^{[n]} + p^m \cdot B^+)$ on the topos $\mathfrak{X}^\circ$ introduced in Chapter II by associating to a small affine $U = \operatorname{Spec}(R) \to \hat{X}$ the locally constant étale sheaf on $U^\circ \otimes_V K$ defined by the Γ-module $B^+(\hat{R})/(I^{[n]} + p^m \cdot B^+(\hat{R}))$. If $\mathbf{L}$ is any locally constant finite sheaf on $X^\circ \otimes_V K$ annihilated by p^m we can consider $H^*(\mathfrak{X}^\circ \otimes_V \bar{K}, \mathbf{L} \otimes B^+/(I^{[n]} + p^m \cdot B^+)) = H^*(\mathfrak{X}^\circ \otimes_V \bar{K}, \mathbf{L} \otimes B^+/I^{[n]})$, which by the results of Chapter II is almost (up to terms annihilated by the preimage of the maximal ideal $\mathfrak{m} \subset \bar{V}^\wedge$ under $B^+(V) \to \bar{V}^\wedge$) isomorphic to $H^*(X^\circ \otimes_V \bar{K}, \mathbf{L}) \otimes B^+(V)/I^{[n]}$ (Filter $B^+(\hat{R})$ with quotients isomorphic to $\bar{R}$ or $\bar{R}$/some p-power).

Now assume that $V = V_0$. If $R + I \subset B^+(\hat{R})$ denotes the preimage of $R \subset B^+(\hat{R})/I$, the map $R + I \to (R + I)/I = R$ is a PD-thickening, hence defines an object of the crystalline site of X/V. This also holds in the logarithmic context: If $f \in R$ is the equation for one of the components D_a of D, choose roots $f^{p^{-n}}$ in $\bar{R}$. Their limit defines an element $\tilde{f}$ in $S = \text{proj.lim}(\bar{R}/p\bar{R})$, unique up to multiplication with a unit. Then $[\tilde{f}, 0, 0, \ldots] \in W(S)$ gives an element of $B^+(R)$ lifting f, again unique up to units. The line bundle defined by $\tilde{f}^{-1} \cdot B^+(R)$ and its global section 1 lift $\mathcal{O}(D_i)$, respectively its section 1 on X. Note also that under Frobenius $\Phi(\tilde{f}) = \tilde{f}^p$, so that the Frobenius on $B^+(R)$ lifts the Frobenius on R/pR even in the logarithmic sense.

Now a crystal $\mathcal{E}$ on X/V defines a sheaf $\mathcal{E} \otimes_R B^+/I^{[n]}$ on $\mathfrak{X}^\circ$ by associating to $U = \operatorname{Spec}(R)$ $\mathcal{E} \otimes_R B^+(\hat{R})/I^{[n]}$, and any parallel global section of $\mathcal{E}$ defines a global section of $\mathcal{E} \otimes_R B^+(\hat{R})/I^{[n]}$ over $\mathfrak{X}^\circ \otimes_V \bar{K}$. If we resolve $\mathcal{E}$

by suitable acyclic crystals (see for example below) we obtain an exact resolution of $\mathcal{E} \otimes_R B^+/I^{[n]}$ on $\mathfrak{X}^\circ \otimes_V \bar{K}$, which however need not by acyclic for Galois-cohomology. Mapping to an acyclic resolution and taking global sections now defines a transformation $H^*{}_{\mathrm{crys}}(X^\circ/V, \mathcal{E}) \to H^*(\mathfrak{X}^\circ \otimes_V \bar{K}, \mathcal{E} \otimes_R B^+/I^{[n]})$ independent of all choices. This also works in the filtered context, where however we have a small difficulty because for $n \geqslant p$ the F^n on B^+ do not coincide with the divided powers $I^{[n]}$. In applications either $F^p(H^*{}_{\mathrm{crys}}(X^\circ/V, \mathcal{E}))$ vanishes or we use $\mathbf{Q}_p$-coefficients, so this does not hurt us.

Finally for $\mathcal{E} \in \mathfrak{M}\mathfrak{F}^\nabla{}_{[0,p-2]}(X^\circ)$ we can form $\mathbf{L} = \mathbf{D}(\mathcal{E})$, and in $\mathfrak{X}^\circ$ we get maps $\mathcal{E} \otimes_R B^+(\hat{R})/I^{[n]} \to \mathbf{L}^* \otimes B^+(\hat{R})/I^{[n]}$, which in cohomology induce $H^*{}_{\mathrm{crys}}(X^\circ/V, \mathcal{E}) \otimes_V B^+(V)/I^{[n]} \to H^*(X^\circ \otimes_V \bar{K}, \mathbf{L}^*) \otimes B^+(V)/I^{[n]}$ (only almost defined), $H^*{}_{\mathrm{crys}}(X^\circ/V, \mathcal{E}) \otimes_V B^+(V) \to H^*(X^\circ \otimes_V \bar{K}, \mathbf{L}^*) \otimes B^+(V)$ and $\mathrm{Hom}(H^*{}_{\mathrm{crys}}(X^\circ/V, \mathcal{E}), B^+(V) \otimes \mathbf{Q}_p/\mathbf{Z}_p) \leftarrow H^*(X^\circ \otimes_V \bar{K}, \mathbf{L}^*)^*$. On the left we mean almost defined morphisms respecting filtrations and the φ_i (as far as these are defined). However one checks that all such maps are actual homomorphisms if $H^*{}_{\mathrm{crys}}(X^\circ/V, \mathcal{E})$ happens to lie in $\mathfrak{M}\mathfrak{F}_{[0,p-2]}(V)$ (Reduce to objects annihilated by p, and use the description as solutions of certain equations in $\bar{R}/p\bar{R}$), and then we have obtained transformations $H^*(X^\circ \otimes_V \bar{K}, \mathbf{L}^*)^* \to \mathbf{D}(H^*{}_{\mathrm{crys}}(X^\circ/V, \mathcal{E}))$. We intend to show that these are isomorphisms. There is also an analogue for cohomology with compact support, or compact support along only a part of the boundary.

We first consider the transformation $\alpha : H^*{}_{\mathrm{crys}}(X^\circ/V, \mathcal{E}) \otimes_V B^+(V) \to H^*(X^\circ \otimes_V \bar{K}, \mathbf{L}^*) \otimes B^+(V)$. This is a natural transformations of cohomology-theories, respecting pullbacks and cup products. As usual we have to consider the behaviour under direct images:

b) Suppose $f : Y \subset X$ is a closed immersion of pure codimension e, transversal to the divisor at infinity. We obtain transformations $f_* : H^*{}_{\mathrm{crys}}(V^\circ/V, \mathcal{E}) \otimes_V B^+(V)\{e\} \to H^*{}_{\mathrm{crys}}(X^\circ/V, \mathcal{E}) \otimes_V B^+(V)$ and $f_* : H^*(Y^\circ \otimes_V \bar{K}, \mathbf{L}^*) \otimes B^+(V) \to H^*(X^\circ \otimes_V \bar{K}, \mathbf{L}^*) \otimes B^+(V)(e)$. We claim that $\alpha \cdot f_*$ and $f_* \cdot \alpha$ differ by the map from $B^+(V)\{e\}(e)$ to $B^+(V)$ induced by $\beta^{\otimes e} : \mathbf{Z}_p(e) \to F^e(B^+(V))$. There is a corresponding result about Chern-classes of vector bundles, which we state first:

Lemma 5.1. *Suppose $\mathcal{E}$ is a vectorbundle on X, with Chern classes $c_e(\mathcal{E})$ in $H^{2e}(X \otimes_V \bar{K}, \underline{\mathbf{Z}/n\mathbf{Z}})(e)$ and $H^{2e}{}_{\mathrm{crys}}(X/V, \mathcal{O}_X/n\mathcal{O}_X)\{-e\}$. Then α maps the crystalline Chern-class $c_e(\mathcal{E})$ to the image under $\beta^{\otimes e}$ of the étale class.*

Proof. By the splitting principle it suffices to consider $c_1(\mathcal{L})$, $\mathcal{L}$ a line bundle on X. We also may assume that X is geometrically irreducible. Choose an open affine covering $X = \cup_{i \in I} U_i$ which trivializes $\mathcal{L}$, so that $\mathcal{L}$ can be represented by a cocycle $\{f_{ij}\} \in \Gamma(U_{ij}, \mathcal{O}_X^*)$. Now the crystalline Chern-class is given by the cocycle $\{d \log(f_{ij})\}$, while the étale class can be described as follows: If $\Delta_{ij} = \pi_1(U_{ij} \otimes_V \bar{K})$ denotes the geometric fundamental group of U_{ij}, the coverings defined by the n-th roots of f_{ij} define homomorphisms $\Delta_{ij} \to \mathbf{Z}/n\mathbf{Z}(1)$ and thus Δ_{ij}-invariant 1-cocycles in the canonical complex $D^*(\Delta_{ij}, \mathbf{Z}/n\mathbf{Z}(1))$ which computes the cohomology $H^*(\Delta_{ij}, \mathbf{Z}/n\mathbf{Z}(1))$. They also behave as 1-cocycles with respect to the indices ij, and so define an invariant 2-cocycle in the total complex made up from the $D^*(\Delta_I, \mathbf{Z}/n\mathbf{Z})(1))$. This cocycle represents the étale Chern-class $c_1(\mathcal{L})$.

To describe α we need the resolution of $\mathcal{O}_{X/V}$ underlying the isomorphism of crystalline and de Rham cohomology (compare [B1], Chapter V, 2). If $D^n(X \times X)$ denotes the n-th infinitesimal PD-neighbourhood of the diagonal in $X \times X$, with its projections pr_{1n} and pr_{2n} onto X, this is given by $L(\Omega_{X/V}^*) = \text{proj.lim}\, pr_{1n*}(pr_{2n}^*(\Omega_{X/V}^*)$, with the differentials induced from $\Omega_{X/V}^*$, and the inclusion $\mathcal{O}_{X/V} \to L(\mathcal{O}_{X/V})$ associates to any regular function on X its Taylor series. The map α uses the complexes $L(\Omega_{X/V}^*)(B^+(R_{ij}))$ (R_{ij} the affine ring of U_{ij}), on which Δ_{ij} operates. They are quasiisomorphic to the total complexes associated to the double complexes (using continuous cochains everywhere) $D^*(\Delta_{ij}, L(\Omega_{X/V}^*)(B^+(R_{ij})))$, which in turn are quasiisomorphic to $D^*(\Delta_{ij}, B^+(R_{ij}))$. Now in $L(\Omega_{X/V}^*)(B^+(R_{ij}))$ $d \log(f_{ij})$ is the differential of $\log(pr_2^*(f_{ij})/pr_1^*(f_{ij})) \in L(\mathcal{O}_{X/V})(B^+(R_{ij}))$, and we have to study the action of Δ_{ij} on this logarithm. For this choose p-power roots of f_{ij} in $\bar{R}_{ij}$ (this can be done in such a way that these roots satisfy the cocycle condition: The second Cech-cohomology of $\mathbf{Z}_p(1)$ vanishes, by restricting to the generic point of X). Then $pr_1^*(f_{ij})$ stands for the limit of those roots in $S_{ij} = \text{lim.proj}(\bar{R}_{ij}/p\bar{R}_{ij})$, which is an Δ_{ij}-eigenvector with character given by $\Delta_{ij} \to \mathbf{Z}_p(1) \to B^+(R_{ij})^*$. So on the logarithm we obtain the composition of $\Delta_{ij} \to \mathbf{Z}_p(1)$ (defined by f_{ij}) with $-\beta$. That is the Δ_{ij}-invariant class defined by $\log(pr_2^*(f_{ij})/pr_1^*(f_{ij}))$ in the simple complex associated to $D^*(\Delta_{ij}, L(\Omega_{X/V}^*)(B^+(R_{ij})))$ has as differential the difference of $\{d \log(f_{ij})\}$ and the image (under β) of the cocycle in $D^*(\Delta_{ij}, \mathbf{Z}_p(1))$ which represents the étale first Chern-class of $\mathcal{L}$. This proves the assertion.

c) We now can show the announced result about direct images:

LEMMA 5.2. *Suppose $f: Y \subset X$ is an inclusion of pure codimension e (X and Y proper and smooth over $V = V_0$), transversal to the boundary. If $\mathcal{E} \in \mathcal{MF}^{\nabla}_{[0,p-2]}(X)$ then on $H^*_{\text{crys}}(Y^\circ/V, f^*(\mathcal{E})) \otimes_V B^+(V)\{e\}$ we have $\beta^{\otimes e} \cdot f_* \cdot \alpha = \alpha \cdot f_*$. The same holds for cohomology with compact support, or compact support along part of the boundary.*

Proof. We use deformation to the normal bundle (Compare [Fu], Chapter V). Let $Z \to X \times \mathbf{P}^1$ denote the blow-up along $Y \times \{\infty\}$, and $\tilde{f}: Y \times \mathbf{P}^1 \subset Z$ the inclusion. If for any $t \in \mathbf{P}^1$ Z_t denotes the fibre of Z over t, and $f_t : Y \subset Z_t$ the inclusion, then $Z_\infty = Z_\infty' \cup Z_\infty''$ is the union of two smooth components meeting transversally. Here Z_∞'' is the blow-up of X in Y, $Z_\infty' = \mathbf{P}_Y(\mathcal{N}^* \oplus \mathcal{O}_Y)$ is the projective completion of the normal bundle of Y in X, and the intersection $Z_\infty' \cap Z_\infty'' = \mathbf{P}_Y(\mathcal{N}^*)$ does not meet Y. If H^* denotes any of the theories $H^*_{\text{crys}}(?^\circ/V, \mathcal{E})$ or $H^*(?^\circ \otimes_V \bar{K}, \mathbf{L}^*)$ (even with compact support, or partial compact support) then the map $H^*(Z_\infty')$ $\to H^*(Z_\infty' \cap Z_\infty'')$ is surjective, so we get exact sequences $H_!^*(Z - Z_\infty)$ $\to H^*(Z) \to H^*(Z_\infty') \oplus H^*(Z_\infty'')$ (the first term denotes cohomology with compact support along Z_∞, in addition to the other boundary conditions chosen in the beginning). In both theories f_* commutes with transverse pullbacks. It follows that for $x \in H^*(Y)$ $\tilde{f}_*(x)$ vanishes on Z_∞''. If we know its restriction to Z_∞' we know $\tilde{f}_*(x)$ up to terms in $H_!^*(Z - Z_\infty)$, which are annihilated when restricting to Z_0.

All in all we derive that it suffices to prove the assertion of Lemma 2 for the embedding $Y \subset Z_\infty'$. As this embedding has a section we only have to consider the direct image $f_*(1)$ (in cohomology with constant coefficients). This is given by Chern-classes and everything follows from Lemma 1.

It also follows that α respects up to $\beta^{\otimes d}$ the trace-maps from $H^{2d}_{\text{crys}}(X/V, \mathcal{O}_X)$ to $V\{d\}$ respectively from $H^{2d}(X \otimes_V \bar{K}, \mathbf{Z}_p)$ to $\mathbf{Z}_p(-d)$ (check on the characteristic class of a point).

d) We now can show the main result.

THEOREM 5.3. *Assume X is proper and smooth over V, of pure dimension d. If $a + d \leqslant p - 2$ and if $\mathcal{E} \in \mathcal{MF}^{\nabla}_{[0,a]}(X)$ is associated to* $\mathbf{L} = \mathbf{D}(\mathcal{E})$, *then the transformations α above induce natural isomorphisms* $H^*(X^\circ \otimes_V \bar{K}, \mathbf{L}^*)^* \cong \mathbf{D}(H^*_{\text{crys}}(X^\circ/V, \mathcal{E}))$.

The analogue result holds for cohomology with compact support, or partial compact support. The isomorphism is functorial and respects cup-products and Chern-classes.

Proof. Both sides satisfy Poincaré-duality pairing cohomology of $\mathcal{E}$ and $\mathcal{E}^t = \underline{\mathrm{Hom}}(\mathcal{E}, \mathcal{O}_X/n\mathcal{O}_X\{a\})$, respectively $\mathbf{L}$ and $\mathbf{L}^t = \underline{\mathrm{Hom}}(\mathbf{L}, \underline{\mathbf{Z}/n\mathbf{Z}}(a)) = \mathbf{D}(\mathcal{E}^t)$, into $\mathbf{Z}/n\mathbf{Z}(a + d)$. This is described by the characteristic class of the diagonal and the cup-product,

$$\mathbf{Z}/n\mathbf{Z}(a + d) \to H^*(X^\circ \otimes_V \bar{K}, \mathbf{L}^*)^* \otimes H^*_!(X^\circ \otimes_V \bar{K}, \mathbf{L}^{t*})^* \to \mathbf{Z}/n\mathbf{Z}(a + d),$$

as well as the analogues in crystalline cohomology.

Now α respects these structures (use the immersion of X into the blow-up of $X \times X$ along the $D_i \times D_i$, and Lemma 3.1, Chapter III, as well as its crystalline version). The assertion follows by the usual methods.

e) We might ask whether the hypotheses "$a + d \leqslant p - 2$" might be relaxed to "$a + b \leqslant p - 2$" when considering cohomology in degree b. For projective varieties something be done using a hyperplane argument: Suppose X is smooth and projective over $V = V_0$, of pure relative dimension d, $D \subset X$ a divisor with relative normal crossings, $\mathcal{O}(1)$ an ample line bundle on X, and $Y \subset X$ a smooth hypersurface of degree $r \gg 0$ meeting all strata transversally. By the usual Lefschetz-argument ($X - Y$ is affine) for any locally constant $\mathbf{L}$ the restriction maps $H^i(X^\circ \otimes_V \bar{K}, \mathbf{L}) \to H^i(Y^\circ \otimes_V \bar{K}, \mathbf{L})$ and $H_!^i(X^\circ \otimes_V \bar{K}, \mathbf{L}) \to H_!^i(Y^\circ \otimes_V \bar{K}, \mathbf{L})$ are bijective for $i < d - 1$, and injections for $i = d - 1$. There is also a corresponding result on crystalline cohomology, i.e. $H^i(X^\circ/V, \mathcal{E}) \to H^i(Y^\circ, \mathcal{E})$ and $H_!^i(X^\circ/V, \mathcal{E}) \to H_!^i(Y^\circ, \mathcal{E})$ are bijective respectively injective for any $\mathcal{E} \in \mathcal{MF}(X^\circ)$: Let us assume for simplicity that $X = X^\circ$ (otherwise we have to replace differentials by differentials with logarithmic poles, etc.). Using devissage we may assume that $\mathcal{E}$ is annihilated by p, and from the Hodge spectral sequence and the exact sequences $0 \to \Omega^q{}_{X/V}(-r) \to \Omega^q{}_{X/V} \to \Omega^q{}_{X/V}|Y \to 0$ and $0 \to \Omega^{q-1}{}_{Y/V}(-r) \to \Omega^q{}_{X/V}|Y \to \Omega^q{}_{Y/V} \to 0$ we derive that everything follows from the fact that $H^i(X, \mathcal{E} \otimes \Omega^q{}_{X/V}(-j \cdot r))$ vanishes for $r \gg 0, j > 0, i < d$ (Show by induction on q that $H^i(Y, \mathcal{E} \otimes \Omega^q{}_{Y/V}(-j \cdot r))$ vanishes for $i + q < d - 1$).

Now by induction over $\dim(X)$ we derive:

Proposition 5.4. *Suppose X is projective and smooth over V, of pure relative dimension d. Then if $\mathcal{E} \in \mathcal{MF}^\nabla{}_{[0,a]}(X^\circ)$, $\mathbf{L} = \mathbf{D}(\mathcal{E})$ and $a + b < p - 2$, the maps $H^b(X^\circ \otimes_V \bar{K}, \mathbf{L}^*)^* \to \mathbf{D}(H^b{}_{\mathrm{crys}}(X^\circ/V, \mathcal{E})$ are bijective. They are still surjective for $a + b = p - 2$. The same holds for cohomo-*

logy with compact support, or compact support along part of the boundary.

f) If $V \neq V_0$ or if the restriction on degrees is not fulfilled we still can show a result about $\mathbf{Q}_p$-coefficients. Let $\bar{X} = X \otimes_V k$ denote the special fibre of X over V. We start with a convergent F-isocrystal $\mathcal{E}$ on $\bar{X}/V_0$, whose evaluation on the enlargement $\hat{X}$ (= formal completion of X) has a filtration satisfying Griffiths-transversality. For a small affine $U = \mathrm{Spec}(R)$ of X we have constructed $B^+(\hat{R})$ as the completion of a divided power-hull $D_I(W(S)$. Then the I-adic completion of $D_I(W(S))[I^p/p]$ defines a V_0-enlargement of $\bar{X}$, and as it maps to $B^+(\hat{R})$ and $B(\hat{R})$ we can form $\mathcal{E}(B(\hat{R}))$. This is a $B(\hat{R})$-module with Frobenius and Galois-action, and $\mathcal{E}(B(\hat{R})) \otimes_{K_0} K$ is filtered. If $\mathcal{E}$ is finitely generated it has a natural p-adic topology, and the Galois-action is continuous. We then form $H^*(\mathfrak{X}^\circ, \mathcal{E}(B(\hat{R})))$ (using continuous Galois-cohomology) and obtain as before a transformation $H^*_{\mathrm{cris}}(\bar{X}/V_0, \mathcal{E}) \to H^*(\mathfrak{X}^\circ, \mathcal{E}(B(\hat{R})))$: Locally write $X = \mathrm{Spec}(R/I)$ and resolve (J = ideal of diagonal embedding) $\mathcal{E}$ by $L(\mathcal{E} \otimes \Omega^*_{R/V}) = \mathrm{proj.lim}\ \mathcal{E}((R \otimes_V R)[J^n/p)^\wedge) \otimes_R \Omega^*_{R/V}$. This is exact (even in the topological sense, using the projective limit of p-adic topologies) as the de Rham complex for the open unit ball is, just as in proving (Chapter IV) that the cohomology of the de Rham complex is independent of the smooth embedding. Now these resolutions behave functorially, and by comparing to acyclic (for Galois) resolutions we obtain the desired transformation just as above.

We say that $\mathcal{E}$ is associated to a smooth $\mathbf{Q}_p$-adic étale sheaf $\mathbf{L}$ on $X^\circ \otimes_V K$ if functorially in R $\mathcal{E}(B(\hat{R})) \cong B(\hat{R}) \otimes \mathbf{L}$ (preserving Galois-action, Frobenius and filtrations). This implies that $\mathcal{E}$ is locally free, that is $\mathcal{E}(A)$ is projective of finite rank over $A \otimes \mathbf{Q}_p$, for any enlargement A: It suffices to consider enlargements constructed from local smooth V_0-lifts of $\bar{X}$ or, by using descent, from $\hat{R}$. The assertion now follows by looking at graded pieces, using that $gr_F(B(\hat{R}) \otimes_{K_0} K)$ is faithfully flat over $\hat{R} \otimes_V K$. In fact it follows that even the associated graded is locally free. We can also use [O] if there is no divisor at infinity.

LEMMA 5.5.

i) *"Associated" preserves kernels, cokernels, tensor-products and internal* $\underline{\mathrm{Hom}}$*'s.*

ii) *If $\mathcal{E}$ is associated to some $\mathbf{L}$, then $\mathcal{E}$ and $gr_F(\mathcal{E})$ (over V) are locally free of finite rank.*

iii) *If* $\mathcal{E}_1$ *and* $\mathcal{E}_2$ *are associated to* $\mathbf{L}_1$ *respectively* $\mathbf{L}_2$, *then* $\mathrm{Hom}(\mathcal{E}_1, \mathcal{E}_2) = \mathrm{Hom}(\mathbf{L}_1, \mathbf{L}_2)$ (*Homomorphisms of F-crystals respectively Galois-homomorphisms*).

iv) *Any homomorphism between such* $\mathcal{E}$*'s* (*associated to some* $\mathbf{L}$) *is strict for the F-filtrations.*

Proof. i) is obvious, and ii) has been shown above. For iii) the question is local, so we may assume that X_0 lifts to a smooth formal affine V_0-scheme given by R_0, the p-adic completion of a smooth V_0-algebra such that there exists elements $t_1, \ldots, t_d$ defining an étale map (respecting stratifications) from $V_0[T_1, \ldots, T_d]$ into R_0. Then the formal completion $\hat{X}$ is defined by $R = R_0 \otimes_{V_0} V$. We also chose the Frobenius-lift Φ for R_0 which raises the t_i to their p-th power. Now $\mathcal{E}$ is given by an $R_0 \otimes \mathbf{Q}_p$-module M_0 with an integrable connection ∇ and a parallel Frobenius-automorphism Φ, such that $M = M_0 \otimes_{V_0} V$ has a filtration satisfying Griffiths-transversality. Furthermore we have a $\mathbf{Q}_p$-$\mathrm{Gal}(\bar{R}/R)$-module $\mathbf{L}$ and an isomorphism $M_0 \otimes_{R_0} B(\hat{R}) \cong \mathbf{L} \otimes B(\hat{R})$ respecting all structures. Note that for defining the left hand side we have to chose a lifting $R_0 \to B^+(\hat{R})$ of the natural homomorphism into $B^+(\hat{R})/F^1$, and this cannot be done in a Galois-invariant way. However because of the connection the tensor product is up to canonical isomorphism independent of this choice. If $R_\infty \supset R_0$ denotes the extension defined by adjoining p-power roots of the t_i, we get a canonical $\mathrm{Gal}(\bar{R}/R_\infty \otimes_{V_0} V)$-invariant lift $\gamma : R_\infty \to B^+(\hat{R})$, which we shall use in the following. However under $\mathrm{Gal}(R_\infty \otimes_{V_0} \bar{V}/R_0 \otimes_{V_0} \bar{V}) \cong \mathbf{Z}_p(1)^d$ this homomorphism transforms by the rule $\sigma(\gamma(t_i)) = \alpha(\sigma_i) \cdot \gamma(t_i)$, $\alpha : \mathbf{Q}_p(1) \to B^+(R)^*$ the canonical morphism defined in Chapter II (On $\mathbf{Z}_p(1)$ $\alpha = \exp(\beta)$). It follows that such a σ operates on $M_0 \otimes_{R_0} B(\hat{R})$ by the usual rule $\sigma(m \otimes b) = \Sigma_I \nabla(\partial)^I(m) \otimes (\beta(\sigma)^I \cdot \sigma(b)/I!)$. The sum is over all multiindices $I = (i_1, \ldots, i_d)$, and ∂ denotes the vector consisting of the derivations $t_i \cdot \partial/\partial t_i$. Especially it follows that $m \otimes 1$ is Galois-invariant if and only if m is annihilated by ∇.

It is also clear that $\mathrm{Hom}(\mathcal{E}_1, \mathcal{E}_2)$ injects into $\mathrm{Hom}(\mathbf{L}_1, \mathbf{L}_2)$ (as $\mathbf{L} =$ Frobenius-invariants in $\mathcal{E}(B(R) \cap F^0)$, but it is harder to show that any Galois-homomorphism is induced from one of F-crystals. Using internal $\underline{\mathrm{Hom}}$'s we reduce to showing that Galois-invariants in $\mathbf{L}$ come from (Φ, ∇)-invariants in $F^0(M)$. So assume that l is such an invariant, so that $x = l \otimes 1 \in M_0 \otimes_{R_0} B(R)$ is Φ- and Galois-invariant and lies in F^0. We have to show that x is of the form $m \otimes 1$ (then m is (Φ, ∇)-invariant). Filter $B(\hat{R})$ by its natural filtration. Then $gr_F(B(\hat{R})) = \oplus_n \bar{R}^\wedge \otimes \mathbf{Q}_p(n)$ has only Galois-invariants in degree 0 (use the technique of almost unramified extensions), and the continuous cohomology $H^1(\mathrm{Gal}(\bar{R}^\wedge/\hat{R} \otimes_V \bar{V}, gr^1{}_F(B(\hat{R})) \cong (\hat{R} \otimes_V \bar{K})^{\wedge d}$, with a basis h_i given by $\mathrm{Gal}(R_\infty \otimes_{V_0} \bar{V}/R \otimes_V \bar{V}) = \mathbf{Z}_p(1)^d \to \mathbf{Z}_p(1) \to gr^1{}_F(B(\hat{R}))$ (maps given by projections pr_i and β). The Galois-action on $gr_F(M_0 \otimes_{R_0} B(\hat{R})) \cong M \otimes_R gr_F(B(\hat{R}))$ is induced from the action on $gr_F(B(\hat{R}))$, so its only Galois-invariants live in degree zero, and are isomorphic to M. The obstruction to lifting an $m \in M$ to a Galois-invariant element of $F^0(M_0 \otimes_{R_0} B(\hat{R}))/F^2(M_0 \otimes_{R_0} B(\hat{R}))$ is easily computed to be $\Sigma_i \nabla(\partial_i)m \cdot h_i \in M \otimes_R H^1(\mathrm{Gal}(\bar{R}^\wedge/\hat{R} \otimes \bar{V}, gr^1{}_F(B(\hat{R}))$. It follows that $m \otimes 1$ lifts if and only if m is ∇-parallel, and that the Galois-invariants in $M_0 \otimes_{R_0} B(\hat{R})$ are contained in $M_0 \otimes_{V_0} B(V)$. By [Fo1], Proposition 2.3.3 they are all of the form $m \otimes 1$.

Finally assertion iv) follows from the fact that $gr_F(M)$ is isomorphic the Galois-invariants in $\mathbf{L} \otimes gr_F((B(\hat{R}) \otimes_{V_0} V)$. So an injective map on M's induces an injection on the $gr_F(M)$'s, hence is strict. The same for surjective maps (pass to duals). This concludes the proof of Lemma 5.5.

Remark. We could not prove that associated defines a one to one relation between subobjects. Let us note that this was already the most difficult part of the corresponding theorem for the category $\mathfrak{MF}$.

g) Now employing characteristic classes and Poincaré-duality we derive:

THEOREM 5.6. *Suppose $\mathcal{E}$ and $\mathbf{L}$ are associated. Then $H^*{}_{\mathrm{cris}}(\bar{X}^\circ/V_0, \mathcal{E}) \otimes_{V_0} B(V)$ and $H^*(X^\circ \otimes_V \bar{K}, \mathbf{L}) \otimes B(V)$ are associated (in the sense of [Fo1], Chapter 3), that is $H^*{}_{\mathrm{cris}}(\bar{X}^\circ/V_0, \mathcal{E}) \otimes_{V_0} B(V) \cong H^*(X^\circ \otimes_V \bar{K}, \mathbf{L}) \otimes B(V)$, the functorial isomorphism preserving Frobenius, Galois-actions, filtrations (after tensoring with K), and Chern-classes (up to suitable powers of β). This also holds for cohomology with compact support, or compact support along part of the boundary.*

COROLLARY. *The conjectures C_{cris}, C'_{cris} and $C_{\mathrm{cris}}{}^{\mathrm{pot}}$ of Fontaine ([Fo2] appendix) are true.*

Remark. Theorems 5.3 and 5.6 are compatible in the following way: Suppose $V = V_0$, and $\mathcal{E}$ is such that for each n $\mathcal{E}/p^n\mathcal{E}$ is in $\mathfrak{MF}^\nabla{}_{[0,a]}(X)$. Then $\mathcal{E}$ is associated to $\mathbf{L} = \lim \mathrm{proj}\, \mathbf{D}(\mathcal{E}/p^n\mathcal{E})^* \otimes \mathbf{Q}_p$, and the isomorphisms between $H^*(X^\circ \otimes_V \bar{K}, \mathbf{D}(\mathcal{E}/p^n\mathcal{E})^*)$ and $\mathbf{D}(H^*{}_{\mathrm{crys}}(X^\circ/V, \mathcal{E}/p^n\mathcal{E}))^*$ given by Theorem 5.3 (if it applies) induce the isomorphism given by Theorem 5.6.

VI. The relative case. a) We intend to consider the relative case. So assume that X and Y are smooth over $V = V_0$, that $D \subset X$ and $E \subset Y$ are relative divisors with simple normal crossings, and that $f : X \to Y$ is a proper map with $f^{-1}(E) \subset D$ which is smooth in the logarithmic sense, that is $df : f^*(\Omega_{Y/V}(d \log \infty) \to \Omega_{X/V}(d \log \infty)$ is locally a direct summand. We denote its cokernel by $\Omega_{X/Y}(d \log \infty)$. We intend to show that under suitable restrictions the crystalline direct images of an $\mathcal{E} \in \mathfrak{MF}^\nabla(X^\circ)$ lie in $\mathfrak{MF}^\nabla(Y^\circ)$. Besides conditions about degrees these restrictions require that f is smooth in codimension one, which is equivalent to the fact that any component of D has multiplicity $\leqslant 1$ in $f^*(E)$, or that $D \geqslant f^*(E)$ as divisors. This is sufficient (and in fact also necessary) to assure the validity of a relative Cartier isomorphism. It also assures that we have a good local model. After that the proof follows the previous arguments.

b) So suppose that $f : X \to Y$ is a log-smooth map of smooth V-schemes. Furthermore we assume that for the divisors at infinity we have $E \leqslant f^{-1}(D)$. Then locally in X and Y there exist étale maps from X to $\operatorname{Spec}(V[S_1, \ldots, S_d])$ and from Y to $\operatorname{Spec}(V[T_1, \ldots, T_e])$ such that the divisors at infinity are pullbacks of $\{\Pi\, S_i = 0\}$ respectively $\{\Pi\, T_j = 0\}$, and such that in these local coordinates the map f sends T_j to a product of some S_i: To get local parameters at $x \in X$ first choose equations T_i for the components of D through x, and then add some other T_i such that their differentials form a basis of the image of $\Omega_{X/V} \otimes k(x) \to \Omega_{X/V}(d \log \infty) \otimes k(x)$. Similar for the image $y = f(x) \in Y$. As $\Omega_{Y/V}(d \log \infty) \otimes k(x)$ is a direct summand in $\Omega_{X/V}(d \log \infty) \otimes k(x)$ the assertion now follows easily.

Now let us explain the relative Cartier-isomorphism: We are dealing with a local problem, so assume $f : X = \operatorname{Spec}(B) \to Y = \operatorname{Spec}(A)$ is a log-smooth map of affine schemes over V, corresponding to a homomorphism of rings $A \to B$ of finite type over V. Furthermore we assume that on the p-adic completions $\hat{A}$ and $\hat{B}$ we have chosen compatible Frobenius-lifts Φ even in the logarithmic sense, that is $\Phi^*(D) = p \cdot D$ and $\Phi^*(E) = p \cdot E$ (These exist because of log-smoothness). The induced maps on differentials are divisible by p, and so for the relative de Rham complex $\Omega^*_{B/R}(d \log \infty)$ we obtain Φ-semilinear transformations $\Omega^q{}_{B/R}(d \log \infty) \to (\Omega^q{}_{B/R}(d \log \infty) \otimes_B B/pB)^{d=0}$ induced by $d\Phi_*/p$. The induced map into cohomology is independent of all choices.

Proposition 6.1. *If f is smooth in codimension* 1 *these transformations induce isomorphisms* $\Omega^q{}_{B/A}(d \log \infty) \otimes_{R\ \Phi} A/pA \to H^q(\Omega^*_{B/A}(d \log \infty) \otimes_B B/pB)$.

Proof. The assertion is invariant under étale localisation, so it suffices to treat the standard local model, that is a tensor product of maps $A = V[S] \to B = V[T_1, \ldots, T_d]$, sending S to $\Pi\, T_i$. A convenient Frobenius-lift is obtained by raising all coordinates to their p-th power, and an explicit calculation gives the assertion.

Remark. The relative Cartier-isomorphism does not hold for general log-smooth maps. A counterexample is given by the n-th power map ($n > 1$ and prime to p) from $\mathbf{A}^1$ (with $D = \{0\}$) onto itself.

c) Let us note a corollary: *For $M \in \mathcal{MF}^\nabla(B)$ the semilinear maps φ : $\tilde{M} \to M$ and $d\Phi_*/p : \Omega_{B/A}(d \log \infty) \to \Omega_{B/A}(d \log \infty)$ induce a quasi-isomorphism $(M \otimes_B \Omega^*_{B/A}(d \log \infty))^\sim \to M \otimes_B \Omega^*_{B/A}(d \log \infty)$.*

Use devissage to reduce to M annihilated by p, and then to M trivial as in Chapter III. There is also a variant with compact support (using the ideal defining $D - f^*(E)$). This already gives us the necessary local facts to prove the main result.

THEOREM 6.2. *Suppose $f : X \to Y$ is proper, log-smooth and smooth in codimension one, of relative dimension d. If $\mathcal{E} \in \mathcal{MF}^\nabla{}_{[0,a]}(X^\circ)$ and if $a + \min(b, d) \leqslant p - 2$, then $\mathbf{R}^b f^\circ{}_{*\mathrm{crys}}(\mathcal{E})$ lies in $\mathcal{MF}^\nabla{}_{[0,a+b]}(Y^\circ)$. If $a + d \leqslant p - 2$ then $\mathbf{D}(\mathbf{R}^b f^\circ{}_{*\mathrm{crys}}(\mathcal{E}))$ is isomorphic to $\mathbf{R}^b f^\circ{}_*(\mathbf{D}(\mathcal{E})^*)^*$. The same holds for direct images with compact support.*

Proof. The crystalline direct image is represented by the filtered complex $\mathcal{E} \otimes \Omega^*_{X/Y}(d \log \infty)$. We may assume that $Y = \mathrm{Spec}(A)$ is affine, and that we have chosen a Frobenius-lift Φ on $\hat{A}$. Covering X by open affines and choosing compatible local Frobenius-lifts we obtain (as in Chapter IV) for $i \leqslant p - 1$ semilinear transformations $\varphi_i : H^*(F^i(\mathcal{E} \otimes \Omega^*_{X/Y}(d \log \infty)) \otimes_{A\,\Phi} A \to H^*(\mathcal{E} \otimes \Omega^*_{X/Y}(d \log \infty))$, induced by the local transformations above, which define an object in $\mathcal{MF}_{\mathrm{big}}(B)$. If we assume for the moment that p annihilates $\mathcal{E}$ we get for $a + b \leqslant p - 2$ isomorphisms $H^b(gr_F(\mathcal{E} \otimes \Omega^*_{X/Y}(d \log \infty))) \otimes_{A\,\Phi} A \cong H^b(\mathcal{E} \otimes \Omega^*_{X/Y}(d \log \infty))$, and an injection $H^b(\oplus_{0 \leqslant i \leqslant p-2}\, gr^i{}_F(\mathcal{E} \otimes \Omega^*_{X/Y}(d \log \infty))) \otimes_{A\,\Phi} A \oplus H^b(F^{p-1}(\mathcal{E} \otimes \Omega^*_{X/Y}(d \log \infty))) \otimes_{A\,\Phi} A \to H^b(\mathcal{E} \otimes \Omega^*_{X/Y}(d \log \infty))$ for $a + b = p - 1$ (as in Chapter IV). It follows that in the spectral sequence associated to the filtration all differentials entering or leaving $H^b(gr_F(\mathcal{E} \otimes \Omega^*_{X/Y}(d \log \infty)))$ with $a + b \leqslant p - 2$ vanish, and that we obtain an object in $\mathcal{MF}(B)$: Prove by induction over the height of the prime ideal $\mathfrak{p} \subset B$ (containing pB) that this holds after localisation in $\mathfrak{p}$. If $\mathfrak{p}$ has height one everything has finite length over $B_\mathfrak{p}$, and the assertion follows as in Chapter IV by length-estimates. In general the assertion already holds up to $\mathfrak{p}$-torsion, so especially the spectral sequence degenerates up to torsion. On the other hand the length of $\mathfrak{p}$-torsion in $H^b(gr_F(\mathcal{E} \otimes \Omega^*_{X/Y}(d \log \infty)))$ is not less than the length of $\mathfrak{p}$-torsion in $H^b(\mathcal{E} \otimes \Omega^*_{X/Y}(d \log \infty))$, which contradicts the isomorphism $H^b(gr_F(\mathcal{E} \otimes \Omega^*_{X/Y}(d \log \infty))) \otimes_{A, \Phi} A \cong H^b(\mathcal{E} \otimes \Omega^*_{X/Y}(d \log \infty))$ (respectively the injection above for $a + b = p - 1$), as base extension by Φ multiplies the length of $\mathfrak{p}$-torsion by a factor bigger than one. This proves the assertion for $\mathcal{E}$ annihilated by p, and the general case follows by devissage (and Chapter II, Proposition 2.2).

Concerning the relation between the crystalline and the étale direct image we obtain transformations ($\mathcal{H}^*$ now denotes a relative version of Galois-cohomology) $H^*_{\text{crys}}(X^\circ/A, \mathcal{E}) \otimes_A B^+(\hat{A}) \to \mathcal{H}^*(X^\circ/\bar{A}, \mathcal{E} \otimes B^+) \to \mathcal{H}^*(X^\circ/\bar{A}, \mathbf{L}^* \otimes B^+) \leftarrow H^*(X^\circ \otimes_A K(A)^-, \mathbf{L}^*) \otimes B^+(\hat{A})$, and similar for cohomology with compact support. As in Chapter IV we try to show that the second arrow is an almost isomorphism. However we do not quite succeed:

Locally we can find compatible étale homomorphisms $A \leftarrow V[T_1, \ldots, T_e]$ and $B \leftarrow V[S_1, \ldots, S_d]$ respecting strata, such that f sends each T_j to a monomial in the S_i. Then replace $\bar{A}$ and $\bar{B}$ by A_∞ and B_∞ in calculations about Galois-cohomology, and redo the theory of Chapter II. For example we get induced maps of topoi from $\mathfrak{X}^\circ$ to $\mathfrak{Y}^\circ$ and from $\mathfrak{X}$ to $\mathfrak{Y}$, and we have to study direct images. If locally the map is represented by a ring homomorphism $A \to B$, the direct image of $\mathbf{L}$ corresponds to Galois-cohomology $H^*(\Delta_X, \text{Map}(\Delta_Y, \mathbf{L}))$ (= Galois-cohomology of the kernel of $\Delta_X \to \Delta_Y$, if this maps happens to be surjective). This Galois-cohomology is represented by functorial complexes, etc. All in all we see that the theory $\mathcal{H}^*$ satisfies Poincaré-duality (feasible as any finitely presented A_∞-module has finite projective dimension), has characteristic classes, and so on. However there is no nice geometric construction of the characteristic class of the diagonal, except over the open stratum Y° (where the previous construction "with additional parameters" applies).

It follows that $H^*(X^\circ \otimes_A K(A)^-, \mathbf{L}^*) \otimes B^+(\hat{A})$ is almost isomorphic to a direct summand in $\mathcal{H}^*(X^\circ/\bar{A}, \mathbf{L}^* \otimes B^+)$, and to all of $\mathcal{H}^*(X^\circ/\bar{A}, \mathbf{L}^* \otimes B^+)$ if $\text{Spec}(A) \subset Y^\circ$. It follows that in general the complement of $H^*(X^\circ \otimes_A K(A)^-, \mathbf{L}^*) \otimes B^+(\hat{A})$ is torsion (for the ideal defining the divisor at infinity), and this suffices to obtain a Galois-linear map from $H^b(X^\circ \otimes_A K(A)^-, \mathbf{L}^*)^*$ to $\mathbf{D}(H^b_{\text{crys}}(X^\circ/A, \mathcal{E}))$. Under suitable restrictions on b (as in the theorem) this is an isomorphism, as can be seen for example by restricting to the fibre over a V-rational point of Y°.

Remark. If f is projective the condition "$a + d \leqslant p - 2$" can be replaced by "$a + b \leqslant p - 2$", as in Chapter V, Proposition 5.4, except that for the highest possible value of b we only obtain a surjection instead of an isomorphism. However sometimes we know that both sides have the same order. For example if X is a projective abelian scheme over Y and if $p > 2$ then we derive that its Tate-module is the crystalline representation associated to its first crystalline cohomology group.

d) For general V we can still make assertions about $\mathbf{Q}_p$-cohomology. For a smooth proper map $f: X \to Y$ we define the direct images $R^i f_{*\mathrm{cris}}(\mathcal{E})$ of a convergent isocrystal $\mathcal{E}$ on X/V by associating to any V-enlargement A of Y the crystalline cohomology of $\mathcal{E}$ on X/A, computed by the relative de Rham complexes of smooth morphisms $A \to B$ lifting f (that is B is smooth and of finite type over A, but A itself can be arbitrary). It is straightforward to check that this defines a crystal in the derived sense (see [O], Lemma 3.3), and the cohomology-groups itself are crystals if they are finitely generated, which happens for finitely generated convergent F-isocrystals (same idea as in Chapter IV: Choose integral structures and look at fibres modulo p). Also higher direct images of finitely generated convergent F-isocrystals are again F-isocrystals (again choose integral structures, and some local Cartier-isomorphism). As usual everything works in the logarithmic context.

After these remarks it is straightforward to generalise the theory above and obtain:

THEOREM 6.3. *Assume X and Y are smooth over V, with relative normal crossings divisors D and E, and that $f: X \to Y$ with $f^{-1}(E) \subset D$ is proper, log-smooth and smooth in codimension one. Furthermore assume that $\mathcal{E}$ is a filtered convergent F-isocrystal on $\bar{X}^\circ / V_0$ associated to the smooth $\mathbf{Q}_p$-adic sheaf $\mathbf{L}$ on $X^\circ \otimes_V \bar{K}$. Then for each i the crystalline direct image $\mathbf{R}^i f_*(\mathcal{E})$ is a filtered convergent F-isocrystal on $\bar{Y}^\circ / V_0$ associated to the étale i-th direct image on $\mathbf{L}$. The same holds for direct images with compact support.*

VII. Applications. a) As a first application let us note that we have the degeneracy of various Hodge-spectral sequences (compare [DI]). By standard methods this gives degeneracy over any field of characteristic 0: Everything is defined over a subring of the field which is finitely generated over $\mathbf{Z}$. This subring has many maximal ideals of positive characteristic, and making base-change into p-adic fields we get the assertion. For example this shows degeneration of the Hodge-spectral sequence for any proper smooth X over a field of characteristic 0. It holds also in the logarithmic context, or in the relative case for proper and log-smooth maps which are smooth in codimension one.

b) We obtain assertions about Hodge-Tate decompositions: If X is proper and smooth over $V = V_0$, $\mathcal{E} \in \mathcal{MF}^{\nabla}_{[0,a]}(X)$ and $\mathbf{L} = \mathbf{D}(\mathcal{E})$, we have (for $a + \dim(X/V) \leqslant p - 2$) obtained filtered maps $H^b{}_{\text{crys}}(X/V, \mathcal{E}) \otimes_V B^+(V) \to H^b(X \otimes_V \bar{K}, \mathbf{L}^*) \otimes B^+(V)$. Duality gives an inverse up to $\beta^{\otimes a+d}$, which on the graded pieces gives an isomorphism up to $p^{(a+d)/(p-1)}$ between $\bigoplus_{0 \leqslant i \leqslant b} gr^i{}_F(H^b{}_{\text{crys}}(X/V, \mathcal{E})) \otimes_V \bar{V}(-i)$ and $H^b(X \otimes_V \bar{K}, \mathbf{L}^*) \otimes \bar{V}$. The same again in the logarithmic context, or for rings $V \neq V_0$ and $\mathbf{Q}_p$-coefficients. However in a recent preprint ([H]) O. Hyodo has obtained Hodge-Tate decompositions for more general $\mathbf{Q}_p$-adic sheaves.

c) It is known that the first cohomology-groups can be described by group-schemes. It is thus not surprising that there is a relation between finite flat group-schemes of p-power rank and $\mathcal{MF}^{\nabla}_{[0,1]}$:

THEOREM 7.1. *Suppose X is smooth over $V = V_0$ and that $p > 2$. Then there is an anti-equivalence of categories between finite flat group-schemes of p-power rank on the formal completion $\hat{X}$, and $\mathcal{MF}^{\nabla}_{[0,1]}(X)$. If such a group-scheme G corresponds to $\mathcal{E}$, then $\mathbf{D}(\mathcal{E})$ is the étale sheaf associated to $G(K(\hat{X})^-)$.*

Proof. The assertion is local, so we may assume that X is affine. First we show that for any G the associated $\mathrm{Gal}(\hat{R}^-/\hat{R})$-representation is dual-crystalline and associated to an $M(G) \in \mathcal{MF}^{\nabla}_{[0,1]}(R)$. This holds if $G = A[p^n]$ consists of the p^n-torsion-points of a projective abelian scheme A over $\hat{R}$, by Chapter IV, Theorem 6.2 applied to $H^1(A/R)$. The assertion follows as locally in the étale topology any G can be embedded in such an A, and as any subrepresentation of a dual-crystalline representation is again dual-crystalline. Furthermore we obtain that $F^1(\mathcal{E}(G))$ is functorially isomorphic to $t^*{}_{G/R}$ = translation-invariant sections in $\Gamma(G, \Omega_{G/R})$: Locally in the étale topology choose an exact sequence $0 \to G \to A \to B \to 0$ with abelian schemes A and B. Then $M(G)$ is the cokernel of $H^1_{\text{crys}}(B/R, \mathcal{O}_B) \to H^1_{\text{crys}}(A/R, \mathcal{O}_A)$, so $F^1(M(G))$ is the cokernel $t^*{}_{G/R}$ of the pullback $t^*{}_{B/R} \to t^*{}_{A/R}$. This identification is independent of the choice of A (compare two embeddings via the diagonal) and functorial. It also follows that $M(\ )$ is fully faithful: A map between $M(G_2)$ and $M(G_1)$ induces a map $G_1 \to G_2$ over $\hat{R} \otimes_V K$. It extends to $\hat{R}$ if this holds after localisation in height one primes of R, as G_1 and G_2 are finite and flat (consider the induced map on affine algebras). So we may replace R by such a localisation. Now the closure G in $G_1 \times G_2$ of the graph of the generic map is finite and flat, and the first projection $G \to G_1$ induces an isomorphism on Galois-representations, hence also $M(G_1) \cong M(G)$, and $t^*_{G_1/R} \cong t^*_{G/R}$. Finally the projection is unramified and must be an isomorphism, and everything follows.

d) It remains to construct for each $M \in \mathcal{MF}^{\nabla}_{[0,1]}(R)$ a G with $M = M(G)$, or equivalently a finite flat Hopf-algebra $\mathfrak{A}$ over $\hat{R}$ with $G = \mathrm{Spec}(\mathfrak{A})$. The question is local in R, so we may assume that there exists an étale map $V[T_1^{\pm 1}, \ldots, T_d^{\pm 1}] \to R$. We denote by $R_1 \supset R$ the extension obtained by adjoining p-power roots of the T_i. We shall first assume that M is annihilated by p, and construct a group-scheme $G_1 = \mathrm{Spec}(\mathfrak{A}_1)$ over $\hat{R}_1$ with $M(G_1) = M_1 = M \otimes_R R_1$.

After localisation we choose a basis $m_1, \ldots, m_h$ of M over R/pR, and such that $F^1(M)$ is generated by a subset of this basis. If $e_i \in \{0, 1\}$ denotes the degree on m_i, we have $\varphi_{e_i}(m_i) = \Sigma\, a_{ij} \cdot m_j$, $a_{ij} \in R$. Choose $b_{ij} \in R_1$ with $b_{ij}{}^p \equiv a_{ij}$ modulo p, and consider the finite flat $\hat{R}_1$-algebra $\mathfrak{A}_1 = \hat{R}_1[X_1, \ldots, X_h]/(X_i^p - (-p)^{ei} \cdot \Sigma\, b_{ij} \cdot X_j)$ (already occurring in Chapter II). It has as basis the monomials X^J with $J \in [0, p-1]^h$, and we filter it by defining $F^r(\mathfrak{A}_1)$ as the $\hat{R}_1$-submodule generated by $p^a \cdot X^J$ with $p \cdot a + \Sigma\, e_i \cdot j_i \geqslant r$. This makes $\mathfrak{A}_1$ into a complete filtered $\hat{R}_1$-algebra, and for any $x \in F^1(\mathfrak{A}_1)$ x^p is divisible by p. If $\mathfrak{B}$ is any such algebra, and if in addition for any $b \in \mathfrak{B}$ $p \cdot b$ in $F^{r+p}(\mathfrak{B})$ implies that $b \in F^r(\mathfrak{B})$ one derives that giving a filtered $\hat{R}_1$-homomorphism from $\mathfrak{A}_1$ to $\mathfrak{B}$ corresponds to choosing elements $x_i \in F^{e_i}(\mathfrak{B})/F^2(\mathfrak{B})$ such that $x_i^p \equiv p^{e_i} \cdot \Sigma\, b_{ij} \cdot x_j$ modulo $F^{2+pe_i}(\mathfrak{B})$: These solutions lift uniquely, as in Chapter I where we considered the special case of $\mathfrak{B} = \hat{R}^-$ (with $F^r(\mathfrak{B})$ the ideal generated by $p^{r/p}$). Apply this to $\mathfrak{B} = \mathfrak{A}_1 \otimes \mathfrak{A}_1$, and the approximate solutions $x_i = X_i \otimes 1 + 1 \otimes X_i$: This defines a comultiplication $\mathfrak{A}_1 \to \mathfrak{A}_1 \otimes \mathfrak{A}_1$ which makes $G_1 = \mathrm{Spec}(\mathfrak{A}_1)$ into a finite flat group-scheme over $\hat{R}_1$ (check co-associativity using $\mathfrak{B} = \mathfrak{A}_1 \otimes \mathfrak{A}_1 \otimes \mathfrak{A}_1$, etc.). Also $G_1(\hat{R}^-) = \mathbf{D}(M_1)$, by the theory in Chapter I.

Note that $\mathfrak{A}_1$ is uniquely determined once we have chosen the b_{ij}. Changing them modulo p gives an isomorphic $\mathfrak{A}_1$, by the previous arguments using filtrations and liftings, and up to this ambiguity everything is unique: Use the Frobenius-lift Φ which sends the T_i to their p-th power, and note that Φ induces an isomorphism from $\hat{R}_1$ to $\hat{R}$. It also follows that up to canonical isomorphism $\mathfrak{A}_1$ is determined by the object $\Phi^*(M)$ in $\mathcal{MF}(R_1)$ obtained via Φ from $M \in \mathcal{MF}(R)$.

Now to define a $G/\hat{R}$ we need a descent-datum on G_1 (or $\mathfrak{A}_1$), relative to the faithfully flat map $\hat{R} \to \hat{R}_1$. This is an isomorphism between the two pullbacks of G_1 to $\hat{R}_1 \otimes_R \hat{R}_1$, satisfying some cocycle condition. We note that although $\hat{R}_1 \otimes_R \hat{R}_1$ is no more smooth over V, it still has a Frobenius-lift (namely $\Phi \otimes \Phi$), and that we might speak of $\mathfrak{MF}(\hat{R}_1 \otimes_R \hat{R}_1)$. Now one checks that the two pullbacks of $\mathfrak{A}_1$ to $\hat{R}_1 \otimes_R \hat{R}_1$ are determined by the two pullbacks of $\Phi^*(M)$, and because of the connection ∇ (which we have not used so far) these two pullbacks are naturally isomorphic: The kernel of the multiplication-map $\hat{R}_1 \otimes_R \hat{R}_1 \to \hat{R}_1$ admits divided powers.

This defines the descent-datum and thus the group-scheme G over $\hat{R}$. That it is indeed associated to M is easily seen on the level of Galois-representations.

e) This concludes the proof for objects annihilated by p. It follows that for any G annihilated by p, and any projective abelian scheme A over $\hat{R}$, there is an isomorphism $\mathrm{Ext}(G, A) \cong \mathrm{Ext}(M(A), M(G))$, where the Ext's are computed in group-schemes respectively in $\mathfrak{MF}^{\nabla}(R)$, and we set $M(A) = \text{proj.lim}\, M(A[p^n])$: As $M(\ )$ is faithfully flat there is an injection, and we have to show that any extension $0 \to M(G) \to M \to M(A) \to 0$ is given by some $0 \to A \to E \to G \to 0$. However the first extension splits as extension of $\hat{R}$-modules (as $M(A)$ is locally free), so it is the pullback of an extension $0 \to M(G) \to M/pM \to M(A)/pM(A) \to 0$, which (by the result about objects annihilated by p) comes from an extension of G by $A[p]$. The pushout via $A[p] \subset A$ now defines E.

Finally we prove Theorem 7.1 in full generality, using devissage: By induction M/pM and $pM \subset M$ are associated to a finite flat group-scheme G_1 respectively G_2. Localising in the étale topology we embed G_2 into a projective abelian scheme A. This defines an extension of $M(A)$ by M/pM which is given by $0 \to G_1 \to E \to A \to 0$. Then the preimage $G \subset E$ of $G_2 \subset A$ is finite flat with $M(G) = M$.

Remark. For Cartier-duals we have $M(G^t) = M(G)^t = \mathrm{Hom}(M(G), R[1/p]/R\{1\})$ (internal Hom in $\mathfrak{MF}$), and the filtered module with connection underlying $M(G)$ is of course equal to the dual of the crystal associated to G by the universal vector-extension $0 \to t^*_{G^t} \to E(G) \to G \to 0$, as defined in [Me].

VIII. The de Rham conjecture. a) We intend to sketch how our methods allow us to prove the so called de Rham conjecture. To explain it assume as always that V is a complete discrete valuation ring of uneven characteristic, with perfect residue field, and that K is its field of fractions. We may form $B^+{}_{DR}(V) = \text{proj.lim}(B(V)/F^n(B(V))[1/p])$, and $B_{DR}(V) = B^+{}_{DR}(V)[\beta(1)^{-1}]$. These are complete filtered K-algebras with $\text{Gal}(\bar{K}/K)$-action, and $gr^n{}_F(B_{DR}(V))$ is isomorphic to $\bar{V}^\wedge[1/p](n)$. The conjecture now asserts that for any X which is smooth and proper over K, there is a natural filtered Galois-linear isomorphism $H^*(X, \Omega^*{}_{X/K}) \otimes_K B_{DR}(V) \cong H^*(X \otimes_K \bar{K}, \mathbf{Q}_p) \otimes B_{DR}(V)$. To prove this we extend X to a proper normal V-scheme (denoted by X again), and try to construct rings $B_{DR}(R)$ for sufficiently small affine pieces $\text{Spec}(R) \subset X$. We then show that both sides are isomorphic to the Galois-cohomology with values in $B_{DR}(R)$.

The construction of $B_{DR}(R)$ is not quite straightforward as we cannot define the crystalline $B(R)$ in our context. We only know that R is normal and that $R[1/p]$ is smooth over K of relative dimension d. We proceed as follows:

b) Assume that we are given units $u_1, \ldots, u_d$ in R^* such that their logarithmic derivatives $\{d \log(u_i)\}$ generate $\Omega_{R/V} \otimes_V K$. Then we have good control over the normalisation A of R in the étale extension of $R[1/p]$ generated by adjoining $\bar{K}$ and p-power roots of the u_i: Up to some finite p-power it is isomorphic to the tensor product (over $V[T_1^{\pm 1}, \ldots, T_d^{\pm 1}]$ of R with $\bar{V}[T_i{}^{\pm p^{-\infty}}]$ (see [Fa], Chapter III, Lemma 1.1).

We say that the u_i are good for a normal Galois-extension $R \subset C \subset \bar{R}$ if C contains A, and if for the exists a p-power p^e such that for any $A \subset B \subset C$ of finite degree over A $p^e \cdot e_{B/A}$ is integral, and the image of the trace map $\text{tr}_{B/A}$ contains $p^e \cdot A$ (This is stronger than the corresponding condition in [Fa], Chapter III). It then follows easily (using [Fa], Chapter III, Theorem 1.3) that the continuous Galois-cohomology $H^i(\text{Gal}(C/R), \hat{C}[1/p](n))$ vanishes for $i \neq n$, while $H^n(\text{Gal}(C/R), \hat{C}[1/p](n))$ is canonically isomorphic to $\Omega^n{}_{R/V} \otimes_R \hat{R}[1/p]$. Now if C admits a good system of units we can construct a flat filtered and complete $B^+{}_{DR}(V)$-algebra $B^+{}_{DR}(R, C)$ as follows:

If $R = R_0 = V[T_1^{\pm 1}, \ldots, T_d^{\pm 1}]$ and $C = C_0 = \bar{V}[T_i^{\pm p^{-\infty}}]$ we take $B^+{}_{DR}(R_0, C_0)$ = completion of $B^+(V)[1/p][X_i^{\pm p^{-\infty}}]$ in the (p, F^1)-adic topology. On it $\mathbf{Z}_p(1)^d = \Delta = \text{Gal}(C_0/R_0 \otimes_V \bar{V}) \subset \Gamma = \text{Gal}(C_0/R_0)$ operates via $\sigma(X^I) = \alpha(\langle I, \sigma\rangle) \cdot X^I$, $I \in \mathbf{Z}[1/p]^d$ a multiindex, and we can extend this to Γ (= a semidirect product of $\text{Gal}(\bar{V}/V)$ with Δ) by requiring that $\text{Gal}(\bar{V}/V)$ fixes the X_i. This operation is continuous and respects the filtration. Furthermore each $B^+{}_{DR}(C_0, R_0)/F^n$ has a natural p-adic topology, that is a p-adic norm unique up to equivalence. Equivalently it has a submultiplicative p-adic norm, or a subring (of integral elements) which is a lattice. We use the abbreviation "integral structure" for this.

For general C we have by assumption an $A \subset C$ obtained by adjoining p-powers roots of units u_i, and this defines a map $C_0 \to A$. For any n $B^+{}_{DR}(R_0, C_0)/F^n$ lifts the $\bar{K}^\wedge$-algebra $\hat{C}_0[1/p]$ to a flat $B^+{}_{DR}(V)/F^n$-algebra. We can extend this uniquely first to a lifting of $\hat{A}[1/p]$ with p-adic topology: As a ring with p-adic topology we take the completion of $R \otimes_{V[X_1^{\pm 1},\dots,X_d^{\pm 1}]} B^+{}_{DR}(R_0, C_0)$. Before completion Galois operates, as $R[1/p]$ is étale over $R_0[1/p]$, and this action extends to the completion as it is continuous and respects the p-adic topology (if we know this modulo F^{n-1}, the obstruction of it being true modulo F^n amounts to a derivation from R to $C[1/p]/C(n)$ which is trivial on R_0, hence annihilated by a fixed p-power). After that we extend to a lifting of $\hat{C}[1/p]$ (because of the boundedness of ramification, as in [Fa], Chapter I, Theorem 2.3). We define $B^+{}_{DR}(R, C)$ as the projective limit of these liftings. It is a flat filtered complete $B^+{}_{DR}(V)$-algebra and has the following properties:

i) $gr_F(B^+{}_{DR}(R, C)) \cong \hat{C}[1/p] \otimes gr_F(B^+{}_{DR}(V))$.

ii) each $B^+{}_{DR}(R, C)/F^n$ has an integral structure.

iii) $\Gamma = \mathrm{Gal}(C/R)$ operates on $B^+{}_{DR}(R, C)$, respecting all structures.

iv) The derivation $R \to H^1(\Gamma, \hat{C}[1/p](1)) \cong \Omega_{R/V} \otimes_R \hat{R}[1/p]$ given by the long exact cohomology-sequence associated to the Γ-extension $gr^1{}_F \to B^+{}_{DR}(R, C)/F^2 \to gr^0{}_F$ is equal to the canonical derivation.

(Property iv) follows by explicit calculation, and the rest is clear).

We show that these properties uniquely determine $B^+{}_{DR}(R, C)$ and even better than any inclusion $C_1 \subset C_2$ extends to a unique continuous equivariant map $B^+{}_{DR}(R, C_1) \to B^+{}_{DR}(C_2)$: If $\{u_1, \dots, u_d\}$ are units in R^* good for C_1, inducing a map $C_0 \to C_1$, it suffices to extend to $B^+{}_{DR}(R_0, C_0) \to B^+{}_{DR}(R, C_2)$ (The rest follows from bounded ramification).

We first construct units $\{x_1, \dots, x_d\}$ in $B^+{}_{DR}(R, C_2)$ which lift $\{u_1, \dots, u_d\}$, and which transform the right way under Galois, that is $\sigma(x_i) = \alpha(\sigma_i) \cdot x_i$. If we have found them modulo F^{n-1}, we get a 1-cocycle for Γ with values in $gr^n{}_F(B^+{}_{DR}(R, C_2))$ by sending $\sigma \in \Gamma$ to $(\sigma(x_i) - \alpha(\sigma_i) \cdot x_i)/x_i$. Changing the lifting amounts to changing this cocycle by a coboundary, and so it suffices to check that its class in $H^1(\Gamma, gr^n{}_F(B^+{}_{DR}(R, C_2)))$ vanishes. This is true for $n = 1$ by assumption iv), and for $n > 1$ because then the cohomology itself is zero. Similarly one checks that the lifts x_i are unique. As the $gr^n{}_F(B^+{}_{DR}(R, C_2))$ are p-divisible it follows that there are unique p^n-th roots of the x_i which lift the p^n-th roots of the u_i. All in all we have defined a Galois-linear homomorphism from $B^+(V)[1/p][X_i^{\pm p^{-\infty}}]$ into $B^+{}_{DR}(R, C_2)$. It remains to check that this is p-adically continuous. If we already know this modulo F^{n-1} the obstruction for this to be true modulo F^n is (up to bounded p-power) given by a Γ-invariant derivation from $B^+(V)[1/p][X_i^{\pm p^{-\infty}}]$ into $C_2 \otimes (K/V)(n)$. The corresponding map on differentials sends the logarithmic differential of each $X_p^{p^{-n}}$ into the Γ-invariants in $C_2 \otimes (K/V)(n)$, which are annihilated by a fixed p-power. This easily implies the assertion.

The rest is now comparatively easy, except that we have to correct a mistake in [Fa], Chapter III: The proof of Proposition 1.2 there is unfortunately incomplete, which forces us to replace the assertion of Theorem 3.3 by the fact that for any R there exist maps $R \to R_1$ forming a rigid étale covering, and $C_1 \subset \bar{R}_1$ which admit a good system of units, such that $\bar{R} \otimes_R R_1$ is contained in C_1. It follows that for any proper V-scheme X with smooth generic fibre the pairs (R, C), where R is rigid étale over X and $C \subset \bar{R}$ admits a good system of units, form a filtering system, and we can build hypercoverings from them. Using $H^*(\mathrm{Gal}(C/R \otimes_V \bar{V}), B_{DR}(R, C))$ as building blocks we can form a theory $\mathcal{H}^*_{DR}(X)$, and using our previous methods and [Fa], Chapter III, Theorem 1.3 one checks that it is isomorphic to $H^*(X, \Omega^*_{X/V}) \otimes_V B_{DR}(V)$. Finally p-adic étale cohomology of the geometric generic fibre maps to $\mathcal{H}^*_{DR}(X)$, and as Chern-classes are preserved we obtain an isomorphism $H^*(X \otimes_V \bar{K}, \mathbf{Q}_p) \otimes B_{DR}(V) \cong \mathcal{H}^*_{DR}(X)$. The same holds for open varieties (over K they always have good compactifications, by resolution of singularities), and we have shown the conjecture C_{DR} of Fontaine ([Fo2], A.6]:

THEOREM 8.1. *Suppose X is smooth and separated over K. Then there is a natural* $\mathrm{Gal}(\bar{K}/K)$*-linear filtered isomorphism* $H^*(X, \Omega^*_{X/K}) \otimes_K B_{DR}(V) \cong H^*(X \otimes_V \bar{K}, \mathbf{Q}_p) \otimes B_{DR}(V)$. It respect Chern-classes.

PRINCETON UNIVERSITY

REFERENCES

[B1] P. Berthelot, Cohomologie Cristalline des Schémas de Characteristic $p > 0$, *Springer Lecture Notes* **407** (1974).

[B2] ______, Géométrie rigide et cohomologie des variétés algébriques on characteristic p, *Soc. Math. de France, Mémoire* **23** (1986), 7-32.

[BO1] ______ and A. Ogus, *Notes on crystalline cohomology*, Princeton University Press, Princeton 1978.

[BO2] ______ and ______, F-isocrystals and de Rham cohomology I, *Invent. Math.* **72** (1983), 159-199.

[DI] P. Deligne and L. Illusie, Relèvements modulo p^2 et decomposition du complexe de de Rham, *Invent. Math.* **89** (1987), 247-270.

[Fa] G. Faltings, p-adic Hodge-theory, *Journal of the AMS* **1** (1988), 255-299.

[Fo1] J. M. Fontaine, Modules galoisiens, modules filtrés et anneaux de Barsotti-Tate, *Astérisque* **65** (1979), 3-80.

[Fo2] ______, Sur certains types de représentations p-adiques du groupe de Galois d'un corps local, construction d'un anneau de Barsotti-Tate, *Ann. of Math.* **115** (1982), 529-577.

[Fo3] ______, Cohomologie de de Rham, cohomologie crystalline et représentations p-adiques, *Springer Lecture Notes* **1016** (1983), 86-108.

[FM] ______ and W. Messing, p-adic periods and p-adic etale cohomology, prepring MSRI 1987.

[FL] ______ and G. Lafaille, Construction de représentations p-adiques, *Ann. Sci. Ec. Norm. Sup.* **15** (1982), 547-608.

[Fu] W. Fulton, *Intersection Theory*, Springer Verlag, Berlin 1984.

[H] O. Hyodo, On variation of Hodge-Tate structures, prepring Nara (Japan) Oct. 17th 1987.

[Ma] B. Mazur, Frobenius and the Hodge filtration (estimates), *Ann. of Math.* **98** (1973), 58-95.

[Me] W. Messing, The crystals associated to Barsotti-Tate groups, *Lecture Notes in Math.* **264** (1972).

[O] A. Ogus, F-isocrystals and de Rham cohomology II—convergent isocrystals, *Duke Math J.* **51** (1984), 765-850.

[T] J. Tate, P-divisible groups in Conference on local fields (Driebergen) 158-183 Springer-Verlag, Berlin 1967.

[SGA4] M. Artin, A. Grothendieck, and J. L. Verdier, Théorie des Topos et Cohomologie Etale des Schémas (SGA4), Springer Lecture Notes **269**, **270**, **305** (1972/73).

ON FOURIER COEFFICIENTS OF EISENSTEIN SERIES

By M. Furusawa and J. A. Shalika

0. Introduction and notations.

0.1. Many authors have addressed the problem of calculating Fourier coefficients of Eisenstein series. For treatment of this problem in the setting of holomorphic forms, we refer the reader to the work of S. Böcherer [Bo], N. Kurokawa [K-M], S. Mizumoto [Mi], T. Orloff [O-S], G. Shimura [Shi1,2] and J. Sturm [O-S].

Earlier ideas of Hecke and Siegel were elaborated in the framework of arithmetic cohomology by G. Harder [Ha].

Very detailed and systematic results on the topic of this paper were obtained by F. Shahidi [G-S], [Sha] in a long series of papers in which he studies the "normalizing factor" $L(s, \pi, r)$ of Langlands' Eisenstein series. Many deep results have been obtained about these L-functions by Shahidi by interpreting them as Fourier coefficients of Eisenstein series.

We refer finally to the striking results of I. I. Piatetskii-Shapiro and S. Rallis [PS-R1,2] on integral representations of automorphic L-functions.

We believe that it is important to set the known results on this subject in a general framework. We shall see, in a very elementary way, how coefficients of Eisenstein series with respect to "multiplicity one subgroups" can be expressed as infinite products (of explicit integrals). In all the cases that have been treated concretely, and to which we have referred, these infinite products are actually ratios of automorphic L-functions. In this paper, we will, for the most part, only concern ourselves with the simple formal aspects of this relationship, ignoring entirely the difficult question of the calculation of Kloosterman integrals (c.f. Section 2).

It will be clear to the reader that in order to make progress we will need to find more examples of multiplicity one subgroups. We can only

Manuscript received 8 November 1988.

hope that as more information becomes available we may obtain new results on automorphic L-functions or their special values.

Finally we believe that the work of S. Gelbart, I. I. Piatetskii-Shapiro and S. Rallis [G-PS-R] will provide new examples of multiplicity one subgroups. We also believe that their work will also provide examples with which to test the method of the double trace formula. One can hope to eventually obtain new special cases of Langlands' functoriality as a consequence.

We also refer to the results of Jacquet and one of the authors [J-S] on exterior square L-functions for GL_n (see Section 4 for other references).

We refer to H. Jacquet [J-1,2,3], N. V. Kuznietsov [K] and Y. Ye [Y] for the methods and applications of the double trace formula.

Finally we would like to thank Y. Nisnevich for his kind assistance.

We would also like to mention the beautiful results of Bump, Friedberg and Hoffstein [B-F-H].

0.2. We now list our principal notations. The ground field F is a global field. (We will assume in Section 3 that F has characteristic zero, esp. in Proposition 3.2.) Denote by $F_{\mathbf{A}}$ or $\mathbf{A}$ the ring of adeles of F. The completion of F at a place v will be denoted by F_v.

If G is an algebraic group defined over F, we write G_F, $G_v = G_{F_v}$, $G_{\mathbf{A}} = G(\mathbf{A})$ resp. for the points of G over F, F_v and $\mathbf{A}$. We will denote by $Z = Z_G$ the center of G.

We will denote by $d_r g$ (resp. $d_l g$) the right (resp. left) Haar measure on G_v (or $G_{\mathbf{A}}$). We have then $d_r(xg) = \delta_G(x) d_r(g)$, $d_l(gx^{-1}) = \delta_G(x) d_l(g)$, for $x, g \in G_v$ (or $G_{\mathbf{A}}$). In this case of $G_{\mathbf{A}}$ we take for $d_r(g)$ and $d_l(g)$ the corresponding Tamagawa measure (defined by a right or left invariant volume form).

Suppose H is an algebraic subgroup of G. Even though the space $H_{\mathbf{A}} \backslash G_{\mathbf{A}}$ may not carry an invariant measure, we do have an invariant positive functional, denoted by $\int_{H_{\mathbf{A}} \backslash G_{\mathbf{A}}} f(g) d_r^* g$, on the space of functions f on $G_{\mathbf{A}}$ satisfying

$$f(hg) = \delta_G^{-1}(h) \delta_H(h) f(g),$$

for $g \in G_{\mathbf{A}}$, $h \in H_{\mathbf{A}}$. One then has, say for $f \in C_c(G_{\mathbf{A}})$,

$$\int_{G_{\mathbf{A}}} f(g) d_r(g) = \int_{H_{\mathbf{A}} \backslash G_{\mathbf{A}}} d_r^* g \int_{H_{\mathbf{A}}} \delta_G(h) \delta_H^{-1}(h) f(hg) d_r(h).$$

We shall refer, by abuse of language, to $d_r^*(g)$ as the invariant measure on $H_{\mathbf{A}}\backslash G_{\mathbf{A}}$. We shall normalize this measure by taking for $d_r(g)$ and $d_r(h)$ the Tamagawa measures on $G_{\mathbf{A}}$ and $H_{\mathbf{A}}$ resp. We shall use similar notation in the local setting.

Next let (π, V) be an admissible, irreducible representation of the group $G_{\mathbf{A}}$. In the current context we will say that π is automorphic if V is a space of automorphic functions on $G_{\mathbf{A}}$ and π is realized by right translations:

$$(\pi(g)f)(x) = (R_g f)(x) = f(xg),$$

with $x, g \in G_{\mathbf{A}}$. We also have the notion of a smooth, irreducible, automorphic representation π of $G_{\mathbf{A}}$. Suppose for example that ω is a character of $Z_{\mathbf{A}}$ trivial on Z_F and that π acts irreducibly on a closed subspace $\mathcal{H}$ of ${}^0L_2(\omega, G_F\backslash G_{\mathbf{A}})$, the corresponding space of cusp forms. Then the space $V = \mathcal{H}^\infty$ of smooth vectors in $\mathcal{H}$ is stable under right translations and affords a smooth representation of $G_{\mathbf{A}}$. In either case we let ω_π denote the central (quasi-) character of π.

If V is a locally-convex space we denote by $C_c^\infty(G_{\mathbf{A}}; V)$ the space of smooth maps

$$f : G_{\mathbf{A}} \to V,$$

of compact support. We set for $x \in G_{\mathbf{A}}$, $g \in G_{\mathbf{A}}$, $(R_x f)(g) = f(gx)$, $(L_x f)(g) = f(xg)$. We refer to [W] (vol. I) for details concerning Bruhat theory (c.f. (3.20) below).

If V_1 and V_2 are locally convex spaces, $V_1 \hat{\otimes} V_2$ will denote the projective tensor product (c.f. [W]).

1. Eisenstein series. We review the basic definition of Eisenstein series.

Let G be a (connected) reductive group defined over F. Let P be a maximal parabolic subgroup of G (defined over F). Let $P = MU$ be a Levi-decomposition.

Next let (π, V) be a smooth, irreducible, automorphic representation of the group $M_{\mathbf{A}}$ (with central quasi-character ω_π). We suppose that V is a space of smooth automorphic functions on $M_{\mathbf{A}}$ on which π operates by right translations (c.f. (0.2)). We may suppose for example that π occurs in ${}^0L_2(\omega, M_F\backslash M_{\mathbf{A}})$ for an appropriate character ω of $Z_{M,\mathbf{A}}$, trivial on $Z_{M,F}$.

We may regard π as a representation of $P_{\mathbf{A}}$ via the canonical identification $P_{\mathbf{A}}/U_{\mathbf{A}} \simeq M_{\mathbf{A}}$. For $\phi \in V$ we set $\lambda(\phi) = \phi(e)$.

Next let $f : G_{\mathbf{A}} \to V$ be a smooth map of compact support. We set, for $s \in \mathbf{C}$, $g \in G_{\mathbf{A}}$,

$$f^{\#}(s, g) = \int_{P_{\mathbf{A}}} \delta_P^{1/2+s}(p)\lambda\{\pi(p)f(p^{-1}g)\}d_l(p). \tag{1.1}$$

The Eisenstein series we wish to consider are the functions on $G_{\mathbf{A}}$ the form

$$E(s, g) = E(s, f, g) = \sum f^{\#}(s, \gamma g), \tag{1.2}$$

the sum over γ in $P_F \backslash G_F$.

Next let H be an algebraic subgroup of G (defined over F). We assume that H contains $Z_G = Z$, the center of G. *We assume throughout that the quotient $Z_{\mathbf{A}}H_F\backslash H_{\mathbf{A}}$ is compact.* (One expects completely similar results in the case when $Z_{\mathbf{A}}H_F\backslash H_{\mathbf{A}}$ has finite volume.) Next let (σ, W) be an irreducible, automorphic representation of $H_{\mathbf{A}}$. As above we assume that σ is a smooth representation acting by right translations on a space W of smooth automorphic functions. We will assume that $\omega_\pi\omega_\sigma = 1$ on $Z_{\mathbf{A}}$.

Our object of study are the integrals

$$Z(s) = Z(s, \pi, \sigma) = Z(s, f) = \int_{H_F Z_{\mathbf{A}}\backslash H_{\mathbf{A}}} E(s, h)\phi(h)d_r(h), \tag{1.3}$$

where ϕ varies in the space W of the automorphic representation σ. As we have indicated in the introduction, such integrals have been extensively studied. In all the cases we have reviewed, in the context of this paper, the functions $Z(s)$ are ratios of automorphic L-functions. In fact, if $L(s, \pi, r)$ is the "normalizing factor" for the Eisenstein series (c.f. [L], [G-S]), then $L(s, \pi, r)Z(s, \pi, \sigma)$ is (essentially) an automorphic L-function associated functorially to the pair (π, σ). A common feature in many of the cases we have referred to is the fact that *the induced representation* $\mathrm{Ind}(G_A, H_A; \sigma)$ *has multiplicity one*. In what follows we will indicate how this hypothesis, in actuality a much weaker variation of this hypothesis, leads naturally to a factorization

$$Z(s) = \prod_v Z_v(s)$$

of $Z(s)$ into a product of explicit local integrals (Kloosterman integrals), (c.f. Theorem 4.3). We do not know if our hypotheses on multiplicities (see below) always imply that the factors $Z_v(s)$ are Eulerian. Our experience is however that this is certainly true in many cases.

2. Kloosterman integrals. For the sake of clarity we choose to formulate our problem in a more general context. We let then H_1 and H_2 be algebraic subgroups of G (over F). We suppose that H_1 and H_2 contain the center Z of G. Let Δ denote the subgroup of $H_1 \times H_2$ consisting of all pairs (z, z) with $z \in Z$. Let (π_i, V_i) $(i = 1, 2)$ be smooth, irreducible, automorphic representations of $H_{1,\mathbf{A}}$ and $H_{2,\mathbf{A}}$ resp. Again we suppoose that V_i $(i = 1, 2)$ is a space of smooth automorphic functions and that π_i $(i = 1, 2)$ acts by right translations. Let ω_i $(i = 1, 2)$ be the central quasi-character of π_i. We suppose $\omega_1\omega_2 = 1$ on $Z_{\mathbf{A}}$. We also let χ_1 and χ_2 be quasi-characters of $H_{1,\mathbf{A}}$ and $H_{2,\mathbf{A}}$ (resp.) trivial on $H_{1,F}$ and $H_{2,F}$ (resp.). We suppose $\chi_1\chi_2 = 1$ on $Z_{\mathbf{A}}$. Let $V = V_1 \hat{\otimes} V_2$ (projective tensor product) regarded as a space of functions on $H_{1,\mathbf{A}} \times H_{2,\mathbf{A}}$. Next, for $\phi \in V$, we set $\lambda(\phi) = \phi(e)$. We set, for $f \in C_c^\infty(G_{\mathbf{A}}; V)$,

$$(2.1)\quad T(f) = T_{\chi_1,\chi_2}(f) = \int_{\Delta_{\mathbf{A}}(H_{1,F}\times H_{2,F})\backslash H_{1,\mathbf{A}}\times H_{2,\mathbf{A}}} \sum_{\gamma\in G_F} \chi_1(h_1)\chi_2(h_2) \cdot \lambda\{\pi_1(h_1) \otimes \pi_2(h_2) f(h_1^{-1}\gamma h_2)\} d_r h_1 d_r h_2.$$

Integrals of this type have been extensively studied by Jacquet [J-1,2,3] and Kuznetzov [K]. We want to know in general under what conditions we can express such an integral as a (generally) infinite sum of infinite products of local integrals.

To proceed we transform the integral (2.1) in the standard fashion. We write then for $\gamma \in G_F$, $\gamma = x^{-1}\xi y$, with $x \in H_{1,F}$, $y \in H_{2,F}$ and ξ varying in the set of double cosets $H_{1,F}\backslash G_F/H_{2,F}$. Integrating in stages we get at once the identity

$$(2.2)\qquad T(f) = \sum T_\xi(f),$$

the sum over ξ in $H_{1,F}\backslash G_F/H_{2,F}$. The distribution $T_\xi = T_{\xi,\chi_1,\chi_2}$ is given by the integral

(2.3) $$T_\xi(f) = \int_{\Delta_{\mathbf{A}}(\xi)\backslash H_{1,\mathbf{A}}\times H_{2,\mathbf{A}}} \chi_1(h_1)\chi_2(h_2) \cdot \mu_\xi\{\pi_1(h_1)\otimes\pi_2(h_2)f(h_1^{-1}\xi h_2)\}d_r^*(h_1, h_2).$$

Here $\Delta(\xi)$ is the subgroup of $H_1 \times H_2$ consisting of all pairs (x, y) satisfying $x^{-1}\xi y = \xi$. The continuous linear form $\mu_\xi = \mu_{\xi,\chi_1,\chi_2}$ on the space V is given by the integral

(2.4) $$\mu_\xi(\phi) = \int_{\Delta_{\mathbf{A}}\Delta_F(\xi)\backslash\Delta_{\mathbf{A}}(\xi)} \chi_1(x)\chi_2(y)\phi(x, y)(\delta_{H_1\times H_2}\delta_\xi^{-1})(x, y)d_r(x, y).$$

Here δ_ξ is the modulus for the group $\Delta_{\mathbf{A}}(\xi)$. (All of our measures are Tamagawa measures). The linear form $\mu = \mu_\xi = \mu_{\xi,\chi_1,\chi_2}$ satisfies the following condition:

(2.5) $$\mu(\pi_1(x)\otimes\pi_2(y)v) = (\delta_{H_1\times H_2}^{-1}\delta_\xi)(x, y)\chi_1^{-1}(x)\chi_2^{-1}(y)\mu(v),$$

for all $v \in V$ and $(x, y) \in \Delta_{\mathbf{A}}(\xi)$. (The form μ is a generalization of the Whittaker functional). Let us denote, for $\xi \in G_F$, by $V^*(\xi) = V^*(\pi_1, \pi_2, \chi_1, \chi_2; \xi)$ the space of continuous linear forms on V satisfying (2.5). Since π_1 and π_2 are factorizable: $\pi_i = \otimes\,\pi_{i,v}$, $V_i = \otimes_v V_{i,v}$ $(i = 1, 2)$, we may define, in a completely analogous way, the local analogue of this space. We denote this latter space by $V_v^*(\xi) = V_v^*(\pi_1, \pi_2, \chi_1, \chi_2, \xi)$, $\xi \in G_v$. We now make the following hypothesis:

(2.6). *Hypothesis* A. *For all places v of F, and all $\xi \in G_v$*, dim $V_v^*(\xi) \leq 1$. This hypothesis implies of course that

(2.7) $$\dim V^*(\xi) \leq 1.$$

Moreover each linear form μ_ξ is then a product

(2.8) $$\mu_\xi = \bigotimes_v \mu_{\xi,v}.$$

We remark here that the distribution $T_\xi = 0$ if and only if $\mu_\xi = 0$.

If then the test function f is a pure tensor:

$$f = \bigotimes_v f_v,$$

with f_v of the form

$$f_v = \alpha_v \cdot v_{1,v} \otimes v_{2,v}, \qquad \alpha_v \in C_c^\infty(G_v), \qquad v_{1,v} \in V_{1,v}, \qquad v_{2,v} \in V_{2,v},$$

then the integral $T_\xi(f)$ in (2.3) may be expressed as a product over the places v of F:

$$T_\xi(f) = \prod T_{\xi,v}(f_v), \tag{2.9}$$

where each factor $T_{\xi,v}(f_v)$ is expressed as an integral

$$T_{\xi,v}(f_v) = \int_{\Delta_v(\xi)\backslash H_{1,v}\times H_{2,v}} \chi_1(h_1)\chi_2(h_2)W_v(h_1, h_2) \cdot \alpha_v(h_1^{-1}\xi h_2)d_r^*(h_1, h_2). \tag{2.10}$$

Here the function $W_v(h_1, h_2)$ (generalized Whittaker function) is given by

$$W_v(h_1, h_2) = \mu_{\xi,v}(\pi_1(h_1) \otimes \pi_2(h_2)v_{1v} \otimes v_{2v}), \tag{2.11}$$

for $h_1 \in H_{1,v}$, $h_2 \in H_{2,v}$. Integrals of this type have been extensively studied in many interesting cases by Jacquet [J-1,2,3] and Ye [Y]. They have also been studied by Gelbart, Piatetskii-Shapiro and Rallis [G-PS-R], [PS-R1,2,3] in the context of their important work on automorphic L-functions.

3. Infinite products. Recall our basic identity

$$T(f) = \sum T_\xi(f), \tag{3.1}$$

the sum over $\xi \in H_{1,F}\backslash G_F/H_{2,F}$. In this section we will formulate conditions under which this sum essentially reduces to a single term.

Before proceeding we recall a basic result of Borel-Serre [B-S].

PROPOSITION 3.2. *Suppose F is a number field. Then the canonical map*

$$j : H_{1,F}\backslash G_F/H_{2,F} \to H_{1,\mathbf{A}}\backslash G_{\mathbf{A}}/H_{2,\mathbf{A}}$$

is finite to one.

The actual statement proved in [B-S] is that the Hasse-kernel for H^1 is finite. It is not difficult however to deduce Proposition 3.2 from this result. We are indebted to Y. Nisnevich for this observation.

We will say in what follows that two elements ξ and $\eta \in G_F$ are weakly equivalent if and only if there exist elements $x \in H_{1,\mathrm{A}}, y \in H_{2,\mathrm{A}}$ such that $\xi = x\eta y^{-1}$. The proposition states then that there are finitely many rational double cosets in a given weak-equivalence class.

Next let us denote by $Bil(\pi_1, \pi_2) = Bil(\pi_1, \pi_2, \chi_1, \chi_2)$ the space of V-distributions (or distributions) on $G_{\mathbf{A}}$ satisfying

$$T(L_{h_1}R_{h_2}^{-1}f) = \chi_1(h_1)\chi_2(h_2)T(\pi_1(h_1) \otimes \pi_2(h_2)f), \tag{3.3}$$

for all $f \in C_c^\infty(G; V)$, $h_1 \in H_{1,\mathbf{A}}$, $h_2 \in H_{2,\mathbf{A}}$ (c.f. (0.2)).

It is clear that each distribution T_ξ (defined by (2.3)) belongs to the space $Bil(\pi_1, \pi_2, \chi_1, \chi_2)$. Let us denote by T_ξ^* the distribution defined by

$$T_\xi^* = \sum_\eta T_\eta, \tag{3.4}$$

the sum over all η in $H_{1,F}\backslash G_F/H_{2,F}$ which are weakly-equivalent to ξ. We clearly have

$$T = \sum T_\xi^*, \tag{3.5}$$

the sum over weak-equivalence classes of elements of G_F.

(3.6) It is easy to see that any set of nonzero distributions of the form T_ξ, with ξ varying in a fixed set of representatives for the weak-equivalence classes, is linearly independent. On the other hand we shall now show, assuming our hypothesis (2.8), that terms T_η appearing in (3.4) are proportional.

Let us suppose then that ξ and $\eta \in G_F$ are weakly-equivalent. Choose $x \in H_{1,\mathbf{A}}, y \in H_{2,\mathbf{A}}$ such that $\xi = x\eta y^{-1}$. If we set $z = (x, y) \in H_{1,\mathbf{A}} \times H_{2,\mathbf{A}}$, we see at once that

$$\Delta_{\mathbf{A}}(\xi) = z\Delta_{\mathbf{A}}(\eta)z^{-1}.$$

Let us momentarily denote by $d_r{}^*(\xi; h_1, h_2)$ the invariant measure on the quotient space $\Delta_{\mathbf{A}}(\xi)\backslash H_{1,\mathbf{A}} \times H_{2,\mathbf{A}}$ (c.f. (0.2)). We have then

$$(3.7) \qquad d_r{}^*(\eta; x^{-1}h_1x, y^{-1}h_2y) = c(x, y)d_r{}^*(\xi; h_1, h_2),$$

for a suitable positive constant $c(x, y)$.

Next an elementary computation, starting with the definition of T_η, (c.f. (2.3)) leads at once to the identity

$$(3.8) \quad T_\eta(f) = c(x, y)\chi_1^{-1}(x)\chi_2^{-1}(y)$$

$$\cdot \int_{\Delta_{\mathbf{A}}(\xi)\backslash H_{1,\mathbf{A}}\times H_{2,\mathbf{A}}} \nu_\eta\{\pi_1(h_1) \otimes \pi_2(h_2)f(h_1^{-1}\xi h_2)\}d_r^*(h_1, h_2).$$

Here ν_η is the linear form on V given by

$$(3.9) \qquad \nu_\eta(v) = \mu_\eta(\pi_1(x^{-1}) \otimes \pi_2(y^{-1})v), \qquad \text{for} \quad v \in V.$$

It follows easily from (3.9) that the linear form ν_η belongs to the space $V^*(\xi) = V^*(\pi_1, \pi_2, \chi_1, \chi_2; \xi)$ (c.f. (2.5)).

Let us suppose that the space $V^*(\xi)$ is not zero. Our hypothesis (2.8) implies then that $\dim V^*(\xi) = 1$. Hence $\nu_\eta = c^*(x, y)\mu_\xi$, for some constant $c^*(x, y)$. Thus from (3.8) we have

$$(3.10) \qquad T_\eta = c^*(\eta)T_\xi,$$

with

$$(3.11) \qquad c^*(\eta) = c^*(\eta; \chi_1, \chi_2) = c(x, y)c^*(x, y)\chi_1^{-1}(x)\chi_2^{-1}(y).$$

Next let

$$(3.12) \qquad c(\xi) = c(\xi; \chi_1, \chi_2) = \Sigma\, c^*(\eta),$$

the sum over η in $H_{1,F}\backslash G_F/H_{2,F}$ weakly-equivalent to ξ. We have then from (3.4)

(3.13) $$T_\xi^* = c(\xi; \chi_1, \chi_2) T_\xi,$$

and finally

(3.14) $$T_{\chi_1,\chi_2} = \sum_\xi c(\xi; \chi_1, \chi_2) T_{\xi,\chi_1,\chi_2},$$

the sum over weak-equivalence classes of elements of G_F.

We have seen that the distributions T_ξ (or T_ξ^*) belong to the space $Bil(\pi_1, \pi_2, \chi_1, \chi_2)$. In a completely analogous way we may define the local analogue $Bil(\pi_{1,v}, \pi_{2,v}, \chi_{1,v}, \chi_{2,v})$ of this space. We now make the following additional assumption on the quadruple $(\pi_1, \pi_2, \chi_1, \chi_2)$.

(3.15). *Hypothesis* B. *For all places v of F,*

$$\dim Bil(\pi_{1,v}, \pi_{2,v}, \chi_{1,v}, \chi_{2,v}) \leq 1.$$

Since $C_c^\infty(G_{\mathbf{A}}; V)$ is a restricted tensor product, this assertion implies that

(3.16) $$\dim Bil(\pi_1, \pi_2, \chi_1, \chi_2) \leq 1.$$

(3.17) Before proceeding we make the following definition. We say that the quadruple $(\pi_1, \pi_2, \chi_1, \chi_2)$ is *generic* if and only if there exists an element ξ of G_F such that the corresponding linear form $\mu_\xi = \mu_{\xi,\chi_1,\chi_2}$ is nonzero.

Suppose then that $(\pi_1, \pi_2, \chi_1, \chi_2)$ is generic. *Choose* $\xi \in G_F$ *such that* $\mu_\xi = \mu_{\xi,\chi_1,\chi_2}$ *is nonzero.* Then the corresponding distribution $T_\xi = T_{\xi,\chi_1,\chi_2}$ is also nonzero. If we assume then Hypothesis B, it follows from (3.6) that the weak-equivalence class of ξ is uniquely determined. We denote this class by $\xi(\pi_1, \pi_2, \chi_1, \chi_2)$. We summarize our conclusions in the following proposition.

PROPOSITION 3.18. *Suppose that the quadruple* $(\pi_1, \pi_2, \chi_1, \chi_2)$ *is generic. Suppose further that* $(\pi_1, \pi_2, \chi_1, \chi_2)$ *satisfies the hypotheses* (2.8) *and* (3.15). *Let* $\xi = \xi(\pi_1, \pi_2, \chi_1, \chi_2)$ *be the associated weak-equivalence class in* G_F. *Then the distribution* T_{ξ,χ_1,χ_2} *is nonzero. Further we have*

(3.19) $$T_{\chi_1,\chi_2} = c(\xi, \chi_1, \chi_2) T_{\xi,\chi_1,\chi_2}$$

with the constant $c(\xi, \chi_1, \chi_2)$ *given by* (3.12).

Remark (3.20). In some of the concrete cases we have examined, the notions of weak and strong equivalence coincide, i.e. the map j of Proposition 3.2 is injective. In that case of course the constant $c(\xi, \chi_1, \chi_2) = 1$.

Remark 3.21. The reader who is familiar with Bruhat theory will realize, at least when F is p-adic, that we have the inequality

$$(3.21)\ \dim Bil(\pi_{1,v}, \pi_{2,v}, \chi_{1,v}, \chi_{2,v}) \leq \sum \dim V^*(\pi_{1,v}, \pi_{2,v}, \chi_{1,v}, \chi_{2,v}; \xi),$$

the sum taken over double cosets of $H_{1,v}\backslash G_v/H_{2,v}$ (assuming the latter set is finite). One has a similar statement when F is real or complex. In many concrete cases the sum on the right is equal to one (for all v). Of course in that case both of our hypotheses A and B will hold. See [W] for more details.

4. Fourier coefficients. We apply the results of Sections 2 and 3 to our main problem which is the calculation of the integral (1.3).

For that purpose we take $H_1 = P, H_2 = H$ (c.f. Section 1). Recall that we are assuming that the quotient $Z_{\mathbf{A}}H_F\backslash H_{\mathbf{A}}$ is compact. We let $\pi_1 = \pi$, $\pi_2 = \sigma$, $V_1 = V$, $V_2 = W$, $\chi_1 = \delta_P^{s-1/2}$, $\chi_2 = 1$. We have then, for $f_1 \in C_c^\infty(G_{\mathbf{A}}; V_1)$,

$$(4.1) \qquad f^{\#}{}_1(s, g) = \int_{P_{\mathbf{A}}} \chi_1(p)\lambda_1\{\pi(p)f_1(p^{-1}g)\}d_r(p).$$

Here $\lambda_1(\phi_1) = \phi_1(e)$, $\phi_1 \in V_1$. (Recall that V_1 is a space of functions on $M_{\mathbf{A}}$.)

Next let $\phi_2 \in V_2$. Set $f = f_1 \otimes \phi_2$. We regard f as an element of $C_c(G_{\mathbf{A}}; V)(V = V_1 \hat{\otimes} V_2)$. Finally let $\lambda = \lambda_1 \otimes \lambda_2$: for $\phi = \phi_1 \otimes \phi_2 (\phi_1 \in V_1, \phi_2 \in V_2)$, $\lambda(\phi) = \phi(e)$.

With these identifications we find at once the equality

$$(4.2) \qquad Z(s) = \int_{H_F Z_{\mathbf{A}}\backslash H_A} E(s, f_1, h)\phi_2(h)d_r(h) = T_{\chi_1,\chi_2}(f).$$

Next our assumption that $Z_{\mathbf{A}}H_F\backslash H_{\mathbf{A}}$ is compact implies that, for all $\xi \in G_F$, the quotient $\Delta_{\mathbf{A}}\Delta_F(\xi)\backslash\Delta_{\mathbf{A}}(\xi)$ is compact. It follows that the (quasi-) character $\chi_1 \times \chi_2$ of $H_{1,\mathbf{A}} \times H_{2,\mathbf{A}} = P_{\mathbf{A}} \times H_{\mathbf{A}}$ has a trivial restriction to the group $\Delta_{\mathbf{A}}(\xi)$ (for all $\xi \in G_F$). *Thus the linear form* $\mu_\xi = \mu_{\xi,\chi_1,\chi_2}$ *does not depend on* (χ_1, χ_2). Similarly the space $V^*(\xi) =$

$V^*(\pi_1, \pi_2, \chi_1, \chi_2; \xi)$ does not depend on χ_1 or χ_2 (c.f. (2.4), (2.5)). The same is true for Hypothesis A ((2.6)).

Let us assume now that the pair (π, σ) is generic, i.e. there is a $\xi \in G_F$ such that the corresponding linear form μ_ξ is nonzero ((3.17)). We also assume (2.6) for the pair (π, σ) and (3.15) for (π, σ) and all s. Let $\xi = \xi(\pi, \sigma)$ be the associated weak-equivalence class. This class then does not depend on s. Applying Proposition 3.18 we obtain at once the following theorem.

THEOREM 4.3. *Let the notations and assumptions be as above. Suppose* $\mathrm{Re}(s)$ *is large. Then* $Z(s, f) = c(s)\, T_{\xi,s}(f)$. *Here* $c(s) = c(\xi, \chi_1, \chi_2)$ *is the elementary function of s (a linear combination of exponentials) given by* (3.11) *and* (3.12) *and* $T_{\xi,s} = T_{\xi,\chi_1,\chi_2}$ *is given by* (2.3). *Moreover, if* $f = \otimes_v f_v$ *is a pure-tensor, then* $T_{\xi,s}(f)$ *may be expressed as an infinite product of local integrals (Kloosterman integrals) given explicitly by* (2.10):

$$Z(s, f) = c(s) \prod_v Z(s, f_v),$$

with

$$Z(s, f_v) = \int_{\Delta_v(\xi)\backslash P_v \times H_v} \delta_P^{s-1/2}(h_1) W_v(h_1, h_2) \alpha_v(h_1^{-1}\xi h_2) d_r^*(h_1, h_2).$$

Remark (4.4). We refer the reader again to Remark (3.21). We also add the following. Suppose F is local. Write G for G_v, H_i for $H_{i,v}$ $(i = 1, 2)$. Let π_i be an admissible representation (smooth if F is archimedean) of H_i $(i = 1, 2)$. Let

$$I(\pi_1) = \mathrm{Ind}(G, H_1, \pi_1 \otimes \delta_{H_1}^{1/2}).$$

Then we have the canonical identification (Frobenius reciprocity)

$$Bil(\pi_1, \pi_2) = Bil(I(\pi_1)|H_2, \pi_2)^{H_2}, \tag{4.5}$$

the latter space being the space of H_2-invariants in $Bil(I(\pi_1)|H_2, \pi_2)$. In our case the representations $I(\pi \otimes \delta_P^s)$ are typically irreducible. Thus, outside of an exceptional set of values of s, our hypothesis B is implied by the *multiplicity-one hypothesis* for the pair (π_2, H_2), i.e. the assertion that

(4.6) $$\dim Bil(\pi | H_2, \pi_2)^{H_2} \leq 1,$$

for all admissible irreducible representations π of G_ν. The reader will find examples in the next section which actually satisfy this latter condition (c.f. [N-PS], [Ro], [So]).

5. Examples. We discuss very briefly the results of the authors we have referred to in the introduction (0.1).

(5.1). (Shahidi). We take G to be a split group (defined over F) with center Z. Let N be the unipotent radical of a Borel subgroup of G (over F). Let $\psi \neq 1$ be a basic-additive character of $\mathbf{A}$ (trivial on F). We define a character θ of $N_{\mathbf{A}}$ by setting

$$\theta(n) = \psi(\sum_{\alpha} x_\alpha).$$

Here α is a simple root and x_α is the corresponding morphism into the additive group. Let P be any maximal parabolic in G (defined over F). Let π be a cuspidal, automorphic representation of $M_{\mathbf{A}}$ (or $P_{\mathbf{A}}$). We assume that π is generic [Sha]. Next let $H = Z \cdot N$. Let σ be the representation of $H_{\mathbf{A}}$ defined by setting, for $h = zn$, $z \in Z_{\mathbf{A}}$, $n \in N_{\mathbf{A}}$, $\sigma(zn) = \omega_\pi^{-1}(z)\theta(n)$.

Let $L(s, \pi, r)$ be the "normalizing factor" attached by Langlands [L] to (G, P, π). Shahidi has proved that $Z(s)$ is the product of $L(s, r, \pi)^{-1}$ times a product of finitely many "elementary" factors. Of course this involves a detailed study of the local integrals (2.11) [Sha]. Similar results are true more generally for quasi-split groups. See also [G-S].

(5.2). (Böcherer, Kurokawa, Mizumoto, Orloff, Sturm). Here we take $G = GSP_4$. Let P be the stabilizer of the line determined by the vector $e_i = {}^t[1, 0, 0, 0]$. We may take the group $M = M_0 M_1$ as the Levi-factor. Here M_0 is the group of (diagonal) matrices of the form $m_0 = \mathrm{diag}(\alpha, 1, \alpha^{-1}, 1)$ and M_1 the group of matrices of the form

$$m_1 = \begin{bmatrix} 1 & 0 & 0 & 0 \\ 0 & a & 0 & b \\ 0 & 0 & \lambda & 0 \\ 0 & c & 0 & d \end{bmatrix},$$

with $\lambda = ad - bc \neq 0$.

Let π_1 be a cuspidal automorphic representation of $\mathrm{GL}_2(\mathbf{A})$. We take for π the representation of $M_{\mathbf{A}}$ defined by setting

$$\pi(m) = \pi(m_0 m_1) = \nu(\alpha)\pi_1\left(\begin{bmatrix} a & b \\ c & d \end{bmatrix}\right), \tag{5.2.1}$$

with ν a quasi-character of $F^\times \backslash \mathbf{A}^\times$.

Next let S be a 2×2 definite symmetric matrix with coefficients in F. Let $G(S)$ be the corresponding similitude group: the set of $g \in \mathrm{GL}_2$ such that ${}^t g S g = \mu(g) S$. We may identify the connected component of the identity $G^0(S)$ of the group $G(S)$ with the multiplicative group $K^\times$ of a quadratic extension K of Q. Next let T denote the subgroup of G consisting of all matrices of the form

$$t = \begin{bmatrix} g & 0 \\ 0 & \lambda {}^t g^{-1} \end{bmatrix},$$

with $g \in G^0(S)$, $\lambda = \mu(g)$.

Let V be the group of all matrices of the form

$$v = \begin{bmatrix} 1 & C \\ 0 & 1 \end{bmatrix},$$

where C is a 2×2 symmetric matrix. Let H be the (semi-direct) product

$$H = T \cdot V.$$

Define a (one-dimensional) representation of $H_{\mathbf{A}}$ as follows. Let χ be a character of $T_F \backslash T_{\mathbf{A}} \simeq K^\times \backslash K_{\mathbf{A}}^\times$. If $\psi \neq 1$ is a character of $F \backslash \mathbf{A}$, we set $\theta_S(v) = \psi(\mathrm{tr}(CS))$, for $v \in V_{\mathbf{A}}$. Finally set, for $h = tv$, $t \in T_{\mathbf{A}}$, $v \in V_{\mathbf{A}}$,

$$\sigma(h) = \chi(t)\theta_S(v). \tag{5.2.2}$$

We assume $\chi | F_{\mathbf{A}}^\times \cdot \omega_{\pi_1} \nu = 1$. In this case the authors have calculated the local integrals in the unramified situation. We have

$$Z(s) = L(2s + 1/2, \pi_1 \otimes \pi(\chi) \otimes \nu)/L(4s + 1, \mathrm{Sym}^2(\pi_1) \otimes \omega_{\pi_1}^{-1} \nu),$$

up to multiplication by finitely many local factors. Here $\pi(\chi)$ is the "Weil-representation" of $\mathrm{GL}_2(\mathbf{A})$ associated to χ. For references see (0.1). We refer especially to the important work of Piatetskii-Shapiro and Soudry on generalized Bessel models ([PS-S1,2]).

(5.3). (Maeda). In this case $G = GSU(2, 2)$, the unitary similitude group of type (2, 2) associated to a (fixed) quadratic extension K of F. Thus G is the group of matrices in $\mathrm{GL}_4(K)$ satisfying

$$g\begin{pmatrix} 0 & 1_2 \\ -1_2 & 0 \end{pmatrix}{}^t g^\sigma = \mu(g)\begin{pmatrix} 0 & 1_2 \\ -1_2 & 0 \end{pmatrix},$$

with $\mu(g) \in F$. Here σ is the nontrivial automorphism of K over F. Again we take P to be the stabilizer of the line determined by the vector $e_1 = {}^t[1, 0, 0, 0]$.

The group H is defined in a fashion analogous to that of (5.2), taking for S a 2×2 definite Hermitian matrix.

We have not yet calculated the integrals (2.11) in this case. However in special cases, T. Maeda has calculated the integral $Z(s)$ by other methods [M]. His results are consistent with our expectations, i.e. $Z(s)$ is a ratio of L-functions, the numerator in this case being the standard L-function associated to an appropriate cusp form on GL_2.

We remark that in both cases (5.2) and (5.3), our results apply to either of the two maximal parabolics.

(5.4) Rankin-Selberg convolutions. There are examples of multiplicity-one subgroups H which are associated with the various known integral representations of L-functions, following the method of Rankin-Selberg. For example let $G = \mathrm{GL}_{2n}$. Let H be the group of matrices of the form

$$h = \begin{pmatrix} g & 0 \\ 0 & g \end{pmatrix}\begin{pmatrix} 1 & u \\ 0 & 1 \end{pmatrix},$$

with $g \in \mathrm{GL}_n$, $u \in M_n$. We take for σ the one-dimensional representation of $H_{\mathbf{A}}$ given by

$$\sigma(h) = \chi(\det g)\psi(\mathrm{tr}(u)),$$

with χ a character of $F^\times \backslash \mathbf{A}^\times$. D. Soudry [So] has proved that when $n = 2$, the induced representation $\mathrm{Ind}(G_v, H_v; \sigma_v)$ decomposes with multiplicity one. We suspect that this is the case for any n. The pair (σ, H) arises in the integral representation for the exterior-square L-function $L(s, \Lambda^2(\pi))$, with π a cusp form on GL_{2n} [J-S2]. There is a great deal of evidence to suggest that the same pair (σ, H) may be used in the double-trace formula to give a new proof of the converse theorem for GSP_4 [J-PS-S].

A "compact version" of this example may be obtained as follows: let $G = \mathrm{GL}_2(D)$, D a division algebra over F. Let H be the subgroup of G consisting of all matrices of the form

$$h = \begin{pmatrix} g & 0 \\ 0 & g \end{pmatrix} \begin{pmatrix} 1 & u \\ 0 & 1 \end{pmatrix},$$

with $g \in D^\times$, $u \in D$. This time let, for $h \in H_{\mathbf{A}}$,

$$\sigma(h) = \chi(\nu(g))\psi(\mathrm{tr}(u)),$$

with ν the reduced norm, tr the reduced trace. Again we expect that the representation $\mathrm{Ind}(G_v, H_v; \sigma_v)$ decomposes with multiplicity-one.

Another example of this type is provided by the 4×2 integral of Piatetskii-Shapiro and Soudry [PS-S1,2] for the group $G = GSP_4$. Here the representation σ is infinite-dimensional. One should be able to use the double trace formula (Section 2) to pass from certain Eisenstein series on GL_4 to cusp forms on GSP_4.

Finally the integral representations of Asai [A] for the Asai L-function and those of Gelbart and Piatetskii-Shapiro and Rallis [G-PS-R] for the classical groups should provide us with many other possible examples with which to test the theme of this paper.

THE JOHNS HOPKINS UNIVERSITY

REFERENCES

[A] T. Asai, On certain Dirichlet series associated with the Hilbert modular forms and Rankin's method, *Math. Ann.* **221** (1977), 81–94.

[Bo] S. Böcherer, Über gewisse Siegelsche Modulformen zweiten grades, *Math. Ann.* **261** (1982), 23-41.

[B] A. Borel, Some finiteness properties of adele groups over number fields, *Publ. Math. Inst. Hautes Etudes Sci.* **16** (1963), 5-30.

[B-S] ______ and J.-P. Serre, Théorèmes de finitude en cohomologie galoisienne, *Comm. Math. Helvetici* **39** (1964), 111-164.

[B-F-H] D. Bump, S. Freidberg and J. Hoffstein, Eisenstein series on the metaplectic group and nonvanishing theorems for automorphic L-functions and their derivatives.

[G-PS-R] S. Gelbart, I. I. Piatetskii-Shapiro and S. Rallis, Explicit constructions of automorphic L-functions, Lecture Notes in Mathematics, Vol. 1254, Springer-Verlag, New York, 1987.

[G-S] ______ and F. Shahidi, Analytic properties of automorphic L-functions, *Perspectives in Mathematics* **6** (1988), Academic Press.

[H] G. Harder, Period integrals of cohomology classes which are represented by Eisenstein series, in *Automorphic Forms, Representation Theory and Arithmetic*, Tata Inst. Jan. 1979.

[H-L-R] ______, R. P. Langlands and M. Rapoport, Algebraische Zykeln auf Hilbert-Blumenthal-Flächen, *Journal Für Math.* **366** (1986), 53-120.

[Ha] M. Harris, Special values of zeta functions attached to Siegel modular forms, *Ann. Scient. Ec. Norm. Sup. 4e series*, t.14, 1981, 77-120.

[J1] H. Jacquet, Sur un résultat de Waldspurger, *An. Scient. Ec. Norm. Sup. 4e series*, t.19 (1986), 185-229.

[J2] ______, Sur un résultat de Waldspurger II, *Comp. Math.* **63** (1987), 315-389.

[J3] ______, On the nonvanishing of some L-functions, *Proc. of Indian Acad. Sci. (Math. Sci.)* **97** (1987), 117-155.

[J-L] ______ and K. F. Lai, A relative trace formula, *Comp. Math.* **54** (1985), 234-310.

[J-S1] ______ and J. A. Shalika, A non-vanishing theorem for zeta-functions of GL_n, *Inventiones Math.* **38** (1976), 1-16.

[J-S2] ______ and ______, Exterior square L-functions, Proceedings of the Ann Arbor Conference on Automorphic forms, Shimura varieties and L-functions.

[J-PS-S] ______, I. I. Piatetskii-Shapiro and J. A. Shalika, The converse theorem for GSP_4, to appear.

[K] N. V. Kuznietsov, Petersson hypothesis for parabolic forms of weight zero and Linnik hypothesis, *Math. Sbornik* **111** (1980), 334-383.

[K-M] N. Kurokawa and S. Mizumoto, On Eisenstein series of degree two, *Proc. Jap. Ac.* **57**A2 (1981).

[L] P. R. Langlands, *Euler Products*, Yale Mathematical Monographs 1, Yale University Press (1971).

[M] T. Maeda, private communication.

[Mi] S. Mizumoto, Fourier coefficients of generalized Eisenstein series of deg two. I, *Invent. Math.* **65** (1981), 115-135.

[N-PS] M. Novodvorsky and I. I. Piatetskii-Shapiro, Generalized Bessel models for the symplectic group of rank 2, *Math. Sb.* **90**(2) (1973), 245-256.

[O-S] T. Orloff and J. Sturm, On the Fourier expansions of certain Eisenstein series of genus two, preprint.

[PS-R1] I. I. Piatetskii-Shapiro and S. Rallis, L-functions of automorphic forms on simple classical groups, in *Modular Forms* (editor R. Rankin), Ellis Harwood, (1984), 251-262.

[PS-R2] ______ and ______, Rankin triple L-functions, *Compositio Mathematica* **64** (1987), 31-115.

[PS-S1] ______ and D. Soudry, L and ϵ factors for $GSP(4)$, *Jour. Fac. Sci. Univ. of Tokyo*, **28** (1982), 505–530.

[PS-S2] ______ and ______, Automorphic forms on the symplectic group of order four, *Inst. des Hautes Etude Sci.*, Bures-sur-Yvette, 1983.

[PS-S3] ______ and ______, Special representations of rank one orthogonal groups, *School of Mathematical Sciences*, Tel Aviv Univ.

[Ro] F. Rodier, Les répresentations de $GSP(4, k)$ où k est un corps local, *C.R. Acad. Sci. Paris*, **282** (1976), Sèr. A, 429–431.

[Sha] F. Shahidi, On certain L-functions, *Amer. J. Math.* **103**(2) (1981), 297–355.

[Shi 1] G. Shimura, Confluent hypergeometric functions on tube domains, *Math. Ann.* **260** (1982), 259–302.

[Shi 2] ______, On Eisenstein series, *Duke Math. J.* **50** (1983), 417–476.

[So] D. Soudry, A uniqueness theorem for representations of $GSO(6)$ and the strong multiplicity one theorem for generic representations of $GSP(4)$, *Israel Journ. of Math.* **58**(3) (1987), 257–287.

[St] J. Sturm, The critical values of zeta functions associated to the symplectic group, *Duke Math. J.* **48** (1981), 327–350.

[Wa] J.-L. Waldspurger, Sur les valeurs de certaines fonctions L automorphes en leur centre de symetrie, *Comp. Math.* **54** (1985), 173–242.

[W] G. Warner, *Harmonic Analysis on Semi-Simple Lie Groups* 1, Springer-Verlag, **188** (1972).

[Y] Y. Ye, Kloosterman integrals and base change, to appear.

QUASICHARACTERS OF CONGRUENCE GROUPS

By DORIAN GOLDFELD*

---

1. Modular forms that transform by a quasicharacter. Let Γ be a congruence subgroup of $SL(2, \mathbf{Z})$, let $\psi : \Gamma \to \mathbf{C}^{\times}$ be a quasicharacter of Γ satisfying

$$\psi(\alpha\alpha') = \psi(\alpha)\psi(\alpha')$$

for $\alpha, \alpha' \in \Gamma$, and let $\mathcal{H}$ denote the upper half plane. Our aim is to study the arithmetic properties of holomorphic modular forms $g : \mathcal{H} \to \mathbf{C}$ which transform by a quasicharacter ψ: i.e. so that

$$g(\gamma z) = \psi(\gamma)g(z) \tag{1}$$

for all $\gamma \in \Gamma$ and $z \in \mathcal{H}$. Note that there are no holomorphic solutions to (1) if $|\psi| = 1$, (i.e. ψ is a character). This fundamental principle will play a central role in our investigation. By taking logarithmic derivatives, it is easily seen that $(g'/g)(z)$ must be a meromorphic modular form of weight two.

We shall now consider functions of type (1) satisfying the differential equation

$$\frac{g'}{g}(z) = cf(z) \tag{2}$$

where $c \in \mathbf{C}$ and f is a holomorphic Hecke cuspform (newform) of weight two normalized so that

$$f(z) = \sum_{1}^{\infty} a(n)e^{2\pi inz}$$

Manuscript received 30 September 1988.
*Supported in part by NSF grant DMS-87-02169.

with $a(1) = 1$. Explicitly solving the differential equation (2), we obtain for every $\lambda \in \mathbf{C}$ a solution

$$g_\lambda(z) = e^{-2\pi i\lambda \int_z^{i\infty} f(\tau)d\tau} \tag{3}$$

where

$$c = 2\pi i\lambda.$$

It follows that

$$g_\lambda(\gamma z) = \psi_\lambda(\gamma)g_\lambda(z)$$

for all $\gamma \in \Gamma$ where

$$\psi_\lambda(\gamma) = e^{-2\pi i\lambda \int_z^{\gamma z} f(\tau)d\tau}$$

and

$$H(\gamma) = -2\pi i \int_z^{\gamma z} f(\tau)d\tau$$

is independent of z and lies in the homology of $\Gamma\backslash\mathcal{H}$ with rational coefficients. Since ψ_λ is a quasicharacter, it follows that H is a homomorphism of Γ. We shall say that ψ_λ is a quasicharacter for Γ associated to f. This is because ψ_λ really depends on f. Since $g_\lambda(z)$ is never zero, we see that the vector space of holomorphic modular forms which transform by ψ_λ is a one dimensional space. For otherwise, if there were some other $h(z)$ in this space, $h(z)/g(z)$ would be a holomorphic modular form of weight zero, and would have to be a constant.

The function $g_\lambda(z)$ has a Fourier expansion

$$g_\lambda(z) = \sum_{n=0}^{\infty} b_\lambda(n)e^{2\pi inz}$$

where we have

$$b_\lambda(0) = 1$$

because

$$b_\lambda(0) = \lim_{z \to i\infty} e^{-2\pi i \lambda \int_z^{i\infty} f(\tau)d\tau}$$

$$= 1.$$

Let $k = \mathbf{Q}(a_1, a_2, a_3, \ldots)$ be the field generated by the coefficients of f. We can now assert:

PROPOSITION 1. *For $\lambda \in \mathbf{C}$ and $n = 1, 2, \ldots$ we have that $b_\lambda(n) \in k(\lambda)$.*

Proof. It follows from the differential equation (2), after dividing by the common factor of $2\pi i$, that

$$\lambda f(z) g_\lambda(z) = \lambda \left(\sum_{n=1}^{\infty} a(n) e^{2\pi i n z} \right) \left(\sum_{n=0}^{\infty} b_\lambda(n) e^{2\pi i n z} \right)$$

$$= \sum_{n=1}^{\infty} b_\lambda(n) \cdot n e^{2\pi i n z},$$

and this implies the recursion relation

$$b_\lambda(n) = \frac{\lambda}{n} \sum_{j=0}^{n-1} a(n-j) b_\lambda(j),$$

which holds for $n > 1$. Since $b_\lambda(0) = 1$, it easily follows by induction that

$$b_\lambda(n) \in k(\lambda).$$

PROPOSITION 2. *For every $\epsilon > 0$, we have*

$$|b_\lambda(n)| \leq e^{2(n\lambda^2(1+(1/e)))^{(1/3)+\epsilon}}$$

for all $n \gg_\epsilon |\lambda|$ where the implied constant depends at most on ϵ.

Proof. Firstly, we have

$$b_\lambda(n) = e^{2\pi n y} \int_0^1 g_\lambda(z) e^{-2\pi i n x} dx.$$

Hence

$$|b_\lambda(n)| \le e^{2\pi ny} \max_{0\le x\le 1} |g_\lambda(z)|. \tag{4}$$

It follows from (3) and Deligne's estimate [D] (Petersson conjecture)

$$|a(n)| \le d(n)\sqrt{n},$$

where $d(n)$ denotes the number of divisors of n, that

$$|g_\lambda(z)| = |e^{\lambda \sum_1^\infty (a(n)/n)e^{2\pi inz}}|$$

$$\le e^{|\lambda| \sum_1^\infty (d(n)/\sqrt{n})e^{-2\pi ny}}.$$

Now, for y sufficiently small, we have

$$\sum_1^\infty \frac{d(n)}{\sqrt{n}} e^{-2\pi ny} \le \sum_{n\le 1/(2\pi y)} \frac{d(n)}{\sqrt{n}} + \int_{1/(2\pi y)}^\infty e^{-2\pi ty} t^{(1/2)+\epsilon} \frac{dt}{t}$$

$$\le \left(1 + \frac{1}{e}\right)(2\pi y)^{-(1/2)-\epsilon}.$$

This yields

$$\max_{0\le x\le 1} |g_\lambda(z)| \le e^{|\lambda|(1+(1/e))(2\pi y)^{-(1/2)-\epsilon}}.$$

If we choose

$$2\pi ny = |\lambda|\left(1 + \frac{1}{e}\right)(2\pi y)^{-(1/2)-\epsilon}$$

or equivalently

$$y \approx \frac{1}{2\pi}\left(\frac{|\lambda|\left(1 + \frac{1}{e}\right)}{n}\right)^{(2/3)-\epsilon}$$

and put this in equation (4), we obtain the proposition.

2. Properties of arithmetic quasicharacters. We now enumerate some standard properties of quasicharacters for the group

$$\Gamma_o(N) = \left\{ \begin{pmatrix} a & b \\ c & d \end{pmatrix} \in SL(2, \mathbf{Z}) \,|\, c \equiv 0 (\mathrm{mod}\, N) \right\}$$

associated to a newform $f(z) = \Sigma_1^\infty a(n)e^{2\pi inz}$ of weight two. Proofs are omitted since they follow easily from the properties of Shimura maps given in [G]. In the following

$$\begin{pmatrix} a & b \\ c & d \end{pmatrix} \in \Gamma_o(N)$$

and ψ_λ is the quasicharacter.

(5)
$$\psi_\lambda\left(\begin{pmatrix} 1 & 1 \\ 0 & 1 \end{pmatrix}\right) = 1.$$

(6)
$$\psi_\lambda\left(\begin{pmatrix} a & -b \\ -c & d \end{pmatrix}\right) = \overline{\psi_{\bar\lambda}\left(\begin{pmatrix} a & b \\ c & d \end{pmatrix}\right)}.$$

(7) For each N, there exists $\epsilon_N = \pm 1$ such that

$$\psi_\lambda\left(w\begin{pmatrix} a & b \\ c & d \end{pmatrix}w^{-1}\right) = \psi_\lambda\left(\begin{pmatrix} d & -\dfrac{c}{N} \\ -bN & a \end{pmatrix}\right) = \psi_{\epsilon_N\lambda}\left(\begin{pmatrix} a & b \\ c & d \end{pmatrix}\right)$$

where we have put

$$w = \begin{pmatrix} 0 & \dfrac{1}{\sqrt{N}} \\ -\sqrt{N} & 0 \end{pmatrix}.$$

(8) Let p be a prime number not dividing N. Set

$$\sigma_p = \begin{pmatrix} p & 0 \\ 0 & 1 \end{pmatrix},$$

and

$$\sigma_j = \begin{pmatrix} 1 & j \\ 0 & p \end{pmatrix}$$

for $j = 0, 1, \ldots, p-1$. Assume that $\alpha, \sigma_k \alpha \sigma_k^{-1} \in \Gamma_o(N)$ for $k = 0, 1, \ldots, p$. (Note that for

$$\alpha = \begin{pmatrix} a & b \\ c & d \end{pmatrix},$$

this holds if $p|b$, $p|c$ and $p|(d-a)$.) Then we have

$$\prod_{k=0}^{p} \psi_\lambda(\sigma_k \alpha \sigma_k^{-1}) = \psi_{a(p)\lambda}(\alpha).$$

The next proposition gives the analogue of the Hecke operators for modular forms which transform by a quasicharacter. In this situation the Hecke operator takes the form of a product.

PROPOSITION 3. *Let $f(\tau) = \Sigma_1^\infty a(n)e^{2\pi i n\tau}$ be a normalized Hecke newform of weight two for $\Gamma_o(N)$. Let*

$$g_\lambda(z) = e^{-2\pi i\lambda \int_z^{i\infty} f(\tau)d\tau}$$

be a modular form transforming by a quasicharacter of $\Gamma_o(N)$ associated to f. Then we have

$$\prod_{k=0}^{p} g_\lambda(\sigma_k z) = g_{a(p)\lambda}(z)$$

with σ_k as defined in (8) *above.*

Proof. We compute

$$\prod_{k=0}^{p} g_\lambda(\sigma_k z) = e^{-2\pi i\lambda \Sigma_{k=0}^{p} \int_{\sigma_k z}^{i\infty} f(\tau)d\tau}$$

$$= e^{-2\pi i\lambda \int_z^{i\infty} \Sigma_{k=0}^{p} f(\sigma_k^{-1}\tau)d(\sigma_k^{-1}\tau)}$$

$$= e^{-2\pi i a(p)\lambda \int_z^{i\infty} f(\tau)d\tau}$$

since f being an eigenfunction of the Hecke operators satisfies

$$\sum_{k=0}^{p} f(\sigma_k{}^{-1}\tau)d(\sigma_k{}^{-1}\tau) = a(p)\cdot f(\tau)d\tau.$$

3. Shimura maps. Let $f(\tau)$ be a normalized newform of weight two for $\Gamma_o(N)$. We assume that f is defined over $\mathbf{Q}$ so that the Fourier coefficients of f are rational. Then by the theory of Shimura [S] there exists an elliptic curve E, defined over $\mathbf{Q}$, which is modular and of level N. We set $E = \mathbf{C}/\Lambda$, where $\Lambda = \{\Omega_1, \Omega_2\}$ is the period lattice of E.

If $f(z)dz$ is precisely the pullback of the standard differential one-form on E then the Shimura map $H : \Gamma_o(N) \to \mathbf{C}$ defined by

$$H(\gamma) = -2\pi i \int_z^{\gamma z} f(\tau)d\tau$$

satisfies

$$H(\gamma) = m(\gamma)\Omega_1 + n(\gamma)\Omega_2$$

with $m(\gamma), n(\gamma) \in \mathbf{Z}$ for $\gamma \in \Gamma_o(N)$. Let f, f' be two normalized newforms of weight two for $\Gamma_o(N)$ defined over $\mathbf{Q}$. This gives two Shimura maps $H : \Gamma_o(N) \mapsto \{\Omega_1, \Omega_2\}$, $H' : \Gamma_o(N) \mapsto \{\Omega'_1, \Omega'_2\}$ onto the period lattices of two elliptic curves E, E' satisfying

$$H(\gamma) = m(\gamma)\Omega_1 + n(\gamma)\Omega_2, \tag{9}$$

$$H'(\gamma) = m'(\gamma)\Omega'_1 + n'(\gamma)\Omega'_2, \tag{10}$$

with $m(\gamma), n(\gamma), m'(\gamma), n'(\gamma) \in \mathbf{Z}$.

We will now show that these maps are uniquely characterized by their respective m and n components.

PROPOSITION 4. *If there exists two Shimura maps H, H', as above satisfying* (9) *and* (10) *and in addition $m(\gamma) = m'(\gamma)$ or $n(\gamma) = n'(\gamma)$ for all $\gamma \in \Gamma_o(N)$ then $H \equiv H'$.*

Proof. Let f, f' be the two cusp forms for $\Gamma_o(N)$ as above. Without loss of generality, we may assume that Ω_1, Ω_2 are real and Ω_1', Ω_2' are pure imaginary. Define two modular functions

$$g_\lambda(z) = e^{-2\pi i\lambda \int_z^{i\infty} f(\tau)d\tau}$$

$$g_\lambda'(z) = e^{-2\pi i\lambda \int_z^{i\infty} f'(\tau)d\tau}.$$

Note that here the prime does not mean derivative. Since g_λ transforms by the quasicharacter

$$\psi_\lambda(\gamma) = e^{-2\pi i\lambda H(\gamma)}$$

$$= e^{-2\pi i\lambda[m(\gamma)\Omega_1+n(\gamma)\Omega_2]},$$

and similarly g_λ' transforms by

$$\psi_\lambda'(\gamma) = e^{-2\pi i\lambda[m'(\gamma)\Omega_1'+n'(\gamma)\Omega_2']},$$

we easily see that if $m(\gamma) = m'(\gamma)$ then the modular function

$$g_{1/\Omega_1}(z)\cdot g'_{-1/\Omega_2'}(z)$$

must transform by the character

$$e^{-2\pi i[(\Omega_2/\Omega_1)n(\gamma)-(\Omega_2'/\Omega_1')n'(\gamma)]}.$$

But any holomorphic modular function for $\Gamma_o(N)$ which transforms by a character must be a constant. It follows that

$$\frac{\Omega_2}{\Omega_1} = \frac{\Omega_2'}{\Omega_1'} = \kappa$$

for some constant κ, and

$$\frac{\Omega_2}{\Omega_1}\, n(\gamma) = \frac{\Omega_2'}{\Omega_1'}\, n'(\gamma).$$

Consequently $n \equiv n'$. Similarly, we may show that $n \equiv n'$ implies that $m \equiv m'$.

Let ψ_λ be a quasicharacter associated to the aforementioned newform f. We consider the two special quasicharacters ψ_{1/Ω_1} and ψ_{1/Ω_2}. An interesting question is how large or small can these quasicharacters get.

CONJECTURE 1. *If*

$$\alpha = \begin{pmatrix} a & b \\ c & d \end{pmatrix} \in \Gamma_o(N)$$

with $|a|, |b|, |c|, |d| \leq N^2$ *then there exists* $\kappa > 0$ *such that for all* $N > 1$

$$e^{-N^\kappa} \ll |\psi_{1/\Omega_1}(\alpha)| \ll e^{N^\kappa}$$

and

$$e^{-N^\kappa} \ll |\psi_{1/\Omega_2}(\alpha)| \ll e^{N^\kappa}.$$

Now conjecture (1) is equivalent to a well known conjecture of Szpiro for the special case of elliptic modular curves.

CONJECTURE 2. (Szpiro) *Let E be an elliptic curve over* **Q** *in minimal form. If E has conductor N and discriminant D, then there exists an absolute constant κ (independent of N, D) such that*

$$D \ll N^\kappa.$$

To see the equivalence, note that if Ω_1 is real and Ω_2 is pure imaginary, then conjecture (1) says that

$$\left| m(\gamma) \frac{\Omega_1}{\Omega_2} \right| \ll N^\kappa$$

$$\left| n(\gamma) \frac{\Omega_2}{\Omega_1} \right| \ll N^\kappa.$$

But, if we use the well known formula

$$D = \frac{\Delta\left(-\frac{\Omega_1}{\Omega_2}\right)}{2\pi^{12}\Omega_2{}^{12}} = \frac{\Delta\left(\frac{\Omega_2}{\Omega_1}\right)}{2\pi^{12}\Omega_1{}^{12}}$$

where $\Delta(z)$ is the Ramanujan cusp form of weight twelve for the full modular group, we see that the estimate

$$D \ll N^N$$

leads to

$$N^{-1-\epsilon} \ll \left|\frac{\Omega_1}{\Omega_2}\right| \ll N^{1+\epsilon}$$

for some $\epsilon > 0$.

Hence, conjecture (1) is equivalent to

$$|m(\gamma)| \ll N^{\kappa+1}$$

$$|n(\gamma)| \ll N^{\kappa+1}.$$

This is just conjecture (4) of [G]. As shown in [G], conjecture (4) of that paper is equivalent to Szpiro's conjecture in the special case that E is an elliptic modular curve.

4. Modular convolutions. Let f, E and $\Lambda = \{\Omega_1, \Omega_2\}$ be as in Section 3. We assume that Ω_1 is real and Ω_2 is pure imaginary. Let us define two nonholomorphic modular functions.

$$G(z) = g_{\pi i/\Omega_1}(z) \cdot \overline{g_{-\pi i/\Omega_1}(z)} = g_{\pi i/\Omega_1}(z) \cdot g_{\pi i/\Omega_1}(-\bar{z})$$

and

$$G^*(z) = g_{\pi/\Omega_2}(z) \cdot \overline{g_{-\pi/\Omega_2}(z)} = g_{\pi/\Omega_2}(z) \cdot g_{-\pi/\Omega_2}(-\bar{z}).$$

It easily follows that

$$G(\gamma z) = e^{2\pi i m(\gamma)} G(z) = G(z)$$

and

$$G^*(\gamma z) = e^{2\pi in(\gamma)}G^*(z) = G^*(z).$$

If we expand

$$G(z) = \sum_{-\infty}^{\infty} B_n(y)e^{2\pi inx}$$

in a Fourier series, then it is easily seen that

$$B_0(y) = \sum_{n=0}^{\infty} b_{\pi i/\Omega_1}(n)^2 e^{-4\pi ny},$$

where

$$g_\lambda(z) = \sum_{n=0}^{\infty} b_\lambda(n)e^{2\pi inz}.$$

Since $b_\lambda(0) = 1$, it is evident that $B_0(\infty) = 1$. Similarly, we have

$$B_k(y) = \sum_{n=k}^{\infty} b_{\pi i/\Omega_1}(n)b_{\pi i/\Omega_1}(n-k)e^{-2\pi(2n-k)y}.$$

Now,

$$G(z),\ G^*(z) \in \mathcal{L}^2(\Gamma_o(N)\backslash\mathcal{H}).$$

It follows that $G(z)$ has a Selberg spectral expansion in terms of Maass forms and Eisenstein series for $\Gamma_o(N)$.

Let

$$\Delta = -y^2\left(\frac{\partial^2}{\partial x^2} + \frac{\partial^2}{\partial y^2}\right)$$

be the Laplace operator. For $j = 0, 1, 2, \ldots$, let $\eta_j(z)$ denote an orthonormal set of Maass forms for the group $\Gamma_o(N)$ satisfying

$$\Delta\eta_j(z) = \lambda_j\eta_j(z),$$

where η_0 is the constant function, and $\lambda_0 = 0$. These functions span the discrete part of the spectrum of Δ. We also denote by $E_l(z, s)$ the Eisenstein series at the lth cusp, and by h the total number of cusps of the congruence subgroup $\Gamma_o(N)$. Here $\Delta E_l(z, s) = s(1 - s)E_l(z, s)$.

By the Selberg spectral expansion (see [**V**]) we have

$$(11) \quad G(z) = \sum_{j=0}^{\infty} < G, \eta_j > \eta_j(z)$$

$$+ \sum_{l=1}^{h} \frac{1}{4\pi i} \int_{(1/2)-i\infty}^{(1/2)+i\infty} < G, E_l(\ , s) > E_l(z, s) ds,$$

where $\langle\ \rangle$ denotes the Petersson inner product. Since Δ is a self adjoint operator, we see that

$$(12) \qquad \langle G, \Delta E_l(\ , s)\rangle = \overline{s(1-s)}\langle G, E_l(\ , s)\rangle$$

$$= \langle \Delta G, E_l(\ , s)\rangle.$$

But we may write

$$\Delta = -4y^2 \frac{\partial}{\partial z} \cdot \frac{\partial}{\partial \bar{z}}$$

where

$$\frac{\partial}{\partial z} = \frac{1}{2}\left(\frac{\partial}{\partial x} + i\frac{\partial}{\partial y}\right)$$

and it follows that

$$\Delta G = -4y^2 \frac{\partial}{\partial z} \frac{\partial}{\partial \bar{z}} (g_{\pi i/\Omega_1}(z) \cdot \overline{g_{-\pi i/\Omega_1}(z)})$$

$$= -4y^2\left(\frac{\partial}{\partial z} g_{\pi i/\Omega_1}(z)\right) \cdot \overline{\left(\frac{\partial}{\partial z} g_{-\pi i/\Omega_1}(z)\right)}.$$

We compute

$$\frac{\partial}{\partial z} g_\lambda(z) = 2\pi i \lambda f(z) e^{-2\pi i \lambda \int_z^{i\infty} f(\tau)d\tau}.$$

Therefore

$$\Delta G(z) = \frac{16\pi^4}{\Omega_1^2} |f(z)|^2 G(z).$$

It then follows from (12) that

$$\langle G, E_l(\ ,s)\rangle = \frac{16\pi^4}{\Omega_1^2 s(1-s)} \langle y^2 |f|^2 G, E_l(\ ,s)\rangle \tag{13}$$

$$\ll \frac{1}{|s|^2 \Omega_1^2}$$

for $\mathrm{Re}(s) = 1/2$.

In a completely analogous fashion, we obtain the bound

$$\langle G, \eta_j\rangle \ll \frac{1}{|\lambda_j|^2 \Omega_1^2}. \tag{14}$$

In (13) and (14), the implied constants depend at most on the level N. We now see from (13) and (14) that the spectral expansion (11) converges absolutely.

Proposition 5. *Let $c_l(y, s) = \int_0^1 E_l(z, s)dx$ denote the constant term of the Eisenstein series $E_l(z, s)$. Then we have*

$$\sum_{n=0}^{\infty} b_{\pi i/\Omega_1}(n)^2 e^{-4\pi n y} = \langle G, \eta_0\rangle \eta_0$$

$$+ \sum_{l=1}^{h} \frac{1}{4\pi i} \int_{(1/2)-i\infty}^{(1/2)+i\infty} \langle G, E_l(\ ,s)\rangle c_l(y, s) ds.$$

In particular

$$\left| \sum_{n=0}^{\infty} b_{\pi i/\Omega_1}(n)^2 e^{-4\pi n y} \right| \ll 1 + \frac{\sqrt{y}}{|\Omega_1|^2}$$

where the implied constant is independent of y and depends at most on the level N.

Proof. If we apply the integral operator $\int_0^1 dx$ to both sides of (11), we obtain the first part of the proposition. This is because only η_0 and the Eisenstein series have constant terms. On the other hand, it is well known that the constant term of the Eisenstein series has the form

$$c_l(y, s) = y^s + \phi_l(s) y^{1-s}$$

where

$$|\phi_l(s)| \ll_\epsilon |s|^\epsilon$$

for any $\epsilon > 0$ on the line $\mathrm{Re}(s) = 1/2$. This and the estimate (13) gives

$$\int_{(1/2)-i\infty}^{(1/2)+i\infty} \langle G, E_l(\ , s)\rangle c_l(y, s) ds \ll \frac{\sqrt{y}}{|\Omega_1|^2}.$$

This proves the proposition.

Proposition (5) shows that their is enormous cancellation among the sum of the squares of the Fourier coefficients of certain holomorphic modular forms which transform by a quasicharacter. This cancellation is particularly striking in view of the fact that these Fourier coefficients probably do not have a polynomial growth. Because of this there is no notion of the L-function or Rankin-Selberg L-function associated to such a modular form. Nevertheless, it should be possible to give a meaning to such series by use of the theory of distributions.

DORIAN GOLDFELD
DEPARTMENT OF MATHEMATICS
COLUMBIA UNIVERSITY
NEW YORK, NY 10027

REFERENCES

[D] P. Deligne, La conjecture de Weil, *Publ. Math IHES*, **43** (1974), 273–308.

[G] D. Goldfeld, Modular elliptic curves and diophantine problems, to appear in *Proc. Canadian Number Theory Conf.*, Banff.

[S] G. Shimura, *Introduction to the Arithmetic Theory of Automorphic Functions*, Princeton University Press (1971).

[V] Venkov, A. B., Spectral Theory of Automorphic Functions, *Proc. Steklov Inst. of Math.*, (1982) issue 4. (English translation).

NEARLY ORDINARY HECKE ALGEBRAS AND GALOIS REPRESENTATIONS OF SEVERAL VARIABLES

By HARUZO HIDA*

---

0. Introduction. The purpose of this paper is to supplement our previous papers [7] and [8] on Hecke algebras over totally real fields with a result on the canonical Galois representations into GL_2 with coefficients in the total quotient rings of the Hecke algebras. The construction of such Galois representations is automatic (as already done in the case of $\mathbf{Q}$ in [6]) from the known result on Galois representations already available in [11], [12], [13] and [1] in view of the result (or rather the proof of the result) in [8] if the degree of the base totally real field is odd, and even in the remaining case (i.e. the even degree case), after learning the ingenious method of Wiles in [17] (see also Mazur [9, 1.8] and Gouvêa [4, III.5]) of glueing together infinitely many residual representations into the bigger representation, we now know how to construct them from the knowledge of the Hecke algebra studied in [8]. In this paper, adopting the method of Wiles in our nearly ordinary case, we shall construct such representations. These representations give nontrivial (series of) examples of Galois representations into $GL_2(\mathbf{Z}_p[[X_0, \ldots, X_{d+s}]])$ for the degree d of the base field ($s \geq 0$, and if the Leopoldt conjecture holds for the base field and p, then $s = 0$). Moreover, the description similar to [17, Theorem 2] of the restriction to the decomposition group at p (see Theorem I (iv) below) is expected to have an important application in the Iwasawa theory of CM-fields (as already done in works of Mazur and Wiles [10], [16] for the cyclotomic $\mathbf{Z}_p$-extension over totally real fields and of Tilouine [15] for the anti-cyclotomic $\mathbf{Z}_p$-extension over imaginary quadratic fields). The author hopes to return to this problem in a near future. As naturally presumed from what we have already mentioned, our method of constructing Galois representations heavily relies on the Wiles' method in [17], the existence of Galois representations attached to classical cusp forms proven by Wiles [17] and

---

Manuscript received June 22, 1988.

*Research partially supported by NSF grant DMS 8802001.

Taylor [14] and the description of the big nearly ordinary Hecke algebra given in [8]. Actually, we will construct the Galois representation into GL_2 over the full universal Hecke algebra (not only for the nearly ordinary part; see Theorem II below). This result may be of some interest from the point of view of the theory of universal (or the deformation of) Galois representations due to Mazur [9], and we shall formulate our theorems this perspective in mind. In fact, Gouvêa [4, III.5.6] has already constructed the representation as in Theorem II when $F = \mathbf{Q}$ using Mazur's theory of deformation (under some restriction on p).

Now let us give a precise formulation of the result: We fix throughout the paper a rational prime p and a totally real field F of finite degree. We use the notation introduced in [8] without explaining them in detail. Let $\mathfrak{m}$ be an ideal of the integer ring $\mathfrak{r}$ of F. We consider the open subgroups $U_0(\mathfrak{m})$ and $U_1(\mathfrak{m})$ of $GL_2(\hat{\mathfrak{r}})$ defined by

$$U_0(\mathfrak{m}) = \left\{ \begin{pmatrix} a & b \\ c & d \end{pmatrix} \in GL_2(\hat{\mathfrak{r}}) | c \in \mathfrak{m}\hat{\mathfrak{r}} \right\},$$

$$U_1(\mathfrak{m}) = \left\{ \begin{pmatrix} a & b \\ c & d \end{pmatrix} \in GL_2(\hat{\mathfrak{r}}) | c \in \mathfrak{m}\hat{\mathfrak{r}} \text{ and } a \equiv 1 \bmod \mathfrak{m}\hat{\mathfrak{r}} \right\},$$

$$U(\mathfrak{m}) = \left\{ \begin{pmatrix} a & b \\ c & d \end{pmatrix} \in GL_2(\hat{\mathfrak{r}}) | c \in \mathfrak{m}\hat{\mathfrak{r}} \text{ and } a \equiv d \equiv 1 \bmod \mathfrak{m}\hat{\mathfrak{r}} \right\},$$

where $\hat{\mathfrak{r}} = \Pi_{\mathfrak{p}} \mathfrak{r}_{\mathfrak{p}}$ is the product of the completion $\mathfrak{r}_{\mathfrak{p}}$ at $\mathfrak{p}$ over all prime ideals $\mathfrak{p}$ of $\mathfrak{r}$. We denote by $\bar{\mathbf{Q}}$ the field of all numbers algebraic over $\mathbf{Q}$ inside $\mathbf{C}$ and fix an algebraic closure $\bar{\mathbf{Q}}_p$ of the p-adic field $\mathbf{Q}_p$ and an embedding of $\bar{\mathbf{Q}}$ into $\bar{\mathbf{Q}}_p$ once and for all. We suppose throughout the paper that F is contained in $\bar{\mathbf{Q}}$. Let Φ be the subfield of $\bar{\mathbf{Q}}$ generated by all the conjugates of F. Let K be a finite extension of the closure $\hat{\Phi}$ inside $\bar{\mathbf{Q}}_p$ of Φ and $\mathcal{O}$ denote the p-adic integer ring of K. For each open compact subgroup S of $GL_2(\hat{\mathfrak{r}})$ containing $U_1(N)$ for an integral ideal N prime to p, we consider the full Hecke algebra $\mathbf{h}(S; \mathcal{O})$ of infinite p-power level and its nearly ordinary part $\mathbf{h}^{\mathrm{n.ord}}(S; \mathcal{O})$ defined in [8, Section 2]: Let us recall their definition. We denote by I the set of all the embeddings of F into $\bar{\mathbf{Q}}$ and let $\mathbf{Z}[I]$ be the free module generated by the elements of I. We take an element $k \in \mathbf{Z}[I]$ with $k \geq 2t$ for $t = \Sigma_\sigma \sigma \in \mathbf{Z}[I]$. Put $n = k - 2t$ and suppose that there exists

$0 \leq v \in \mathbf{Z}[I]$ such that $n + 2v = \mu t$ with $0 \leq \mu \in \mathbf{Z}$. Then we consider the space $\mathbf{S}_{k,w}(S(p^\alpha); \mathbf{C})$ of holomorphic cusp forms on $GL_2(F_\mathbf{A})$ in the sense of [7, Section 2] with respect to $S(p^\alpha) = S \cap U(p^\alpha)$ and with the automorphic factor at the infinity given by

$$\det\left(\begin{pmatrix} a & b \\ c & d \end{pmatrix}\right)^{-w} (cz + d)^k,$$

where $w = v + k - t$ and $\left(\begin{smallmatrix} a & b \\ c & d \end{smallmatrix}\right) \in GL_2(F_\infty)$ and z is a variable on the product $\mathcal{H}^I$ of copies of the upper half complex planes indexed by I. On $\mathbf{S}_{k,w}(S(p^\alpha);$ $\mathbf{C})$, we have the following three type of operators: For each ideal $\mathfrak{n}$, the Hecke operator $T(\mathfrak{n})$, the action of the group $\mathbf{G}^\alpha = S_0(p^\alpha)\hat{\mathfrak{r}}^\times / S(p^\alpha)\mathfrak{r}^\times$ for $S_0(p^\alpha) = S \cap U_0(p^\alpha)$ and the action of the center $F_\mathbf{A}^\times$. For each ideal $\mathfrak{q}$ prime to Np, we choose $q \in F_\mathbf{A}^\times$ such that $q\mathfrak{r} = \mathfrak{q}$ and $q_{Np} = q_\infty = 1$ and define an operator $\langle \mathfrak{q} \rangle$ on $\mathbf{S}_{k,w}(S(p^\alpha); \mathbf{C})$ by the action of the double coset $S(p^\alpha)qS(p^\alpha)$ as in [7, Section 2]. Then the Hecke algebra $\mathfrak{h}_{k,w}(S(p^\alpha); \mathfrak{r}_\Phi)$ over the integer ring $\mathfrak{r}_\Phi$ of Φ is the subalgebra of $\mathrm{End}_C(\mathbf{S}_{k,w}(S(p^\alpha); \mathbf{C}))$ generated over $\mathfrak{r}_\Phi$ by the action of the group $\mathbf{G}^\alpha$ the Hecke operators $T(\mathfrak{n})$ and $\langle \mathfrak{n} \rangle$ for $\mathfrak{n}$ and $p^{-v}T(p)$. The restriction of the operator in $\mathfrak{h}_{k,w}(S(p^\beta); \mathfrak{r}_\Phi)$ to the subspace $\mathbf{S}_{k,w}(S(p^\alpha); \mathbf{C})$ for $\beta \geq \alpha > 0$ yields a surjective algebra homomorphism over the group algebra $\mathfrak{r}_\Phi[\mathbf{G}^\alpha]$ of $\mathfrak{h}_{k,w}(S(p^\beta); \mathfrak{r}_\Phi)$ onto $\mathfrak{h}_{k,w}(S(p^\alpha); \mathfrak{r}_\Phi)$. After putting $\mathfrak{h}_{k,w}(S(p^\alpha); \mathcal{O}) = \mathfrak{h}_{k,w}(S(p^\alpha); \mathfrak{r}_\Phi) \otimes_{\mathfrak{r}_\Phi} \mathcal{O}$, we define

$$\mathbf{h}(S; \mathcal{O}) = \varprojlim_{\alpha} \mathfrak{h}_{k,w}(S(p^\alpha); \mathcal{O}).$$

We equip on $\mathbf{h}(S; \mathcal{O})$ the topology of the projective limit of the p-adic topology of $\mathfrak{h}_{k,w}(S(p^\alpha); \mathcal{O})$ which makes it into a compact ring. The nearly ordinary part $\mathbf{h}^{\mathrm{n.ord}}(S; \mathcal{O})$ is the maximal algebra direct summand of $\mathbf{h}(S; \mathcal{O})$ on which the image of $p^{-v}T(p)$ is a unit. Then it is known (cf. [8, Theorem 2.3]) that these algebras $\mathbf{h}(S; \mathcal{O})$ and $\mathbf{h}^{\mathrm{n.ord}}(S; \mathcal{O})$ are determined independently of the choice of the weight (k, w). Strictly speaking, the pair of $\mathbf{h}(S; \mathcal{O})$ and specified elements $T(\mathfrak{n})$ is determined independently of weight (k, w) because the isomorphism between the Hecke algebras of two different weights takes $T(\mathfrak{n})$ to $T(\mathfrak{n})$. Naturally $\mathbf{h}(S; \mathcal{O})$ becomes an algebra over the continuous group algebra $\mathcal{A} = \mathcal{O}[[\mathbf{G}]]$ for the profinite group $\mathbf{G} = \varprojlim_{\alpha} \mathbf{G}^\alpha$. Then $\mathbf{G}$ is isomorphic to the product of the following two groups $\mathfrak{r}_p^\times$ and $\bar{Z}_0 = \bar{Z}_0(S)$, which is the image of the center $\hat{\mathfrak{r}}^\times$ of $GL_2(\hat{\mathfrak{r}})$ in $\mathbf{G}$

([8, Lemma 2.1]). Let $\mathbf{W}$ be the torsion free part of $\mathbf{G}$ and write $\mathbf{A}$ for the continuous group algebra $\mathcal{O}[[\mathbf{W}]]$ of $\mathbf{W}$. Then $\mathbf{W}$ is free of finite rank $\geq d+1$ as $\mathbf{Z}_p$-module for $d = [F:\mathbf{Q}]$, and by [8, Theorem 2.4], $\mathbf{h}^{\text{n.ord}}(S;\mathcal{O})$ is of finite type and torsion free as $\mathbf{A}$-module. Let $\mathbf{L}$ be the quotient field of $\mathbf{A}$, and fix an algebraic closure of $\bar{\mathbf{L}}$ of $\mathbf{L}$. We denote by $\mathcal{C}$ the category of complete noetherian local $\mathcal{O}$-algebras with finite residue field. Any object A in $\mathcal{C}$ with maximal ideal $\mathfrak{m}$ is assumed to be complete under the $\mathfrak{m}$-adic topology. Let $A \in Ob(\mathcal{C})$ be an integral domain and Q be the quotient field of A. We say that a representation $\pi: \mathrm{Gal}(\bar{\mathbf{Q}}/F) \to GL_2(Q)$ is *continuous* if there exists a finitely generated A-submodule L of Q^2 stable under π and $L \otimes_A Q = Q^2$. Then for the maximal ideal $\mathfrak{m}$ of A, $L/\mathfrak{m}^r L$ has only finitely many elements, and $\pi: \mathrm{Gal}(\bar{\mathbf{Q}}/F) \to \mathrm{Aut}_A(L) = \varprojlim_{\alpha} \mathrm{Aut}_A(L/\mathfrak{m}^\alpha L)$ is continuous with respect to the $\mathfrak{m}$-adic topology of $\mathrm{Aut}_A(L)$; that is, the topology of projective limit of the finite groups $\mathrm{Aut}_A(L/\mathfrak{m}^\alpha L)$.

THEOREM I. *Let $A \in Ob(\mathcal{C})$ be an integral domain of characteristic different from 2 and $\lambda: \mathbf{h}^{\text{n.ord}}(S;\mathcal{O}) \to A$ be a continuous $\mathcal{O}$-algebra homomorphism. Let Q be the quotient field of A. Then there exists a unique semisimple Galois representation π:* $\mathrm{Gal}(\bar{\mathbf{Q}}/F) \to GL_2(Q)$ *such that:*

(i) *π is continuous;*
(ii) *π is unramified outside Np, where N is the level of S, i.e., the largest ideal prime to p with $S \supset U_1(N)$;*
(iii) *For the Frobenius element $\phi_{\mathfrak{q}}$ for each prime $\mathfrak{q}$ outside Np, we have*

$$\det(1-\pi(\phi_{\mathfrak{q}})X) = 1 - \lambda(T(\mathfrak{q}))X + \lambda(\langle\mathfrak{q}\rangle)\mathfrak{N}_{F/\mathbf{Q}}(\mathfrak{q})X^2;$$

(iv) *Let $\mathfrak{p}$ be a prime factor of p and fix a decomposition group $D_{\mathfrak{p}}$ of $\mathfrak{p}$ in* $\mathrm{Gal}(\bar{\mathbf{Q}}/F)$. *Then there exist two characters ϵ, δ of $D_{\mathfrak{p}}$ with values in A such that the restriction of π to $D_{\mathfrak{p}}$ is, up to equivalence, of the following form:*

$$\pi(\sigma) = \begin{pmatrix} \epsilon(\sigma) & * \\ 0 & \delta(\sigma) \end{pmatrix} \quad \text{for} \quad \sigma \in D_{\mathfrak{p}}.$$

Moreover if A is finite and torsion-free over $\mathbf{A}$ and λ is an $\mathbf{A}$-algebra homomorphism, then π is absolutely irreducible.

This theorem will be proven in Section 3. Here are some remarks about the theorem:

(i) Since $\mathbf{h}^{\text{n.ord}}(S; \mathcal{O})$ is of finite type and torsion-free as $\mathbf{A}$-module, any irreducible component of $\mathrm{Spec}(\mathbf{h}^{\text{n.ord}}(S; \mathcal{O}))$ is isomorphic to $\mathrm{Spec}(\mathbf{I})$ for an integral domain $\mathbf{I}$ finite and torsion-free over $\mathbf{A}$. Writing $\lambda\colon \mathbf{h}^{\text{n.ord}}(S; \mathcal{O}) \to \mathbf{I}$ for the projection morphism, we have the canonical Galois representation $\pi\colon \mathrm{Gal}(\bar{\mathbf{Q}}/F) \to GL_2(\mathbf{K})$ for the quotient field $\mathbf{K}$ of $\mathbf{I}$. Especially, if we fix an isomorphism: $\mathbf{W} \cong \mathbf{Z}_p^r$ with $r \geq d + 1$, we can identify $\mathbf{A}$ with $\mathcal{O}[[X_1, \ldots, X_r]]$. (If the Leopoldt conjecture is true for F and p, $r = d + 1$). Thus if $\mathbf{I}$ as above coincides with $\mathbf{A}$ and $p \neq 2$, we have a Galois representation of several variables:

$$\pi\colon \mathrm{Gal}(\bar{\mathbf{Q}}/F) \to GL_2(\mathcal{O}[[X_1, \ldots, X_r]])$$

(see [17, Remark in Section 2.2]).

(ii) One can even determine exactly the characters ϵ and δ by the values $\lambda(T(\mathfrak{p}))$ and $\lambda(g)$ for $g \in \mathbf{G}$ (see Proposition 2.3 in the text for details).

(iii) Let $\rho\colon \mathbf{G} \to \bar{Z}_0$ be the projection map and consider the induced algebra homomorphism $\rho\colon \mathfrak{A} = \mathcal{O}[[\mathbf{G}]] \to \mathcal{O}[[\bar{Z}_0]]$. We write $\mathfrak{p}$ for the kernel of ρ. As one can easily see, the ordinary Hecke algebra $\mathbf{h}^{\text{ord}}(S; \mathcal{O})$ (which is written for $S = U_1(N)$ as $\mathbf{h}_0^{\text{ord}}(N; \mathcal{O})$ in [7]) is isomorphic to the torsion free part of $\mathbf{h}^{\text{n.ord}}(S; \mathcal{O}) \otimes_{\mathfrak{A}} \mathfrak{A}/\mathfrak{p}$. Thus $\mathbf{h}^{\text{ord}}(S; \mathcal{O})$ is a residue algebra of $\mathbf{h}^{\text{n.ord}}(S; \mathcal{O})$ by an ideal $\mathcal{P}$. If $\lambda\colon \mathbf{h}^{\text{n.ord}}(S; \mathcal{O}) \to A$ factors through $\mathbf{h}^{\text{ord}}(S; \mathcal{O})$, then for the associated Galois representation π, it has been shown by Wiles [17] the restriction of π to $D_\mathfrak{p}$ is reducible and the character δ as in the theorem is unramified.

(iv) Since $\mathbf{G} = \mathfrak{r}_p^\times \times \bar{Z}_0$, we may consider $\mathcal{O}[[\mathbf{G}]] = \mathcal{O}[[\mathfrak{r}_p^\times]] \hat{\otimes}_{\mathcal{O}} \mathcal{O}[[\bar{Z}_0]]$. When $F = \mathbf{Q}$, $\mathbf{h}^{\text{n.ord}}(S : \mathcal{O})$ can be naturally considered as a subalgebra of $\mathbf{h}^{\text{ord}}(S; \mathcal{O}) \hat{\otimes}_{\mathcal{O}} \mathcal{O}[[\mathbf{Z}_p^\times]]$ with the same total quotient ring, because of the nonexistence of the modular forms of multiple weight. When $F = \mathbf{Q}$, we can also identify $\mathbf{Z}_p^\times$ with the Galois group of the cyclotomic extension of all p-power roots of unity by means of the cyclotomic character. Let ι denote the Galois character with values in $\mathcal{O}[[\mathbf{Z}_p^\times]]$ which is the composite of the cyclotomic character with the natural tautological inclusion of $\mathbf{Z}_{p^\times}$ into $\mathcal{O}[[\mathbf{Z}_p]]$. If $\lambda\colon \mathbf{h}^{\text{ord}}(S; \mathcal{O}) \to A$ and $\mu\colon \mathcal{O}[[\mathbf{Z}_p^\times]] \to B$ are continuous $\mathcal{O}$-algebra homomorphisms and $A \hat{\otimes}_{\mathcal{O}} B$ is an integral domain in $\mathcal{C}$, then for the ordinary Galois representation π attached to λ, the Galois representation attached to $\lambda \otimes \mu$ is given by $\pi \otimes (\mu \circ \iota)$ (see also [9, Proposition 15]). Thus, when $F = \mathbf{Q}$, there is nothing essentially new here, and Theorem I follows almost directly from [6, Theorem 2.1].

Now let us present a result for the full Hecke algebra $\mathbf{h}(S; \mathcal{O})$. Let $\mathbf{h}$ be the smallest closed subalgebra of $\mathbf{h}(S; \mathcal{O})$ containing $\mathbf{A}$ and $T(\mathfrak{n})$ for all $\mathfrak{n}$ prime to Np. Then it is known that $\mathbf{h}$ is a reduced compact algebra.

THEOREM II. *Let $A \in Ob(\mathcal{C})$ be an integral domain of characteristic different from 2 and Q be the quotient field of A. For any continuous $\mathcal{O}$-algebra homomorphism $\lambda: \mathbf{h} \to A$, there exists a unique Galois representation $\pi(\lambda)$: $\mathrm{Gal}(\bar{\mathbf{Q}}/F) \to GL_2(Q)$ such that:*

(i) *$\pi(\lambda)$ is continuous and semi-simple;*
(ii) *$\pi(\lambda)$ is unramified outside Np;*
(iii) *For the Frobenius element $\phi_\mathfrak{q}$ for each prime $\mathfrak{q}$ outside Np, we have*

$$\det(1 - \pi(\lambda)(\phi_\mathfrak{q})X) = 1 - \lambda(T(\mathfrak{q}))X + \lambda(\langle \mathfrak{q} \rangle)\mathfrak{N}_{F/\mathbf{Q}}(\mathfrak{q})X^2.$$

This theorem will be proven in Section 3. In view of the duality between Hecke algebra $\mathbf{h}(S; \mathcal{O})$ and the space of p-adic modular forms (cf. [7, Section 5]), this theorem implies that one can associate a canonical Galois representations to any p-adic common eigenform of all Hecke operators including classical cusp forms of weight less than $2t$ (cf. [13] and [17]).

1. Pseudo-representations. Before proving Theorem I in Section 3, we shall explain Wiles' method (in [17] and [14, Section 2]) of patching together the residual representations modulo prime ideals for an integral domain R of characteristic 0 into a representation into R. Let R be a topological commutative ring in which 2 is not a zero divisor. Let G be a compact group, $c \in G$ be a specified element of G of order 2 and $\pi: G \to GL_2(R)$ be a continuous representation with $\det(\pi(c)) = -1$ (such a representation will be called an odd representation with respect to c). Since π is odd, replacing π by a suitable isomorphic representation over the localization $R[2^{-1}]$ of R, we may assume that $\pi: G \to GL_2(R[2^{-1}])$ satisfies the following conditions:

(1.1a) $\mathrm{Tr}(\pi(\sigma))$ and $\det(\pi(\sigma))$ belong to R for all $\sigma \in G$;

(1.1b) $$\pi(c) = \begin{pmatrix} -1 & 0 \\ 0 & 1 \end{pmatrix}.$$

For each $\sigma \in G$, we write $\pi(\sigma) = \left(\begin{smallmatrix} a(\sigma) & b(\sigma) \\ c(\sigma) & d(\sigma) \end{smallmatrix}\right)$ and define continuous functions

$X : G \times G \to R$, $A : G \to R$ and $D : G \to R$ by $X(\sigma, \tau) = 4b(\sigma)c(\tau)$, $A(\sigma) = 2a(\sigma)$ and $D(\sigma) = 2d(\sigma)$. Note that by (1.1b), we see easily that

$$A(\sigma) = (\mathrm{Tr}(\pi(\sigma)) - \mathrm{Tr}(\pi(\sigma c))) \quad \text{and} \quad D(\sigma) = (\mathrm{Tr}(\pi(\sigma)) + \mathrm{Tr}(\pi(\sigma c)))$$

$$\text{and} \quad X(\sigma, \tau) = 2A(\sigma\tau) - A(\sigma)A(\tau).$$

This shows that these functions in fact have values in R. Then this triple $\pi' = (A, D, X)$ of functions satisfies the following properties:

(1.2a) As functions on G or G^2, A, D and X are continuous,

$$\text{(1.2b)} \quad 2A(\sigma\tau) = A(\sigma)A(\tau) + X(\sigma, \tau), \quad 2D(\sigma\tau) = D(\sigma)D(\tau) + X(\tau, \sigma)$$

$$\text{and} \quad 4X(\sigma\tau, \rho\gamma) = A(\sigma)A(\gamma)X(\tau, \rho) + A(\gamma)D(\tau)X(\sigma, \rho)$$
$$+ A(\sigma)D(\rho)X(\tau, \gamma) + D(\tau)D(\rho)X(\sigma, \gamma),$$

$$\text{(1.2c)} \qquad A(1) = D(1) = D(c) = 2, \qquad A(c) = -2,$$

$$\text{and} \quad X(\sigma, \rho) = X(\rho, \tau) = 0 \quad \text{if} \quad \rho = 1 \quad \text{and} \quad c,$$

$$\text{(1.2d)} \qquad X(\sigma, \tau)X(\rho, \eta) = X(\sigma, \eta)X(\rho, \tau),$$

$$\text{(1.2e)} \qquad A(\sigma) + D(\sigma) \in 2R \quad \text{for all} \quad \sigma \in G.$$

The properties (1.2c–d) follow directly from the definition and the first half of (1.2b) can be proven by computing directly the multiplicative formula:

$$\begin{pmatrix} a(\sigma) & b(\sigma) \\ c(\sigma) & d(\sigma) \end{pmatrix} \begin{pmatrix} a(\tau) & b(\tau) \\ c(\tau) & d(\tau) \end{pmatrix} = \begin{pmatrix} a(\sigma\tau) & b(\sigma\tau) \\ c(\sigma\tau) & d(\sigma\tau) \end{pmatrix}.$$

In addition to the two first formulas of (1.2b), we also have

$$b(\sigma\tau) = a(\sigma)b(\tau) + b(\sigma)d(\tau) \quad \text{and} \quad c(\sigma\tau) = c(\sigma)a(\tau) + d(\sigma)c(\tau).$$

Thus we know that

$$4X(\sigma\tau, \rho\gamma) = 16b(\sigma\tau)c(\rho\gamma) = 16(a(\sigma)b(\tau) + b(\sigma)d(\tau))(c(\rho)a(\gamma)$$

$$+ d(\rho)c(\gamma))$$

$$= A(\sigma)A(\gamma)X(\tau, \rho) + A(\gamma)D(\tau)X(\sigma, \rho)$$

$$+ A(\sigma)D(\rho)X(\tau, \gamma) + D(\tau)D(\rho)X(\sigma, \gamma).$$

For any topological algebra R in which 2 is not a zero divisor, we now define, according to [17, Section 2.2], a *pseudo-representation* of G into R to be a triple $\pi' = (A, D, X)$ consisting of continuous functions on G or G^2 satisfying the conditions (1.2a–e) (when 2 is invertible in R, the condition (1.2e) is superfluous). We define the trace $\mathrm{Tr}(\pi')$ (resp. the determinant $\det(\pi')$) of the pseudo representation π' to be a function on G given by

$$\mathrm{Tr}(\pi')(\sigma) = 2^{-1}(A(\sigma) + D(\sigma))$$

$$(\text{resp. } \det(\pi')(\sigma) = 4^{-1}(A(\sigma)D(\sigma) - X(\sigma, \sigma)).$$

They are all continuous functions with values in R. The following propositions due to Wiles describes how to recover the representation into R with the same trace and determinant out of a pseudo-representation and how pseudo-representations behave in much more coherent way than the representations under the specialization process:

PROPOSITION 1.1. *Let $\pi' = (A, D, X)$ be a pseudo-representation of G into an integral domain R with quotient field Q. Suppose that Q is not of characteristic 2. Then there exists a representation $\pi: G \to GL_2(Q)$ with the same trace and determinant as π'. Moreover if R is an object of $\mathcal{C}$, then π is continuous.*

Proof. We repeat here the proof in [17, Section 2.2]. We divide our argument into two cases:

Case 1. The case when there exists $\rho, \gamma \in G$ such that $X(\rho, \gamma) \neq 0$. Then we define $\pi(\sigma) = \begin{pmatrix} a(\sigma) & b(\sigma) \\ c(\sigma) & d(\sigma) \end{pmatrix}$ by

$$a(\sigma) = 2^{-1}A(\sigma), \qquad d(\sigma) = 2^{-1}D(\sigma), \qquad c(\sigma) = 4^{-1}X(\rho, \sigma)$$

$$\text{and} \quad b(\sigma) = X(\sigma, \gamma)/X(\rho, \gamma).$$

Write $x(\sigma, \tau)$ for $4^{-1}X(\sigma, \tau)$. Then

$$b(\sigma)c(\tau) = x(\rho, \sigma)x(\tau, \gamma)/x(\rho, \gamma) = x(\sigma, \tau)$$

by the property (1.2d) of pseudo representation. Thus we know that the entry of $\pi(\sigma)\pi(\tau)$ at the upper left corner is equal to, by (1.2b),

$$a(\sigma)a(\tau) + b(\sigma)c(\tau) = a(\sigma)a(\tau) + x(\sigma, \tau) = a(\sigma\tau).$$

Similarly the lower right corner of $\pi(\sigma)\pi(\tau)$ is equal to

$$d(\sigma)d(\tau) + b(\tau)c(\sigma) = d(\sigma)d(\tau) + x(\tau, \sigma) = d(\sigma\tau).$$

We now compute lower left corner of $\pi(\sigma)\pi(\tau)$, which is given by

$$c(\sigma)a(\tau) + d(\sigma)c(\tau) = x(\rho, \sigma)a(\tau) + d(\sigma)x(\rho, \tau).$$

By applying the last formula in (1.2b) to $(1, \rho, \sigma, \tau)$, we have

$$c(\sigma\tau) = x(\rho, \sigma\tau) = a(\tau)x(\rho, \sigma) + d(\sigma)x(\rho, \tau),$$

since $x(1, \sigma) = x(1, \tau) = 0$ by (1.2c). This shows that

$$c(\sigma\tau) = c(\sigma)a(\tau) + d(\sigma)c(\tau).$$

Similarly by applying the same formula in (1.2b) to $(\sigma, \tau, 1, \gamma)$, we have

$$b(\sigma\tau)x(\rho, \gamma) = x(\sigma\tau, \gamma) = a(\sigma)x(\tau, \gamma) + d(\tau)x(\sigma, \gamma)$$

$$= (a(\sigma)b(\tau) + d(\tau)b(\sigma))x(\rho, \gamma),$$

which finishes the proof of the formula $\pi(\sigma)\pi(\tau) = \pi(\sigma\tau)$. Obviously, by definition, $\pi(1) = \begin{pmatrix} 1 & 0 \\ 0 & 1 \end{pmatrix}$ and hence π is the desired representation.

Case 2. The case when $X(\sigma, \tau) = x(\sigma, \tau) = 0$ for all σ, τ in G. In this case, by (1.2b), we have $a(\sigma)a(\tau) = a(\sigma\tau)$ and $d(\sigma)d(\tau) = d(\sigma\tau)$ for all σ, $\tau \in G$. Then we simply put

$$\pi(\sigma) = \begin{pmatrix} a(\sigma) & 0 \\ 0 & d(\sigma) \end{pmatrix}$$

which does the job. Now assuming that $R \in Ob(\mathcal{C})$, we shall prove the continuity of π in Case 1 because in Case 2, the continuity is obvious. Let $\mathfrak{a}$ be the ideal of R generated by $c(\tau)$ for all $\tau \in G$. Since $b(\sigma)c(\tau) = x(\sigma, \tau) \in 4^{-1}R$ for all τ by our construction, $b(\sigma) \in 4^{-1}\mathfrak{a}^{-1}$ for all $\sigma \in G$. Thus $\mathrm{Im}(\pi)$ is contained in

$$\begin{pmatrix} 2^{-1}R & 4^{-1}\mathfrak{a}^{-1} \\ \mathfrak{a} & 2^{-1}R \end{pmatrix}.$$

Thus we put $L_0 = \left\{\binom{a}{b} \middle| a \in R \text{ and } b \in \mathfrak{a}\right\}$, then $L = \Sigma_{\sigma \in G}\, \pi(\sigma)L_0$ is contained in $4^{-1}L_0$. Since A is noetherian, L is finitely generated over A, and since L contains L_0, $L \otimes_A Q = Q^2$. By definition, L is stable under π. Thus, according to the definition of continuity given above Theorem I, π is continuous.

Remark 1.2. Let the notation be as in Proposition 1.1 and its proof. Assume that R is an object in $\mathcal{C}$ and the residual characteristic of R is different from 2. By the above proof, it is clear that if a pseudo-representation $\rho = (a, d, x)\colon G \to R$ has $\sigma, \tau \in G$ such that $x(\sigma, \tau)$ is a unit of R, then we can construct a representation $\pi\colon G \to GL_2(R)$ out of ρ with the same trace and determinant.

PROPOSITION 1.3. *Suppose that R be a product of finitely many objects in $\mathcal{C}$ (thus, R may not be local). Let $\mathfrak{a}$ and $\mathfrak{b}$ be two ideals of R. Let $\pi(\mathfrak{a})$ and $\pi(\mathfrak{b})$ be pseudo representations of G into $R/\mathfrak{a}$ and $R/\mathfrak{b}$, respectively. Suppose that $\pi(\mathfrak{a})$ and $\pi(\mathfrak{b})$ are compatible; namely, there exist functions* tr *and* det *on a dense subset Σ of G with values in $R/\mathfrak{a} \cap \mathfrak{b}$ such that for all $\sigma \in \Sigma$,*

$$\mathrm{Tr}(\pi(\mathfrak{a})(\sigma)) \equiv \mathrm{tr}(\sigma) \bmod \mathfrak{a} \quad \textit{and} \quad \mathrm{Tr}(\pi(\mathfrak{b})(\sigma)) \equiv \mathrm{tr}(\sigma) \bmod \mathfrak{b}$$

$$\det(\pi(\mathfrak{a})(\sigma)) \equiv \det(\sigma) \bmod \mathfrak{a} \quad \textit{and} \quad \det(\pi(\mathfrak{b})(\sigma)) \equiv \det(\sigma) \bmod \mathfrak{b}.$$

Then there exists a pseudo representation $\pi(\mathfrak{a} \cap \mathfrak{b})$ of G into $R/\mathfrak{a} \cap \mathfrak{b}$ such that $\mathrm{Tr}(\pi(\mathfrak{a} \cap \mathfrak{b})(\sigma)) = \mathrm{tr}(\sigma)$ *and* $\det(\pi(\mathfrak{a} \cap \mathfrak{b})(\sigma)) = \det(\sigma)$ *on* Σ.

Proof. We again repeat the proof given in [17]. We consider the exact sequence:

$$0 \to R/\mathfrak{a} \cap \mathfrak{b} \to R/\mathfrak{a} \oplus R/\mathfrak{b} \xrightarrow{\alpha} R/\mathfrak{a} + \mathfrak{b} \to 0$$

$$a \mapsto a \bmod \mathfrak{a} \oplus a \bmod \mathfrak{b}$$

$$a \oplus b \mapsto a - b \bmod \mathfrak{a} + \mathfrak{b}.$$

We consider the pseudo representation $\pi = \pi(\mathfrak{a}) \oplus \pi(\mathfrak{b})$ with values in $R/\mathfrak{a} \oplus R/\mathfrak{b}$. The function $\alpha \circ \mathrm{Tr}(\pi)$ vanishes constantly on Σ. Since this function is continuous on G and Σ is dense in G, $\alpha \circ \mathrm{Tr}(\pi)$ vanishes on G. Thus $\mathrm{Tr}(\pi)$ has values in $R/\mathfrak{a} \cap \mathfrak{b}$. If we write $\pi = (A, D, X)$, then $A(\sigma) = (\mathrm{Tr}(\pi(\sigma)) - \mathrm{Tr}(\pi(\sigma c)))$ and $D(\sigma) = (\mathrm{Tr}(\pi(\sigma)) + \mathrm{Tr}(\pi(\sigma c)))$ and $X(\sigma, \tau) = 2A(\sigma\tau) - A(\sigma)A(\tau)$. Thus π itself has values in $R/\mathfrak{a} \cap \mathfrak{b}$ and gives the desired pseudo representation.

THEOREM 1.4 (Wiles). *Let R be a topological $\mathcal{O}$-algebra and $\{\mathfrak{p}_i\}_{i=1}^{\infty}$ a countable set of ideals of R such that $R/\mathfrak{p}_i$ belongs to $\mathcal{C}$ for all i. Suppose that* (i) $R = \varprojlim_{\alpha} R/(\bigcap_{i=1}^{\alpha} \mathfrak{p}_i)$ *as a topological algebra and* (ii) *there exist a dense subset Σ of G, functions* $\mathrm{tr}: \Sigma \to R$ *and* $\det: \Sigma \to R$ *and a pseudo representation $\pi_i = (A_i, D_i, X_i): G \to R/\mathfrak{p}_i$ such that*

$$\mathrm{Tr}(\pi_i(\sigma)) = \mathrm{tr}(\sigma) \bmod \mathfrak{p}_i \quad \textit{and} \quad \det(\pi_i(\sigma)) = \det(\sigma) \bmod \mathfrak{p}_i$$

for all $\sigma \in \Sigma$. Then there exists a unique pseudo-representation $\pi = (A, D, X): G \to R$ such that $\pi(\sigma) \bmod \mathfrak{p}_i = \pi_i(\sigma)$ for all i on Σ. Moreover, if $\lambda: R \to A$ is a continuous algebra homomorphism into an integral domain A in $\mathcal{C}$ of characteristic different from 2, then there exists a semi-simple continuous Galois representation $\Pi: G \to GL_2(Q)$ for the quotient field Q of A such that $\det(1 - \Pi(\sigma)X) = 1 - \lambda(\mathrm{Tr}(\pi(\sigma))X + \lambda(\det(\pi(\sigma))X^2$ *for all* $\sigma \in \Sigma$.

Proof. Write $\mathcal{P}_\alpha$ for $\bigcap_{i=1}^{\alpha} \mathfrak{p}_i$. Then by our assumption, we know that $R = \varprojlim_{\alpha} R/\mathcal{P}_\alpha$ and $R/\mathcal{P}_\alpha$ satisfies the assumption of Proposition 1.2. By Proposition 1.3, we can lift inductively the pseudo representations π_i for $i = 1, \ldots, \alpha$ to a pseudo-representation $\pi_\alpha = (A_\alpha, D_\alpha, X_\alpha): G \to R/\mathcal{P}_\alpha$ such that $\mathrm{Tr}(\pi_\alpha(\sigma)) = \mathrm{tr}(\sigma) \bmod \mathcal{P}_\alpha$ and $\det(\pi_\alpha(\sigma)) = \det(\sigma) \bmod \mathcal{P}_\alpha$. Since we have

$$A_\alpha(\sigma) = \mathrm{Tr}(\pi_\alpha(\sigma)) - \mathrm{Tr}(\pi_\alpha(\sigma c)), \qquad D_\alpha(\sigma) = \mathrm{Tr}(\pi_\alpha(\sigma)) + \mathrm{Tr}(\pi_\alpha(\sigma c))$$

and

$$X_\alpha(\sigma, \tau) = 2A_\alpha(\sigma\tau) - A_\alpha(\sigma)A_\alpha(\tau) \quad \text{on} \quad \Sigma,$$

we see that

$$\pi_\alpha(\sigma) \equiv \pi_\beta(\sigma) \bmod \mathcal{P}_\alpha \quad \text{on} \quad \Sigma \quad \text{if} \quad \alpha < \beta.$$

Since Σ is dense, the above identity holds on the whole G, and thus, defining $\pi(\sigma) = \lim_{\overleftarrow{\alpha}} \pi_\alpha(\sigma)$ for each $\sigma \in G$, we obtain the desired pseudo representation π. The last assertion follows from Proposition 1.1 applied to the pseudo representation $\lambda \circ \pi : G \to A$.

2. Representation theoretic preliminaries. In this section, we gather several representation theoretic results necessary to the proof of Theorems I and II.

Let $\mathfrak{p}$ be a prime ideal of $\mathfrak{r}$ and consider an admissible irreducible representation ρ of $GL_2(F_\mathfrak{p})$ on a infinite dimensional $\mathbf{C}$-vector space $\mathbf{U}$. We first quote some well known result on local representations:

PROPOSITION 2.1 (Casselman). *Suppose that ρ is irreducible. Let $c(\rho)$ be the largest ideal of $\mathfrak{r}_\mathfrak{p}$ such that the subspace V of all vectors fixed by*

$$U_0(c(\rho))_\mathfrak{p} = \left\{ \begin{pmatrix} a & b \\ c & d \end{pmatrix} \in GL_2(\mathfrak{r}_\mathfrak{p}) | c \in c(\rho) \right\}$$

is nontrivial. Then this subspace V has dimension 1 and on it the Hecke operator $T(\mathfrak{p})$ acts as follows:

(i) *If ρ is absolutely cuspidal, then $c(\rho)$ is divisible by $\mathfrak{p}^2$ and $T(\mathfrak{p}) = 0$ on V;*

(ii) *Suppose that ρ is a principal series representation $\pi(\xi, \eta)$ with characters ξ and η of $F_\mathfrak{p}^\times$. Then $c(\rho)$ is the product of the conductor of ξ and η. If both the characters ξ and η are unramified, then $T(\mathfrak{p})$ acts on V by the multiplication of $\xi(\bar{\omega}) + \eta(\bar{\omega})$ for a prime element $\bar{\omega}$ of $\mathfrak{r}_\mathfrak{p}$. If ξ is unramified and η is ramified, then $T(\mathfrak{p})$ acts on V by the multiplication of $\xi(\bar{\omega})$ for a prime element $\bar{\omega}$ of $\mathfrak{r}_\mathfrak{p}$. If both the characters are ramified, then $T(\mathfrak{p})$ annihilates V;*

(iii) *Suppose that ρ is a special representation $\sigma(\xi, \xi\alpha^{-1})$ for characters ξ and α of $F_\mathfrak{p}^\times$ with $\alpha(x) = |x|_\mathfrak{p}$. Then, if ξ is unramified $c(\rho) = \mathfrak{p}$ and $T(\mathfrak{p})$ acts on V by the multiplication of $\xi(\bar{\omega})$, and if ξ is ramified, then $c(\rho)$ is the square of the conductor of ξ and $T(\mathfrak{p})$ annihilates V.*

Here $T(\mathfrak{p})$ acts on $\mathbf{U}^U$ for $U = U_0(\mathfrak{p}^r)_\mathfrak{p}$ by $\mathbf{u}|T(\mathfrak{p}) = \Sigma_x\, \rho(x^{-1})\mathbf{u}$ for a representative set $\{x\}$ of left coset decomposition $U\backslash(U\left(\begin{smallmatrix}1 & 0\\ 0 & \varpi\end{smallmatrix}\right)U)$. Usually in the references, the representation $\pi(\xi, \eta)$ here is written as $\rho(\alpha^{-1/2}\xi^{-1}, \alpha^{-1/2}\eta^{-1})$ and is the induced representation of the character: $\left(\begin{smallmatrix}a & b\\ 0 & d\end{smallmatrix}\right) \mapsto \alpha^{-1/2}(ad)\xi^{-1}(a)\eta^{-1}(d)$ from the standard Borel subgroup to $GL_2(F_{\mathfrak{p}})$ (cf. [1, 0.5]). Similar notational convention also applies to $\sigma(\xi, \xi\alpha^{-1})$. Under this convention, the above result is just an interpretation of the result shown in [2]. We now put for each integer r

$$U_r = U(\mathfrak{p}^r)_{\mathfrak{p}} = \left\{\begin{pmatrix} a & b \\ c & d \end{pmatrix} \in U_0(\mathfrak{p}^r)_{\mathfrak{p}} \middle| a \equiv d \equiv 1 \bmod \mathfrak{p}^r\mathfrak{r}_{\mathfrak{p}}\right\}.$$

Then $\mathbf{G}_r = U_0(\mathfrak{p}^r)_{\mathfrak{p}}/U_r$ acts naturally on $V_r = \mathbf{U}^{U_r}$. We normalize the action of $\mathbf{G}_r$ as $\mathbf{u}|g = \rho(g^{-1})\mathbf{u}$, which is a right action. Similarly, by choosing a prime element ϖ of $\mathfrak{r}_{\mathfrak{p}}$, we can define a Hecke operator $T(\varpi)$ by

$$\mathbf{u}|T(\varpi) = \sum_x \rho(x^{-1})\mathbf{u}$$

for a representative set $\{x\}$ of left coset decomposition $U_r\backslash(U_r\left(\begin{smallmatrix}1 & 0\\ 0 & \varpi\end{smallmatrix}\right)U_r)$.

COROLLARY 2.2. *Let the notation be as in Proposition* 2.1 *and suppose that ρ is irreducible and that $V_r \neq 0$ and $T(\varpi)$ on V_r is not nilpotent for some large r. Then,*

(i) *ρ is isomorphic to either a principal series representation $\pi(\xi, \eta)$ or a special representation $\sigma(\xi, \eta)$.*

(ii) *Suppose that $\rho = \pi(\xi, \eta)$ and write $\mathfrak{p}^i$ (resp. $\mathfrak{p}^j$, $\mathfrak{p}^k$) for the conductor of $\xi^{-1}\eta$ (resp. ξ, η). Then we have $r \geq \max(j, k) \geq i$. When $i = j = k = r = 0$, then $T(\varpi)$ acts on V_0 by the multiplication of $\xi(\varpi) + \eta(\varpi)$, $\dim_{\mathbf{C}} V_0 = 1$. When $r > 0$ and $\xi \neq \eta$, then we can decompose $V_r = V(\xi) \oplus V(\eta) \oplus N$ so that $\dim_{\mathbf{C}} V(\xi) = \dim_{\mathbf{C}} V(\eta) = 1$, and on $V(\xi)$ (resp. $V(\eta)$), $U_0(\mathfrak{p}^r)_{\mathfrak{p}}$ acts by $\mathbf{u}|\left(\begin{smallmatrix}a & b\\ c & d\end{smallmatrix}\right) = \xi(a)\eta(d)\mathbf{u}$ (resp. $\mathbf{u}|\left(\begin{smallmatrix}a & b\\ c & d\end{smallmatrix}\right) = \eta(a)\xi(d)\mathbf{u}$), and on N (resp. on $V(\xi)$ and $V(\eta)$), $T(\varpi)$ acts as a nilpotent operator (resp. by the multiplication of $\xi(\varpi)$ and $\eta(\varpi)$). When $r > 0$ and $\xi = \eta$, then we can decompose $V_r = V(\xi) \oplus N$ so that $\dim_{\mathbf{C}} V(\xi) = 2$, and on $V(\xi)$, $U_0(\mathfrak{p}^r)_{\mathfrak{p}}$ acts by $\mathbf{u}|\left(\begin{smallmatrix}a & b\\ c & d\end{smallmatrix}\right) = \xi(ad)\mathbf{u}$ and on $V(\xi)$ (resp. on N), $T(\varpi)$ acts nonsemisimply with unique eigenvalue $\xi(\varpi)$ (resp. as a nilpotent operator).*

(iii) *Suppose that $\rho = \sigma(\xi, \xi\alpha^{-1})$ and write $\mathfrak{p}^j$ for the conductor of ξ. then we have $r \geq \max(j, 1)$, and we can decompose $V_r = V(\xi) \oplus N$ such that on V_r, $U_0(\mathfrak{p}^r)_{\mathfrak{p}}$ acts via $\mathbf{u}|\left(\begin{smallmatrix}a & b\\ c & d\end{smallmatrix}\right) = \xi(ad)\mathbf{u}$ and $T(\varpi)$ acts on N (resp. on*

$V(\xi)$) as a nilpotent operator (resp. by the multiplication of $\xi(\varpi)$). We also have $\dim_{\mathbf{C}} V(\xi) = 1$.

Proof. Define a character ω by $\mathbf{u}|z = \rho(z^{-1})\mathbf{u} = \omega(z)\mathbf{u}$ for $z \in F_{\mathfrak{p}}^{\times}$. Then we can decompose $V_r = \bigoplus_{(\chi,\psi)} V(\chi, \psi)$ so that $\mathbf{G}_r$ acts on $V(\chi, \psi)$ by $\mathbf{u}|\left(\begin{smallmatrix} a & b \\ c & d \end{smallmatrix}\right) = \chi(a)\psi(d)\mathbf{u}$. Then $\chi\psi = \omega$ on $\mathfrak{r}_{\mathfrak{p}}^{\times}$. Extending χ to a character of $F_{\mathfrak{p}}^{\times}$, we now modify the action of ρ and consider $\chi \otimes \rho$ whose action on $\mathbf{u} \in \mathbf{U}$ is given by $\chi \otimes \rho(x)\mathbf{u} = \chi(\det(x))\rho(x)\mathbf{u}$. Then, on $V(\chi, \psi)$, we see $\mathbf{u}|\left(\begin{smallmatrix} a & b \\ c & d \end{smallmatrix}\right) = \chi \otimes \rho\left(\begin{smallmatrix} a & b \\ c & d \end{smallmatrix}\right)^{-1}\mathbf{u} = \psi\chi^{-1}(d)\mathbf{u}$. On the other hand, we have for $z \in F_{\mathfrak{p}}^{\times}$,

$$\mathbf{u}|z = \chi \otimes \rho(z^{-1})\mathbf{u} = \omega\chi^{-2}(z)\mathbf{u} = \psi\chi^{-1}(z)\mathbf{u}.$$

We write the subspace $V(\chi, \psi)$ under the action of ρ as $V_\rho(\chi, \psi)$ to avoid confusion. Then by the above argument, the identity map ι induces an isomorphism $\iota\colon V_\rho(\chi, \psi) \cong V_{\chi\otimes\rho}(1, \psi\chi^{-1})$. We see easily, for the Hecke operator $T_\rho(\varpi)$ under the action of ρ,

$$(*) \qquad \iota(\mathbf{u}|T_\rho(\varpi)) = \chi(\varpi)\iota(\mathbf{u})|T_{\chi\otimes\rho}(\varpi).$$

Thus if ρ is super cuspidal, then $\chi \otimes \rho$ is also super cuspidal for any character χ by definition, and therefore, by Proposition 2.1, $T_{\chi\otimes\rho}(\varpi)$ is nilpotent on $V_{\chi\otimes\rho}(1, \psi\chi^{-1})$ (see [5, Lemma 3.3] and the argument below). Then by (*), $T_\rho(\varpi)$ is nilpotent on $V_\rho(\chi, \psi)$. Thus if $T(\varpi)$ is not nilpotent on V_r for some r, then ρ must be principal or special. Now assume that $\rho = \pi(\xi, \eta)$. Then $\chi \otimes \rho = \pi(\chi^{-1}\xi, \chi^{-1}\eta)$. Thus, by Proposition 2.1, $T_{\chi\otimes\rho}(\varpi)$ is not nilpotent on $V_{\chi\otimes\rho}(1, \psi\chi^{-1})$ if and only if either $\chi^{-1}\xi$ or $\chi^{-1}\eta$ is unramified. Thus we may assume that either $\chi = \xi$ or $\chi = \eta$. Since $\pi(\xi, \eta) \cong \pi(\eta, \xi)$, the argument is the same in either case and thus hereafter, we assume that $\chi = \xi$. Then $\psi = \eta$ and thus we have $r \geq \max(j, k) \geq i$. The assertion in the case $i = j = k = r = 0$ follows directly from Proposition 2.1, (ii). Now we treat the remaining case where $r > 0$. Since $\chi \otimes \rho = \pi(1, \xi^{-1}\eta)$, by Proposition 2.1, $V_{\chi\otimes\rho}(1, \xi^{-1}\eta)$ has dimension $r - i + 1 > 0$ if $\mathfrak{p}^i$ is the conductor of $\xi^{-1}\eta$. Let $\mathbf{u}_0$ be a nontrivial vector in $V_{\chi\otimes\rho}(1, \xi^{-1}\eta)$ fixed by $U_1(\mathfrak{p}^i)$. Then it is well known that

$$\mathbf{u}_n = \xi \otimes \rho\left(\begin{pmatrix} \varpi^{-n} & 0 \\ 0 & 1 \end{pmatrix}\right)\mathbf{u}_0$$

for $n = 0, \ldots, r - i$ gives a basis of $V_{\chi\otimes\rho}(1, \xi^{-1}\eta)$. When $i = 0$, we define inductively

$$\mathbf{w}_0 = \mathbf{u}_0 - \mathbf{u}_1, \qquad \mathbf{w}_1 = \mathbf{u}_0 - \xi\eta^{-1}(\bar{\omega})\mathbf{u}_1,$$

$$\mathbf{w}_2 = \mathbf{w}_1 - \xi \otimes \rho\left(\begin{pmatrix} \bar{\omega}^{-1} & 0 \\ 0 & 1 \end{pmatrix}\right)\mathbf{w}_1 \quad \text{and}$$

$$\mathbf{w}_n = \xi \otimes \rho\left(\begin{pmatrix} \bar{\omega}^{-n+2} & 0 \\ 0 & 1 \end{pmatrix}\right)\mathbf{w}_2 \quad \text{for} \quad 2 \le n \le r - i.$$

Then one see easily (e.g. [5, Lemma 3.3]) that if $\xi \neq \eta$, then $\mathbf{w}_n$'s form a basis of $V_{\chi\otimes\rho}(1, \xi^{-1}\eta)$ and

$$\mathbf{w}_0|T(\bar{\omega}) = \mathbf{w}_0, \qquad \mathbf{w}_1|T(\bar{\omega}) = \xi^{-1}\eta(\bar{\omega})\mathbf{w}_1, \qquad \mathbf{w}_2|T(\bar{\omega}) = 0$$

$$\text{and} \quad \mathbf{w}_n|T(\bar{\omega}) = \mathbf{w}_{n-1} \quad \text{if} \quad n > 2.$$

When $i = 0$ and $\xi = \eta$, then $\mathbf{w}_0, \mathbf{u}_1, \mathbf{w}_2, \ldots, \mathbf{w}_{r-i}$ form a basis and $T(\bar{\omega})$ acts on $\mathbf{C}\mathbf{u}_1 + \mathbf{C}\mathbf{w}_0$ via an unipotent matrix and on $\mathbf{C}\mathbf{w}_2 + \cdots + \mathbf{C}\mathbf{w}_{r-i}$ as a nilpotent operator. When $i > 0$, we similarly define

$$\mathbf{w}_0 = \mathbf{u}_0 - \mathbf{u}_1, \qquad \mathbf{w}_1 = \xi\eta^{-1}(\bar{\omega})\mathbf{w}_0 - \xi \otimes \rho\left(\begin{pmatrix} \bar{\omega}^{-1} & 0 \\ 0 & 1 \end{pmatrix}\right)\mathbf{w}_0$$

$$\text{and} \quad \mathbf{w}_n = \xi \otimes \rho\left(\begin{pmatrix} \bar{\omega}^{-n+2} & 0 \\ 0 & 1 \end{pmatrix}\right)\mathbf{w}_1.$$

Then again $\mathbf{w}_n$'s form a basis of $V_{\chi\otimes\rho}(1, \xi^{-1}\eta)$ and satisfies $\mathbf{w}_0|T(\bar{\omega}) = \mathbf{w}_0$ and $\mathbf{w}_1|T(\bar{\omega}) = 0$ and $\mathbf{w}_n|T(\bar{\omega}) = \mathbf{w}_{n-1}$ if $n > 1$. This shows the desired assertion by (*) for principal series representations. Similar argument shows the assertion also for special representations.

Now consider a common eigenform $\mathbf{f}$ in $\mathbf{S}_{2t,t}(S(p^\alpha); \mathbf{C})$ ($t = \Sigma_\sigma \sigma \in \mathbf{Z}[I]$) of all Hecke operators $T(\mathfrak{n})$ for $\mathfrak{n}$ prime to p and $T(\mathfrak{p})$ for all prime factors $\mathfrak{p}$ of p. The operators $T(\mathfrak{q})$ for primes $\mathfrak{q}$ are defined as follows: We choose a prime element $\bar{\omega}$ of $\mathfrak{r}_\mathfrak{q}$ and consider it as an element of $F_\mathbf{A}^\times$ and define $T(\mathfrak{q})$ by the action of the double coset $S(p^\alpha)\left(\begin{smallmatrix} 1 & 0 \\ 0 & \bar{\omega} \end{smallmatrix}\right)S(p^\alpha)$ on $\mathbf{S}_{2t,t}(S(p^\alpha);$ $\mathbf{C})$; namely, for a representative set $\{x\}$ for $S(p^\alpha)\backslash\left(S(p^\alpha)\left(\begin{smallmatrix} 1 & 0 \\ 0 & \bar{\omega} \end{smallmatrix}\right)S(p^\alpha)\right)$, $\mathbf{f}|T(\mathfrak{q})(y) = \Sigma_x \mathbf{f}(yx^{-1})$. We define the Hecke operator $T(p)$ by $\Pi_{\mathfrak{p}|p}T(\mathfrak{p})^{a(\mathfrak{p})}$ using the prime decomposition $p = \Pi_{\mathfrak{p}|p}\mathfrak{p}^{a(\mathfrak{p})}$ in $\mathfrak{r}$. The Hecke operator $T(\mathfrak{p})$ and $T(p)$ may depend on the choice of $\bar{\omega}$, but $T(\mathfrak{n})$ for $\mathfrak{n}$ prime to p is independent of the choice of $\bar{\omega}$. More generally, one can define a Hecke operator $T(x)$ for each element $0 \neq x \in \mathfrak{r}_{\mathfrak{p}}$ with nonnegative valuation by the action of double coset $S(p^\alpha)\left(\begin{smallmatrix}1 & 0\\ 0 & x\end{smallmatrix}\right)S(p^\alpha)$. The limit $e = e_\alpha = \lim_{n\to\infty} T(p)^{n!} = \lim_{n\to\infty} T(x)^{n!}$ for $0 \neq x \in p\mathfrak{r}_p$ exists in $\mathfrak{h}_{2t,t}(S(p^\alpha); \mathcal{O})$ and is independent of the choice of x. Then by definition,

$$\mathfrak{h}_{2t,t}{}^{\mathrm{n.ord}}(S(p^\alpha); \mathcal{O}) = e_\alpha \mathfrak{h}_{2t,t}(S(p^\alpha); \mathcal{O}).$$

Note that the idempotent e is in fact algebraic and gives an element of $\mathfrak{h}_{2t,t}(S(p^\alpha); \bar{\mathbf{Q}})$. Thus e acts on $\mathbf{S}_{2t,t}(S(p^\alpha); \mathbf{C})$ and we put $\mathbf{S}_{2t,t}{}^{\mathrm{n.ord}}(S(p^\alpha); \mathbf{C}) = e\mathbf{S}_{2t,t}(S(p^\alpha); \mathbf{C})$. Assume that $\mathbf{f} \in \mathbf{S}_{2t,t}{}^{n.\mathrm{ord}}(S(p^\alpha); \mathbf{C})$ and let $\rho = \bigotimes_{\mathfrak{q}} \rho_{\mathfrak{q}}$ be the automorphic representation on the linear span of all right translations of $\mathbf{f}$ by the elements of $GL_2(F_{\mathbf{A}})$. Then by the strong multiplicity one theorem [2, Theorem 2], ρ is irreducible and by Corollary 2.2, $\rho_{\mathfrak{p}}$ for each prime factor $\mathfrak{p}$ of p is principal or special because $T(\mathfrak{p})$ acts nontrivially on $\mathbf{f}$ by near ordinarity. Thus we write $\rho_{\mathfrak{p}} = \pi(\xi_{\mathfrak{p}}, \eta_{\mathfrak{p}})$ if $\rho_{\mathfrak{p}}$ is principal and $\rho_{\mathfrak{p}} = \sigma(\xi_{\mathfrak{p}}, \eta_{\mathfrak{p}})$ if $\rho_{\mathfrak{p}}$ is special. For prime $\mathfrak{q}$ outside Np, $\rho_{\mathfrak{q}}$ is spherical and hence, we can write $\rho_{\mathfrak{q}} = \pi(\xi_{\mathfrak{q}}, \eta_{\mathfrak{q}})$ for all primes $\mathfrak{q}$ outside Np. We may assume by Corollary 2.2 that $|\xi_{\mathfrak{p}}(\bar{\omega}_{\mathfrak{p}})|_p = 1$ in $\bar{\mathbf{Q}}_p$. Then by Corollary 2.2, $\mathbf{f}|T(x) = \xi_{\mathfrak{p}}(x)\mathbf{f}$ for all $0 \neq x \in \mathfrak{p}\mathfrak{r}_{\mathfrak{p}}$. Now we want to find a finite order character χ of $F_{\mathbf{A}}^\times/F^\times$ such that $\chi_{\mathfrak{p}} = \xi_{\mathfrak{p}}$ on $\mathfrak{r}_{\mathfrak{p}}^\times$ for all prime factors $\mathfrak{p}$ of p and unramified at every infinite place. In fact, we consider the restriction $\xi_{\mathfrak{p}}\colon \mathfrak{r}^\times \to \mathbf{C}^\times$ and denote by m the order of the image $\xi_{\mathfrak{p}}(\mathfrak{r}^\times)$. Then by a theorem of Chevalley [3], we can find an integral ideal $\mathfrak{q}(\mathfrak{p})$, which is prime to pN for any given integer N, so that if $\epsilon \equiv 1 \bmod \mathfrak{q}(\mathfrak{p})$ for $\epsilon \in \mathfrak{r}^\times$, then ϵ is an m-th power in $\mathfrak{r}^\times$. Thus the kernel of $\xi_{\mathfrak{p}}$ on $\mathfrak{r}^\times$ contains $\{\delta \in \mathfrak{r}^\times | \delta \equiv 1 \bmod \mathfrak{q}(\mathfrak{p})\}$. Thus we can find a finite order character $\psi_{\mathfrak{p}}$ of $F_{\mathfrak{q}(\mathfrak{p})}^\times$ such that $\psi_{\mathfrak{p}}\xi_{\mathfrak{p}}$ is trivial on $\mathfrak{r}^\times$. Then we can extend $\psi_{\mathfrak{p}}\xi_{\mathfrak{p}}$ on $\mathfrak{r}_{\mathfrak{q}(\mathfrak{p})}^\times \times \mathfrak{r}_{\mathfrak{p}}^\times$ to a finite order character $\chi^{\mathfrak{p}}$ of $F_{\mathbf{A}}^\times/F^\times$ whose conductor is a product of a power of $\mathfrak{p}$ and $\mathfrak{q}(\mathfrak{p})$. We may take $\mathfrak{q}(\mathfrak{p})$'s for $\mathfrak{p}|p$ so that they are mutually prime each other. Then $\chi = \prod_{\mathfrak{p}|p} \chi^{\mathfrak{p}}$ satisfies the required property. Let A be a subring of $\bar{\mathbf{Q}}_p$ generated over $\mathcal{O}$ by the eigenvalue $\lambda(h)$ of $\mathbf{f}$ for all $h \in \mathfrak{h}_{k,w}{}^{n.\mathrm{ord}}(S(p^\alpha); \mathcal{O})$ and Q be the quotient field of A. Let $\sigma = \sigma(\rho)$ be the Galois representation corresponding to ρ (see [14, Theorem 2]) of $GL_2(\bar{\mathbf{Q}}/F)$ into $GL_2(Q)$ such that

(2.1a) *σ is continuous, irreducible and unramified outside Np,*
(2.1b) *For the Frobenius element $\phi_{\mathfrak{q}}$ for prime $\mathfrak{q}$ outside Np,*

$$\det(1 - \sigma(\phi_{\mathfrak{q}})X) = (1 - \xi_{\mathfrak{q}}(\bar{\omega}_{\mathfrak{q}})X)(1 - \eta_{\mathfrak{q}}(\bar{\omega}_{\mathfrak{q}})X).$$

We define an algebra homomorphism $\lambda : \mathfrak{h}_{k,w}{}^{n.\mathrm{ord}}(S(p^\alpha); \mathcal{O}) \to A$ by $\mathbf{f}|h = \lambda(h)\mathbf{f}$. Then we have $\lambda(T(\mathfrak{q})) = \xi_\mathfrak{q}(\bar{\omega}_\mathfrak{q}) + \eta_\mathfrak{q}(\bar{\omega}_\mathfrak{q})$ and $\lambda(\langle \mathfrak{q} \rangle)\mathfrak{N}_{F/\mathbf{Q}}(\mathfrak{q}) = \xi_\mathfrak{q}(\bar{\omega}_\mathfrak{q})\eta_\mathfrak{q}(\bar{\omega}_\mathfrak{q})$. Thus this representation σ is the representation as in Theorem I associated to λ. We consider $\chi \otimes \rho$, which is again a cuspidal automorphic representation. Then its component at $\mathfrak{p}$ is given either $\pi(\chi_\mathfrak{p}{}^{-1}\xi_\mathfrak{p}, \chi_\mathfrak{p}{}^{-1}\eta_\mathfrak{p})$ or $\sigma(\chi_\mathfrak{p}{}^{-1}\xi_\mathfrak{p}, \chi_\mathfrak{p}{}^{-1}\eta_\mathfrak{p})$. Since $\mathbf{f}$ is of weight $2t$, if $\chi_\mathfrak{p}{}^{-1}\eta_\mathfrak{p}$ is unramified and if $\rho_\mathfrak{p}$ is principal, then $\chi_\mathfrak{p}{}^{-2}\xi_\mathfrak{p}\eta_\mathfrak{p}(\bar{\omega}_\mathfrak{p}) = \mathfrak{N}_{F/\mathbf{Q}}(\mathfrak{p})\chi^{-2}\psi(\mathfrak{p})$ for the central character ψ (which is of finite order) of ρ. Thus $\chi_\mathfrak{p}{}^{-1}\xi_\mathfrak{p} \neq \chi_\mathfrak{p}{}^{-1}\eta_\mathfrak{p}$, and without assuming the unramifiedness of $\chi_\mathfrak{p}{}^{-1}\eta_\mathfrak{p}$, the subspace in the representation space of $\rho_\mathfrak{p}$ on which $T(\bar{\omega}_\mathfrak{p})$ has p-adic unit eigenvalue is one dimensional. Since χ is of finite order, its values are roots of unity and hence $|\chi_\mathfrak{p}{}^{-1}\xi_\mathfrak{p}(\bar{\omega}_\mathfrak{p})|_p = 1$ and hence by Proposition 2.1 (and [7, Lemma 12.2]), the primitive form $\mathbf{f}_0$ belonging to $\chi \otimes \rho$ is ordinary; namely, for the global conductor $c(\chi \otimes \rho)$, the subspace of the representation space of $\chi \otimes \rho$ realized on automorphic forms on $GL_2(F_\mathbf{A})$ fixed by $U_1(c(\chi \otimes \rho))$ is one dimensional and on which $T(p)$ acts by a p-adic unit. Note that $\sigma(\chi \otimes \rho) = \chi^{-1} \otimes \sigma(\rho)$. Then, by a theorem of Wiles [17, Theorem 2], on $D_\mathfrak{p}$, we have, up to equivalence

$$\sigma(\chi \otimes \rho)(\tau) = \begin{pmatrix} \epsilon'(\tau) & * \\ 0 & \delta'(\tau) \end{pmatrix} \quad \text{for} \quad \tau \in D_\mathfrak{p},$$

where $\delta' : D_\mathfrak{p} \to \bar{\mathbf{Q}}_p$ is unramified and for the Frobenius element $\phi_\mathfrak{p}$, we have $\delta'(\phi_\mathfrak{p}) = \chi_\mathfrak{p}{}^{-1}\xi_\mathfrak{p}(\bar{\omega}_\mathfrak{p})$. This combined with the formula $\sigma(\chi \otimes \rho) = \chi^{-1} \otimes \sigma(\rho)$ shows that

$$\sigma(\rho)(\tau) = \begin{pmatrix} \epsilon(\tau) & * \\ 0 & \delta(\tau) \end{pmatrix} \quad \text{for} \quad \tau \in D_\mathfrak{p},$$

where δ is a character of $D_\mathfrak{p}$ corresponding to $\xi_\mathfrak{p}$ by local class field theory at $\mathfrak{p}$. Thus δ has values in A by Corollary 2.2, since $\xi_\mathfrak{p}(\bar{\omega}) \in A$ and $\mathfrak{h}_{k,w}{}^{\mathrm{n.ord}}(S(p^\alpha); \mathcal{O})$ contains the action of $S_0(p^\alpha)/S(p^\alpha)$. Since $\epsilon\delta = \det(\sigma(\rho))$ has values in A, ϵ also has values in A over $D_\mathfrak{p}$. Thus we have

PROPOSITION 2.3. *Let A be a subring of $\bar{\mathbf{Q}}_p$ and let $\Lambda : \mathfrak{h}_{k,w}{}^{\mathrm{n.ord}}(S(p^\alpha); \mathcal{O}) \to A$ (for $n \geq 0$ and $v \geq 0$ with $n + 2v \in \mathbf{Z}t$) be an algebra homomorphism. Then there exists A-valued characters ϵ and δ of the decomposition group $D_\mathfrak{p}$ for each prime factor $\mathfrak{p}$ of p such that the restriction of the Galois* *representation attached to λ to the decomposition group $D_{\mathfrak{p}}$ at each prime factor $\mathfrak{p}$ of p is, up to equivalence, of the following form:*

$$\begin{pmatrix} \epsilon(\tau) & * \\ 0 & \delta(\tau) \end{pmatrix} \quad \textit{for} \quad \tau \in D_{\mathfrak{p}}.$$

Moreover if $\xi_{\mathfrak{p}}$ denotes the character of $F_{\mathfrak{p}}^{\times}$ corresponding to δ by local class field theory, then $\lambda(x^{-v}T(x)) = \xi_{\mathfrak{p}}(x)$ for all $0 \neq x \in \mathfrak{r}_{\mathfrak{p}}$.

When $(k, w) = (2t, t)$ (i.e. $n = v = 0$), this proposition follows from the above argument. The general case follows from [8, Theorem 3.3] and the proof of Theorem I in the following section, which only involves the result in the case where $n = v = 0$.

3. Proof of Theorems I and II. We take, as the group G in Section 1, the Galois group over F of the maximal extension of F unramified outside Np, where N is the ideal prime to p which is maximal under the condition $S \supset U_1(N)$. This ideal is called the level of S. Then define a subset Σ of G by

$$\Sigma = \bigcup_{\mathfrak{q}} \{\phi_{\mathfrak{q}} \in G \mid \phi_{\mathfrak{q}} \text{ is a Frobenius element of } \mathfrak{q}\}$$

where $\mathfrak{q}$ runs over all prime ideals outside Np. Then Σ is dense in G by the density theorem of Chebotarev. Let R be either the subalgebra $\mathbf{h}$ of $\mathbf{h}(S; \mathcal{O})$ generated over $\mathbf{A}$ by $T(\mathfrak{n})$'s for $\mathfrak{n}$ prime to Np or the reduced part of $\mathbf{h}^{\text{n.ord}}(S; \mathcal{O})$ (i.e. the quotient of $\mathbf{h}^{\text{n.ord}}(S; \mathcal{O})$ by its nilradical). Now we know from [8, Theorem 2.3] that

$$R = \varprojlim_{\alpha} R_{\alpha} \text{ as a topological algebra,}$$

where R_{α} is a subalgebra of $\mathfrak{h}_{2t,t}(S(p^{\alpha}); \mathcal{O})$ generated over $\mathcal{O}[G^{\alpha}]$ by $T(\mathfrak{n})$'s for $\mathfrak{n}$ prime to Np if $R = \mathbf{h}$ and in the other case, R_{α} is the reduced part of $\mathfrak{h}_{k,w}{}^{\text{n.ord}}(S(p^{\alpha}); \mathcal{O})$ under the notation in [8, Section 2]. Note that R_{α} is a reduced algebra free of finite rank over $\mathcal{O}$. Therefore, R_{α} is a product of finitely many objects in $\mathcal{C}$, and there exists finitely many minimal prime ideals $\mathfrak{p}_{\alpha,i}$ of R_i such that $R_i/\mathfrak{p}_{\alpha,i}$ is free of finite rank over $\mathcal{O}$ and $\bigcap_i \mathfrak{p}_{\alpha,i} = \{0\}$. We lift $\mathfrak{p}_{\alpha,i}$'s to R and consider them as prime ideals of R. If λ denotes the algebra homomorphism of R_i into $\bar{\mathbf{Q}}_p$ whose kernel coincides with $\mathfrak{p}_{\alpha,i}$, then there exists a common eigenform $\mathbf{f}$ of all Hecke operators in

$\mathbf{S}_{2t,t}(U_1(Np^\beta); \mathcal{O})$ for sufficiently large β such that $\mathbf{f}|T(\mathfrak{n}) = \lambda(T(\mathfrak{n}))\mathbf{f}$ (see the proof of Proposition 2.3). Thus by the result of Taylor [14], there exists a Galois representation $\pi(\lambda)$: $\mathrm{Gal}(\bar{\mathbf{Q}}/F) \to GL_2(\bar{\mathbf{Q}}_p)$ such that

(3.1a) $\pi(\lambda)$ is continuous, irreducible and unramified outside Np,
(3.1b) For each $\phi_\mathfrak{q} \in \Sigma$, we have

$$\det(1 - \pi(\lambda)(\phi_\mathfrak{q})X) = 1 - \lambda(T(\mathfrak{q}))X + \mathfrak{N}_{F/\mathbf{Q}}(\mathfrak{q})\lambda(\langle \mathfrak{n} \rangle)X^2.$$

Then by identifying $R/\mathfrak{p}_{\alpha,i}$ with $\mathrm{Im}(\lambda)$, we can construct a pseudo representation $\pi_{\alpha,i}$ into $R/\mathfrak{p}_{\alpha,i}$. As the element $c \in G$ as in Section 1, we take the natural image of complex conjugation. Then it is well known that $\det(\pi(\lambda)(c)) = -1$ and hence one can associate the pseudo representation $\pi_{\alpha,i}$ to each $\mathfrak{p}_{\alpha,i}$. Now define functions tr: $\Sigma \to R$ and det: $\Sigma \to R$ by $\mathrm{tr}(\phi_\mathfrak{q}) = T(\mathfrak{q})$ and $\det(\phi_\mathfrak{q}) = \mathfrak{N}_{F/\mathbf{Q}}(\mathfrak{q})\langle \mathfrak{q} \rangle$. Thus, we have pseudo representations $\{\pi_{\alpha,i}\}$, the set of ideals $\{\mathfrak{p}_{\alpha,i}\}$, the dense subset Σ of G, and functions tr and det satisfying the assumption of Theorem 1.4. Then the assertion of Theorems I (except for (iv)) and Theorem II follows from Theorem 1.4 and the result in [8], especially, Theorems 2.3 and 2.4 and Corollary 2.5. We shall now show the assertion (iv) in Theorem I. Let $\mathfrak{p}$ be a prime factor of p in $\mathfrak{r}$. For each element x with nonnegative valuation in $\mathfrak{r}_\mathfrak{p}$, we consider the Hecke operator $T_\alpha(x)$ corresponding to the double coset action of $S(p^\alpha)\left(\begin{smallmatrix}1 & 0\\ 0 & x\end{smallmatrix}\right)S(p^\alpha)$ in $\mathfrak{h}_{2t,t}(S(p^\alpha); \mathcal{O})$. Then the projection of $T(x) = \lim_{\overleftarrow{\alpha}} T_\alpha(x)$ in $\mathbf{h}^{\mathrm{n.ord}}(S; \mathcal{O})$ is a unit. Since $T(xy) = T(x)T(y)$, we can define a character $\Delta: F_\mathfrak{p}^\times \to \mathbf{h}^{\mathrm{n.ord}}(S; \mathcal{O})$ by $\Delta(x) = T(x)$. By local class field theory, we may regard Δ as a character of $D_\mathfrak{p}$. Similarly, we can define another Galois character ι: $\mathrm{Gal}(\bar{\mathbf{Q}}/F) \to \mathbf{h}^{\mathrm{n.ord}}(S; \mathcal{O})$ so that ι is unramified outside Np for the level N of S and $\iota(\phi_\mathfrak{q}) = \langle \mathfrak{q} \rangle \mathfrak{N}_{F/\mathbf{Q}}(\mathfrak{q})$ for each prime ideal $\mathfrak{q}$ outside Np. Then for each minimal prime ideal $\mathfrak{p}$ of $\mathbf{h}^{\mathrm{n.ord}}(S; \mathcal{O})$ and the projection ρ: $\mathbf{h}^{\mathrm{n.ord}}(S; \mathcal{O}) \to \mathbf{I} = \mathbf{h}^{\mathrm{n.ord}}(S; \mathcal{O})/\mathfrak{p}$, $\det(\pi) = \rho \circ \iota$ for $\pi = \pi(\rho)$. We define a Galois representation π' into $GL_2(\mathbf{K})$ for the quotient field $\mathbf{K}$ of $\mathbf{I}$ by

$$\pi'(\sigma) = \begin{pmatrix} \rho \circ (\iota\Delta^{-1}) & 0 \\ 0 & \rho \circ \Delta \end{pmatrix}.$$

Then, for any prime ideal $\mathfrak{p}_{\alpha,i}$ as above containing $\mathfrak{p}$, the two representations $\pi'|_{D_\mathfrak{p}}$ and $\pi|_{D_\mathfrak{p}}$ modulo $\mathfrak{p}_{\alpha,i}$ have the same trace by Proposition 2.3. Since prime ideals $\mathfrak{p}_{\alpha,i}$ containing $\mathfrak{p}$ are Zariski dense in $\mathrm{Spec}(\mathbf{I})$, the semi-simplification of $\pi|_{D_\mathfrak{p}}$ coincides with $\pi'|_{D_\mathfrak{p}}$. This shows the assertion (iv) because any algebra homomorphism λ: $\mathbf{h}^{n.ord}(S; \mathcal{O}) \to A$ as in Theorem I factors through some ρ as above.

UNIVERSITY OF CALIFORNIA, LOS ANGELES

REFERENCES

[1] H. Carayol, Sur les représentations l-adiques associées aux formes modulaires de Hilbert, *Ann. Scient. Éc. Norm. Sup.*, 4e *série*, **19** (1986), 409-468.

[2] W. Casselman, On some results of Atkin and Lehner, *Math. Ann.* **201** (1973), 301-314.

[3] C. Chevalley, Deux théorèmes d'arithmétique, *J. Math. Soc. Japan*, **3** (1951), 36-44.

[4] F. Q. Gouvêa, Arithmetic of p-adic modular forms, *Lecture notes in Math.* 1304, Springer, 1988.

[5] H. Hida, A p-adic measure attached to the zeta functions associated with two elliptic modular forms, I. *Invent. Math.* **79** (1985), 159-195.

[6] ———, Galois representations into $GL_2(\mathbf{Z}_p[[X]])$ attached to ordinary cusp forms, *Invent. Math.* **85** (1986), 545-613.

[7] ———, On p-adic Hecke algebras for GL_2 over totally real fields, *Ann. of Math.* **128** (1988), 295-384.

[8] ———, On nearly ordinary Hecke algebras for GL_2 over totally real fields, *Advanced Studies in Pure Math.* **17** (1989), 139-169.

[9] B. Mazur, Deforming Galois representations, preprint.

[10] ——— and A. Wiles, Class fields of abelian extension of $\mathbf{Q}$, *Invent. Math.* **76** (1984), 179-330.

[11] M. Ohta, On l-adic representations attached to automorphic forms, *Japan. J. Math.* **8** (1982), 1-47.

[12] ———, On the zeta function of an abelian scheme over the Shimura curve, *Japan. J. Math.* **9** (1983), 1-25.

[13] J. D. Rogawski and J. B. Tunnell, On Artin L-functions associated to Hilbert modular forms of weight one, *Invent. Math.* **74** (1983), 1-42.

[14] R. Taylor, On Galois representations associated to Hilbert modular forms, preprint.

[15] J. Tilouine, Theorie d'Iwasawa classique et de l'algèbre de Hecke ordinaire, *Compositio Math.* **65** (1988), 265-320.

[16] A. Wiles, On p-adic representations for totally real fields, *Ann. of Math.* **123** (1986), 407-456.

[17] ———, On ordinary λ-adic representations associated to modular forms, preprint.

ON THE PRESENTATIONS OF RESOLUTION DATA

By H. HIRONAKA

Noted and Appended by T. T. MOH

Introduction. The purpose of this paper is to examine some of the typical examples of singularities that have suggested us ideas useful toward resolution of singularities. We will deal with the hypersurfaces case. The forms of the equation with respect to various sets of variables are important. From them we may deduce some useful numerical characters for resolutions. We shall examine the behaviors of them under blow-ups.

1. Preliminary. To begin with, let us examine the following 2-dimensional singular surface.

Example 1. Let the ground field k be with characteristic $\neq 2$. Let the surface X be defined by the following equation

$$F = x^2 - y^2 z^2 = 0.$$

It is easy to deduce by the Jacobian criteria that the singular locus of $X = \{x = y = 0\} \cup \{x = z = 0\}$. The origin appears to be the worst singular point of X. We may attempt to blow up the origin. However in one of the affine coordinate system on the transform, we have

$$\begin{cases} \text{coordinate } \left(\dfrac{x}{z}, \dfrac{y}{z}, z\right)(=(x_1, y_1, z_1)) \\ \text{equation } F' = \left(\dfrac{x}{z}\right)^2 - \left(\dfrac{y}{z}\right)^2 z^2 \\ \qquad\qquad = x_1^2 - y_1^2 z_1^2 = 0. \end{cases}$$

Manuscript received October 21, 1988; revised November 11, 1988.

Comparing this with the original equation $F = 0$, we see absolutely no improvement in singularity. □

It is easy to check that if we blow up any one of the two curves in the singular locus, then the singularity is effectively reduced in this case. Thus we come up with an idea: *In the surfaces case we blow up a largest center subject to a certain natural "permissibility" condition.*

The most appropriate "permissibility" condition is proved to be (by all good reasons) that the center should be smooth by itself and the hypersurface in question should have constant multiplicity along the center.

In fact, there is a theorem due to Beppo-Levi and Zariski which states as follows:

THEOREM. *In the case of a surface in* 3*-space over an algebraically closed field k of characteristic* 0*, the following procedure will resolve all singularities. If there are curves, permissible, in the singular locus, then blow up any one of them. If not, blow up any one of those points which are either isolated m-fold points (i.e., all the nearby points have multiplicities* $\leq m - 1$) *or singular points of irreducible curves in the singular locus. Then after a finite number of such blow-ups, the transform of the surface is smooth.*

In the higher dimensional case, the following example is illustrative (cf Spivakovsky [S]).

Example 2. Let a 3-dimensional hypersurface X be defined by the following equation

$$F = x^2 + yzw(yz + y^2w + z^2w) = 0.$$

It is routine to see that the singular locus consists of the following 3 lines:

(1) $x = y = z = 0$
(2) $x = y = w = 0$
(3) $x = z = w = 0$.

By the above discussion of permissibility, we see that each of those 3 lines is permissible. However the transforms of X by blowing up one of the 3 permissible lines are very different. To begin with, let us blow up the curve (2) and look at the coordinate chart $(x/w, y/w, w, z)$. Then the transform X' of X is defined by

$$F' = \left(\frac{x}{w}\right)^2 + \left(\frac{y}{w}\right)zw\left(\left(\frac{y}{w}\right)z + \left(\frac{y}{w}\right)^2 w^2 + z^2\right)$$

$$= x_1^2 + y_1 z_1 w_1 (y_1 z_1 + y_1^2 w^2 + z_1^2).$$

For the hypersurface X', the curve $x_1 = z_1 = w_1 = 0$ is certainly a largest permissible center. After blowing up with this as center then we get in one coordinate chart the following

$$\begin{cases} \text{coordinate } \left(\dfrac{x_1}{w_1}, y_1, \dfrac{z_1}{w_1}, w_1\right) = \left(\dfrac{x}{w^2}, \dfrac{y}{w}, \dfrac{z}{w}, w\right) \\ \qquad (=(x_2, y_2, z_2, w_2)) \\ \text{equation } F'' = \left(\dfrac{x_1}{w_1}\right)^2 + y_1\left(\dfrac{z_1}{w_1}\right)w_1\left(y_1 \dfrac{z_1}{w_1} + y_1^2 w_1 + \left(\dfrac{z_1}{w_1}\right)^2 w_1\right) \\ \qquad = x_2^2 + y_2 z_2 w_2 (y_2 z_2 + y_2^2 w_2 + z_2^2 w_2). \end{cases}$$

Again we see no improvement at this point from the original hypersurface.

Note that the equation $F = 0$ is symmetric with respect to y and z. So, same arguments as above can be applied to blowing up the curve (3). It turns out that the blowing-ups with center (1) are good and effective (i.e., after that, blowing-ups of any largest permissible centers, successively, resolve all the singularities). □

The above example indicates that there are subtle differences between curves (1), (2) and (3). These serve as the topics of the remaining sections of this article.

2. Youssin ideals. At this point, let us cite a beautiful theory of Boris Youssin (in his dissertation, Harvard 1988) in search of a good effective center to blow up, at least in the case of characteristic zero.

First of all, we call for the definition of Youssin polyhedral ideal. Let k be a field and let $R = k[[x_1, \ldots, x_N]]$ be a formal power series ring in N variables. Let us divide the variables $(x_1, \ldots, x_N)$ into blocks as follows:

$$\overbrace{x_1, \ldots, x_{n_1}, \overbrace{x_{n_1+1}, \ldots, x_{n_2}, \ldots, \overbrace{x_{n_{r-1}+1}, \ldots, x_{n_r}, \ldots, x_N}^{y_3'}}^{y_2'}}$$

$$\underbrace{x_1, \ldots, x_{n_1}}_{y_1}, \underbrace{x_{n_1+1}, \ldots, x_{n_2}}_{y_2}, \ldots, \underbrace{x_{n_{r-1}+1}, \ldots, x_{n_r}}_{y_r}, \ldots, x_N.$$

We then consider an ideal I in R associated with an expression of the form

$$(y_1)^{c_1} \oplus \text{mon}_1(y_2')[(y_2)^{c_2} \oplus \text{mon}_2(y_3')[(y_3)^{c_3} \oplus \cdots]]$$

where c_i are nonnegative rational numbers and $\text{mon}(y)$ denotes a monomial in the variables in y with nonnegative rational exponents.

Meaning of the expression. Let b be a positive integer such that bc_i are integers for all i and $(\text{mon}_i(y_{i+1}'))^b$ have integral exponents (i.e., $\in R$) for all i. Let $\tilde{x}_j = x_j^{1/b}$ for all j and $\tilde{R} = k[[\tilde{x}_1, \ldots, \tilde{x}_N]]$. Let $\tilde{y}_i$, $\tilde{y}_i'$ be the systems of variables $\tilde{x}_j$, which are obtained from y_i, y_i' respectively by replacing each x_j by $\tilde{x}_j$. Now, in $\tilde{R}$, $(y_i)^{c_i}$ should be understood to mean $(\tilde{y}_i)^{bc_i}$ and $\text{mon}_i(y_{i+1}')$ to mean a monomial in $\tilde{y}_{i+1}'$ with integral exponents by the relations $x_j = \tilde{x}_j^b$. Thus the above expression defines an ideal J in $\tilde{R}$ if we replace $\oplus$ by $+$. Let J^* be the integral closure of J in $\tilde{R}$ in the sense of Zariski. Finally $I = J^* \cap R$. (We used $\oplus$ instead of $+$ to indicate the operation of taking integral closure.)

Definition. An ideal I as above is called *Youssin polyhedral ideal*, or *Youssin ideal* in short, if the following two conditions are satisfied:

(Y,1) $$\deg_{y_{i+1}}(\text{mon}_i(y_{i+1}')) \geq c_i - c_{i+1}, \qquad \forall_i \geq 1.$$

This means that, in $\tilde{R}$ with interpretations as above,

$$\text{mon}_i(y_{i+1}') \in (y_{i+1})^{c_i - c_{i+1}}, \qquad \forall_i \geq 1.$$

(Y,2) The orders of the terms in the expression, from left to right, is strictly increasing. In other words,

$$c_1 < |\text{mon}_1| + c_2 < |\text{mon}_1| + |\text{mon}_2| + c_3 < \cdots$$

where $|\text{mon}_i|$ = total degree of the fractional monomial mon_i. □

Definition. Given a Youssin polyhedral ideal as above, we associate with it a measurement, called *Youssin measure*, that is

$$(c_1, -n_1, |\mathrm{mon}_1|, c_2, -n_2, |\mathrm{mon}_2|, \ldots). \qquad \square$$

We then consider the lexicographical ordering among various Youssin measures (possibly with different lengths) in the following sense. First, one with bigger c_1 is defined to be bigger; then, among those with the same c_1, one with bigger $-n_1$ (i.e., smaller n_1) is to be bigger; then, among those with the same c_1, and $-n_1$, one with bigger $|\mathrm{mon}_1|$ is bigger, and so on. Here it should be said that, if the two measures in comparison have different lengths, the shorter one should be lengthened by adding ∞'s after its tail end.

Now, when a power series F is given in R (which, in applications, is the Taylor expansion of the equation of a hypersurface in N-space at one of its singular points) we consider all possible choices of parameters for R and all possible Youssin polyhedral ideals with respect to each choice of parameters, in which F is contained.

Definition. Among all these ideals, we pick one, denoted by $Y(F)$, such that

(Y,3) $\qquad$ $Y(F)$ has the biggest Youssin measure.

We call $Y(F)$ *Youssin ideal* of F.

It is a theorem of Boris Youssin that, F being given, the Youssin ideal of F is *unique* provided k has characteristic zero. It should be noted, however, that there are many choices of parameters $(x_1, \ldots, x_N)$ which give us the Youssin ideal of F, unique only as an ideal.

At any rate, whether k has characteristic zero or not, it is often useful to consider a Youssin ideal of F, combined with a choice of parameters which gives it. A combination of such an ideal and parameters will be called *Youssin presentation* of F in R.

Let us go back to Example 2. In this case, a Youssin presentation of F is as follows:

$$F \in (x)^2 \oplus yzw(y, z)^2.$$

In other words, $(x, y, z, w) = (x_1, x_2, x_3, x_4)$, $y_1 = x_1$, $y_2 = (x_2, x_3)$, $c_1 = c_2 = 2$, $\mathrm{mon}_1 = yzw$, and the Youssin measure is $(2, -1, 3, 2, -2)$.

It should be noted that $(x)^2 + ywz(y, w)$ and $(x)^2 + zwy(z, w)$ are Youssin polyhedral ideals, satisfying the conditions (Y,1) and (Y,2), but are not Youssin ideal of F because (Y,3) is not fulfilled. This distinguishes the center (1) above the other centers (2) and (3) of Example 2, in light of the following theorem of B. Youssin.

To explain this theorem of Youssin, we go back to the general expression of Youssin polyhedral ideal given before the conditions (Y,1)-(Y,2).

Note that n_r could be less than N, i.e., the y's may not exhaust the x's. At any rate, an expression of a Youssin polynomial ideal being given as before, we define a *Youssin center* to be as follows:

Case 1. Assume c_r (the last c) > 0. Then Youssin center is defined by

$$y_1 = y_2 = \cdots = y_r = 0.$$

Case 2. Assume $c_r = 0$. In this case, we use a different tactic (i.e., blow-up a largest "permissible" center in a certain sense) and Youssin center may not be unique. Namely, write

$$\mathrm{mon}_{r-1}(y_r') = \prod_j x_j^{a_j}.$$

To each set of j's, say S, we write $|S|$ for its cardinality and

$$a(S) = \sum_{j \in S} a_j.$$

Let $T = \{S \mid a(S) \geq c_{r-1}\}$, $t = \min\{|S| \mid S \in T\}$ and finally $T' = \{S \in T \mid |S| = t\}$. Now a Youssin center is any one defined by picking any $S \in T'$ and letting

$$y_1 = \cdots = y_{r-1} = 0 \quad \text{and} \quad x_j = 0, \qquad \forall j \in S.$$

Definition. Given a Youssin polyhedral ideal as above, we associate with it a *modified Youssin measure*, that is

$$(c_1, -n_1, c_2, -n_2, \ldots, c_r, -n_r) \quad \text{if} \quad c_r \neq 0$$

$$(c_1, -n_1, c_2, -n_2, \ldots, |\mathrm{mon}_{r-1}|) \quad \text{if} \quad c_r \neq 0.$$

It is a theorem of B. Youssin, similar to the algorithm proposed a few years earlier by Mark Spivakovsky, that modified Youssin measure decreases each time we blow up a Youssin center, again provided that the characteristic of k is zero.

We shall illustrate the above discussion by the following example:

Example 3. Let a hypersurface X be defined by the following equation over a characteristic 0 field,

$$F = x^4 - yzw(yz + y^2w + z^2w) = 0.$$

It is easy to see that the singular locus consists of 3 lines as in Example 1.

(1) $x = y = z = 0$
(2) $x = y = w = 0$
(3) $x = z = w = 0$.

Using the constant multiplicity condition implied in the definition of permissibility it is not hard to see that the origin is the only permissible center. On the other hand let us consider the possible *Youssin presentations* of F. In this case, we have

$$\text{(I)} \qquad F \in (x)^4 \oplus yzw(y, z, w)^2$$

or

$$\text{(II)} \qquad F \in (x)^4 \oplus yzw(y, z)^2.$$

In fact (I) is Youssin and the Youssin center is $x = y = z = w = 0$ (the origin). It checks with our previous discussion. □

3. The characteristic p phenomenon. There are other techniques useful in studying singularities, when the base field k has characteristic zero. One of the most important in regards to desingularization is the following existence theorem.

THEOREM. *Say a hypersurface F has multiplicity m at a point ξ. Then there exists a smooth hypersurface L through ξ, which contains all the infinitely near m-fold points of F at ξ, i.e., the m-fold points, corresponding to ξ, on the strict transforms of F after all possible sequence of permissible transformations.*

This is in fact the key to the inductive proof of resolution of singularities in characteristic zero.

The main difficulty, at this point, of resolution in positive characteristics is that such L does not exist in general. Let us examine the following example.

Example 4. Let the ground field be of characteristic 2. Let a hypersurface X be defined by the following equation

$$F = y^2 + x_1^3 x_2 + x_2^3 x_3 + x_3^7 x_1 = 0.$$

This is weighted homogeneous with weight (15, 19, 7, 32) in (x_1, x_2, x_3, y). Its singular locus (i.e. double locus) is parametrized by t as

$$x_1 = t^{15}, \qquad x_2 = t^{19}, \qquad x_3 = t^7, \qquad y = t^{32}.$$

There are at least two simple arguments to show that the above theorem is false in this case. Firstly, any smooth hypersurface, through the origin, containing the double locus must be defined by

$$y + (\text{terms of weight} \geq 32 \text{ in } t) = 0.$$

So if this should contain all the infinitely near double points, the transform of $y = 0$ should also contain the double locus of the quadratic transform of F. But, in the quadratic transform with

$$\begin{cases} \text{coordinate} \left(\dfrac{x_1}{x_3}, \dfrac{x_2}{x_3}, x_3, \dfrac{y}{x_3}\right) = (x_1^*, x_2^*, x_3^*, y^*) \\ \text{equation } F' = y^{*2} + (x_1^{*3} x_2^* + x_2^{*3} + x_3^{*4} x_1^*) x_3^{*2} = 0 \end{cases}$$

the point (1, 1, 1, 1) is a double point, which is not in the strict transform of $y = 0$. Or secondly, by the peculiar property of the weights, it is easy to see that there is no smooth hypersurface L containing the double locus.

Note also that in the case of characteristic zero, the same equation defines a hypersurface with an isolated singularity at the origin. The hypersurface $y = 0$ serves as the smooth hypersurface L stated in the previous theorem. □

Moreover Youssin blowing-up does not in general make Youssin measure decrease in the positive characteristic cases. In other words, there are possibility of "jumps" in Youssin measure. This phenomenon is the central theme of recent investigations of T. T. Moh at Purdue University.

To begin with let us examine the following example.

Example 5. Let a surface X be defined by the following equation over a field of characteristic $p > 0$,

$$F = x^p + y^{p-1}z(z^p - y^p + y^{3p+1}) = 0.$$

Note that the following is a Youssin presentation

$$F \in (x)^p \oplus y^{p-1}z(y, z)^p$$

with modified Youssin measure

$$(p, -1, p, -2)$$

and Youssin center

$$(x, y, z).$$

But, in the quadratic transform with

$$\begin{cases} \text{coordinate } \left(\dfrac{x}{y}, y, \dfrac{z}{y} - 1\right) = (\bar{x}, \bar{y}, \bar{z}) \\ \text{equation } F' = \bar{x}^p + \bar{y}^p(\bar{z} + 1)((\bar{z} + 1)^p - 1 + \bar{y}^{2p+1}) \\ \qquad = \bar{x}^p + \bar{y}^p(\bar{y}^p + \bar{z}^{p+1} + \bar{y}^{2p+1} + \bar{y}^{2p+1}\bar{z}) \\ \qquad = (\bar{x} + \overline{yz})p + \bar{y}^p(\bar{z}^{p+1} + \bar{y}^{2p+1} + \bar{y}^{2p+1}\bar{z}) \\ \qquad = x_1^p + y_1^p(z_1^{p+1} + y_1^{2p+1} + y_1^{2p+1}z_1) = 0 \end{cases}$$

where $(\bar{x} + \overline{yz}, \bar{y}, \bar{z}) = (x_1, y_1, z_1)$. Note that

$$F' \in (x_1)^p \oplus y_1^p(y_1, z_1)^{p+1}$$

and modified Youssin measure is

$$(p, -1, p + 1, -2).$$

Indeed there is a jump of modified Youssin measures after a quadratic transform. □

Although there might be a jump, is there a numerical bound for the increments? Let us mention an interesting theorem of T. T. Moh (RIMS, 1987) in its simplest form.

THEOREM. *Let a hypersurface $X \subset S^{n+1}$ be defined by the following equation over a field k of characteristic $p > 0$,*

$$F = y^{p^e} + \left(\prod_{i=1}^{n} x_i^{m_i}\right)(f_d + f_{d+1} + \cdots) = 0$$

satisfying the conditions (*)

$$(*) \qquad \left(\prod_{i=1}^{n} x_i^{m_i}\right) f_d \notin k[x_1^{p^e}, \ldots, x_n^{p^e}].$$

Then after blowing up a permissible center, the transform of X is either with a multiplicity $<p^e$ or of the following form

$$F' = \bar{y}^{p^e} + \left(\prod_{i=1}^{n} \bar{x}_i^{\bar{m}_i}\right) f'(x_1, \ldots, x_n)$$

satisfying the condition (*) *with*

$$\operatorname{ord} f' \leq d + p^{e-1}.$$

Furthermore all successive quadratic transforms will not exceed the bound $d + p^{e-1}$. □

For simplicity, let us consider the case of $e = 1$. Then, a pre-Moh measure which is a characteristic p analogue of Youssin measure in characteristic 0 can be defined for surface singularities as follows

$$(m, d + \Delta, nw)$$

where m = multiplicity of X at the origin. To avoid complications we will only define the above numerals for the simplest equation

$$F = x^p + y^n z^\ell f(y, z) = 0$$

satisfying the (*) condition in the above theorem as

$$d = \operatorname{ord} f(y, z)$$

$$\Delta = \begin{cases} 0 & \text{if } \ell = 0, \quad\quad\quad\quad\quad e \leq d + 1, \quad d \geq 2 \\ 1 + \epsilon & \text{if } \ell \neq 0, \quad \text{or} \quad \ell = 0, \quad e > d + 1, \quad d \geq 2 \\ 0 & \text{if } 1 \geq d \geq 0 \end{cases}$$

where $e = \operatorname{ord} f(0, z)$, ϵ is a small rational number and

$$nw = \text{the first Newton weight of } f(y, z)$$
$$= -\frac{1}{s}, \qquad s \text{ the first Newton slope of } f(y, z).$$

We define Moh measure as

$$\min_{n,\ell} \max_{x,y,z}(m, d + \Delta, nw).$$

As in the treatment of *Youssin ideal* we may select n, ℓ, x, y, z, so that pre-Moh measure is *Moh measure* and get *Moh presentation* of F. It is a theorem that Moh measure decreases each time we blow up the origin if $d > 1$ in the surface case. If $d \leq 1$ in the surface case we may blow up curves or points to decrease Moh measure.

Generally speaking, a lot has to be done in characteristic p case.

4. Appendix by T. T. Moh. One of the main techniques employed by Zariski in establishing the Uniformization Theorem for characteristic zero is the "killing of f_1." Namely in the following equation

$$F = z^m + \Sigma f_i(x_1, \ldots, x_n)z^{m-i} = 0 \quad \text{with } m = \operatorname{ord} F,$$

a translation of the form

$$z \to z - \frac{1}{m} f_1(x_1, \ldots, x_n)$$

will annihilate the coefficient $f_1(x_1, \ldots, x_n)$. So we may assume $f_1(x_1, \ldots, x_n) = 0$ to begin with.

We consider a valuation ν, throughout this appendix, in the power series $R = k[[z, x_1, \ldots, x_n]]$, which is finite in R and $\nu(M) > 0$ where M is the maximal ideal of R. Then if $\nu(z) = \min \nu(M)$, and for some $\lambda \neq 0$, $\nu(z - \lambda x_j) > \nu(x_j) = \min \nu(M)$, let $z_1 = z - \lambda x_j$, then we have

$$F = z^m + 0z^{m-1} + \sum_{i=2}^{m} f_i(x_1, \ldots, x_n)z^{m-i}$$

$$= z_1^m + m\lambda x_j z_1^{m-1} + \sum_{i=2}^{m} f_i'(x_1, \ldots, x_n)z_1^{m-i}$$

with $\nu(z_1) > \nu(x_j) = \min \nu(M)$. Clearly the proper transform of F is

$$F' = z_1^m + m\lambda z_1^{m-1} + \cdots$$

with multiplicity $<m$ at the origin.

The preceding technique of Zariski shows at once that, at least in the case of residuely rational valuation, we may simply assume $\nu(z) > \min \nu(M) = \nu(x_j)$ and the proper transform of F is

$$F' = z^m + 0z^{m-1} + \sum_{i=2}^{m} f_i' x_j^{m_i - i} z^{m-i}$$

where $m_i = \operatorname{ord} f_i$, $f_i' =$ the proper transform of f_i with respect to the blow-up restricted to $(x_1, \ldots, x_n)$-space. Thus a reduction of dimension is possible.

The above technique can be generalized to characteristic p as follows.

Let the defining equation $F = 0$ of an hypersurface X be given as

$$F = z^{rp^e} + \sum_{i=1}^{rp^e} \left(\prod_{j=1}^{n} x_j^{\alpha_j} \right)^i f_i(x_1, \ldots, x_n) z^{rp^e - i} = 0 \tag{1}$$

where $m = rp^e$, $p \nmid r$, and α_j may be nonnegative rationals. Following Hironaka, we define the Newton diagram $ND(F; z, x_1, \ldots, x_n) =$ all lattice points corresponding to the monomials in F and their nonnegative translations. In short, it will be denoted $ND(F)$. Let pr_z be the projection with center $(rp^e, 0, \ldots, 0)$ (corresponding to the monomial z^{rp^e}) to n dimensional affine space. Then we have the following.

Definition. The equation $F = 0$ of (1) is said to be z-cleaned if $\forall \sigma: z \to z + (\Pi x_j^{\alpha_j})h(x_1, \ldots, x_n)$ we always have $pr_z(ND(F)) \subset pr_z(ND(\sigma F))$.

We shall pause a moment and consider the following example.

Example 6. Let us assume $p \neq 2$. In the following equation for plane curves in terms of variables z, z_1, z_2

$$F = z^{2p} + 2(x^{2p} + x^{3p})z^p + x^{6p+1}z + x^{4p} + 2x^{5p}$$

$$= (z + x^2 + x^3)^{2p} + x^{6p+1}(z + x^2 + x^3) - x^{6p} + \cdots$$

$$= z_1^{2p} + x^{6p+1}z_1 - x^{6p} + \cdots$$

$$= (z + x^2)^{2p} + 2x^{3p}(z + x^2)^p + x^{6p+1}(z + x^2) - x^{6p+3}$$

$$= z_2^{2p} + 2x^{3p}z_2^p + x^{6p+1}z_2 - x^{6p+3}$$

we have

$$pr_z(ND(F)) = \{a : a \geq 4p\}$$

$$pr_{z_1}(ND(F)) = \{a : a \geq 6p\}$$

$$pr_{z_2}(ND(F)) = \{a : a \geq 6p\}.$$

In fact, the equation $F = 0$ is not z-cleaned while z_1-cleaned and z_2-cleaned. In this example the variables z_1 and z_2 are considered to be superior to z. As illustrated by the following discussions, we shall point out that for this example the variable z_1 is even better than the variable z_2. □

In general we have the following theorem (cf. [MZ]).

THEOREM. *Given a hypersurface defined by equation* (1). *Then there exists a variable* $z_1 = z + (\Pi x_j^{\alpha_j})h(x_1, \ldots, x_n)$ *for a suitable* $h(x_1, \ldots, x_n)$ *such that the equation* (1) *is* z_1*-cleaned.*

As indicated by Zariski's "killing of f_1," the term $f_{p^e}(x_1, \ldots, x_n)$ is of interest. Note that in characteristic zero, we take $p^e = 1$. In the positive characteristic case we shall remove all p^e-th power terms from the weighted leading forms of $(\Pi\, x_j^{\alpha_j})^{p^e} f_{p^e}(x_1, \ldots, x_n)$. Thus we define

Definition. The equation $F = 0$ is said to be maximal z-cleaned iff (1) $F = 0$ is z-cleaned and (2) p^e-th power terms in the weighted leading forms of $(\Pi\, x_j^{\alpha_j})^{p^e} f_{p^e}(x_1, \ldots, x_n)$ are all removed.

We have the following generalization of Zariski's "killing of f_1" (cf. [MZL]).

THEOREM. *Given a hypersurface defined by equation* (1). *Then there exists a variable* $z_1 = z + (\Pi\, x_j^{\alpha_j})h(x_1, \ldots, x_n)$ *for a suitable* $h(x_1, \ldots, x_n)$ *such that the equation* (1) *is maximal* z_1*-cleaned.*

Given an equation of the form $F = 0$ of (1), there is a question of choosing the variables x_j, $1 \leq j \leq n$, suitable for our purpose. We now discuss this question.

First of all, let us assume that $F = 0$ is maximal z-cleaned. Then, among the variables x_j in the expression of (1), those with $\alpha_j \neq 0$ are said to be *fixed* while the others with $\alpha_j = 0$ are said to be *free*. Let $V(1)$, whose dimension is denoted by $\tau(1)$, following Hironaka be the k-vector space spanned by those x_j which are either fixed or indispensable (i.e., it actually appears) in the polynomial expression of the homogeneous part of degree $d'i$ of f_i for at least one i, $1 \leq i \leq rp^e$, where $d' = \min_i\{\operatorname{ord} f_i/i\}$.

We then consider *free-variable transformations* in x, i.e., coordinate transformations in the variables x_j, $1 \leq i \leq n$, which leave all the fixed variables x_j's unchanged. If we apply such a transformation in x, the equation $F = 0$ may no longer z-cleaned, or maximal, even if it was originally maximal z-cleaned. But, after the transformation, we can always make it maximal z-cleaned by applying maximal z-cleaning according to the above theorem.

Definition. Assume that $F = 0$ of (1) is maximal z-cleaned. We say that $F = 0$ is x-adjusted if $\tau(1) = \dim V(1)$ cannot be made smaller by any free-variable transformation in x followed by maximal z-cleaning.

THEOREM. *Suppose the equation* (1) *is maximal z-cleaned and x-adjusted. If for some valuation* v, $v(z) = \min v(M)$ *and for some* $\lambda \neq 0$ $v(z - \lambda x_j) > v(x_j) = \min v(M)$, *then the proper transform* F' *of* F *after blow-up is with multiplicity* $< rp^e =$ *the multiplicity of* F *at the origin.*

One of the complications in characteristic p is that the maximal clean variable z is not stable under blow-ups as illustrated in Example 5. The following theorem establishes a partial stability for maximal clean variables.

THEOREM. *Let a permissible center be defined by* $I = (z, x_{i_1}, \ldots, x_{i_s})$. *If for some valuation* v, $v(x_{i_1}) < v(z)$ *and* $v(x_{i_1}) < v(x_{i_j})$ $\forall j \neq 1$. *Then after blowing up* I, *the proper transform* F' *of* F *is either with smaller multiplicity or maximal* z/x_{i_1}*-cleaned.*

It seems to us that one of the tasks in studying characteristic p resolution is to observe the behavior of resolution data expressed in maximal clean variables under blowing ups.

Given an equation (1) we can sometimes continue the process of maximal cleaning and define a sequence of maximal cleaned variables $(z_1, z_2, \ldots, z_t)$ in an inductive way. However, instead of messing up the beautiful lecture of Hironaka by going to the full details of the technical definitions, we shall be satisfied to discuss maximal (z_1, z_2)-clean only (i.e., the case of $t = 2$). Let us assume that the equation F takes the following form $(m = d_1)$.

$$F = z_1^{d_1} + \sum_{i_1=1}^{d_1} (\Pi\, x_j^{\alpha_j})^{i_1} f_{i_1} z_1^{d_1 - i_1}. \tag{2}$$

Assume F is maximal z_1-cleaned. Let

$$d_2 = \min_{i_1} \left\{ \frac{\operatorname{ord} f_{i_1}}{i_1} \right\}, \qquad \text{previously denoted by } d'.$$

$$f_{i_1} = \delta_{i_1} z_2^{d_2 i_1} + \sum_{i_2=1}^{d_2 i_1} (\Pi\, x_j^{\beta_j})^{i_2} f_{i_1 i_2} z_2^{d_2 i_1 - i_2}$$

where $f_{i_1 i_2}$ involves only x_j. Assume δ_{i_1} is a unit for some i_1. Then we may consider (z_1, z_2)-clean as follows.

Definition. The equation $F = 0$ of (2), under the above conditions, is said to be (z_1, z_2)-cleaned iff it is maximal z_1-cleaned and $\forall$ automorphisms σ of the following form

$$\sigma \begin{cases} z_1 \to z_1 + (\Sigma\, x_j^{\alpha_j}) h_1(x_1, \ldots, x_n, z_2) \\ z_2 \to z_2 + (\Pi\, x_j^{\beta_j}) h_2(x_1, \ldots, x_n) \end{cases}$$

we have $pr_{z_2}(pr_{z_1}(ND(F))) \subset pr_{z_2}(pr_{z_1}(ND(\sigma F)))$ where pr_{z_2} is the projection with center $(d_1 d_2, 0, \ldots, 0)$ (corresponding to the monomial $z_2^{d_2 i_1}$ with δ_{i_1} a unit as required above) to n dimensional affine space (corresponding to the variables x_j).

Similarly we may define maximal (z_1, z_2)-clean as

Definition. The equation $F = 0$ is said to be maximal (z_1, z_2)-cleaned iff (1) $F = 0$ is (z_1, z_2)-cleaned and (2) for some i_1, δ_{i_1} is a unit, $p^{e'} | d_2 i_1$ and $p^{e'+1} \nmid d_2 i_1$, $p^{e'}$-th power terms in the weighted leading forms of $(\Pi x_j^{\alpha_{j2}})^{p^{e'}} \cdot f_{i_1 p^{e'}}$ are all removed.

When $F = 0$ of (2) is maximal (z_1, z_2)-cleaned, we can again define the notion of x-adjustedness as follows. In this case, the *free* variables are those x_j such that $\alpha_j = 0$ and $\beta_j = 0$. The other x_j with either $\alpha_j \neq 0$ or $\beta_j \neq 0$ are said to be *fixed*. Let $V(2)$, with $\tau(2) = \dim V(2)$, be the k-vector space spanned by those x_j fixed together with those x_j indispensable in the homogeneous part of degree $d_3 i_2$ of f_{i_1, i_2} for at least one (i_1, i_2), $1 \leq i_1 \leq d_1$ and $1 \leq i_2 \leq d_2 i_1$, where $d_3 = \min_{i_1, i_2}\{\text{ord } f_{i,i_2}/i_2\}$. Once again, we can speak of free-variable transformations in x and maximal (z_1, z_2)-cleaning.

Definition. Assume that $F = 0$ of (2) is maximal (z_1, z_2)-cleaned. We say that $F = 0$ is x-adjusted if $\tau(2)$ cannot be made smaller by any free-variable transformation followed by maximal (z_1, z_2)-cleaning.

Then we have the following theorem (cf [MZ]).

THEOREM. *Suppose: the equation $F = 0$ is maximal (z_1, z_2)-cleaned, x-adjusted and $d_2 > 1/p$. If for some valuation v, $v(z_1) = \min v(M)$ or $v(z_2) = \min v(M)$ and for some λ_i, $v(z_i - \lambda_i x_j) > v(x_j) = \min v(M)$, then the proper transform F' of F after blow up is either with smaller multiplicity d_1 or with nonincreasing d_1 and d_2.*

Note that as indicated by Example 5, the numeral d_2 may increase under blow-ups. The above theorem illustrates that under special conditions there might be some stability phenomena for numeral d_2.

H. HIRONAKA
DEPARTMENT OF MATHEMATICS
HARVARD UNIVERSITY
1 OXFORD
CAMBRIDGE, MA 02138

REFERENCES

[A] S. S. Abhyankar, *Local uniformization on algebraic surfaces over ground fields of characteristic $p \neq 0$*, Annals of Mathematics, **63** (1956), pp. 491-526.

[H1] H. Hironaka, *Resolution of singularities of an algebraic variety over a field of characteristic zero*, Annals of Math, **79** (1964), pp. 109-326.

[H2] ______, *Characteristic polyhedra of singularities*, J. Math. Kyoto Univ., **7** (1964), pp. 251-293.

[H3] ______, *Certain numerical characters of singularities*, J. Math. Kyoto Univ., **10** (1970), pp. 151-187.

[H4] ______, *Introduction to the theory of infinitely near points*, Publications del Instituto "Jorge Juan" de Mathematicas, Univ. of Madrid, 1974.

[H5] ______, *Idealistic exponents of singularity*, Algebraic Geometry, The Johns Hopkins Centennial Lectures, Johns Hopkins Univ. Press (1977).

[M] T. T. Moh, *On a stability theorem for local uniformization in characteristic p*, Journal of RIMS, **23** No. 6 (Nov. 1987).

[MZ] ______ and Y. T. Zhang, *On a Local Uniformization Theorem for 3-fold in characteristic $p > 0$*, Lecture Note, pp. 300+, In preparation.

[S] M. Spivakovsky, *A counterexample to the theorem of Beppo Levi in three dimensions*, to appear in Investiones Math..

[Y] Boris Youssin, *Newton Polyhedra without coordinates*, Dissertation, Harvard 1988.

[Z] O. Zariski, *Local uniformization on algebraic varieties*, Annals of Math, **41** (1940), pp. 852-896.

WHAT COMES NEXT AFTER $\mathbf{Q}(\mu_{\ell^\infty})$?

By YASUTAKA IHARA

This is, basically, a summary of my talk at the JAMI Inaugural Conference held at the Johns Hopkins University (May, 1988). A few results obtained later are also incorporated. The main theme is (the kernel of) the Galois action on the pro-ℓ fundamental groups of $\mathbf{P}^1 - \{0, 1, \infty\}$ and other related varieties. A substantial portion is devoted to (i) a brief account of a joint work with G. W. Anderson (Section 2C), and (ii) work influenced by P. Deligne (Section 3).

Details will appear in [1, 2, 9, 10].

1. The most well-known type of Galois representations is the one arising from Galois actions on etale cohomology groups of algebraic varieties. If we take the rational number field $\mathbf{Q}$ as the base field, it looks like

$$(1)_\ell \qquad G_{\mathbf{Q}} = \mathrm{Gal}(\bar{\mathbf{Q}}/\mathbf{Q}) \to \mathrm{Aut}\, H^i_{et}(X \otimes_{\mathbf{Q}} \bar{\mathbf{Q}}, \mathbf{Q}_\ell)$$

$(i = 0, 1, 2, \ldots)$, where $\bar{\mathbf{Q}}$ is an algebraic closure of $\mathbf{Q}$, X is an algebraic variety over $\mathbf{Q}$, ℓ is a prime number, $\mathbf{Q}_\ell$ is the ℓ-adic field, and H^i_{et} is the i-dimensional etale cohomology group. As is well-known, there is a series of very general and fascinating conjectures as well as deep results in some special cases. They are mainly on:

(i) links connecting arithmetic-geometric properties of X with analytic properties of the L-function associated to $(1)_\ell$, or with arithmetic properties of its special values;

(ii) something hidden behind the ℓ-adic and the Hodge theory (motives, K-theory, . . .).

Among the representations $(1)_\ell$ for various X, the very basic is the one for $X = \mathbf{G}_m = \mathbf{P}^1 - \{0, \infty\}$, $i = 1$. It gives the (inverse of the) ℓ-cyclotomic character χ_ℓ: $G_{\mathbf{Q}} \to Z_\ell^\times$ describing the $G_{\mathbf{Q}}$-action on the group μ_{ℓ^∞} of

---

Manuscript received November 4, 1988; revised November 9, 1988.

ℓ-powerth roots of unity. The powers of χ_ℓ appear everywhere in the general case of $(1)_\ell$, e.g. as the determinant, and in Hodge-Tate theory.

Now there is another type of Galois representations, which has a shorter history of research. Grothendieck is (again) probably the first person who has *explicitly* proposed to study these representations [5], although some other works (e.g. [3][7]) have originated independently. It looks like

$$(2)^* \qquad \phi_X{}^*: G_{\mathbf{Q}} \to \operatorname{Aut} \hat{\pi}_1(X \otimes_{\mathbf{Q}} \bar{\mathbf{Q}}, *),$$

or

$$(2) \qquad \phi_X: G_{\mathbf{Q}} \to \operatorname{Out} \hat{\pi}_1(X \otimes_{\mathbf{Q}} \bar{\mathbf{Q}}),$$

where X is again an algebraic variety over $\mathbf{Q}$, assumed to be smooth and absolutely irreducible, $\hat{\pi}_1(\ )$ is the algebraic (profinite) fundamental group, $*$ a $\mathbf{Q}$-rational point of X (or some substitute), and Out the outer automorphism group. These representations (2)*, (2) are induced directly from the exact sequence

$$1 \to \hat{\pi}_1(X \otimes_{\mathbf{Q}} \bar{\mathbf{Q}}, *) \to \hat{\pi}_1(X, *) \overset{s^*}{\leftrightarrows} G_{\mathbf{Q}} \to 1,$$

where s^* is a section corresponding to $*$. Roughly speaking, the representation (2) describes the $G_{\mathbf{Q}}$-action on the "field of definition" of finite etale coverings of $X \otimes_{\mathbf{Q}} \bar{\mathbf{Q}}$. In [3], V. G. Belyĭ proved, among other things, that (2) is *injective* already for $X = \mathbf{P}^1 - \{0, 1, \infty\}$. This means that $\bar{\mathbf{Q}}$ is the "smallest common field of definition" for all finite etale coverings of $\mathbf{P}^1 - \{0, 1, \infty\}$. He proved this by showing that for any algebraic number $\alpha \in \bar{\mathbf{Q}}$ with $\alpha \neq 0, 1$, there exists a finite branched covering $f: \mathbf{P}^1 \to \mathbf{P}^1$ unramified outside $0, 1, \infty$, such that $f^{-1}(\{0, 1, \infty\})$ contains 4 distinct points having α as their cross-ratio. There are some interesting basic problems related to the representations (2)*, (2) (e.g. Grothendieck [5]), and down-to-earth effective applications to construction of finite Galois extensions over $\mathbf{Q}$ with given Galois group (V. G. Belyĭ, M. D. Fried, J. G. Thompson, W. Feit, B. H. Matzat, etc.). But here, we shall restrict our attention to the representation obtained from ϕ_X by the passage to the maximal pro-ℓ quotient $\pi_1^{\text{pro-}\ell}(X \otimes_{\mathbf{Q}} \bar{\mathbf{Q}})$ of $\pi_1(X \otimes_{\mathbf{Q}} \bar{\mathbf{Q}})$:

$$(2)_\ell \qquad \phi_X{}^{(\ell)}: G_{\mathbf{Q}} \to \operatorname{Out} \pi_1^{\text{pro-}\ell}(X \otimes_{\mathbf{Q}} \bar{\mathbf{Q}}).$$

This is a generalization of $(1)_\ell$ for $i = 1$, in the sense that $(1)_\ell$ for $i = 1$ is the (dual of the) representation obtained from $(2)_\ell$ by abelianization of the representation space (group). Deligne expects that (at least) some of (i) (ii) for $(1)_\ell$ can be generalized to $(2)_\ell$. In other words, a theory essentially independent of ℓ is expected. On the other hand, this has also some similarity to Iwasawa theory (in this sense, each ℓ should be so individual as to be denoted by "p"). Whatever point of view one may choose, it should be emphasized that the representation $(2)_\ell$ is generally very big, very nonabelian and nonsemisimple (in some sense). Now, for one who is less of a geometer than a number theorist, his (or her) interest, related to links connecting geometric properties of X with various properties of the representation $(2)_\ell$, may easily be replaced by a search for something canonical and absolutely number theoretic over $\mathbf{Q}$. Thus we are especially interested in the study of the specific case $X = \mathbf{P}^1 - \{0, 1, \infty\}$, because this case may be as basic among the representation $(2)_\ell$ as χ_ℓ is among the $(1)_\ell$. Note that in this case $\pi_1^{\text{pro-}\ell}(X \otimes_{\mathbf{Q}} \bar{\mathbf{Q}})$ is a free pro-ℓ group of rank 2. Unlike the profinite case, $\phi_X{}^{(\ell)}$ has a big kernel, and the Galois extension $\Omega = \Omega^{(\ell)}$ over $\mathbf{Q}$ corresponding to the kernel is of arithmetic interest relative to the prime ℓ.

2. Now let $\mathbf{P}^1$ be the projective t-line, $X = X_1 = \mathbf{P}^1 - \{0, 1, \infty\}$, and write $\pi_1 = \pi_1^{\text{pro-}\ell}(X \otimes_{\mathbf{Q}} \bar{\mathbf{Q}})$, $\phi = \phi_X{}^{(\ell)}$. If $M = M^{(\ell)}$ is a maximal pro-ℓ extension over $\bar{\mathbf{Q}}(t)$ unramified outside 0, 1, ∞, then π_1 is isomorphic to $\text{Gal}(M/\bar{\mathbf{Q}}(t))$. Let $\Omega = \Omega^{(\ell)}$ denote the Galois extension of $\mathbf{Q}$ corresponding to the kernel of ϕ. Then by the definition of Ω and by the triviality of the center of π_1 it follows directly that Ω is the minimal subfield of $\bar{\mathbf{Q}}$ with which M has a *Galois Ω-model* M_Ω, i.e., a Galois extension $M_\Omega/\Omega(t)$ in M such that $M_\Omega \cap \bar{\mathbf{Q}} = \Omega$ and $M_\Omega.\bar{\mathbf{Q}} = M$. Such an M_Ω is unique. Now we shall briefly review the main known results related to the minimal common field of definition Ω and the minimal model field M_Ω.

(A) The field Ω is an infinite pro-ℓ extension of $\mathbf{Q}(\mu_{\ell^\infty})$ unramified outside ℓ [4a] [7]. It is not known whether Ω is maximal under these conditions.

(B) Using the lower central series $\{\pi_1(m)\}$ of π_1 starting with $\pi_1 = \pi_1(1)$, we can equip the extension $\Omega/\mathbf{Q}$ with the following natural filtrations. Define $\mathbf{Q}(m)$ $(m \geqslant 1)$ to be the Galois extension of $\mathbf{Q}$ corresponding to the kernel of the quotient representation

$$\phi^{(m)}: G_{\mathbf{Q}} \to \text{Out}\, \pi_1 \to \text{Out}(\pi_1/\pi_1(m + 1)).$$

Then

$$\mathbf{Q}(\mu_{\ell^\infty}) = \mathbf{Q}(1) \subseteq \mathbf{Q}(2) \subseteq \cdots \subseteq \cup \mathbf{Q}(m) = \Omega,$$

and this ascending tower of Galois extensions satisfies the following properties. For each $m \geqslant 1$, $\mathcal{G}^{(m)} = \mathrm{Gal}(\mathbf{Q}(m+1)/\mathbf{Q}(m))$ is isomorphic to $Z_\ell^{r_m}$ for some nonnegative integer r_m, $\mathcal{G}^{(m)}$ is central in $\mathrm{Gal}(\mathbf{Q}(m+1)/\mathbf{Q}(1))$, and as a $\mathrm{Gal}(\mathbf{Q}(1)/\mathbf{Q})$-module, has the Tate twist of order m. Moreover,

$$\mathcal{G} = \bigoplus_{m \geqslant 1} \mathcal{G}^{(m)}$$

is naturally equipped with the structure of graded Lie algebra over Z_ℓ. For small m, the value of the rank r_m is given by $r_1 = r_2 = r_4 = r_6 = 0$, $r_3 = r_5 = r_7 = 1$. Deligne conjectures that the structure of the Lie algebra $\mathcal{G} \otimes \mathbf{Q}_\ell$ is "independent of ℓ" and, moreover, that there exist $x_m \in \mathcal{G}^{(m)}$ for each odd integer $m \geqslant 3$, such that $\mathcal{G} \otimes \mathbf{Q}_\ell$ is generated by $x_3, x_5, \ldots$ as Lie algebra over $\mathbf{Q}_\ell$. It is known that $\mathcal{G}^{(m)}$, for m odd $\geqslant 3$, has a quotient isomorphic to Z_ℓ arising from some *abelian* extension of $\mathbf{Q}(1)$ (with Tate twist m), studied by Soulé [11], Deligne [4b] (cf. also [6] [9]). As to "how nonabelian $\mathcal{G}$ is," we have [9]:

$$[\mathcal{G}^{(m)}, \mathcal{G}^{(n)}] \neq 0,$$

and, in fact, moreover,

$$[\mathcal{G}^{(m)}[\mathcal{G}^{(m)}, \mathcal{G}^{(n)}]] \neq 0$$

for $m > n \geqslant 3$, m, n odd. Can one determine the structure of $\mathcal{G}$ in the near future? Questions related to $\mathcal{G} \otimes \mathbf{F}_\ell$ seem to be quite different in nature from those related to $\mathcal{G} \otimes \mathbf{Q}_\ell$. (As different as p and ℓ.)

(C) Joint work with G. W. Anderson, [1], [2].

This actually deals with the case of $\mathbf{P}^1 - S_0$ for more general finite set of points S_0, but we shall restrict our attention to the case $S_0 = \{0, 1, \infty\}$. This work consists of:

(i) Explicit generation of the Galois extensions $\Omega/\mathbf{Q}$ and $M_\Omega/\mathbf{Q}(t)$ in terms of "higher circular ℓ-units" [1], and

(ii) Comparison of the Galois action; on π_1 and on the group of higher circular ℓ-units [2].

We begin by defining the group of higher circular ℓ-units. For any finite subset $S_0 \subseteq \mathbf{P}^1(k)$ (k: any algebraically closed field) with $S_0 \supseteq \{0, 1, \infty\}$, and a prime number ℓ, we associate a subgroup $E(S_0) = E^{(\ell)}(S_0)$ of $k^\times$ as follows. Call $\mathcal{S}_k$ the set of all finite subsets $S \subseteq \mathbf{P}^1(k)$ such that $S \supseteq \{0, 1, \infty\}$. Define $\mathcal{S}(S_0)$ to be the smallest subset of $\mathcal{S}_k$ such that

(i) $S_0 \in \mathcal{S}(S_0)$;

(ii) if $S \in \mathcal{S}(S_0)$, a, b, c are three distinct points of S, and $T_{a,b,c}$ is the linear fractional transform of $\mathbf{P}^1$ which maps a, b, c to $0, 1, \infty$, respectively, then $T_{a,b,c}(S) \in \mathcal{S}(S_0)$;

(iii) if $S \in \mathcal{S}(S_0)$, then $S^{1/\ell} = \{s \in \mathbf{P}^1(k); s^\ell \in S\} \in \mathcal{S}(S_0)$.

Note that this definition is *constructive*, and that $\mathcal{S}(S_0)$ is independent of the choice of the ambient field k. Now the group $E(S_0)$ is, by definition, the subgroup of $k^\times$ generated by $s \in S - \{0, \infty\}$ for all $S \in \mathcal{S}(S_0)$.

For example, $E(\{0, 1, \infty\})$ is a subgroup of $\bar{\mathbf{Q}}^\times$ consisting only of ℓ-units, and contains all elements of the form ζ, $1 - \zeta$, $1 - (1 - \zeta)^{1/\ell}$, etc., where $\zeta \in \mu_{\ell^\infty}$, $\zeta \neq 1$. If t is a variable, then the elements of $E(\{0, 1, \infty, t\})$ are algebraic functions over $\mathbf{Q}(t)$, unramified outside $0, 1, \infty$, and are all contained in a pro-ℓ Galois extension over $\mathbf{Q}(\mu_\ell, t)$. This group $E(\{0, 1, \infty, t\})$ contains, in addition to elements of $E(\{0, 1, \infty\})$, such elements as $1 - t$, $1 - (1 - t)^{1/\ell}$, $1 - (1 - (1 - t)^{1/\ell})^{1/\ell}$, . . . etc.

THEOREM (Anderson-Ihara). (i) $\Omega = \mathbf{Q}(E(\{0, 1, \infty\}))$ [1]; (ii) $M_\Omega = \mathbf{Q}(E(\{0, 1, \infty, t\}))$ (*see Corollary in Section* 3).

As in the profinite case (cited above in Section 1), the key point in proving the theorem is to look at those branched coverings $f: \mathbf{P}^1 \to \mathbf{P}^1$ unramified outside $0, 1, \infty$, having Galois closure of ℓ-power degree. There are "sufficiently many" such f, and $E(\{0, 1, \infty\})$ is the group generated by the cross ratio of (various) 4 points from $f^{-1}(\{0, 1, \infty\})$ for various choices of f.

Remark. Recall that the smallest common field of definition has played a significant role in number theory; namely, the complex multiplication theory, generalizations by Shimura-Taniyama and further generalizations by Shimura. There, the relevant extension fields are (essentially) abelian, and the results and methods are deep. Here, the extensions are essentially nonabelian (and infinite), but the proof is essentially elementary in nature.

As a corollary, we obtain the $\mathbf{Q}(E(\{0, 1, \infty\}))$-rationality of ℓ-power division points of the Jacobian $\mathrm{Jac}(X)$ of various branched coverings f:

$X \to \mathbf{P}^1$ of ℓ-power degree [1] Section 3. They are, in general, *nonabelian* over $\mathbf{Q}(\mu_{\ell^\infty})$. Thus, one wants to compare the Galois actions on the Tate module $T_\ell(\mathrm{Jac}(X))$ and $E(\{0, 1, \infty\})$, or even more, those on π_1 and $E(\{0, 1, \infty\})$. Roughly speaking, the comparison is described in the following way. First, for each $\sigma \in G_{\mathbf{Q}}$, the outer action of σ on π_1 is appropriately "lifted" to a genuine action on π_1 (Belyĭ's lifting [3], Deligne's base point [4b]). The latter action is described by a two-variable (noncommutative) formal power series $\psi_\sigma(u_1, u_2)$ over Z_ℓ [8]. Knowing the coefficients of degree $\leqslant m$ in $\psi_\sigma(u_1, u_2)$ is equivalent to knowing the action of σ on $\pi_1/\pi_1(m+1)$. For each monomial $u_{(i)} = u_{i_m} \ldots u_{i_1}$ of degree m, we can express the coefficient of $u_{(i)}$ in $\psi_\sigma(u_1, u_2)$ module ℓ^n explicitly in terms of the action of σ on elements of $E(\{0, 1, \infty\})$ of "depth (m, ℓ^n)," i.e., those elements obtained by extracting ℓ^n-th roots m times, alternating with taking linear fractional transforms, starting from $\{0, 1, \infty\}$. Cf. [2] for details.

3. The study below was motivated by discussion with Deligne. Put $X_1 = X = \mathbf{P}^1 - \{0, 1, \infty\}$, and for each $n \geqslant 1$, consider the configuration space

$$X_n = F_{0,n}(X_1) = \{(x_1, \ldots, x_n) \in (X_1)^n;\ x_i \neq x_j \text{ if } i \neq j\}.$$

By investigating a special type of automorphisms of the pro-ℓ completion of the pure sphere braid groups, and using some properties of the braid monodromy, we obtain [10]:

THEOREM. $\mathrm{Ker}\,\phi_{X_n}{}^{(\ell)} = \mathrm{Ker}\,\phi_{X_1}{}^{(\ell)}$ $(n \geqslant 1)$.

The pure sphere braid group is relevant, because $X_n \approx F_{0,n+3}\mathbf{P}^1/PGL_2$ and $\pi_1(X_n) \cong P_{n+3}/C_2$. Here, P_n is the pure braid group of the 2-sphere with n strings and $C_2 = \pi_1(PGL_2(C)) \cong Z/2$ is its center.

The above theorem may be viewed as saying: every etale pro-ℓ covering of X_n has a canonical Ω-model, or negatively, as: one cannot obtain an extension larger than Ω by considering X_n instead of X_1. If one can prove that Ω is the maximal pro-ℓ extension of $\mathbf{Q}(\mu_{\ell^\infty})$ unramified outside ℓ, then this will also imply the above theorem, because such $\phi_{X_n}{}^{(\ell)}$ is obviously unramified outside ℓ. At any rate, this shows that $\phi_{X_1}{}^{(\ell)}$ is, to some extent, "universal."

Finally, let M_n denote the union of the function fields of all etale pro-ℓ coverings of $X_n \otimes \bar{\mathbf{Q}}$, considered as a Galois extension of the function field

$\bar{\mathbf{Q}}(t_1, \ldots, t_n)$ of $X_n \otimes \bar{\mathbf{Q}}$. Let $(M_n)_\Omega$ be its canonical Ω-model. Then the above theorem, combined with [1] gives

COROLLARY. $(M_n)_\Omega = \mathbf{Q}(E\{0, 1, \infty, t_1, \ldots, t_n\})$.

FACULTY OF SCIENCE
UNIVERSITY OF TOKYO

REFERENCES

[1] G. W. Anderson and Y. Ihara, Pro-ℓ branched coverings of $\mathbf{P}^1$ and higher circular ℓ-units, *Ann. of Math.* **128** (1988), 271-293.
[2] ———, ibid, Part II, in preparation.
[3] V. G. Belyĭ, On Galois extensions of a maximal cyclotomic field, *Izv. Akad. Nauk USSR* **43** (1979) 2; (*Math. USSR Izv.* **14** (1980) 2, 247-256).
[4] P. Deligne, (a) Letters to A. Grothendieck (Nov. 19, 1982, and an earlier one undated?) (b) to S. Bloch (Feb. 2, 1984).
[5] A. Grothendieck, Esquisse d'un programme, Mimeographed Note 1984.
[6] H. Ichimura and K. Sakaguchi, The non-vanishing of a certain Kummer character χ_m (after Soulé) and some related topics, *Adv. St. in Pure Math.* **12** (1987), 53-64.
[7] Y. Ihara, Profinite braid groups, Galois representations and complex multiplications, *Ann. of Math.* **123** (1986) 43-106.
[8] ———, On Galois representations arising from towers of coverings of $\mathbf{P}^1 - \{0, 1, \infty\}$, *Invent. Math.* **86** (1986), 427-459.
[9] ———, The Galois representation arising from $\mathbf{P}^1 - \{0, 1, \infty\}$ and Tate twists of even degree, to appear in the Proceedings of the Microprogram on "Galois groups over $\mathbf{Q}$" (March 1987), MSRI publications, Springer.
[10] ———, Automorphisms of pure sphere braid groups and Galois representations, to appear in Grothendieck Festschrift, Birkhäuser.
[11] C. Soulé, On higher p-adic regulators, Springer Lecture Notes in Math. **854** (1981), 372-401.

THE FLAG MANIFOLD OF KAC-MOODY LIE ALGEBRA

By M. Kashiwara

0. Introduction. In this paper, we shall construct the flag variety of a Kac-Moody Lie algebra as an infinite-dimensional scheme. There are several constructions by Kac-Peterson ([K-P]), Kazhdan-Lusztig ([K-L]), S. Kumar ([Ku]), O. Mathieu ([M]), P. Slodowy ([S]), J. Tits ([T]), but there the flag variety is understood as a union of finite-dimensional varieties.

We give here two methods of construction of the flag variety. For a Kac-Moody Lie algebra g, let $\hat{g}$ be the completion of g. The first construction is to realize the flag variety as a subscheme of Grass$(\hat{g})$, the Grassmann variety of $\hat{g}$. More precisely, taking the Borel subalgebra $b_- \subset \hat{g}$ and regarding this as a point of Grass$(\hat{g})$, we define the flag variety as its orbit by the infinitesimal action of $\hat{g}$ in Grass$(\hat{g})$.

The other construction is to realize the flag variety as G/B_-. Of course, in the Kac-Moody Lie algebra case, we cannot expect that there is a group scheme whose Lie algebra is g. But we can construct a scheme G on which g acts infinitesimally from the left and the right. Then we define the flag variety G/B_-, where B_- is the Borel subgroup. More precisely, we consider the ring of regular functions as in [K-P]. Then its spectrum admits an infinitesimal action of g. But its action is not locally free. Roughly speaking, G is the open subscheme where g acts locally freely (Proposition 6.3.1).

The flag variety of a Kac-Moody algebra shares the similar properties to the finite-dimensional ones, such as Bruhat decompositions.

I would like to acknowledge mathematicians I saw at Tata Institute, especially S. Kumar, Moody, Verma. I also thank for hospitalities of the staffs in The Johns Hopkins University during my preparing this article.

Manuscript received 5 December 1988.

1. Scheme of countable type.

1.1. In this paper, we treat infinite-dimensional schemes such as $\mathbf{A}^\infty$, $\mathbf{P}^\infty$, etc.. We shall discuss their local properties briefly.

Let k be a commutative ring.

Definition 1.1.1. *A k-algebra A is called of countable type over k, if A is generated by k and countable numbers of elements.*

The following is easily proven just as in EGA.

LEMMA 1.1.2. *Let X be a scheme over k. Assume that there is an open affine covering $X = \cup U_j$ of X such that $\Gamma(U_j; \mathcal{O}_X)$ is of countable type. Then, for any open affine subset U of X, $\Gamma(U; \mathcal{O}_X)$ is of countable type.*

Definition 1.1.3. *A scheme X over k is called of countable type if for any open affine subset U of X, $\Gamma(U; \mathcal{O}_X)$ is a k-algebra of countable type.*

LEMMA 1.1.4. *Let k be a noetherian ring. Then any ideal of a k-algebra A of countable type is generated by countable elements.*

Proof. Assume A is generated by x_i ($i = 1, 2, \ldots$). Then for any ideal I of A, $I \cap k[x_1, \ldots, x_n]$ is generated by finitely many elements.

LEMMA 1.1.5. *Let k be an algebraically closed field such that k is not a countable set, and let X be a k-scheme of countable type. If X has no k-valued point, then X is empty.*

Proof. We may assume $X = \mathrm{Spec}(A)$ and $A \cong k[T_n; n \in \mathbf{N}]/I$, where T_n are indeterminates. Then I is generated by countably many elements f_j. Let k' be the subring of k generated by the coefficients of the f_j. Set $A' = k'[T_n; n \in \mathbf{Z}]/I'$ where I' is the ideal generated by f_j. Then $A \cong k \otimes_{k'} A'$. If $A \neq 0$, there is a homomorphism $A' \to K'$ from A' to a field K'. We may assume K' is generated by the image of A' as a field. Then K' has at most countable transcendental dimension over the prime field. Hence $k' \to k$ splits $k' \to K' \xrightarrow{\varphi} k$ for some φ. Therefore X has a k-valued point.

PROPOSITION 1.1.6. *Let k be a noetherian ring, and $A \cong \varinjlim_n A_n$, where $\{A_n\}_{n \in \mathbf{N}}$ is an inductive system of k-algebra of finite type and $A_n \to A_{n+1}$ is flat. Then $\mathcal{O}_{\mathrm{Spec}(A)}$ is a coherent ring.*

Proof. Any homomorphism $\varphi : A^{\oplus m} \to A$ comes from some φ' :

$A^{\oplus m} \to A_n$. Then Ker φ' is finitely generated over A_n and hence Ker $\varphi \cong A \otimes_{A_n}$ Ker φ' is also finitely generated over A.

Let us give an example.

Example 1.1.7. *Infinite-dimensional affine space*: $A^\infty = \operatorname{Spec} k[X_i; i \in \mathbf{N}]$. The set of k-valued points of A^∞ is $\{(x_i)_{i\in\mathbf{N}}; x_i \in k\}$. The structure ring is coherent by Proposition 1.1.6, since $k[X_i; i \in \mathbf{N}] = \bigcup_m k[X_1, \ldots, X_m]$.

2. Grassmann variety.

2.1. Let k be a base field.

Definition 2.1.1. *An l.c. k-vector space V is a k-vector space with a topology satisfying*

(i) *The addition map* $V \times V \to V$ *is continuous.*
(ii) *V is Hausdorff and complete.*
(iii) *The open k-vector subspaces form a neighborhood system of* 0.

Let V_1 and V_2 be two *l.c.* vector spaces. We set

$$V_1 \hat{\otimes} V_2 = \varprojlim_{U_1, U_2} (V_1/U_1) \otimes (V_2/U_2) \tag{2.1.1}$$

where U_j ranges over open linear subspaces of 0 in V_j ($j = 1, 2$). We endow $V_1 \hat{\otimes} V_2$ with the structure of *l.c.* vector space such that $\operatorname{Ker}(V_1 \hat{\otimes} V_2 \to (V_1/U_1) \otimes (V_2/U_2))$ form a neighborhood system of 0.

Definition 2.1.2. *An l.c. k-vector space V is called a c.l.c. k-vector space if V is an l.c. k-vector space and it satisfies furthermore*

(iv) *There is a decreasing sequence* $\{W_n\}_{n\in\mathbf{Z}}$ *of open vector subspaces forming a neighborhood system of* 0 *such that* $V = \bigcup_{n\in\mathbf{Z}} W_n$ *and* dim $W_n/W_m < \infty$ for $n \leq m$.

Remark that in this case the family $\mathfrak{F}(V)$ of open vector subspace W of V which is contained by some W_n is independent from the choice of $\{W_n\}$. In fact, $\mathfrak{F}(V)$ is the family of open vector subspaces W of V such that $\dim(W/W') < \infty$ for any open subspace $W' \subset W$.

2.2. For a *c.l.c.* vector space V, define the Grassmann variety as follows.

For a k-scheme S, set $\mathcal{O}_S \hat{\otimes} V = \varinjlim_{W \in \mathfrak{F}(V)} \mathcal{O}_S \otimes (V/W)$ and consider the functor

(2.2.1) $\mathrm{Grass}(V) : S \mapsto \{\mathfrak{F};$ $\mathfrak{F}$ is a sub-$\mathcal{O}_S$-module of $\mathcal{O}_S \hat{\otimes} V$ such that locally in the Zariski topology there exists a $W \in \mathfrak{F}(V)$ such that $\mathfrak{F} \to \mathcal{O}_S \otimes (V/W)$ is an isomorphism$\}$.

For $W \in \mathfrak{F}(V)$, we set

(2.2.2) $\mathrm{Grass}_W(V) : S \mapsto \{\mathfrak{F};$ $\mathfrak{F}$ is a sub-$\mathcal{O}_S$-module of $\mathcal{O}_S \hat{\otimes} V$ such that $\mathfrak{F} \to \mathcal{O}_S \otimes (V/W)$ is an isomorphism$\}$.

Hence $\mathrm{Grass}(V) = \bigcup_W \mathrm{Grass}_W(V)$ in the Zariski topology.

PROPOSITION 2.2.1. Grass(V) *is represented by a separated scheme.*

Proof. This proposition follows from the following two statements

(2.2.3) $\mathrm{Grass}_W(V)$ is represented by an affine scheme of countable type.

(2.2.4) For $W, W' \in \mathfrak{F}(V)$, there exists $f \in \Gamma(\mathrm{Grass}_W(V); \mathcal{O})$ and $f' \in \Gamma(\mathrm{Grass}_{W'}(V); \mathcal{O})$ such that $\mathrm{Grass}_W(V) \cap \mathrm{Grass}_{W'}(V)$ is represented by the open subscheme defined by $f \neq 0$ of $\mathrm{Grass}_W(V)$ and that we have $ff' = 1$ on $\mathrm{Grass}_W(V) \cap \mathrm{Grass}_{W'}(V)$.

We shall prove first (2.2.3). Let us take $\{e_i\}_{i \in I}$ in V such that $\{e_i\}$ forms a base of V/W. Take $\{u_j\}_{j \in J}$ in W such that u_j tends to 0 and any element of W is uniquely written as $\Sigma\, a_j u_j$ $(a_j \in k)$. Then for a scheme S and $\mathfrak{F} \in \mathrm{Grass}_W(V)(S)$, there exist $a_{ij} \in \mathcal{O}(S)$ such that $\mathfrak{F}$ is generated by $e_i + \Sigma_j\, a_{ij} u_j$. Hence $\mathrm{Grass}_W(V)$ is represented by $\mathrm{Spec}(k[T_{ij};\, i \in I, j \in J])$.

Now, we shall prove (2.2.4).

For $\mathfrak{F} \in \mathrm{Grass}(V)(S)$, let $\mathcal{G}$ be the cokernel of $\mathfrak{F} \to \mathcal{O}_S \otimes V/(W \cap W')$, and consider the diagram

$$
\begin{array}{ccccccccc}
 & & 0 & \longrightarrow & \mathcal{O}_S \otimes W/(W \cap W') & \overset{\sim}{\longrightarrow} & \mathcal{O}_S \otimes W/(W \cap W') & \longrightarrow & 0 \\
 & & \downarrow & & \downarrow & & \downarrow & & \\
0 & \longrightarrow & \mathfrak{F} & \longrightarrow & \mathcal{O}_S \otimes V/(W \cap W') & \longrightarrow & \mathcal{G} & \longrightarrow & 0 \\
 & & \downarrow & & \downarrow & & & & \\
 & & \mathfrak{F} & \longrightarrow & \mathcal{O}_S \otimes V/W & & & &
\end{array}
$$

Hence if $\mathfrak{F} \in \mathrm{Grass}_W(V)(S)$, $\mathcal{G}$ is isomorphic to $\mathcal{O}_S \otimes W/(W \cap W')$. The similar diagram obtained by exchanging W and W' shows that $\mathcal{O}_S \otimes W'/(W' \cap W') \to \mathcal{G}$ and $\mathfrak{F} \to \mathcal{O}_S \otimes V/W'$ has the same kernel and the cokernel. Hence if we denote by f the determinant of $\psi : \mathcal{O}_S \otimes W'/(W \cap W') \to \mathcal{G} \rightleftarrows \mathcal{O}_S \otimes W/(W \cap W')$, then $\mathrm{Grass}_W(V) \cap \mathrm{Grass}_{W'}(V)$ is defined by $f \neq 0$. On $\mathrm{Grass}_{W'}(V)$, we define f' as the determinant of ψ' : $\mathcal{O}_S \otimes W/(W \cap W') \to \mathcal{G} \rightleftarrows \mathcal{O}_S \otimes W'/(W \cap W')$. Then since ψ and ψ' are inverse to each other on $\mathrm{Grass}_W(V) \cap \mathrm{Grass}_{W'}(V)$, we have $ff' = 1$ there.

COROLLARY 2.2.2. $\mathrm{Grass}_W(V)$ *is open in* $\mathrm{Grass}(V)$ *and isomorphic to* $\mathbf{A}^\infty$ (*if* $\dim V = \infty$).

COROLLARY 2.2.3. (i) *For* $W, W' \in \mathfrak{F}(V)$, $\mathrm{Grass}_W(V) \cap \mathrm{Grass}_{W'}(V) = \emptyset$ *if* $\dim W/(W \cap W') \neq \dim W'/(W \cap W')$.

(ii) *Fix* $W \in \mathfrak{F}(V)$. *Then*

$$\mathrm{Grass}(V) = \bigcup_{d \in \mathbf{Z}} \mathrm{Grass}^d(V) \quad \text{and} \quad \mathrm{Grass}^d(V) = \bigcup_{W'} \mathrm{Grass}_{W'}(V)$$

where W' *ranges over* $\mathfrak{F}(V)$ *with* $\dim W/(W \cap W') - \dim W'/(W \cap W') = d$.

2.3. Let G be an affine group scheme over a field k. We say that G acts on a k-vector space (or V is a G-module) if V is an $\mathcal{O}(G)$-comodule; i.e. there is a comultiplication $\mu : V \to \mathcal{O}(G) \otimes V$ such that

$$
\text{(2.3.1)} \qquad
\begin{array}{ccc}
V & \longrightarrow & \mathcal{O}(G) \otimes V \\
 & \searrow & \downarrow \\
 & & k \otimes V
\end{array}
\quad \text{and} \quad
\begin{array}{ccc}
V & \longrightarrow & \mathcal{O}(G) \otimes V \\
 & & \mathcal{O}(G) \otimes \mu \downarrow \downarrow \mu_G \otimes V \\
 & & \mathcal{O}(G) \otimes \mathcal{O}(G) \otimes V
\end{array}
$$

commutes, where $\mathcal{O}(G) \to k$ is the evaluation map at the identity and μ_G :

$\mathcal{O}(G) \to \mathcal{O}(G) \otimes \mathcal{O}(G)$ is the comultiplication. As well-known, in this case, V is a union of finite-dimensional sub-G-modules.

Now, let V be an *l.c.* k-vector space. We endow $\mathcal{O}(G)$ with the discrete topology. We say that V is a (*l.c.*) G-module if there is given a continuous comultiplication $V \to \mathcal{O}(G) \hat{\otimes} V$ such that

$$
\begin{array}{ccc}
V & \longrightarrow & \mathcal{O}(G) \hat{\otimes} V \\
 & \searrow & \downarrow \\
 & & k \otimes V
\end{array}
\quad \text{and} \quad
\begin{array}{ccc}
V & \longrightarrow & \mathcal{O}(G) \hat{\otimes} V \\
 & & {\scriptstyle \mathcal{O}(G) \hat{\otimes} \mu} \downarrow \downarrow {\scriptstyle \mu_G \otimes V} \\
 & & \mathcal{O}(G) \hat{\otimes} \mathcal{O}(G) \hat{\otimes} V
\end{array}
\tag{2.3.2}
$$

commute. In this case, there exists a neighborhood system of 0 by linear subspaces U_i $(i \in I)$ such that V/U_i is a G-module and $V/U_i \to V/U_i'$ is a morphism of G-modules if $U_i \subset U_i'$.

PROPOSITION 2.3.1. *If V is a c.l.c. G-module, then G acts on* Grass(V).

Proof. It is enough to construct

$$G(S) \times \text{Grass}(V)(S) \to \text{Grass}(V)(S)$$

functorially in S. An S-valued point of G gives $\mathcal{O}(G) \xrightarrow{a} \mathcal{O}(S)$.

Then we obtain

$$g : \mathcal{O}_S \hat{\otimes} V \xrightarrow{\mathcal{O}_S \hat{\otimes} \mu} \mathcal{O}_S \hat{\otimes} \mathcal{O}(G) \hat{\otimes} V \xrightarrow{a} \mathcal{O}_S \hat{\otimes} V.$$

This is an isomorphism. Hence for $F \subset \mathcal{O}_S \otimes V$, $\varphi(F) \subset \mathcal{O}_S \hat{\otimes} V$ and it gives the map Grass$(V)(S) \to$ Grass$(V)(S)$.

3. Kac-Moody Lie algebra.

3.1. Following Kac, Moody, Mathieu, we start by the following data: a free $\mathbf{Z}$ module P, at most countably generated, and $\alpha_i \in P$ and $h_i \in \text{Hom}_{\mathbf{Z}}(P, \mathbf{Z})$ indexed by an index set I.

We set $t^0 = \mathbf{C} \otimes_{\mathbf{Z}} P$, $t = \text{Hom}_{\mathbf{C}}(t^0, \mathbf{C}) \cong \text{Hom}_{\mathbf{Z}}(P, \mathbf{C})$ with the structure of *l.c.* vector space induced from the discrete topology of t^0. We assume the following conditions:

(3.1.1) $\{\langle \alpha_i, h_j \rangle\}_{i,j}$ is a generalized Cartan matrix, i.e. $\langle \alpha_i, h_j \rangle \in \mathbf{Z}$, $\langle \alpha_i, h_i \rangle = 2$, $\langle \alpha_i, h_j \rangle \leq 0$ for $i \neq j$ and $\langle \alpha_i, h_j \rangle = 0$ iff $\langle \alpha_j, h_i \rangle = 0$.

(3.1.2) For any i, there is $\lambda \in P$ such that $\langle \lambda, h_i \rangle > 0$ and $\langle \lambda, h_j \rangle = 0$ for any $j \neq i$.

(3.1.3) $\{\alpha_i\}_{i \in I}$ is linearly independent.

(3.1.4) For any $\lambda \in P$, $\langle h_i, \lambda \rangle = 0$ except finitely many i.

Let $\mathcal{G}$ be the Lie algebra generated by t and symbols e_i, f_i ($i \in I$) with the following recover relations:

(3.1.5) $[h, e_i] = \alpha_i(h)e_i$ and $[h, f_i] = -\alpha_i(h)f_i$ for $h \in t$.

(3.1.6) $[e_i, f_j] = \delta_{ij} h_i$.

(3.1.7) $(ade_i)^{1-\alpha_j(h_i)} e_j = 0$ and $(adf_i)^{1-\alpha_j(h_i)} f_j = 0$ for $i \neq j$.

Let n (resp. n_-) be the Lie subalgebra generated by e_i (resp. f_i), $i \in I$. Then we have (e.g. [K])

(3.1.8) $$\mathcal{G} = n \oplus t \oplus n_-.$$

Set

(3.1.9) $$b = t \oplus n, \qquad b_- = t \oplus n_-$$

(3.1.10) $$\mathcal{G}_i = t \oplus \mathbf{C}e_i \oplus \mathbf{C}f_i, \qquad p_i = \mathcal{G}_i + n, \qquad p_i^- = \mathcal{G}_i + n^-.$$

Let Δ be the set of roots of $\mathcal{G}$ and Δ_+ and Δ_- the set of roots of n and n_-, respectively, and let $\mathcal{G}_\alpha$ be the root space with root $\alpha \in \Delta$. We set

(3.1.11) $$n_i = \bigoplus_{\substack{\alpha \in \Delta_+ \\ \alpha \neq \alpha_i}} \mathcal{G}_\alpha, \qquad n_i^- = \bigoplus_{\substack{\alpha \in \Delta_- \\ \alpha \neq -\alpha_i}} \mathcal{G}_\alpha.$$

Let W be the Weyl group, i.e. the subgroup of $\mathrm{GL}(t^0)$ generated by the simple reflections s_i ($i \in I$), where

(3.1.12) $$s_i(\lambda) = \lambda - \langle h_i, \lambda\rangle\alpha_i.$$

We also denote by W' the braid group generated by s_i' $(i \in I)$ with the fundamental relation

(3.1.13)
$$\begin{aligned}
s_i's_j' &= s_j's_i' && \text{if} \quad \langle h_i, \alpha_j\rangle = 0\\
s_i's_j's_i' &= s_j's_i's_j' && \text{if} \quad \langle h_i, \alpha_j\rangle = \langle h_j, \alpha_i\rangle = -1\\
(s_i's_j')^2 &= (s_j's_i')^2 && \text{if} \quad \langle h_i, \alpha_j\rangle\langle h_j, \alpha_i\rangle = 2\\
(s_i's_j')^3 &= (s_j's_i')^3 && \text{if} \quad \langle h_i, \alpha_j\rangle\langle h_j, \alpha_i\rangle = 3
\end{aligned}$$

Then as is well-known, W is isomorphic to the quotient of W' by the subgroup generated by $ws_i'^2w^{-1}$ $(i \in I)$.

For $w \in W$, we denote by $l(w)$ the length of w, i.e. the smallest number l such that w is the product of a sequence of length l in $\{s_i\}$. Recall that

(3.1.14) $$l(w) = \#(\Delta_+ \cap w\Delta_-).$$

Also recall that $l(s_iw) < l(w)$ if and only if $w^{-1}\alpha_i \in \Delta_-$. Note also there exists a unique injection $\iota : W \to W'$ such that

(3.1.15) $$\iota(1) = 1, \qquad \iota(s_i) = s_i' \quad \text{and} \quad \iota(ww') = \iota(w)\iota(w') \quad \text{if} \quad l(ww') = l(w) + l(w').$$

By this, we sometimes embed W into W'.

An element h of t is called regular if $\langle h, \alpha\rangle \neq 0$ for any $\alpha \in \Delta$. Such an element always exists. We set

(3.1.16) $$P_+ = \{\lambda \in P;\ \langle \lambda, h_i\rangle \geq 0 \text{ for any } i\}.$$

For any finite set J of I, we set

(3.1.17) $$P_{J^+} = \{\lambda \in P_+;\ \langle \lambda, h_i\rangle = 0 \text{ for } i \in I\backslash J\}.$$

If we set $P_0 = \{\lambda \in P;\ \langle \lambda, h_i\rangle = 0 \text{ for } i \in I\}$ then P_0 is a free $\mathbf{Z}$-module and P_{J^+}/P_0 is a finitely generated semigroup.

3.2. Now, we shall define a completion of $\mathcal{G}$. For a subset S of Δ_+, we set

$$n_S = \bigoplus_{\alpha \in S} \mathcal{G}_\alpha. \tag{3.2.1}$$

We set

$$\hat{\mathcal{G}} = \varprojlim_S \mathcal{G}/n_S = b_- \oplus \prod_{\alpha \in \Delta_+} \mathcal{G}_\alpha \tag{3.2.2}$$

where S ranges over the subsets of Δ_+ such that $\Delta_+ \backslash S$ is finite. We define the subalgebras $\hat{P}_i$, $\hat{n}_i$, $\hat{b}$, $\hat{n}$ of $\hat{\mathcal{G}}$, similarly. We set also

$$\hat{U}_l(\mathcal{G}) = \varprojlim_S U_l(\mathcal{G})/U_{l-1}(\mathcal{G})n_S$$

(3.2.3)

$$\hat{U}(\mathcal{G}) = \bigcup_l \hat{U}_l(\mathcal{G})$$

Then $\hat{U}(\mathcal{G})$ is an algebra containing $U(\mathcal{G})$ as a subalgebra.

3.3. In general, let $\mathcal{G}$ be a Lie algebra. A vector v of a $\mathcal{G}$-module V is called $\mathcal{G}$-finite if v is contained in a finite-dimensional sub-$\mathcal{G}$-module of V. We call a $\mathcal{G}$-module V is locally finite if any element of V is $\mathcal{G}$-finite.

Let us define a ring homomorphism

$$\delta : U(\mathcal{G}) \to U(\mathcal{G}) \otimes U(\mathcal{G}) \tag{3.3.1}$$

by $\delta(A) = A \otimes 1 + 1 \otimes A$ for $A \in \mathcal{G}$, and an anti-ring automorphism

$$a : U(\mathcal{G}) \to U(\mathcal{G}) \tag{3.3.2}$$

by $A^a = -A$ for $A \in G$. Then δ defines $U(\mathcal{G})^* \otimes U(\mathcal{G})^* \to (U(\mathcal{G}) \otimes U(\mathcal{G}))^* \to U(\mathcal{G})^*$ and this gives a commutative ring structure on $U(\mathcal{G})^*$.

The right and left multiplication of $\mathcal{G}$ on $U(\mathcal{G})$ induces the two $\mathcal{G}$-module structures on $U(\mathcal{G})^*$:

$$(R(A)f)(P) = f(PA), \qquad (L(A)f)(P) = f(a(A)P) \tag{3.3.3}$$

for $A \in U(\mathcal{G})$, $f \in U(\mathcal{G})^*$ and $P \in U(\mathcal{G})$. Then $R(A)$ and $L(A)$ are derivations of the ring $U(\mathcal{G})^*$ for any $A \in \mathcal{G}$.

Now let $\mathfrak{A}$ be an abelian Lie algebra acting on the Lie algebra $\mathcal{G}$ semi-simply, t an abelian subalgebra of $\mathcal{G}$ stable by $\mathfrak{A}$, and $P \subset t^*$ a sub-**Z**-module stable by $\mathfrak{A}$. We assume that t acts semi-simply on $\mathcal{G}$ by the adjoint action and its weights belong to P.

Then, we set

(3.3.5) $A(\mathcal{G}, t, P, \mathfrak{A}) = \bigoplus_{\lambda \in P} \{ f \in U(\mathcal{G})^*; f$ satisfies the following conditions (3.3.6), (3.3.7) and (3.3.8)$\}$.

(3.3.6) f is $\mathcal{G}$-finite with respect to L and R.

(3.3.7) f is a weight vector with weight λ with respect to the left action of t.

(3.3.8) f is $\mathfrak{A}$-finite.

Then $f \in U(\mathcal{G})^*$ belongs to $A(\mathcal{G}, t, P, \mathfrak{A})$ if and only if there exists a two-sided ideal I of $U(\mathcal{G})$ such that

(3.3.9) $f(U(\mathcal{G})/I) = 0$,

(3.3.10) $\dim U(\mathcal{G})/I < \infty$,

(3.3.11) I is $\mathfrak{A}$-invariant,

(3.3.12) t acts semi-simply on $U(\mathcal{G})/I$ by the left multiplication and its weights belong to P.

Then one can see easily that $A(\mathcal{G}, t, P, \mathfrak{A})$ is a subring of $U(\mathcal{G})^*$ and the multiplication map $\mu : U(\mathcal{G}) \otimes U(\mathcal{G}) \to U(\mathcal{G})$ induces the homomorphism

$$\begin{array}{ccc} A(\mathcal{G}, t, P, \mathfrak{A}) & \longrightarrow & A(\mathcal{G}, t, P, \mathfrak{A}) \otimes A(\mathcal{G}, t, P, \mathfrak{A}) \\ \cap & & \cap \\ U(\mathcal{G})^* & \longrightarrow & (U(\mathcal{G}) \otimes U(\mathcal{G}))^* \end{array} \tag{3.3.13}$$

With this, $\operatorname{Spec}(A(\mathcal{G}, t, P, \mathfrak{A}))$ becomes an affine group scheme (see [M]).

We write

$$G(\mathcal{G}, t, P, \mathfrak{A}) = \mathrm{Spec}(A(\mathcal{G}, t, P, \mathfrak{A})). \tag{3.3.14}$$

Remark that $g \mapsto g^{-1}$ is given by $a : U(\mathcal{G}) \to U(\mathcal{G})$.

When $\mathfrak{A} = 0$, we write $G(\mathcal{G}, t, P)$ for $G(\mathcal{G}, t, P, \mathfrak{A})$ for short.

3.4. Coming back to the situation in Section 3.1, we define the affine group schemes B, B_-, T, U, U_-, G_i, U_i, U_i^-, P_i, P_i^- as follows. This construction is due to Mathieu [M].

$$B = G(b, t, P),$$

$$B_- = G(b_-, t, P),$$

$$T = G(t, t, P),$$

$$U = G(n, 0, 0, t),$$

$$U_- = G(n_-, 0, 0, t),$$

$$G_i = G(\mathcal{G}_i, t, P),$$

$$U_i = G(n_i, 0, 0, t),$$

$$U_i^- = G(n_i^-, 0, 0, t),$$

$$P_i = G(p_i, t, P),$$

$$P_i^- = G(p_i^-, t, P),$$

$$G_i^+ = G(t \oplus \mathbf{C}e_i, t, P),$$

$$G_i^- = G(t \oplus \mathbf{C}f_i, t, P).$$

Then we have ([M])

$$B = T \ltimes U = G_i^+ \times U_i,$$

$$B_- = T \ltimes U_- = G_i^- \times U_i^-,$$

$$P_i = G_i \ltimes U_i \supset B \supset T,$$

$$P_i^- = G_i \ltimes U_i^- \supset B_- \supset T,$$

$$T = \operatorname{Spec} \mathbf{C}[P],$$

$$U \cong \operatorname{Spec} S(\bigoplus_{\alpha \in \Delta_+} \mathfrak{g}_\alpha^*),$$

$$U_- \cong \operatorname{Spec} S(\bigoplus_{\alpha \in \Delta_-} \mathfrak{g}_\alpha^*),$$

$$G_i^+ = G_i \cap B, \qquad G_i^- = G_i \cap G_-.$$

More generally, for a subset S of Δ_+ such that $(S + S) \cap \Delta_+ \subset S$, we set $n_S = \oplus \mathfrak{g}_\alpha$ and $U_S = G(n_s, 0, 0, t)$.

Then for $S \supset S'$ such that $S \setminus S'$ is a finite set and that $(S + S') \cap \Delta_+ \subset S'$, $n_S/n_{S'}$ is a finite-dimensional nilpotent Lie algebra and if we denote by $\exp(n_S/n_{S'})$ the associated unipotent group, we have

$$U_S \cong \varprojlim_{S'} \exp(n_S/n_{S'}).$$

3.5. The group P_i acts on the *c.l.c.* space $\hat{\mathfrak{g}}$ by the adjoint action. In fact, $ad : p_i \to \operatorname{End}(\mathfrak{g})$ extends to $ad : U(p_i) \to \operatorname{End}(\mathfrak{g})$. Moreover, for any ideal $\mathfrak{a}$ of p_i with $\operatorname{codim} p_i/\mathfrak{a} < \infty$, $\mathfrak{g}/\mathfrak{a}$ is locally p_i-finite. Hence, for any $A \in \mathfrak{g}$, there is a two-sided ideal I of $U(p_i)$ with $\dim U(p_i)/I < \infty$ and $ad(I)A \subset \mathfrak{a}$. Hence the morphism $P \mapsto ad(P)A$ from $U(p_i)$ to $\mathfrak{g}/\mathfrak{a}$ splits as $U(p_i)/I \to \mathfrak{g}/\mathfrak{a}$. Hence this gives an element of $(U(p_i)/I)^* \otimes \mathfrak{g}/\mathfrak{a} \subset U(p_i)^* \otimes (\mathfrak{g}/\mathfrak{a})$. This element clearly belongs to $A(p_i, t, P) \otimes (\mathfrak{g}/\mathfrak{a})$. Thus we obtained $\mathfrak{g}/\mathfrak{a} \to A(p_i, t, P) \otimes (\mathfrak{g}/\mathfrak{a})^*$. Since $\varprojlim \mathfrak{g}/\mathfrak{a} = \hat{\mathfrak{g}}$, we obtain $\hat{\mathfrak{g}} \to \mathcal{O}(P_i) \hat{\otimes} \hat{\mathfrak{g}}$. This gives an action of P_i on $\hat{\mathfrak{g}}$.

Clearly the action of B on $\hat{\mathfrak{g}}$ obtained from the action of P_i does not depend on $i \in I$.

Especially, P_i acts on the Grassmann variety $\operatorname{Grass}(\hat{\mathfrak{g}})$ by Proposition 2.3.1.

4. The first construction of the flag variety.

4.1. In this section, for a Kac-Moody Lie algebra $\mathfrak{g}$, we construct its flag variety as a subscheme of Grass$(\hat{\mathfrak{g}})$. We keep the notations in Section 3.

4.2. Since $\hat{\mathfrak{g}}$ is a *c.l.c.* vector space, Grass$(\hat{\mathfrak{g}})$ is a separated scheme. Since $\hat{\mathfrak{g}} = b_- \oplus \hat{n}$, b_- gives a $\mathbf{C}$-valued point of Grass$(\hat{\mathfrak{g}})$. We denote this point by x_0. By Section 3.5, P_i and B act on Grass$(\hat{\mathfrak{g}})$.

4.3. Set $s_i' = \exp(-e_i)\exp(f_i)\exp(-e_i) \in G_i \subset P_i$. Then $s_i'^4 = 1$ and s_i' acts on $\hat{\mathfrak{g}}$. This extends to the group homomorphism:

$$W' \to \mathrm{Aut}(\hat{\mathfrak{g}}). \tag{4.3.1}$$

In order to see this, it is enough to prove the braid relation (3.1.13) when the Lie algebra generated by e_i, e_j, f_i, f_j is finite-dimensional. Then the braid condition holds in the corresponding simply connected semi-simple group.

The morphism (4.3.1) induces

$$W' \to \mathrm{Aut}(\mathrm{Grass}(\hat{\mathfrak{g}})). \tag{4.3.2}$$

We have also

(4.3.3) The image of $\mathrm{Ker}(W' \to W)$ in Aut$(\hat{\mathfrak{g}})$ belongs to the image of T in Aut$(\hat{\mathfrak{g}})$.

In fact, $\mathrm{Ker}(W' \to W)$ is generated by the $ws_i'^2w^{-1}$, which belongs to T.

Since $[t, b_-] \subset b_-$, we have

$$Tx_0 = x_0. \tag{4.3.4}$$

Hence for $w \in W$, $w'x_0$ does not depend on the choice of a representative w' of w in W'. We denote it by wx_0.

4.4. As in (2.2.2), we set

$$\mathrm{Grass}_{\hat{n}}(\hat{\mathfrak{g}}) = \{W \in \mathrm{Grass}(\hat{\mathfrak{g}});\ W \oplus \hat{n} \xrightarrow{\sim} \hat{\mathfrak{g}}\}. \tag{4.4.1}$$

This is an affine open subscheme of $\mathrm{Grass}(\hat{\mathfrak{g}})$.

LEMMA 4.4.1. *The morphism* $U \to \mathrm{Grass}(\hat{\mathfrak{g}})$ *given by* $U \ni g \mapsto gx_0$ *is an embedding.*

Proof. First we shall show $Ux_0 \subset \mathrm{Grass}_{\hat{n}}(\hat{\mathfrak{g}})$. For this, it is enough to show, for any $g \in U$,

$$gb_- \oplus \hat{n} = \hat{\mathfrak{g}}. \tag{4.4.2}$$

But this is obvious because $\hat{n}$ is stable by U. Hence it is enough to show that $U \to Y = \mathrm{Grass}_{\hat{n}}(\hat{\mathfrak{g}})$ is a closed embedding. In order to see this, let us take a regular element h of t (i.e. $\langle h, \alpha \rangle \neq 0$ for any $\alpha \in \Delta$). Then for any $F \in \mathrm{Grass}_{\hat{n}}(\hat{\mathfrak{g}})$, $F \oplus \hat{n} = \hat{\mathfrak{g}}$, and hence there exists $\psi(F) \in \hat{n}$ with $h - \psi(F) \in F$. This defines a morphism

$$\psi : Y \to \hat{n}.$$

If we combine $U \to Y \xrightarrow{\psi} \hat{n}$, this is given by

$$U \ni g \mapsto h - g^{-1}h \in \hat{n}.$$

Hence it is enough to show the following lemma.

LEMMA 4.4.2. *Let h be a regular element of t. Then, the morphism* $U \to h + \hat{n}$ *given by* $g \mapsto gh$ *is an isomorphism.*

Proof. Let S be a subset of Δ_+ such that $(S + \Delta_+) \cap \Delta_+ \subset S$ and $\Delta_+ \backslash S$ is finite. Then $U \to h + \hat{n}$ induces $U/U_S \to (h + n)/n_S$, and it is enough to show that this is an isomorphism. Now U/U_S acts on b/n_S. For $A \in n/n_S$, the isotropy group at $h + A$ is the identity. In fact this follows from

$$\{E \in n; [h + A, E] \in n_S\} = n_S. \tag{4.4.3}$$

Since $\dim(h + n)/n_S = \dim U/U_S$, $(U/U_S)(h + A)$ is open in $(h + n)/n_S$. Thus $(U/U_S)(h + A)$ and $(U/U_S)h$ intersect. This shows $U/U_S \xrightarrow{\sim} (U/U_S)h = (h + n)/n_S$.

4.5. We have

$$Bx_0 = Ux_0 \tag{4.5.1}$$

because $Tx_0 = x_0$ and $B = UT$. For $w \in W$, let us denote

$$B \cap {}^wB = A(t \oplus \bigoplus_{\alpha\in\Delta_+\cap{}^w\Delta_+} \mathcal{G}_\alpha, t, P) \tag{4.5.2}$$

$$B \cap {}^wB_- = A(t \oplus \bigcap_{\alpha\in\Delta_+\cap{}^w\Delta_-} \mathcal{G}_\alpha, t, P).$$

They are subgroups of B. Similarly, we define $U \cap {}^wU$ and $U \cap {}^wU_-$. Then we have

$$U \simeq (U \cap {}^wU) \times (U \cap {}^wU_-) \simeq (U \cap {}^wU_-) \times (U \cap {}^wU). \tag{4.5.3}$$

We have also

$$(B \cap {}^wB_-)wx_0 = x_0. \tag{4.5.4}$$

LEMMA 4.5.1. *For $w \in W$, $Bs_i'Bwx_0 \subset Bwx_0 \cup Bs_iwx_0$.*

Proof. We have $Bs_i'Bwx_0 \subset P_iwx_0$. Since $P_i = BG_i \subset B(G_i \cap {}^wB_-) \cup Bs_i'(G_i \cap {}^wB_-)$, $P_iwx_0 \subset B(G_i \cap {}^wB_-)wx_0 \cup Bs_i'(G_i \cap {}^wB_-)wx_0 \subset Bwx_0 \cup Bs_iwx_0$.

Note that for $w_1, w_2 \in W$, $w_1Bw_2x_0$ does not depend on the representatives in W' of $w_1, w_2 \in W$. Hence we denote $w_1Bw_2x_0$ for it.

LEMMA 4.5.2. *Let $w \in W$.*

(i) *If $l(w) > l(s_iw)$, $Bs_iBwx_0 = Bs_iwx_0$.*
(ii) *If $l(w) < l(s_iw)$, $Bs_iBwx_0 = P_iwx_0 = Bwx_0 \cup Bs_iwx_0$.*

Proof. If $l(s_iw) < l(w)$, then $w^{-1}\alpha_i \in \Delta_-$. Hence $G_i^+ = G_i \cap B \subset {}^wB_-$ and $s_iB \subset s_iU_iG_i^+ \subset Bs_iG_i^+$. Hence we have $Bs_iBwx_0 = Bs_iG_i^+wx_0 = Bs_iwx_0$.

If $l(s_iw) > l(w)$, then we have $Bs_iBs_iwx_0 = Bwx_0$ since $l(s_is_iw) < l(s_iw)$. Hence $Bs_iBwx_0 = Bs_iBs_iBwx_0$. Since $Bs_iBs_iB = P_i$, $Bs_iBwx_0 = P_iwx_0$ and it contains wx_0 and s_iwx_0.

LEMMA 4.5.3. *$wBx_0 \subset \bigcup_{w' \leq w} Bw'x_0$, where $\leq$ is the Bruhat order (the order generated by $s_{i_1} \cdots s_{i_{k-1}} s_{i_{k+1}} \cdots s_{i_l} \leq s_{i_r}$ for a reduced expression $s_{i_1} \cdots s_{i_r}$).*

Proof. We shall prove by the induction of $l(w)$. If $l(w) = 0$, it is trivial. Otherwise, set $w = s_i w'$ with $l(w) = 1 + l(w')$. Then by the hypothesis of the induction, $wBx_0 \subset \bigcup_{w'' \leq w'} s_i Bw''x_0 \subset \bigcup_{w'' \leq w'} Bs_i w''x_0 \cup Bw''x_0 \subset \bigcup_{w'' \leq w} Bw''x_0$.

LEMMA 4.5.4.

(i) $Bwx_0 \cap \mathrm{Grass}_{\hat{n}}(\hat{\mathfrak{G}}) = \emptyset$ *if* $w \neq 1$.
(ii) $wBx_0 \cap \mathrm{Grass}_{\hat{n}}(\hat{\mathfrak{G}}) \subset Bx_0$.

Proof. (i) Let $g \in B$ and assume that $gwb_- \rightrightarrows \hat{\mathfrak{G}}/\hat{n}$. Then $wb_- \rightrightarrows \hat{\mathfrak{G}}/\hat{n}$. Hence $w\Delta_- = \Delta_-$, which implies $w = 1$.

(ii) follows from (i) and the preceding lemma.

COROLLARY 4.5.5. *$X = \bigcup_{w \in W'} wBx_0$ is a subscheme of* $\mathrm{Grass}(\hat{\mathfrak{G}})$ *and wBx_0 is open in X for any $w \in W$.*

This easily follows from $X \cap \mathrm{Grass}_{\hat{n}}(\hat{\mathfrak{G}}) = Bx_0$.

Definition 4.5.6. We call X the flag variety of $\mathfrak{G}$.

Since $\mathrm{Grass}(\hat{\mathfrak{G}})$ is a separated scheme, X is also a separated scheme, and $\{wBx_0\}$ is an open affine covering of X. Note that X is not quasi-compact if W is an infinite group. I do not know whether X is a closed subscheme of $\mathrm{Grass}(\hat{\mathfrak{G}})$ or not.

LEMMA 4.5.7. *Bwx_0 is a closed subscheme of wBx_0 and we have a commutative diagram:*

$$\begin{array}{ccc} Bwx_0 & \hookrightarrow & wBx_0 \\ \uparrow\!\wr & & \uparrow\!\wr \\ \hat{n} \cap {}^{w^{-1}}\hat{n} & \hookrightarrow & \hat{n} \end{array} \tag{4.5.5}$$

Proof. We have $U = (U \cap^w U) \times (U \cap^w U_-)$. Since $(U \cap^w U_-)x_0 = x_0$, we have $Uwx_0 = (U \cap^w U)wx_0 = w({}^{w^{-1}}U \cap U)x_0$. Then the lemma follows from Lemma 4.4.1.

COROLLARY 4.5.8. *Bwx_0 is affine and codimension $l(w)$ in X.*

PROPOSITION 4.5.9. *$X(\mathbf{C}) = \bigsqcup_{w \in W} Bwx_0$.*

Proof. By Lemma 4.5.3, it is enough to show $Bwx_0 = Bw'x_0$ implies $w = w'$.

We have $wx_0 \in Bw'x_0 \subset w'Bx_0$. Hence $w'^{-1}wx_0 \subset Bw'^{-1}wx_0 \cap Bx_0$. Then Lemma 4.5.4 implies $w' = w$.

LEMMA 4.5.10. *Let $w_1, w_2 \in W$ and assume $l(w_1s_iw_2) = l(w_1) + l(w_2) + 1$. Then $Bw_1s_iw_2x_0 \subset \overline{Bw_1w_2x_0}$.*

Proof. Since $l(w_1s_i) > l(w_1)$, we have $w_1\alpha_i \in \Delta_+$, and hence $G_i \cap {}^{w_1^{-1}}B \subset G_i \cap B$. Since $l(s_iw_2) > l(w_2)$, $w_2^{-1}\alpha_i \in \Delta_+$ and hence $G_i \cap {}^{w_2}B_- \subset G_i \cap B_-$. Since $(G_i \cap B) \cdot (G_i \cap B_-)$ is dense in G_i, we obtain

$$Bw_1s_{io}w_2x_0 \subset Bw_1G_iw_2x_0 \subset \overline{Bw_1(G_i \cap {}^{w_1^{-1}}B)(G_i \cap {}^{w_2}B_-)w_2x_0}$$
$$= \overline{Bw_1w_2x_0}.$$

PROPOSITION 4.5.11. $\overline{Bwx_0} = \bigcup_{w' \geq w} Bw'x_0$.

Proof. We shall prove first $\overline{Bwx_0} \supset Bw'x_0$ if $w' \geq w$ by the induction of $l(w')$. If $l(w') = 0$, then $w = w' \doteq e$ and this is evident. If $l(w') > 0$, there is $w_1, w_2 \in W$ and i such that $w' = w_1s_iw_2$, $w_1w_2 \geq w$ and $l(w') = l(w_1) + l(w_2) + 1$. Hence $Bw'x_0 \subset \overline{Bw_1w_2x_0} \subset \overline{Bwx_0}$.

Now, we shall prove the converse inclusion.

In order to see this, we shall prove that $\overline{Bwx_0} \supset Bw'x_0$ implies $w \leq w'$ by the induction of $l(w')$. If $l(w') = 0$, $w \neq 1$ implies $Bwx_0 \cap Bx_0 = \emptyset$. Hence $\overline{Bwx_0} \cap Bx_0 = \emptyset$. Assume that $l(w') > 0$. Then there is i such that $l(s_iw') < l(w')$. Thus we have $\overline{Bs_iBwx_0} \supset Bs_iBw'x_0 = Bs_iw'x_0$ by Lemma 4.5.2.

If $l(s_iw) < l(w)$, then by Lemma 4.5.2, $\overline{Bs_iwx_0} = \overline{Bs_iBwx_0} \supset Bs_iw'x_0$ and hence $s_iw' \geq s_iw$, which implies $w' \geq w$.

If $l(s_iw) > l(w)$, then $\overline{Bs_iBwx_0} = \overline{Bwx_0} \supset Bs_iw'x_0$ and hence $w' \geq s_iw' \geq w$.

PROPOSITION 4.5.12. $BwBx_0 = \bigcup_{w' \leq w} Bw'x_0$.

Proof. By Lemma 4.5.3, it is enough to show $BwBx_0 \supset Bw'x_0$ implies $w \geq w'$, or equivalently

$$wBx_0 \cap Bw'x_0 \neq \emptyset \quad \text{implies} \quad w \geq w'. \tag{4.5.8}$$

We shall prove this by the induction on $l(w)$. If $l(w) = 0$, this is already proven. Assume $l(w) > 0$. Then there exists i such that $w'' = s_i w$ satisfies $l(w'') < l(w)$. Then $wBx_0 \cap Bw'x_0 \neq \emptyset$ implies $w''Bx_0 \cap Bs_iBw'x_0 \neq \emptyset$.

If $l(s_iw') < l(w')$, Lemma 4.5.2 implies $w''Bx_0 \cap Bs_iw'x_0 \neq \emptyset$. Hence the hypothesis of the induction implies $w'' \geq s_iw'$, which gives $w \geq w'$. If $l(s_iw') > l(w')$, then $w''Bx_0 \cap (Bs_iw'x_0 \cup Bw'x_0) \neq \emptyset$.

Hence $w' \geq s_iw'$ or $w'' \geq w'$. Hence in the both cases, we have $w \geq w'$.

Corollary 4.5.13. $BwBx_0 = \bigcup_{w' \leq w} w'Bx_0$.

Proof. If $w' \leq w$, $w'Bx_0 \subset \bigcup_{w'' \leq w'} Bw''x_0 \subset BwBx_0$. The inverse inclusion follows from $w'Bx_0 \supset Bw'x_0$ (Lemma 4.5.7).

Remark 4.5.14. For $w, w' \in W$, we have

$$\overline{Bwx_0} \cap w'Bx_0 \cong (U \cap {}^{w'}U) \times (\overline{Bwx_0} \cap w'(B \cap {}^{w'^{-1}}B_-)x_0)$$

because $w'Bx_0 = (U \cap {}^{w'}U) \times w'(B \cap {}^{w'^{-1}}B_-)x_0$ and $\overline{Bwx_0}$ is invariant by $U \cap {}^{w'}U$. Then $\overline{Bwx_0} \cap w'(B \cap {}^{w'^{-1}}B_-)x_0$ is a finite-dimensional variety. Thus, $\overline{Bwx_0}$ is locally finite-dimensional or the product of a finite-dimensional variety and $\mathbf{A}^\infty$.

Proposition 4.5.15. *X is irreducible.*

Proof. Since $X = \bigcup wBx_0$ is an open covering by irreducible subsets, it is enough to show $wBx_0 \cap w'Bx_0 \neq \emptyset$ for any w, w'. This follows from $Bw'^{-1}wBx_0 \supset Bx_0$ (Proposition 4.5.12).

5. The second construction of the flag variety.

5.1. Following Kac-Peterson [K-P], we shall first define the ring of regular functions. Recall that $U(\mathcal{G})^*$ has the structure of two-sided $\mathcal{G}$-modules (Section 3.3).

Definition 5.1.1. $A(\mathcal{G}, P) = \bigoplus_{\mu \in P} \{\varphi \in U(\mathcal{G})^*; \varphi$ satisfies the following conditions (5.1.1) and (5.1.2)$\}$.

(5.1.1) φ is finite with respect to the left action of p_i and the right action of p_i for all i.

(5.1.2) φ is a weight vector of weight μ with respect to the left action of t.

LEMMA 5.1.2. *$A(\mathcal{G}, P)$ is a subring of $U(\mathcal{G})^*$.*

This easily follows from the fact that $\delta : U(\mathcal{G}) \to U(\mathcal{G}) \otimes U(\mathcal{G})$ is p_i-linear with respect to the left and right actions.

Definition 5.1.3. We define G_∞ as $\mathrm{Spec}(A(\mathcal{G}, P))$.

LEMMA 5.1.4. *Let V be a p_i-module, and $v \in V$.*

(i) *If v is b-finite, then $f_i v$ is also b-finite.*
(ii) *If v is b-finite and $f_i^N v = 0$ for $N \gg 0$, then v is p_i-finite.*

Proof. Since $[b, f_i] \subset p_i = b + \mathbf{C} f_i$, we have

$$U(b) f_i \subset U(b) + f_i U(b). \tag{5.1.3}$$

This shows (i). If $f_i^N v = 0$, then $U(p_i)v = \Sigma_{k<N} U(b) f_i^k v$, which shows (ii).

LEMMA 5.1.5. *Let V be a $\mathcal{G}$-module. Then, for any $i \in I$, the set of p_i-finite vectors is a sub-$\mathcal{G}$-module.*

Proof. It is enough to show that if v is a p_i-finite vector then $f_j v$ is also p_i-finite vector for $j \neq i$. By the preceding lemma, $f_j v$ is b-finite. Hence it is enough to show $f_i^N f_j v = 0$ for $N \gg 0$. But this follows from (3.1.7) and $f_i^N f_j v = \Sigma_k \binom{N}{k}((ad f_i)^k f_j) f_i^{N-k} v$.

LEMMA 5.1.6. *For any $\lambda \in t^0$, $\lambda + N\alpha_i$ is not a weight of $U(n_i)$ except finitely many $N \in \mathbf{Z}$.*

Proof. We may assume that λ is a weight of $U(n_i)$ and I is finite. For $\lambda = \Sigma\, m_j \alpha_j \in \bigoplus_j \mathbf{Z}\alpha_j$, set $|\lambda|' = \Sigma_{j \neq i} m_j$. Then if α is a weight of n_i, then $|\alpha|' > 0$. Now assume $\lambda + N\alpha_i$ is a weight of $U(n_i)$. Then

$$\lambda + N\alpha_i = \sum_{\nu=1}^{r} \gamma_\nu$$

where γ_ν are weights of n_i. Hence $|\lambda|' = \Sigma_{\nu=1}^r |\gamma_\nu|'$. Hence $r \le |\lambda|'$ and $|\gamma_\nu|' \le |\lambda|'$. Since for any root β, there is only finitely many roots of the form $\beta + N\alpha_i$, there are only finitely many possibilities for γ_ν. Thus we obtain the result.

LEMMA 5.1.7.

(i) $[n_i, f_i] \subset n_i$.

(ii) $(\mathrm{ad} f_i)$ *acts locally nilpotently on* $U(n_i)$.

(iii) *For any two-sided ideal* I *of* $U(n_i)$ *such that* $[t, I] \subset I$ *and* $\dim(U(n_i)/I) < 0$, *there exists* N *such that*

(a) $(\mathrm{ad} f_i)^m U(n_i) \in I$ *for* $m \geq N$.

(b) $f_i^{N+m} U(n_i) \subset I\mathbf{C}[f_i] + U(n_i)\mathbf{C}[f_i]f_i^m$ *for* $m \geq 0$.

Proof.

(i) follows from $(\Delta_+ - \alpha_i) \cap \Delta \subset \Delta_+ \backslash \{\alpha_i\}$.

(ii) follows from the fact that weights of $U(n_i)$ belong to $\Sigma\, \mathbf{Z}_{\geq 0}\alpha_j$.

(iii) In order to see (a), it is enough to show, for any weight β of $U(n_i)$, $\beta + N\alpha_i$ is not a weight of $U(n_i)$ if $N \gg 0$. This follows from Lemma 5.1.6. (b) follows from (a) and $f_i^{N+m} U(n_i) \subset \Sigma((\mathrm{ad} f_i)^k U(n_i)) f_i^{N+m-k}$.

LEMMA 5.1.8. *If* $\varphi \in U(\mathcal{G})^*$ *is left* b*-finite and right* p_i^-*-finite, then* φ *is left* p_i*-finite.*

Proof. By Lemma 5.1.4, it is enough to show

$$L(f_i)^N \varphi = 0 \quad \text{for} \quad N \gg 0. \tag{5.1.5}$$

There exists a two-sided ideal I of $U(b)$ such that $\varphi(IU(\mathcal{G})) = 0$ and $\dim U(b)/I < \infty$. Then by the preceding lemma, there exists N such that

$$f_i^{N+m} U(n_i) \subset IU(\mathcal{G}) + U(n_i) f_i^m U(p_i^-) \quad \text{for} \quad m \geq 0.$$

Since $U(\mathcal{G}) = U(n_i)U(p_i^-)$, we have

$$\varphi(f_i^{N+m} U(\mathcal{G})) \subset \varphi(IU(\mathcal{G}) + U(n_i) f_i^m U(p_i^-))$$

$$\subset \{R(f_i)^m R(U(p_i^-))\varphi\}(U(\mathcal{G})) = 0$$

for $m \gg 0$.

PROPOSITION 5.1.9. $\mathcal{O}(G_\infty)$ *is a two-sided sub-*$\mathcal{G}$*-module of* $U(\mathcal{G})^*$.
This follows immediately from Lemma 5.1.5.
Let $e \in G_\infty$ be the point given by $U(\mathcal{G}) \to U(\mathcal{G})/U(\mathcal{G})\mathcal{G} \cong \mathbf{C}$.

THEOREM 5.1.10.

(i) P_i *acts on* G_∞ *from the left and* P_i^- *acts on* G_∞ *from the right.*

(ii) *The action of* B *on* G_∞ *induced from the one of* P_i *does not depend on* i.

(iii) *For* $g \in G_i$, $ge = eg$.

Proof. The multiplication homomorphism $\mu_i : U(p_i) \otimes U(\mathcal{G}) \to U(\mathcal{G})$ gives a $\varphi : U(\mathcal{G})^* \to (U(p_i) \otimes U(\mathcal{G}))^*$. We shall show that

$$\varphi(\mathcal{O}(G_\infty)) \subset \mathcal{O}(P_i) \otimes \mathcal{O}(G_\infty). \tag{5.1.6}$$

Then φ is a ring homomorphism and defines $P_i \times G_\infty \to G_\infty$. It is easy to check this is an action of P_i. Similarly $U(\mathcal{G}) \otimes U(p_i^-) \to U(\mathcal{G})$ defines $G_\infty \times P_i^- \to G_\infty$ and it gives the right action of P_i^- on G_∞. The rest is easy to check. Now, we shall show (5.1.6).

Let $f \in \mathcal{O}(G_\infty)$. Then by the definition, there exists a two-sided ideal I of $U(p_i)$ such that $f(IU(\mathcal{G})) = 0$, $U(p_i)/I$ is finite-dimensional and that t acts semisimply and the weights belong to P.

Hence $f \circ \mu_i : U(p_i) \otimes U(\mathcal{G}) \to \mathbf{C}$ splits to $U(p_i) \otimes U(\mathcal{G}) \to (U(p_i)/I) \otimes U(\mathcal{G})$. Hence f belongs to $(U(p_i)/I)^* \otimes U(\mathcal{G})^* \subset \mathcal{O}(P_i) \otimes U(\mathcal{G})^*$. Write $f = \Sigma\, \varphi_k \otimes \psi_k$ with $\varphi_k \in \mathcal{O}(P_i)$ and $\psi_k \in U(\mathcal{G})^*$, such that $\{\varphi_k\}$ is linearly independent. Then there are $R_k \in U(p_i)$ such that $\varphi_k(R_{k'}) = \delta_{kk'}$. Then $\psi_k(P) = f(R_k P)$ for any $P \in U(\mathcal{G})$. Hence $\psi_k \in \mathcal{O}(G_\infty)$ by Proposition 5.1.9.

5.2. For $\Lambda \in t^0$, let us denote $K_\Lambda \in U(\mathcal{G})^*$ given by

$$K_\Lambda : U(\mathcal{G}) \xleftarrow{\sim} U(n) \otimes U(t) \otimes U(n_-) \longrightarrow U(t) \xrightarrow{-\Lambda} \mathbf{C} \tag{5.2.1}$$

where the middle arrow is given by $U(n) \to U(n)/U(n)n \xleftarrow{\sim} \mathbf{C}$ and $U(n_-) \to U(n_-)/U(n_-)n_- \xleftarrow{\sim} \mathbf{C}$ and the last arrow is given by $h \mapsto -\Lambda(h)$. We have in the ring $U(\mathcal{G})^*$

$$K_{\Lambda_1} \cdot K_{\Lambda_2} = K_{\Lambda_1 + \Lambda_2} \quad \text{for} \quad \Lambda_1, \Lambda_2 \in t^0. \tag{5.2.2}$$

$$L(h)K_\Lambda = \langle \Lambda, h \rangle K_\Lambda \quad \text{and} \quad R(h)K_\Lambda = -\langle \Lambda, h \rangle K_\Lambda \quad \text{for} \quad h \in t, \quad \Lambda \in t^0. \tag{5.2.3}$$

LEMMA 5.2.1. *Let* $\varphi \in U(\mathcal{G})^*$ *be a left* b*-finite and right* b_-*-finite element*, a, b *nonnegative integers. Assume that*

$$R(f_i)^{1+a} R(U(n_-))\varphi = 0. \tag{5.2.4}$$

(5.2.5) *Either* $R(e_i)^{1+b}(R(U(n_-))\varphi|_{U(b)}) = 0$ *or* $L(e_i)^{1+b}L(U(n))\varphi = 0$.

(5.2.6) *Assume that* t *acts, by* R, *semisimply on* $(R(U(b_-))\varphi)|_{U(b)} \subset U(b)^*$ *and its weight* Λ *satisfies* $\Lambda(h_i) \leq -a - b$ *and* $\Lambda(h_i) \in \mathbf{Z}$.

Then φ *is* p_i*-finite.*

Proof. Let N be an integer such that $N \geq 1 - \Lambda(h_i)$ for any weight Λ of $R(U(b_-))\varphi|_{U(b)}$. By Lemma 5.1.4, it is enough to show

$$L(f_i)^{N+m}\varphi = 0 \quad \text{if} \quad m \gg 0. \tag{5.2.7}$$

Let I be the ideal of $U(b)$ given by $\{P \in U(b); L(P)\varphi = 0\}$. Then by Lemma 5.1.7 we have $f_i^{N+m}U(\mathfrak{g}) \subset U(n_i)f_i^N\mathbf{C}[e_i]U(b_-) + IU(\mathfrak{g})$. We have

$$f_i^N e_i^k = \sum \frac{N!k!}{(N-\nu)!(k-\nu)!} e_i^{k-\nu}(-h_i - N - k + 2\nu; \nu)f_i^{N-\nu} \tag{5.2.8}$$

where $(x; n) = x(x-1)\cdots(x-n+1)/n!$.

We obtain

$$\varphi(f_i^{N+m}U(\mathfrak{g})) \subset \sum_{0\leq\nu\leq k,N} \varphi(U(n_i)e_i^{k-\nu}(-h_i - N - k + 2\nu; \nu)f_i^{N-\nu}U(b_-)). \tag{5.2.9}$$

Hence it is enough to show

$$\varphi(U(n_i)e_i^{k-\nu}(-h_i - N - k + 2\nu; \nu)U(t)f_i^{N-\nu}U(n_-)) = 0 \quad \text{for} \quad 0 \leq \nu \leq k, N. \tag{5.2.10}$$

If $N - \nu \geq 1 + a$, (5.2.10) holds by (5.2.4). If $k - \nu \geq 1 + b$, (5.2.10) holds by (5.2.5). Hence we may assume $0 \leq N - \nu \leq a$ and $0 \leq k - \nu \leq b$. Then in this case, it is enough to show

$$(R((-h_i - N - k - 2\nu; \nu))R(U(b_-))\varphi)|_{U(b)} = 0. \tag{5.2.11}$$

This is true, if for any weight Λ of $R(U(b_-))\varphi|_{U(b)}$ satisfies

$$0 \leq -\Lambda(h_i) - N - k + 2\nu \leq \nu - 1.$$

This is true if $N \geq 1 - \Lambda(h_i)$, $0 \leq N - \nu \leq a$ and $0 \leq k - \nu \leq b$.

COROLLARY 5.2.2. *$K_\Lambda \in \mathcal{O}(G_\infty)$ if $\Lambda \in P_+$.*
In fact, we can apply the preceding lemma with $a = b = 0$.

5.3. For a subset J of I, we set

$$\Delta_J = \Delta \cap (\sum_{j \in J} \mathbf{Z}\alpha_j) \quad \text{and} \quad \Delta_J^{\pm} = \Delta^{\pm} \cap \Delta_J, \tag{5.3.1}$$

$$\mathcal{G}_J = t \oplus \bigoplus_{\alpha \in \Delta_J} \mathcal{G}_\alpha; \qquad n_J^{\pm} = \bigoplus_{\alpha \in \Delta_{\pm} \setminus \Delta_J} \mathcal{G}_\alpha. \tag{5.3.2}$$

Then $\mathcal{G} = n_J^+ \oplus \mathcal{G}_J \oplus n_J^-$ and $U(\mathcal{G}) \leftrightarrows U(n_J^+) \otimes U(\mathcal{G}_J) \otimes U(n_J^-)$.
We have

$$[\mathcal{G}_J + n_J^+, n_J^+] \subset n_J^+. \tag{5.3.3}$$

Since $\mathcal{G}_J$ is also a Kac-Moody algebra, we set $G_{J\infty}$ the corresponding variety Spec$(A(\mathcal{G}_J, P))$. We also set U_J, U_J^+ the subgroups of U and U^- with the Lie algebra $\hat{n}_J^+$ and $\hat{n}_J^-$. Set

(5.3.4) $A_J = \bigoplus_{\mu \in P} \{\varphi \in U(\mathcal{G})^*; \varphi$ is a weight vector of weight μ with respect to the left action of t and φ is left p_j-finite and right p_j^--finite for any $j \in J$ and φ is left b-finite and right b_--finite$\}$.

Then we can easily show that

(5.3.5) A_J is a subring of $U(\mathcal{G})^*$ and a two-sided sub-$\mathcal{G}$-module of $U(\mathcal{G})^*$.

LEMMA 5.3.4. *$A_J \cong \mathcal{O}(U_J) \otimes \mathcal{O}(G_J) \otimes \mathcal{O}(U_J^-)$.*

Proof. We have

$$\mathcal{O}(U_J) \otimes \mathcal{O}(G_J) \otimes \mathcal{O}(U_J^-) \subset (U(n_J^+) \otimes U(\mathcal{G}_J) \otimes U(n_J^-))^* \cong (U(\mathcal{G}))^*. \tag{5.3.6}$$

We shall show first $A_J \subset \mathcal{O}(U_J) \otimes \mathcal{O}(G_J) \otimes \mathcal{O}(U_J^-)$. For $f \in A_J$, let $\mathfrak{A}$ be the annihilator in $U(b)$ of $L(U(b))f$. Then $f : U(\mathcal{G}) \to \mathbf{C}$ splits into $U(\mathcal{G}) \leftrightarrows U(n_J) \otimes U(\mathcal{G}_J) \otimes U(n_J^-) \to (U(n_J)/(\mathfrak{A} \cap U(n_J))) \otimes U(\mathcal{G}_J) \otimes U(n_J^-)$. Hence f belongs to $\mathcal{O}(U_J) \otimes (U(\mathcal{G}_J) \otimes U(n_J^-))^*$. Similarly f belongs to

$(U(n_J) \otimes U(\mathcal{G}_J)) \otimes \mathcal{O}(U_J)$, and hence to the intersection $\mathcal{O}(U_J) \otimes U(\mathcal{G}_J)^* \otimes \mathcal{O}(U_J^-)$. Write $f = \Sigma_{k=1}^N \varphi_k \otimes \psi_k \otimes \xi_k$ with $\varphi_k \in \mathcal{O}(U_J)$, $\psi_k \in U(\mathcal{G}_J)^*$, $\xi_k \in \mathcal{O}(U_J^-)$. We take an expression such that N is minimal among them. Then there are $S_k^\nu \in U(n_J)$ and $R_k^\nu \in U(n_J^-)$ such that $\varphi_k(S_{k'}^\nu)\psi_k(R_{k'}^\nu) = \delta_{kk'}$. Hence $\psi_k(P) = f(S_k^\nu P R_k^\nu)$. Since A_J is a two-sided $\mathcal{G}$-module, ψ_k belongs to $\mathcal{O}(G_J)$.

We shall prove the converse inclusion $A_J \supset \mathcal{O}(U_J) \otimes \mathcal{O}(G_J) \otimes \mathcal{O}(U_J)$. In order to see this, it is enough to show that any element in $\mathcal{O}(U_J) \otimes \mathcal{O}(G_J) \subset (U(n_J \oplus \mathcal{G}_J))^*$ is b-finite and p_j-finite for any $j \in J$. For any $\varphi \in \mathcal{O}(U_J)$, there exists a two-sided ideal $\mathcal{a}$ of $U(n_J)$ such that $[b, \mathcal{a}] \subset \mathcal{a}$, dim $U(n_J)/\mathcal{a}$ and $\varphi(\mathcal{a}) = 0$. For any $\psi \in \mathcal{O}(G_J)$, there exists an ideal k of $U(\mathcal{G}_J \cap b)$ such that $\dim(U(\mathcal{G}_J \cap b)/k) < \infty$ and $\psi(k) = 0$. Since $bU(n_J) \subset U(n_J) + U(n_J)(b \cap \mathcal{G}_J)$, $U(n_J) \otimes k + \mathcal{a} \otimes U(\mathcal{G}_J)$ is a left b-module. Since $\varphi \otimes \psi$ decomposes into

$$U(n_J) \otimes U(\mathcal{G}_J) \to U(n_J + \mathcal{G}_J)/(U(n_J) \otimes kU(\mathcal{G}_J) + \mathcal{a} \otimes U(\mathcal{G}_J))$$

$$\cong (U(n_J)/\mathcal{a}) \otimes (U(\mathcal{G}_J)/kU(\mathcal{G}_J)),$$

$\varphi \otimes \psi$ is b-finite.

We have

$$(adf_i)^N U(n_J) \subset \mathcal{a} \quad \text{for} \quad N \gg 0 \quad \text{for} \quad i \in J.$$

In fact, this follows from the fact that for any $\lambda \in t^0$, $\lambda + m\alpha_i$ is a weight of $U(n_J)$ except finitely many integer m (Lemma 5.1.6). Hence $\varphi \otimes \psi$ is f_i-finite. Thus, $\varphi \otimes \psi$ is p_i-finite for any $i \in J$. Since $\varphi \otimes \psi$ is b_--finite, we obtain $\varphi \otimes \psi \in A_J$.

Proposition 5.3.5. ([K-P]). *$A_J = \mathcal{O}(G_\infty)[K_\Lambda^{-1}; \Lambda \in P_+, h_j(\Lambda) = 0$ for $j \in J]$.*

Proof. Since K_Λ is invertible in $\mathcal{O}(G_{J\infty})$ if $h_j(\Lambda) = 0$ for $j \in \Lambda$, we have

$$A_J \supset \mathcal{O}(G_\infty)[K_\Lambda^{-1}; \Lambda \in P_+, h_j(\Lambda) = 0 \text{ for } j \in J].$$

Now, we shall show the converse inclusion.

Let $\varphi \in A_J$. Then there exists $a > 0$ such that $R(n_-)^{1+a}\varphi = L(n)^{1+a}\varphi = 0$. Let S be the set of weights of $R(U(b_-))\varphi$ with respect to the right action of t. Taking a sufficiently large, we may assume that $\langle \lambda, h_i \rangle \leq a$ for any $i \in I$ and $\lambda \in S$. Moreover, there exists a finite set K of I such that $R(e_i)\varphi = L(e_i)\varphi = 0$, $\langle \lambda, h_i \rangle = 0$ for any $i \in I \backslash K$ and $\lambda \in S$.

Now, let $\Lambda \in P_+$ be such that $h_j(\Lambda) = 0$ for $j \in J$ and $h_j(\Lambda) \geq a$ for $j \in K \backslash J$. Then $\varphi \cdot K_\Lambda$ is p_j-finite for $j \in J$ and p_j-finite for $j \in I \backslash J$ by Lemma 5.2.1. Hence $\varphi K_\Lambda \in \mathcal{O}(G_\infty)$.

5.4. By Proposition 5.3.5, for finite subsets J and J' with $J \subset J'$, $\mathrm{Spec}(A_J)$ is an open subscheme of $\mathrm{Spec}(A_{J'})$. We set $G_{\infty f} = \bigcup_J U_J \times G_J \times U_J^-$ where J ranges through finite subsets of I. Then $G_{\infty f}$ is an irreducible separated scheme, and $U \times T \times U_-$ is an open subscheme of $G_{\infty f}$. The groups P_i and P_i^- act on $G_{\infty f}$ from the left and the right, respectively.

Definition 5.4.1. Let G be the smallest open subset of $G_{\infty f}$ containing $U \times T \times U_-$ closed by the left and right actions of G_i $(i \in I)$.

5.5. Hence G is invariant by the left action of P_i, and the right action of P_{i-}. Since $G_{\infty f}$ is irreducible, G is also irreducible. In Section 6, we shall study more precisely the structure of $G_{\infty f}$ in the symmetrisable case.

5.6. Since G_i acts on G_∞, $G_{\infty f}$ and G, $s_i' \in G_i$ acts on them. Then we have the braid condition (3.1.13). In fact, if $i, j \in I$ satisfies $\langle h_i, \alpha_j \rangle \langle h_j, \alpha_i \rangle \leq 3$, then the semisimple part of $G_{\{i,j\}}$ is a finite-dimensional group. Thus we can apply the braid condition for finite-dimensional Lie group and hence s_i' and s_j' satisfy the braid condition in $G_{\{i,j\}}$. Since we can check easily that $G_{\{i,j\}}$ acts on G_∞, $G_{\infty f}$ and G, we obtain (3.1.13). Thus the braid group W' acts on G, $G_{\infty f}$ and G_∞.

Let us embed W into W' by $w \mapsto s_{i_1}' \cdots s_{i_l}'$ where $w = s_{i_1} \cdots s_{i_l}$ is a reduced expression of w.

LEMMA 5.6.1. $G = \bigcup_{w \in W} w(U \times T \times U_-)$
$= \bigcup_{w \in W} (U \times T \times U_-)w$.

In fact, we have $P_i^- = G_i \quad U_-$, and $(U \times T \times U_-)P_i^- = Ue \cdot P_i^- = UG_ie \cdot U_- = P_ie \cdot U_-$. Since $P_i \subset s_iBG_i^- \cup BG_i^-$, we have $P_ie \cdot U_- \subset s_iBeU_- \cup Be \cdot U_-$. Thus $\bigcup_{w \in W} w(U \times T \times U_-)$ is invariant by P_{i-}. Hence if A (resp. A') is the smallest open subset containing $U \times T \times U_-$ and invariant by P_i (resp. P_{i-}) for any i, we have $A \supset \bigcup_{w \in W} w(U \times T \times U_-) \supset A'$. Similarly $A \subset A'$. Hence $A = A' = \bigcup_{w \in W} w(U \times T \times U_-)$.

5.7. In general, let X be a scheme and G a group scheme acting on X. We say that G acts locally freely on X if any point has a G-stable open neighborhood which is isomorphic to $G \times U$ for some scheme U. In this case, the quotient X/G in the Zariski topology is representable by a scheme. Note that X/G is not necessarily separated even if X is separated.

5.8. Now, B_- acts on G locally freely. Hence G/B_- is a scheme and covered by open affine subsets $wU \times B_-/B_-$. Note that we have not yet shown that G/B_- is a separated scheme.

PROPOSITION 5.8.1. *$X \cong G/B_-$. Here X is the flag variety defined in Section* 4.

Proof. We have $G/B_- = \bigcup_{w \in W} wUB_-/B_-$ and $X = \bigcup_{w \in W} wUx_0$. We define for $w \in W'$, the morphism

$$\varphi_w : wUB_- \to wUx_0 \quad \text{by} \quad wgb_- \mapsto wg.$$

We shall show

$$\varphi_w = \varphi_{w'} \quad \text{on} \quad wUB_- \cap w'UB_-. \tag{5.8.1}$$

This follows from the case where $w' = 1$. If $w = 1$, this is trivial. If $w = s_i'^{\pm 1}$, then this is trivial because φ_w and φ_1 are the restrictions of $P_i e U_i^- \to X$ given by $geg' \mapsto gx_0$ $(g \in P_i, g' \in U_i^-)$.

Arguing by induction on the length of w, we may assume $w = s_i'^{\pm 1}w''$ and

$$\varphi_{w''}|_{w''UeB_- \cap UeB_-} = \varphi_1|_{w''UeB_- \cap UeB_-}$$

and hence

$$\varphi_w|_{wUeB_- \cap s_i'^{\pm 1}UeB_-} = \varphi_{s_i'^{\pm 1}}|_{wUeB_- \cap s_i'^{\pm 1}UeB_-}.$$

Hence φ_w and φ_1 coincide on $wUeB_- \cap s_i'^{\pm 1}UeB_- \cap UeB_-$. Since $wUeB_- \cap s_i'^{\pm 1}UeB_- \cap UeB_-$ is open dense in $wUB_- \cap w'UB_-$ and X is separated, we have (5.8.1).

Thus, we can construct $\varphi : G \to X$ such that $\varphi|_{wUeB_-} = \varphi_w$. Taking the quotient, we obtain $\tilde{\varphi} : G/B_- \to X$.

By the definition, $\tilde{\varphi}$ is W'-equivariant. Also, $\tilde{\varphi}$ is B-equivariant. This is because $\varphi|_{BeB_-}$ is B-equivariant and BeB_- is open dense in G.

Since $\tilde{\varphi}$ is clearly a local isomorphism and surjective, it is enough to show that $\tilde{\varphi}$ is injective. In order to see this, we shall prove that, for two $\mathbf{C}$-valued points g, g' of G/B_-, $\varphi(g) = \varphi(g')$ implies $g = g'$. Since φ is W'-equivariant, we may assume $g \in BeB_-/B_-$. Since φ is B-equivariant, we may assume $g = e \bmod B_-$. Assume $g' \in wUeb_-/B_-$ for $w \in W$. Write $g' = wuB_-/B_-$ for $u \in U$. Then $\varphi(g) = \varphi(g')$ implies $x_0 = wux_0$. Hence Proposition 4.5.9 implies $w = 1$ and Lemma 4.4.1 implies $u = 1$. Hence $g = g'$.

6. Symmetrisable case.

6.1. In Section 6, we shall assume that the set I of simple roots is finite and the Kac-Moody Lie algebra is symmetrisable. Then by Gabber-Kac [G-K], any integrable $U(\mathfrak{g})$-module generated by a highest weight vector is semisimple. For $\Lambda \in P_+$, let L_Λ be the irreducible $\mathfrak{g}$-module with highest weight Λ. Then we have

LEMMA 6.1.1. ([K-P]). $A(\mathfrak{g}, P) = \mathcal{O}(G_\infty) \cong \bigoplus_{\Lambda \in P_+} L_\Lambda \otimes L_\Lambda^*$.

6.2. We shall assume further that any irreducible finite-dimensional representation of $\mathfrak{g}$ is one-dimensional. This is equivalent to saying that any connected component of the Dynkin diagram of $\mathfrak{g}$ is not finite-dimensional. In this case, letting $P_0 = \{\Lambda \in P;\ \langle \Lambda, h_j \rangle = 0 \text{ for any } j\}$, any irreducible finite-dimensional representation is $\mathbf{C}$ with weight $\Lambda \in P_0$.

LEMMA 6.2.1. *$\bigoplus_{\Lambda \in P_+ \setminus P_0} (L_\Lambda \otimes L_\Lambda^*)$ is an ideal of $A(\mathfrak{g}, P)$.*

Proof. For $\Lambda_1, \Lambda_2 \in P_+ \setminus P_0$,

$$(L_{\Lambda_1} \otimes L_{\Lambda_1}^*) \cdot (L_{\Lambda_2} \otimes L_{\Lambda_2}^*) \subset \sum_{\Lambda} L_\Lambda \otimes L_\Lambda^*$$

where Λ ranges over the set Λ with $L_\Lambda \subset L_{\Lambda_1} \otimes L_{\Lambda_2}$. If $\Lambda \in P_0$ and $L_\Lambda \subset L_{\Lambda_1} \otimes L_{\Lambda_2}$, then we have a homomorphism $L_{\Lambda_1}^* \otimes L_\Lambda \to L_{\Lambda_2}$. Therefore L_{Λ_2} has a lowest weight vector, which implies L_{Λ_2} is finite-dimensional. Hence $\Lambda_2 \in P_0$, which is a contradiction.

Definition 6.2.2. Let us define $\infty \in G_\infty$ by

$$A(\mathcal{G}, P) \to A(\mathcal{G}, P)/(\sum_{\Lambda \in P_+ \setminus P_0} L_\Lambda \otimes L_\Lambda^*) \rightleftarrows \bigoplus_{\Lambda \in P_0} \mathbf{C}K_\Gamma \to \mathbf{C}$$

where the last arrow is given by $K_\Lambda \mapsto 1$.

Note that

$$T \cdot \infty \cong \operatorname{Spec}(\mathbf{C}[K_\Lambda; \Lambda \in P_0]) \tag{6.2.1}$$

$$P_i \infty = \infty P_i^- = T \cdot \infty \quad \text{for any } i. \tag{6.2.2}$$

6.3. PROPOSITION 6.3.1.

$$G_\infty \setminus T \cdot \infty = \bigcup_{\substack{w \in W' \\ J \neq I}} w(U_J \times G_J \times U_J^-) = \bigcup_{\substack{w \in W' \\ J \neq I}} (U_J \times G_J \times U_J^-)w.$$

Proof. The last identity can be proven as in the proof of Lemma 5.6.1. For $v \in L_\Lambda$, $w \in L_\Lambda^*$, let us denote by $\langle v, gw \rangle$ the corresponding function on $g \in G_\infty$. Now, let g be an element of $G_\infty \setminus T \cdot \infty$. Let us denote by G_f the subgroup of $\operatorname{Aut}(L_+)$ generated by the G_i. By the assumption, there is $\Lambda \in P_+ \setminus P_0$ and $v \in L_\Lambda$, $w \in L_\Lambda^*$ such that $\langle v, gw \rangle \neq 0$. Then $\{v' \in L_\Lambda, \langle G_f v', gw \rangle = 0\}$ is a $\mathcal{G}$-module. Hence, it is zero. Therefore, if we denote by v_Λ the highest weight vector of L_Λ, then $\langle G_f v_\Lambda, gw \rangle \neq 0$. Hence there exists $g_0 \in G_f$ such that $\langle v_\Lambda, g_0^{-1} gw \rangle \neq 0$. Since $\bigcup w(U_J \times G_J \times U_J^-)$ is invariant by G_f, we may assume from the beginning $\langle v_\Lambda, gw \rangle \neq 0$.

Similarly, $\{w'; \langle v_\Lambda, gG_f w' \rangle = 0\}$ is $\mathcal{G}$-invariant and hence it is zero. Therefore if $v_{-\Lambda}$ is the lowest weight vector of L_Λ^* such that $\langle v_\Lambda, v_{-\Lambda} \rangle = 1$, then $\langle v_\Lambda, gG_f v_{-\Lambda} \rangle \neq 0$. Hence replacing g with an element in gG_f, we may assume $\langle v_\Lambda, gv_{-\Lambda} \rangle \neq 0$. Since $K_\Lambda(g) = \langle v_\Lambda, gv_{-\Lambda} \rangle \neq 0$, g belongs to $U_{I \setminus \{j\}} \times G_{I \setminus \{j\}} \times U_{I \setminus \{j\}}^-$ for $j \in I$ with $\langle h_j, \Lambda \rangle \neq 0$, by Proposition 5.3.5.

7. Example.

7.1. We shall give here one example $A_\infty^{(1)}$. Let I be $\mathbf{Z}$, $P = \bigoplus_{i \in I} \mathbf{Z}\Lambda_i$, $\alpha_i = 2\Lambda_i - \Lambda_{i+1} - \Lambda_{i-1}$ and $h_i \in t$ is given by $\langle h_i, \Lambda_j \rangle = \delta_{i,j}$.

Let $V' = \mathbf{C}^{\mathbf{Z}} = \Pi_{i \in \mathbf{Z}} \mathbf{C} v_i$, $V_{\leq q} = \Pi_{i \leq q} \mathbf{C} v_i \subset V'$ for $q \in \mathbf{Z}$ and $V = \bigcup V_{\leq q}$. Let us define $g \to \operatorname{End}(V)$ by

$$t \ni h : \sum a_i v_i \mapsto \sum (\Lambda_i(h) - \Lambda_{i-1}(h)) a_i v_i$$

$$e_i : \sum a_j v_j \mapsto a_{i+1} v_i$$

$$f_i : \sum a_j v_j \mapsto a_i v_{i+1}.$$

For $p \leq q$, let $\mathrm{GL}_{p,q}(\infty)$ be the subgroup of $\mathrm{GL}(V)$ given by

$$\{g \in \mathrm{End}(V); g|_{V_{\leq k}} \subset V_{\leq k} \text{ for } k < p \text{ or } k \geq q \text{ and } g|_{V_{\leq k}/V_{\leq k-1}} \text{ is invertible for } k < p \text{ or } k > q \text{ and } g|_{V_{\leq q}/V_{\leq p-1}} \text{ is invertible}\}.$$

This is an affine group scheme. With matrix expression, $\mathrm{GL}_{p,q}(\infty) = \{(g_{ij}); g_{ij} = 0 \text{ for } j < i \text{ and } j < p, j < i \text{ and } i \geq q, g_{ii} \text{ invertible for } i < p \text{ or } i > q \text{ and } \det((g_{ij})_{p \leq i, j \leq q}) \text{ is invertible}\}$. We define the affine group scheme $\mathrm{GL}_{p,q}(\infty)$ by

$$\widetilde{\mathrm{GL}}_{p,q}(\infty) = \mathrm{GL}_{p,q}(\infty) \times \mathbf{C}^*.$$

We define for $p' \leq p \leq q \leq q'$ $\widetilde{\mathrm{GL}}_{p,q}(\infty) \to \widetilde{\mathrm{GL}}_{p',q'}(\infty)$ by

$$(g, c) \mapsto (g, c \det(g|_{V_{\leq q'}/V_{\leq q}})).$$

Then for $p'' \leq p' \leq p \leq q \leq q' \leq q''$,

$$\begin{array}{ccc} \widetilde{\mathrm{GL}}_{p,q}(\infty) & \longrightarrow & \widetilde{\mathrm{GL}}_{p',q'}(\infty) \\ & \searrow & \downarrow \\ & & \widetilde{\mathrm{GL}}_{p'',q''}(\infty) \end{array}$$

commutes. We set

$$\widetilde{\mathrm{GL}}(\infty) = \varinjlim_{(p,q)} \widetilde{\mathrm{GL}}_{p,q}(\infty), \qquad \mathrm{GL}(\infty) = \varinjlim_{(p,q)} \mathrm{GL}_{p,q}(\infty).$$

Then $\widetilde{\mathrm{GL}}(\infty)$ and $\mathrm{GL}(\infty)$ are ind-objects in the category of schemes with group structure. The group $\widetilde{\mathrm{GL}}_{p,q}(\infty)$ coincides with $U_J \times G_J$ where $J = \{i \in \mathbf{Z}; p \leq i \leq q\}$. Note that we have an exact sequence

$$1 \to \mathbf{C}^* \to \widetilde{\mathrm{GL}}(\infty) \to \mathrm{GL}(\infty) \to 1,$$

which does not split.

In this case, the flag variety is, under the notation in Corollary 2.2.3, $\{(W_i)_{i \in \mathbf{Z}};\ W_i \in \text{Grass}^i(V),\ W_i \subset W_{i+1}\}$.

R.I.M.S., KYOTO UNIVERSITY

REFERENCES

[EGA] A. Grothendieck and J. Diendouné, Éléments de géométrie algébrique, *Publ. Math. IHES*, **4** (1960), **8** (1961).

[G-K] O. Gabber and V. G. Kac, On defining relations of certain infinite dimensional Lie algebras, *Bull. Amer. Math. Soc.* **5** (1981), 185–189.

[K] V. G. Kac, *Infinite Dimensional Lie Algebras*, Second ed., Cambridge Univ. Press, 1985.

[K-L] D. A. Kazhdan and G. Lusztig, Schubert varieties and Poincaré duality, *Proc. Symp. in Pure Math. of AMS*, **36** (1980), 185–203.

[K-P] V. G. Kac and D. H. Peterson, [1]Regular functions on certain infinite dimensional groups, in Arithmetic and Geometry (ed. M. Artin and J. Tate), *Progress in Math.* **36**, Birkhäuser, Boston (1983), 141–166. [2]Infinite flag varieties and conjugacy theorems, *Proc. Nat'l Acad. Sci. USA*, **80** (1983), 1778–1782.

[Ku] S. Kumar, Demazure character formula in arbitrary Kac-Moody setting, *Inventiones Math.* **89** (1987), 395–423.

[M] O. Mathieu, Formule de Weyl et de Demazure et Théorème de Borel-Weil-Bott pour les algèbres de Kac-Moody générales (preprint).

[S] P. Slodowy, A character approach to Looijenga's invariant theory for generalized root systems, *Compositio Mat.* (1984).

[T] J. Tits, Groups and group functors attached to Kac-Moody data, in Lecture Notes in Math. **1111**, Springer-Verlag, 1985.

LOGARITHMIC STRUCTURES OF FONTAINE-ILLUSIE

By Kazuya Kato

Introduction. In this note, we present a general formulation of "logarithmic structure" on a scheme found by J. M. Fontaine and L. Illusie. Following their plan, we develop the theory of crystals with logarithmic poles using this logarithmic structure.

The logarithmic structure is "something" which gives rise to differentials with logarithmic poles, crystals and crystalline cohomology with logarithmic poles, . . . etc. For example, a reduced divisor with normal crossings on a regular scheme is such "something," and the logarithmic structure of Fontaine and Illusie is a natural generalization of this example to arbitrary schemes. Their logarithmic structure is defined to be a sheaf of commutative monoids M on the etale site X_{et} of a scheme X, endowed with a homomorphism $M \to \mathcal{O}_X$ satisfying a certain condition. (Cf. Section 1.) For X regular and D a reduced divisor with normal crossings on X, the corresponding M is the sheaf of regular functions on X which are invertible outside D. In general, the homomorphism $M \to \mathcal{O}_X$ is not assumed to be injective.

Algebraic geometry works especially well with smooth morphisms. We can regard the theory of toroidal embeddings as a theory of varieties with smooth logarithmic structures over a field (cf. (3.7)(1)). The logarithmic structure introduces a new range of smoothness, and we expect to have

Manuscript received November 8, 1988

good algebraic geometry for smooth morphisms between logarithmic structures. In subsequent papers [HK] [K′], as was the motivation of Fontaine and Illusie, we apply our theory to schemes with semi-stable reduction, which are examples of schemes with smooth logarithmic structures over discrete valuation rings. (Cf. Complement 2 at the end of this note.)

I am very thankful to Fontaine and Illusie for the original definition of the logarithmic structure, their permission for me to develop their theory in this paper, advice and discussions. I was studying originally crystals with log. poles for regular schemes and reduced divisors with normal crossings, and I wished to write a note on log. str.'s of Fontaine-Illusie to know the best formulation of crystals with logarithmic poles.

Discussions between Illusie and M. Raynaud, and between Illusie and P. Deligne gave good influences to the theory.

The theory of logarithmic structures and crystals with logarithmic poles was developed independently by G. Faltings, and some parts of his papers [Fa_1] [Fa_2] overlap with our study. Our formulation is different from his (cf. Complement 1) and not covered by his theory. The theory of de Rham-Witt complex with logarithmic poles was considered by O. Hyodo [H_1] [H_2] and by M. Gros (unpublished). The theory of N. Katz [K] on connections with logarithmic poles was the guide for our theory.

I thank Université de Paris-Sud and Institut des Haute Etude Scientifique for the supports and hospitality during my study and writing.

1. Logarithmic structures. In this note, a monoid (resp. a ring) means a commutative monoid (resp. ring) with a unit element. A homomorphism of monoids (resp. rings) is required to preserve the unit elements.

For a monoid M, M^{gp} denotes the associated group $\{ab^{-1}; a, b \in M\}$; $ab^{-1} = cd^{-1} \Leftrightarrow sad = sbc$ for some $s \in M$.

For a scheme X and $x \in X$ and for a sheaf $\mathfrak{F}$ on the etale site X_{et}, $\mathfrak{F}_{\bar{x}}$ denotes the stalk of $\mathfrak{F}$ at the separable closure $\bar{x}$ of x. In particular, $\mathcal{O}_{X,\bar{x}}$ denotes the strict henselization of $\mathcal{O}_{X,x}$.

(1.1). Pre-log. structures. Let X be a scheme. A pre-logarithmic structure on X is a sheaf of monoids M on the etale site X_{et} endowed with a homomorphism $\alpha : M \to \mathcal{O}_X$ with respect to the multiplication on $\mathcal{O}_X$.

A morphism $(X, M) \to (Y, N)$ of schemes with pre-log. str.'s is defined to be a pair (f, h) of a morphism of schemes $f : X \to Y$ and a homomorphism $h : f^{-1}(N) \to M$ such that the diagram

$$\begin{array}{ccc} f^{-1}(N) & \xrightarrow{h} & M \\ \downarrow & & \downarrow \\ f^{-1}(\mathcal{O}_Y) & \longrightarrow & \mathcal{O}_X \end{array}$$

is commutative. (We use the notation f^{-1}, not f^*, for the inverse image of a sheaf, for we shall make a special use of the notation f^*, cf. (1.4).)

(1.2). Log. structures. A pre-logarithmic structure (M, α) is called a logarithmic structure if

$$\alpha^{-1}(\mathcal{O}_X^*) \cong \mathcal{O}_X^* \quad \text{via} \quad \alpha$$

where $\mathcal{O}_X^*$ denotes the group of invertible elements of $\mathcal{O}_X$. (We shall often identify $\alpha^{-1}(\mathcal{O}_X^*) \subset M$ with $\mathcal{O}_X^*$ via this isomorphism.) A morphism of schemes with log. str.'s is defined as a morphism of schemes with pre-log. str.'s.

(1.3). The log. str. associated to a pre-log. str. For a pre-log. str. (M, α) on X, we define its associated log. str. M^a to be the push out of

$$\begin{array}{ccc} \alpha^{-1}(\mathcal{O}_X^*) & \longrightarrow & M \\ \downarrow & & \\ \mathcal{O}_X^* & & \end{array}$$

in the category of sheaves of monoids on X_{et}, endowed with

$$M^a \to \mathcal{O}_X; \qquad (a, b) \mapsto \alpha(a)b \quad (a \in M,\ b \in \mathcal{O}_X^*).$$

Then, M^a is universal for homomorphisms of pre-log. str.'s from M to log. str.'s on X.

(*Remark*. If $G \overset{s}{\leftarrow} H \overset{t}{\rightarrow} M$ is a diagram of monoids with G a group, its push out is described as $(M \oplus G)/\sim$, where $(m, g) \sim (m', g') \Leftrightarrow$ there exist $h_1, h_2 \in H$ such that $mt(h_1) = m't(h_2)$, $gs(h_2) = g's(h_1)$.)

(1.4). The direct image and the inverse image. Let $f : X \to Y$ be a morphism of schemes. For a log. str. M on X, we define the log. str. on Y called the direct image of M, to be the fiber product of sheaves

$$\begin{array}{ccc} & & f_*(M) \\ & & \downarrow \\ \mathcal{O}_Y & \longrightarrow & f_*(\mathcal{O}_X)\,. \end{array}$$

For a log. str. M on Y, we define the log. str. on X called the inverse image of M and denoted by $f^*(M)$, to be the log. str. associated to the pre-log. str. $f^{-1}(M)$ endowed with the composite map $f^{-1}(M) \to f^{-1}(\mathcal{O}_Y) \to \mathcal{O}_X$. For a log. str. M on X and for a log. str. N on Y, the following three sets are canonically identified: The set of homomorphisms from N to the direct image of M, the set of homomorphisms from the inverse image of N to M, and the set of extensions of f to a morphism $(X, M) \to (Y, N)$.

The following facts concerning inverse images will be used frequently. Let M be a log. str. on Y.

(1.4.1). $f^{-1}(M/\mathcal{O}_Y^*) \cong (f^*M)/\mathcal{O}_X^*$.

(1.4.2). If M is the log. str. associated to a pre-log. str. M' on Y, $f^*(M)$ coincides with the log. str. associated to the pre-log. str. $f^{-1}(M') \to \mathcal{O}_X$.

(1.5). Examples of log. str.'s. (1) A standard example which we keep in mind is (X, M) where X is a regular scheme with a fixed reduced divisor D with normal crossings, and M is the log. str. on X defined as

$$M = \{g \in \mathcal{O}_X;\ g \text{ is invertible outside } D\} \subset \mathcal{O}_X.$$

The reason why we preferred the etale topology to the Zariski topology in this note is that the definition of "normal crossings" is etale local.

(2) For any scheme X, we call $M = \mathcal{O}_X^* \subset \mathcal{O}_X$ the trivial log. str. on X. This is the initial object in the category of log. str.'s on X. On the other hand, $M = \mathcal{O}_X$ is the final object in this category. The example (1.5)(1) is interpreted to be the direct image of the trivial log. str. on the open sub-scheme $X - D$.

(3) Let P be a monoid, X a scheme, and assume we are given a homomorphism $P \to \Gamma(X, \mathcal{O}_X)$, or equivalently $P_X \to \mathcal{O}_X$ where P_X denotes the constant sheaf on X corresponding to P. Then, let M be the log. str. associated to the pre-log. str. $P_X \to \mathcal{O}_X$. The log. str. of this type will play important roles in this paper. An interpretation of M is the following. For

a ring R, if $R[P]$ denotes the monoid ring on P over R, $\mathrm{Spec}(R[P])$ has a canonical log. str. associated to the canonical map $P \to R[P]$. The above log. str. M on X is the inverse image of the canonical log. str. on $\mathrm{Spec}(\mathbf{Z}[P])$ under the morphism $X \to \mathrm{Spec}(\mathbf{Z}[P])$. We mention what this M is, under a certain assumption.

Claim. In the above, if P has the property "$ab = ac \Rightarrow b = c$" and if X is a scheme over a ring R such that the induced morphism $X \to \mathrm{Spec}(R[P])$ is flat, then M is identified with the sub-monoid sheaf of $\mathcal{O}_X$ generated by $\mathcal{O}_X^*$ and P.

The proof of this claim will be given at the end of this section.

(1.6). Finite inverse limits. The category of schemes with log. str.'s has finite inverse limits. If (X_λ, M_λ) is a finite inverse system, the inverse limit is (X, M) where X is the inverse limit of the system of schemes X_λ, and M is obtained as follows. Let $p_\lambda : X \to X_\lambda$ be the projection, take the inductive limit M' of the inductive system of sheaves of monoids $p_\lambda^{-1}(M_\lambda)$, and then let M be the log. str. associated to the pre-log. str. $M' \to \mathcal{O}_X$.

(1.7). Logarithmic differentials. Let $\alpha : M \to \mathcal{O}_X$ and $\beta : N \to \mathcal{O}_Y$ be pre-log. str.'s, and let $f : (X, M) \to (Y, N)$ be a morphism. Then, we define the $\mathcal{O}_X$-module $\Omega^1_{X/Y}(\log(M/N))$, which is denoted simply by $\omega^1_{X/Y}$ for simplicity when there is no risk of confusion about the pre-log. str.'s, to be the quotient of

$$\Omega^1_{X/Y} \oplus (\mathcal{O}_X \otimes_{\mathbf{Z}} M^{gp})$$

($\Omega^1_{X/Y}$ is the usual relative differential module) divided by the $\mathcal{O}_X$-submodule generated locally by local sections of the following forms.

(i) $(d\alpha(a), 0) - (0, \alpha(a) \otimes a)$ with $a \in M$.
(ii) $(0, 1 \otimes a)$ with $a \in \mathrm{Image}(f^{-1}(N) \to M)$.

The class of $(0, 1 \otimes a)$ for $a \in M$ in $\omega^1_{X/Y}$ is denoted by $d\log(a)$.

It is easily seen that if M^a and N^a denote the associated log str.'s, respectively, we have

$$\Omega^1_{X/Y}(\log(M/N)) = \Omega^1_{X/Y}(\log(M^a/N)) = \Omega^1_{X/Y}(\log(M^a/N^a)).$$

If M and N are log. str.'s, we have a surjection

$$\mathcal{O}_X \otimes_{\mathbf{Z}} M^{gp} \to \omega^1_{X/Y}; \qquad a \otimes b \to a.d\log(b),$$

and the kernel is the $\mathcal{O}_X$-submodule generated locally by local sections of the forms

(i) $\alpha(a) \otimes a - \Sigma_i u_i \otimes u_i$ with $a \in M$ and $u_i \in \mathcal{O}_X^*$ such that $\alpha(a) = \Sigma_i u_i$,

(ii) $1 \otimes a$ with $a \in \mathrm{Image}(f^{-1}(N) \to M)$.

If we have a cartesian diagram of schemes with log. str.'s

$$\begin{array}{ccc} (X', M') & \xrightarrow{f} & (X, M) \\ \downarrow & & \downarrow \\ (Y', N') & \longrightarrow & (Y, N), \end{array}$$

we have an isomorphism

$$f^*\omega^1_{X/Y} \cong \omega^1_{X'/Y'}.$$

(1.8). For example, let P and Q be monoids, $Q \to P$ a homomorphism, R a ring, $X = \mathrm{Spec}(R[P])$, $Y = \mathrm{Spec}(R[Q])$, and endow X and Y with the canonical log. str.'s (1.5)(3), respectively. Then,

$$\mathcal{O}_X \otimes_{\mathbf{Z}} (P^{gp}/\mathrm{Image}(Q^{gp})) \cong \omega^1_{X/Y}; \qquad a \otimes b \mapsto a.d\log(b).$$

(1.9). In the situation (1.7), we define $\omega^{\cdot}_{X/Y}$ to be the exterior algebra on the $\mathcal{O}_X$-module $\omega^1_{X/Y}$. It becomes a complex of $f^{-1}(\mathcal{O}_Y)$-modules in the natural way.

(1.10) Proof of the Claim in (1.5)(3). The problem is the injectivity of $M \to \mathcal{O}_X$. By the description of the push out in Remark in (1.3), it suffices to prove the following: If $x \in X$, $a, b \in P$ and $ab^{-1} \in \mathcal{O}^*_{X,\bar{x}}$, then there exists $c, d \in P$ such that $c, d \in \mathcal{O}^*_{X,\bar{x}}$ and $ac = bd$. (Note an element of P is a nonzero-divisor on X by the flatness assumption, and hence the expression $ab^{-1} \in \mathcal{O}^*_{X,\bar{x}}$ makes sense.) Let $\mathfrak{p}$ be the image of x in $\mathrm{Spec}(R[P])$. Then, since $R[P]_{\mathfrak{p}} \to \mathcal{O}_{X,\bar{x}}$ is faithfully flat, we have $ab^{-1} \in R[P]^*_{\mathfrak{p}}$. Hence $\exists f, g \in R[P]$ which are not contained in the prime ideal $\mathfrak{p}$ such that $af = bg$ in

$R[P]$. Write $f = \Sigma_c f_c c$, $g = \Sigma_c g_c c$, where $c \in P$, f_c, $g_c \in R$. Take $c \in P$ such that f_c, $c \notin \mathfrak{p}$. Then, the equation $af = bg$ shows that there exists $d \in P$ such that $ac = bd$. Since $c \notin \mathfrak{p}$ and $ab^{-1} \in R[P]^*_{\mathfrak{p}}$, we have $d \notin \mathfrak{p}$.

2. Fine log. structures.

(2.1). A log. str. M on a scheme X is called quasi-coherent (resp. coherent) if etale locally on X, there exists a monoid (resp. finitely generated monoid) P and a homomorphism $P_X \to \mathcal{O}_X$ whose associate log. str. is isomorphic to M.

If $(X, M) \to (Y, N)$ is a morphism of schemes with quasi-coherent log. str.'s, $\omega^1_{X/Y}$ (1.7) is a quasi-coherent $\mathcal{O}_X$-module. If furthermore M is coherent and X is noetherian and locally of finite type over Y, it is a coherent $\mathcal{O}_X$-module.

(2.2). A monoid is called integral if "$ab = ac \Rightarrow b = c$" holds. A log. str. M on a scheme X is called integral if M is a sheaf of integral monoids.

(2.3). We call a log. str. "fine" if it is coherent and integral.

In this note, we consider mainly fine logarithmic structures. (To my experience, nonintegral log. str.'s are too much pathological.)

(2.4). The following facts are proved easily.

(2.4.1). If $f : X \to Y$ is a morphism and M is a quasi-coherent (resp. coherent, resp. integral) log. str. on Y, so is $f^*(M)$.

(2.4.2). A quasi-coherent (resp. coherent) log. str. M on a scheme X is integral if and only if etale locally on X, M is isomorphic to the log. str. associated to the pre-log. str. $P_X \to \mathcal{O}_X$ for some integral (resp. finitely generated integral) monoid P.

(2.4.3). If M is coherent (resp. integral), the stalk $M_{\bar{x}}/\mathcal{O}^*_{X,\bar{x}}$ is a finitely generated (resp. integral) monoid for any $x \in X$.

Example (2.5). (1) The log. str. on a regular scheme X corresponding to a reduced divisor with normal crossings (1.5)(1) is fine. Indeed, etale locally on X, write $D = \cup_i$ "$\pi_i = 0$" where "$\pi_i = 0$" are regular closed subschemes of X. Then, M is associated to the pre-log. str.

$$N^r \to \mathcal{O}_X; \qquad (n_i)_{1 \leqq i \leqq r} \mapsto \prod_i \pi_i^{n_i}$$

and the monoid $\mathbf{N}^r$ is clearly finitely generated and integral.

(2) If $X = \mathrm{Spec}(k)$ for an algebraically closed field k, there is a bijection between the two sets

{isomorphism classes of integral log. str.'s on X}

{isomorphism classes of integral monoids having no invertible element other than the unit element}

given in the following way. For an integral monoid P as above, the corresponding log. str. is $M = \mathcal{O}_X^* \oplus P$ with

$$M \to \mathcal{O}_X; \qquad (a, b) \to a \quad (\text{resp. } 0) \quad \text{if} \quad b = 1 \quad (\text{resp. } b \neq 1).$$

Proposition (2.6). *The inverse limit of a finite inverse system of schemes with* log. str.'s (1.6) *is coherent if each member of the system is coherent.*

Proof. It is enough to consider finite direct products and equalizers. By the description of finite inverse limits in (1.6), the case of finite direct products is clear, and for equalizers, it is enough to prove the following. Assume we are given two homomorphisms $g, h : M \to N$ of log. str.'s on X, let L' be the co-equalizer of (g, h) in the category of sheaves of monoids, and let L be the log. str. associated to L'. Then, if M, N are coherent, L is coherent. To see this, take finitely generated monoids P, Q and homomorphisms $s : P_X \to M$, $t : Q_X \to N$ which induce $(P_X)^a \cong M$, $(Q_X)^a \cong N$. If we have homomorphisms $g', h' : P \to Q$ compatible with g and h, respectively, then L is associated to the coequalizer of (g', h') in the category of monoids (which is endowed with the induced homomorphism to $\mathcal{O}_X$), and hence is coherent. In general we may not have such (g', h'), but considering the commutative diagram

$$\begin{array}{ccc} P & \xrightarrow{i_1 (\text{resp. } i_2)} & Q' = P \oplus P \oplus Q \\ \downarrow & & \downarrow {\scriptstyle t' = (g,h,t)} \\ M & \xrightarrow{g (\text{resp. } h)} & N \end{array}$$

($i_1(x) = (x, 0, 0)$, $i_2(x) = (0, x, 0)$), we see that it is enough to construct etale locally a finitely generated monoid Q'' and a factorization of t' as $Q' \to Q'' \to N$ such that $(Q''_X)^a \cong N$. Fix $x \in X$ and take a system of generators $(a_i)_{1 \leqq i \leqq r}$ of Q'. Then $t'(a_i)_{\bar{x}} = t'(b_i)_{\bar{x}} u_i$ for some $b_i \in Q$ and $u_i \in \mathcal{O}^*_{X,\bar{x}}$. Let Q'' be the monoid $(Q' \oplus \mathbf{N}^r)/\sim$ where $\sim$ is the relation generated by the relations $a_i = b_i e_i (1 \leqq i \leqq r)$ with $(e_i)_i$ the canonical base of $\mathbf{N}^r$. On an etale neighbourhood U of $\bar{x}$, we extend t' to $t'' : Q'' \to M$ by $e_i \mapsto u_i$. Then, $(Q_U)^a \to (Q''_U)^a$ is surjective and the composite $(Q_U)^a \to (Q''_U)^a \to M|_U$ is an isomorphism. Hence $(Q''_U)^a \cong M|_U$.

PROPOSITION (2.7). *The inclusion functor from the category of schemes with fine* log. str.'s *to the category of schemes with coherent* log. str.'s *has a right adjoint.*

Proof. Let (X, M) be a scheme with a coherent log. str. We construct (X', M') over (X, M) with M' fine which is universal for morphisms from schemes with fine log. str.'s. We may work etale locally, and hence assume that we have $X \to \mathrm{Spec}(\mathbf{Z}[P])$ which induces M. Let

$$X' = X \times_{\mathrm{Spec}(\mathbf{Z}[P])} \mathrm{Spec}(\mathbf{Z}[P^{int}])$$

where $P^{int} = \mathrm{Image}(P \to P^{gp})$, and let M' be the log. str. on X' induced by $X' \to \mathrm{Spec}(\mathbf{Z}[P^{int}])$. It is easy to see that this (X', M') is universal.

We shall denote the above universal (X', M') by $(X, M)^{int}$.

(2.8). If (X_λ, M_λ) is a finite inverse system of schemes with fine log. str.'s and (X, M) is its inverse limit (1.6) in the category of schemes with log. str.'s, $(X, M)^{int}$ (2.7) is the inverse limit of (X_λ, M_λ) in the category of schemes with fine log. str.'s. Various properties of morphisms between schemes with fine log. str.'s defined in later sections (smoothness, etaleness, etc.) are preserved by base changes using the fiber products in the category of schemes with fine log. str.'s.

Definition (2.9). (1) For a scheme X with a fine log. str. M, a chart of M is a homomorphism $P_X \to M$ for a finitely generated integral monoid P which induces $(P_X)^a \cong M$.

A chart of M exists etale locally.

(2) For a morphism $f : (X, M) \to (Y, N)$ of schemes with fine log. str.'s, a chart of f is a triple $(P_X \to M, Q_Y \to N, Q \to P)$ where $P_X \to M$,

$Q_Y \to N$ are charts of M and N, respectively, and $Q \to P$ is a homomorphism for which

$$\begin{array}{ccc} Q_X & \longrightarrow & P_X \\ \downarrow & & \downarrow \\ f^{-1}(N) & \longrightarrow & M \end{array}$$

is commutative.

A chart of f also exists etale locally. This fact is deduced easily from

LEMMA (2.10). *Let X be a scheme with a fine* log. str. *M, let $x \in X$, G a finitely generated abelian group, and let $h : G \to M_{\bar{x}}^{gp}$ be a homomorphism such that $G \to M_{\bar{x}}^{gp}/\mathcal{O}_{X,\bar{x}}^*$ is surjective. Let $P = (h^{gp})^{-1}(M_{\bar{x}})$. Then, $P \to M_{\bar{x}}$ is extended to a chart $P_U \to M|_U$ for an etale neighbourhood U of $\bar{x}$.*

Proof. First, P is finitely generated since

$$(*) \qquad P/(\text{a subgroup}) \cong M_{\bar{x}}/\mathcal{O}_{X,\bar{x}}^*$$

and $M_{\bar{x}}/\mathcal{O}_{X,\bar{x}}^*$ is finitely generated. When we extend $P \to M_{\bar{x}}$ to a homomorphism $P_U \to M|_U$ for an etale neighbourhood U of $\bar{x}$, (*) proves $((P_U)^a)_{\bar{x}} \cong M_{\bar{x}}$. This shows that $(P_{U'})^a \cong M|_{U'}$ for an etale neighbourhood U' of $\bar{x} \to U$.

3. Smooth morphisms.

(3.1). We call a morphism of schemes with log. str.'s $i : (X, M) \to (Y, N)$ a closed immersion (resp. an exact closed immersion) if the underlying morphism of schemes $X \to Y$ is a closed immersion and $i^*N \to M$ is surjective (resp. an isomorphism).

(3.2). We shall often consider a commutative diagram of schemes with fine log. str.'s

$$\begin{array}{ccc} (T', L') & \xrightarrow{s} & (X, M) \\ {\scriptstyle i}\downarrow & & \downarrow{\scriptstyle f} \\ (T, L) & \xrightarrow{t} & (Y, N) \end{array}$$

such that i is an exact closed immersion (3.1) and T' is defined in T by an ideal I such that $I^2 = (0)$.

(3.3). Smoothness and etaleness. A morphism $f : (X, M) \to (Y, N)$ of schemes with fine log. str.'s is called smooth (resp. etale) if the underlying morphism $X \to Y$ is locally of finite presentation and if for any commutative diagram as in (3.2), there exists etale locally on T (resp. there exists a unique) $g : (T, L) \to (X, M)$ such that $gi = s$ and $fg = t$.

A standard example of a smooth (resp. etale) morphism is given by the following (3.4). In (3.5), which is the main result of this section, we shall see that all smooth (resp. etale) morphisms are essentially of the type of this standard example.

PROPOSITION (3.4). *Let P, Q be finitely generated integral monoids, $Q \to P$ a homomorphism, R a ring, such that the kernel and the torsion part of the cokernel (resp. the kernel and the cokernel) of $Q^{gp} \to P^{gp}$ are finite groups whose orders are invertible in R. Let*

$$X = \mathrm{Spec}(R[P]), \qquad Y = \mathrm{Spec}(R[Q])$$

and endow them with the canonical log. str.'s *M and N, respectively. Then*, the morphism *$(X, M) \to (Y, N)$ is smooth (resp. etale).*

Proof. Consider a commutative diagram as in (3.2). Then, if we embed I in L via the injective homomorphism

$$I \to \mathcal{O}_T^* \subset L; \qquad x \mapsto 1 + x,$$

we have a cartesian diagram

(3.4.1)
$$\begin{array}{ccc} L & \longrightarrow & L/I = L' \\ \downarrow & & \downarrow \\ L^{gp} & \longrightarrow & L^{gp}/I = (L')^{gp} \end{array}$$

By the assumption on $Q^{gp} \to P^{gp}$, we have the following dotted arrow etale locally (resp. uniquely) which makes the diagram commutative;

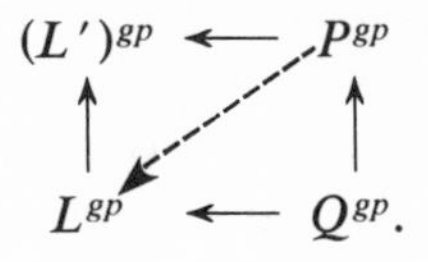

By the cartesian diagram (3.4.1), we obtain $P \to L$ which induces the desired morphism $(T, L) \to (X, M)$.

THEOREM (3.5). *Let $f : (X, M) \to (Y, N)$ be a morphism of schemes with fine* log. str.'s. *Assume we are given a chart* (2.9) $Q_Y \to N$ *of* N. *Then the following conditions* (3.5.1) *and* (3.5.2) *are equivalent.*

(3.5.1). f is smooth (resp. etale).

(3.5.2). Etale locally on X, there exists a chart $(P_X \to M, Q_Y \to N, Q \to P)$ of f (2.9) extending the given $Q_Y \to N$ satisfying the following conditions (i)(ii).

(i) The kernel and the torsion part of the cokernel (resp. The kernel and the cokernel) of $Q^{gp} \to P^{gp}$ are finite groups of orders invertible on X.

(ii) The induced morphism from $X \to Y \times_{\mathrm{Spec}(\mathbf{Z}[Q])} \mathrm{Spec}(\mathbf{Z}[P])$ is etale (in the classical sense).

Remark (3.6). The proof of (3.5) will show the following facts. We can require in the condition (3.5.2)(i) that $Q^{gp} \to P^{gp}$ is injective, without changing the conclusion of (3.5). In the part concerning the smoothness of f, we can replace the etaleness of the morphism from X to the fiber product in (3.5.2), by the smoothness (also in the classical sense), without changing the conclusion of (3.5).

Examples (3.7). (1) Let k be a field and let X be a scheme over k locally of finite type with a fine log. str. M. Then, by (3.5), (X, M) is smooth over $\mathrm{Spec}(k)$ if and only if etale locally on X, there exists a finitely generated integral monoid P and an etale morphism $X \to \mathrm{Spec}(k[P])$ satisfying the following conditions; $M = P\mathcal{O}_X^* \subset \mathcal{O}_X$, the torsion part of P^{gp} is of order invertible in k. Thus (X, M) corresponds to a toroidal embedding [KKMS] which is locally given by the open immersion

$$X \times_{\mathrm{Spec}(k[P])} \mathrm{Spec}(k[P^{gp}]) \subset X.$$

We assume P^{gp} is torsion free in the usual theory of toroidal embeddings, but essentially, the theory of toroidal embeddings is nothing but the theory of schemes with smooth fine log. str.'s over a field (with respect to the trivial log. str. on the base field).

(2) Let A be a discrete valuation ring, X a regular scheme over A such that etale locally on X, there is a smooth morphism $X \to$

$\mathrm{Spec}(A[T_1, \ldots, T_r]/(T_1 \cdots T_r - \pi))$ for $r \geqq 1$ and a prime element π of A. (In this situation, X is called of semi-stable reduction over A.) Then, if M (resp. N) denotes the log. str. on X (resp. $\mathrm{Spec}(A)$) corresponding to the special fiber of X (resp. the closed point of $\mathrm{Spec}(A)$), which is a reduced divisor with normal crossings on a regular scheme, the morphism $(X, M) \to (\mathrm{Spec}(A), N)$ is smooth.

For the proof of (3.5), we use the following facts.

PROPOSITION (3.8). *Let $f : (X, M) \to (Y, N)$ be a morphism of schemes with fine* log. str.'s *such that $f^*N \xrightarrow{\cong} M$. Then f is smooth (resp. etale) if and only if the underlying morphism $X \to Y$ is smooth (resp. etale).*

Proof. Exercise.

PROPOSITION (3.9). *In* (3.2), *assume we are given one morphism $g : (T, L) \to (X, M)$ such that $gi = s$ and $fg = t$. Then there exists a bijection*

$$\{h : (T, L) \to (X, M);\ hi = s,\ fh = t\} \to \mathrm{Hom}_{\mathcal{O}_{T'}}(s^*\omega^1_{X/Y}, I)$$

which sends h to the homomorphism

$$da \to h^*(a) - g^*(a) \quad \text{for} \quad a \in \mathcal{O}_X,$$

$$d\log(a) \to u(a) - 1 \quad \text{for} \quad a \in M,$$

where $u(a)$ is the unique local section of $\mathrm{Ker}(\mathcal{O}_T^* \to \mathcal{O}_{T'}^*) \subset L$ *such that $h^*(a) = g^*(a)u(a)$.*

Proof. Exercise.

The proofs of the following (3.10) (3.12) are reduced to (3.9) in the same way as in the theory of the classical smoothness.

PROPOSITION (3.10). *Let $f : (X, M) \to (Y, N)$ be a smooth morphism of schemes with fine* log. str.'s. *Then the $\mathcal{O}_X$-module $\omega^1_{X/Y}$ is locally free of finite type.*

COROLLARY (3.11). *If $f : (X, M) \to (Y, N)$ is smooth in the diagram* (3.2), *a morphism $g : (T, L) \to (X, M)$ such that $gi = s$ and $fg = t$ exists whenever T is affine.*

Indeed, the obstruction to glueing local g lies in $H^1(T', \mathrm{Hom}_{\mathcal{O}_{T'}}(s^*\omega^1_{X/Y}, I)) = (0)$.

PROPOSITION (3.12). *Let* $(X, M) \xrightarrow{f} (Y, N) \xrightarrow{g} (S, L)$ *be morphisms of schemes with fine* log. str.'s, *and let*

$$f^*\omega^1_{X/S} \xrightarrow{s} \omega^1_{Y/S} \to \omega^1_{X/Y} \to 0$$

be the associated exact sequence. Consider the following conditions.

(i) f *is smooth* (*resp. etale*).

(ii) s *is injective and the image of* s *is locally a direct summand* (*resp.* s *is an isomorphism*).

Then, we have the implication (i) $\Rightarrow$ (ii). If gf is smooth, we have (ii) $\Rightarrow$ (i).

(3.13). *Proof of* (3.5). The implication (3.5.2) $\Rightarrow$ (3.5.1) follows from (3.4) (3.8).

We prove the converse. We construct P etale locally as follows.

Fix $x \in X$. Take elements $t_1, \ldots, t_r$ of $M_{\bar{x}}$ such that $(d\log(t_i))_{1 \leq i \leq r}$ is a basis of $\omega^1_{X/Y,\bar{x}}$ (3.10). Consider

$$\mathbf{N}^r \oplus Q \to M_{\bar{x}}$$

induced by $\mathbf{N}^r \to M_{\bar{x}}$; $(m_i) \to \Pi_i\, t_i^{m_i}$ and by $Q \to f^{-1}(N)_{\bar{x}} \to M_{\bar{x}}$. Note $M_{\bar{x}}/\mathcal{O}^*_{X,\bar{x}}$ is finitely generated. The map

$$\omega^1_{X/Y,\bar{x}} \to \kappa(\bar{x}) \otimes_{\mathbf{Z}} (M^{gp}_{\bar{x}}/(\mathcal{O}^*_{X,\bar{x}}\ \mathrm{Image}(f^{-1}(N)^{gp}_{\bar{x}}))$$

$$d\log(a) \mapsto 1 \otimes a \quad (a \in M_{\bar{x}})$$

shows that

$$\kappa(\bar{x}) \otimes_{\mathbf{Z}} (\mathbf{Z}^r \oplus Q^{gp}) \to \kappa(\bar{x}) \otimes_{\mathbf{Z}} (M^{gp}_{\bar{x}}/\mathcal{O}^*_{X,\bar{x}})$$

is surjective, and hence the cokernel of

$$\mathbf{Z}^r \oplus Q^{gp} \to M^{gp}_{\bar{x}}/\mathcal{O}^*_{X,\bar{x}}$$

is a finite group annihilated by an integer which is invertible in $\mathcal{O}_{X,\bar{x}}$. By using the fact that $\mathcal{O}^*_{X,\bar{x}}$ is n-divisible, we can easily construct a finitely generated abelian group $G \supset \mathbf{Z}^r \oplus Q^{gp}$ such that $G/(\mathbf{Z}^r \oplus Q^{gp})$ is annihilated by n and such that the map $\mathbf{Z}^r \oplus Q^{gp} \to M_{\bar{x}}^{gp}$ is extended to $h: G \to M_{\bar{x}}^{gp}$ which induces a surjection $G \to M_{\bar{x}}^{gp}/\mathcal{O}_{X,\bar{x}}^*$. Let $P = h^{-1}(M_{\bar{x}})$. Then $Q^{gp} \to P^{gp} = G$ is injective and the torsion part of P^{gp}/Q^{gp} is annihilated by n. We have

$$\mathcal{O}_{X,\bar{x}} \otimes_{\mathbf{Z}} (P^{gp}/Q^{gp}) \cong \omega^1_{X/Y,\bar{x}}.$$

By (3.10), by replacing X by an etale neighbourhood of $\bar{x}$, we have

$$(*) \qquad \mathcal{O}_X \otimes_{\mathbf{Z}} (P^{gp}/Q^{gp}) \cong \omega^1_{X/Y}.$$

Furthermore, by (2.10), $P \to M_{\bar{x}}$ is extended to a chart of $M|_U$ for some etale neighbourhood U of $\bar{x}$. By replacing X with U, we have a morphism g from X to the fiber product $X' = Y \times_{\mathrm{Spec}(\mathbf{Z}[Q])} \mathrm{Spec}(\mathbf{Z}[P])$. It remains to prove g is etale. To see this, endow X' with the inverse image of M' of the canonical log. str. of $\mathrm{Spec}(\mathbf{Z}[P])$. Since the inverse image of M' on X is M, it is sufficient (3.8) to show that $(X, M) \to (X', M')$ is etale. But this follows from (3.12) and $g^*\omega^1_{X'/Y} \cong \omega^1_{X/Y}$ ((1.8) and (*) above).

The theory of infinitesimal liftings for smooth morphisms hold in the logarithmic situation as follows. This (3.14) and the related theorem (4.12) were obtained following faithfully suggestions of L. Illusie.

Proposition (3.14). *Let $f: (X, M) \to (Y, N)$ be a smooth morphism between schemes with fine* log. str.'s, *and let $i: (Y, N) \to (\tilde{Y}, \tilde{N})$ with N fine be an exact closed immersion* (3.2) *such that Y is defined in $\tilde{Y}$ by a nilpotent ideal I of $\mathcal{O}_Y$. Then we have:*

(1) *If X is affine, a smooth lifting of (X, M, f) exists and is unique up to isomorphism. Here, by a smooth lifting of (X, M, f), we mean a scheme $\tilde{X}$ with a fine* log. str. *$\tilde{M}$ endowed with a smooth morphism $\tilde{f}: (\tilde{X}, \tilde{M}) \to (\tilde{Y}, \tilde{N})$ and with an isomorphism*

$$g: (X, M) \cong (\tilde{X}, \tilde{M}) \times_{(\tilde{Y},\tilde{N})} (Y, N) \quad \textit{over} \quad (Y, N).$$

(2) *Assume $I^2 = (0)$. Then, for a smooth lifting $(\tilde{X}, \tilde{M}, \tilde{f}, g)$ of (X, M, f), there exists a canonical isomorphism*

$$\mathrm{Aut}(\tilde{X}, \tilde{M}, \tilde{f}, g) = \mathrm{Hom}_{\mathcal{O}_X}(\omega^1_{X/Y}, I\mathcal{O}_{\tilde{X}}).$$

(3) *Assume* $I^2 = (0)$. *If we are given one fixed smooth lifting of* (X, M, f), *we have a bijection from the set of all isomorphism classes of smooth liftings of* (X, M, f) *to*

$$H^1(X, \mathcal{H}om_{\mathcal{O}_X}(\omega^1_{X/Y}, I\mathcal{O}_{\tilde{X}})).$$

(4) *A smooth lifting of* (X, M, f) *exists if* $I^2 = (0)$ *and if*

$$H^2(X, \mathcal{H}om_{\mathcal{O}_X}(\omega^1_{X/Y}, I\mathcal{O}_{\tilde{X}})) = (0).$$

Proof. Once we prove that a smooth lifting exists etale locally on X, the statements in (3.14) are deduced from it by the classical arguments as in SGA I Exposé 3. Etale locally we have a chart ($P_X \to M$, $Q_Y \to N$, $Q \to P$) of f satisfying the condition (3.5.2) such that $Q_Y \to N$ factors through a chart $Q_Y \to \tilde{N}$ of $\tilde{N}$. Let

$$X' = Y \times_{\mathrm{Spec}(\mathbf{Z}[Q])} \mathrm{Spec}(\mathbf{Z}[P]), \qquad \tilde{X}' = \tilde{Y} \times_{\mathrm{Spec}(\mathbf{Z}[Q])} \mathrm{Spec}(\mathbf{Z}[P]).$$

Lift the etale scheme X over X' to an etale scheme $\tilde{X}$ over $\tilde{X}'$ (this is classical; SGA I Exposé 1), and endow $\tilde{X}$ with the inverse image $\tilde{M}$ of the canonical log. str. on $\mathrm{Spec}(\mathbf{Z}[P])$. Then, $(\tilde{X}, \tilde{M}, \tilde{f}, g)$ with the evident definitions of $\tilde{f}$, g is a smooth lifting.

4. Several types of morphisms. We define integral morphisms (4.3), exact morphisms (4.6), and morphisms of Cartier type (4.8), describe their properties, and prove a Cartier isomorphism (4.12).

PROPOSITION (4.1). (1) *Let* $h : Q \to P$ *be a homomorphism of integral monoids. Then, the following conditions* (i) *and* (iv) *are equivalent* (*resp.* (ii), (iii) *and* (v) *are equivalent*).

(i) *For any integral monoid* Q' *and for any homomorphism* $g : Q \to Q'$, *the push out of* $P \leftarrow Q \to Q'$ *in the category of monoids is integral.*

(ii) *The homomorphism* $\mathbf{Z}[Q] \to \mathbf{Z}[P]$ *induced by* h *is flat.*

(iii) *For any field* k, *the homomorphism* $k[Q] \to k[P]$ *induced by* h *is flat.*

(iv) *If* $a_l, a_2 \in Q$, $b_1, b_2 \in P$ *and* $h(a_1)b_1 = h(a_2)b_2$, *there exist* $a_3, a_4 \in Q$ *and* $b \in P$ *such that* $b_1 = h(a_3)b$, $b_2 = h(a_4)b$, *and* $a_1a_3 = a_2a_4$.

(v) *The condition* (iv) *is satisfied and* h *is injective.*

(2) *Let* $f : (X, M) \to (Y, N)$ *be a morphism of schemes with integral* log. str.'s. *Then, for* $x \in X$, *the conditions* (i)–(v) *in* (1) *for* $Q = (f^*N)_{\bar{x}}$ *and* $P = M_{\bar{x}}$ *are equivalent, and they are equivalent to each of* (i)–(v) *for* $Q = f^{-1}(N/\mathcal{O}_Y^*)_{\bar{x}}$ *and* $P = (M/\mathcal{O}_X^*)_{\bar{x}}$.

(4.2). *Proof of* (4.1). (1) We omit the proof of (4.1)(2) since it is easy by considering the conditions (4.1)(1)(iv) and (v).

(i) $\Rightarrow$ (iv). Let $a_1, a_2 \in Q$, $b_1, b_2 \in P$ and $h(a_1)b_1 = h(a_2)b_2$. Define $Q' = (Q \oplus \mathbf{N}^2)/\sim$, where $\sim$ is the equivalence relation

$$(c, m, n) \sim (c', m', n') \Leftrightarrow$$

$$m + n = m' + n' \quad \text{and} \quad ca_1^m a_2^n = c' a_1^{m'} a_2^{n'},$$

and let P' be the push out of $P \leftarrow Q \to Q'$. Since Q' is integral and (i) is satisfied, P' is also integral and we see from this that $(b_1, 1, 0)$ and $(b_2, 0, 1)$ coincide in P'. It follows that there exists a sequence $v_0, \ldots, v_r$ of elements of $P \oplus \mathbf{N}^2$ such that $v_0 = (b_1, 1, 0)$, $v_r = (b_2, 0, 1)$ and such that for each $i = 1, \ldots, r$, there exist $c, c' \in Q$, $m, n, m', n' \in \mathbf{N}$ and $w \in P \oplus \mathbf{N}^2$ satisfying $v_{i-1} = (h(c), m, n)w$, $v_i = (h(c'), m', n')w$, $m + n = m' + n'$, $ca_1^m a_2^n = c' a_1^{m'} a_2^{n'}$. As is easily seen, this implies that there exist $a_3, a_4 \in Q$ and $b \in P$ such that $b_1 = h(a_3)b$, $b_2 = h(a_4)b$ and $a_1 a_3 = a_2 a_4$.

(iv) $\Rightarrow$ (i). Let P' be the push out. We prove the surjection $P' \to (P')^{int}$ is bijective. Let $b_1, b_2 \in P$, $c_1, c_2 \in Q'$ and assume $b_1 c_1$ and $b_2 c_2$ coincide in $(P')^{int}$. Then, an easy observation on the push out of $P^{gp} \leftarrow Q^{gp} \to (Q')^{gp}$ shows that there exist $a_1, a_2 \in Q$ such that $h(a_1)b_1 = h(a_2)b_2$ in P and $g(a_2)c_1 = g(a_1)c_2$ in Q'. By the condition (iv), we have $b_1 = h(a_3)b$, $b_2 = h(a_4)b$, $a_1 a_3 = a_2 a_4$ for some $a_3, a_4 \in Q$, $b \in P$. We have $g(a_3)c_1 = g(a_4)c_2$ in Q', and hence we have in P' (not only in $(P')^{int}$),

$$b_1 c_1 = (h(a_3)b)c_1 = b(g(a_3)c_1) = b(g(a_4)c_2) = (h(a_4)b)c_2 = b_2 c_2.$$

(iii) $\Rightarrow$ (v). We show first h is injective. Let $a_1, a_2 \in Q$ and let k be any field. As is easily seen, the kernel of the multiplication by $a_1 - a_2$ on $k[Q]$ is generated, as an ideal, by elements of the form $\Sigma_{1 \leq i \leq n} c_i$ ($n \geq 1$, $c_i \in Q$) such that $a_1 c_i = a_2 c_{i+1}$ for $i = 1, \ldots, n-1$ and $a_1^n = a_2^n$. By the flatness of $k[Q] \to k[P]$, the images of these elements in $k[P]$ generates as an ideal, the kernel of the multiplication by $h(a_1) - h(a_2)$ on $k[P]$. If $h(a_1) = h(a_2)$, the above elements satisfy $h(c_1) = \cdots = h(c_n)$ and hence $k[P]$ is generated as an ideal by the elements $h(\Sigma_{1 \leqq i \leqq n} c_i) = nh(c_1)$. Hence n is invertible in any field k and hence $n = 1$, that is, $a_1 = a_2$.

Next assume $h(a_1)b_1 = h(a_2)b_2$, $a_1, a_2 \in Q$ and $b_1, b_2 \in P$. Let S be the kernel of

$$k[Q] \oplus k[Q] \to k[Q]; \qquad (f, g) \mapsto a_1 g - a_2 f.$$

By the flatness, the kernel T of

$$k[P] \oplus k[P] \to k[P]; \qquad (f, g) \mapsto h(a_1)g - h(a_2)f$$

is generated as a $k[P]$-module, by the image of S. Since $(b_1, b_2) \in T$, we can write

$$(*) \qquad b_1 = \sum_{1 \leqq i \leqq r} h(c_i) f_i, \qquad b_2 = \sum_{1 \leqq i \leqq r} h(d_i) f_i, \qquad a_1 c_i = a_2 d_i$$

for some $c_i, d_i \in k[Q]$, $f_i \in k[P]$ $(1 \leqq i \leqq r)$. The expression of b_1 in (*) shows that there are $a_3 \in Q$ and $b \in P$ and i such that a_3 appears in c_i, b appears in f_i, and $b_1 = h(a_3)b$. By $a_1 c_i = a_2 d_i$, there exist $a_4 \in Q$ which appear in d_i such that $a_1 a_3 = a_2 a_4$. We have $b_2 = h(a_4)b$ by

$$h(a_2)b_2 = h(a_1)b_1 = h(a_1 a_3)b = h(a_2)h(a_4)b.$$

(v) $\Rightarrow$ (ii). The $\mathbf{Z}[Q]$-module $\mathbf{Z}[P]$ becomes a filtered inductive limit of free $\mathbf{Z}[Q]$-modules which are direct sums of $\mathbf{Z}[Q]$-modules of the form $\mathbf{Z}[Q]b$ with $b \in P$.

Definition (4.3). Let $f : (X, M) \to (Y, N)$ be a morphism of schemes with integral log. str.'s. We say f is integral if for any $x \in X$, the equivalent conditions in (4.1)(2) are satisfied.

That f is integral (Resp. In the case M and N are fine, that f is integral) is equivalent to the following

(4.3.1). For any scheme Y' with an integral (resp. a fine) log. str. N' and for any $(Y', N') \to (Y, N)$, the log. str. of the fiber product $(X, M) \times_{(Y,N)} (Y', N')$ is integral.

(The implication "(4.1)(i) for $Q = (f^*N)_{\bar{x}}$ and $P = M_{\bar{x}}$ holds for any $x \in X \Rightarrow$ (4.3.1)" is proved easily. The implication "(4.3.1) $\Rightarrow$ (4.1)(iv) for $Q = (f^*N)_{\bar{x}}$ and $P = M_{\bar{x}}$ holds for any $x \in X$" is proved by the method of the proof of (iv) $\Rightarrow$ (i) given in (4.2).)

COROLLARY (4.4). *A morphism $f : (X, M) \to (Y, N)$ of schemes with integral* log. str.'s *is integral in each of the following cases*:

(i) *M is isomorphic to the inverse image of N.*
(ii) *For any $y \in Y$, the monoid $(N/\mathcal{O}_Y^*)_{\bar{y}}$ is generated by one element.*

Proof. For the case (ii), consider the condition (iv) in (4.1)(1).

For example, the morphism $(X, M) \to (\mathrm{Spec}(A), N)$ in (3.7)(2) (the semi-stable reduction situation) is integral.

COROLLARY (4.5). *If a morphism f of schemes with fine* log. str.'s *is smooth and integral, the underlying morphism $X \to Y$ is flat.*

Proof. By using (ii) of (4.1)(1), we can find etale locally a chart $(P_X \to M, Q_Y \to N, Q \to P)$ satisfying (3.5.2) such that $\mathbf{Z}[Q] \to \mathbf{Z}[P]$ is flat.

Definition (4.6). (1) We say a homomorphism of integral monoids $h : Q \to P$ is exact if $Q = (h^{gp})^{-1}(P)$ in Q^{gp} where $h^{gp} : Q^{gp} \to P^{gp}$.

(2) We say a morphism of schemes with integral log. str.'s $f : (X, M) \to (Y, N)$ is exact if the homomorphism $(f^*N)_{\bar{x}} \to M_{\bar{x}}$ is exact for any $x \in X$.

Following facts are proved easily. Exact morphisms are stable under composition. For fine log. str.'s, exact morphisms are stable under base changes in the category of schemes with fine log. str.'s in the sense of (2.8). An integral morphism is exact. If f is exact, the homomorphism $f^*N \to M$ is injective. (This last fact shows that a closed immersion (3.1) between schemes with integral log. str.'s is exact if and only if it is an exact closed immersion in the sense of (3.1).)

Now we consider characteristic p.

Definition (4.7). Let p be a prime number. For a scheme X over $\mathbf{F}_p = \mathbf{Z}/p\mathbf{Z}$ and a log. str. M on X, we define the absolute frobenius $F_{(X,M)} : (X, M) \to (X, M)$ as follows. The morphism of schemes underlying $F_{(X,M)}$ is the usual absolute frobenius $F_X : X \to X$, and the homomorphism $F_X^{-1}(M) \to M$ is the multiplication by p on M under the canonical identification of $F_X^{-1}(M)$ with M.

Definition (4.8). Let $f:(X, M) \to (Y, N)$ be a morphism of schemes with integral log. str.'s. Assume Y is a scheme over $\mathbf{F}_p$ with p a prime number. We say that f is of Cartier type if f is integral and the morphism $(f, F_{(X,M)})$ from (X, M) to the fiber product of

$$\begin{array}{ccc} & & (X, M) \\ & & \downarrow f \\ (Y, N) & \xrightarrow{F_{(Y,N)}} & (Y, N) \end{array}$$

is exact.

For example, f is of Cartier type if f has locally a chart of the form $Q = \mathbf{N}$, $P = \mathbf{N}^r (r \geqq 1)$, and $Q \to P$ is the diagonal map. (This happens in the semi-stable reduction situation (3.7)(2).)

Morphisms of Cartier type are stable under compositions and base changes.

(4.9). Let p be a prime number and let $f:(X, M) \to (Y, N)$ be a morphism of schemes with integral log. str.'s over $\mathbf{F}_p$. We say f is weakly purely inseparable if the following (i)–(iii) are satisfied.

(i) The map $X \to Y$ of underlying topological spaces is a homeomorphism.

(ii) For $x \in X$ and $a \in M_{\bar{x}}$, there exists $n \geqq 0$ such that $a^{p^n} \in \mathrm{Image}(f^{-1}(N)_{\bar{x}})$.

(iii) If $x \in X$ and if $a, b \in f^{-1}(N)_{\bar{x}}$ have the same image in $M_{\bar{x}}$, $a^{p^n} = b^{p^n}$ for some $n \geqq 0$.

We say f is purely inseparable if it is exact and weakly purely inseparable.

PROPOSITION (4.10). *Let $f:(X, M) \to (Y, N)$ be a morphism of schemes with fine* log. str.'s.

(1) *Assume that for any $x \in X$ and $a \in M_{\bar{x}}$, there exists $n \geqq 1$ such that $a^n \in \mathrm{Image}((f^*N)_{\bar{x}} \to M_{\bar{x}})$. Then, etale locally on X, f has a factorization $f = f'f''$ such that f' is an etale morphism of schemes with fine* log. str.'s *and f'' is exact.*

(2) *Assume Y is a scheme over $\mathbf{F}_p$ for a prime number p and f is weakly purely inseparable* (4.9). *Then f has a unique factorization $f = f'f''$ such that f' is an etale morphism between schemes with fine* log. str.'s *and f'' is purely inseparable* (4.9).

Proof. In the situation of (1) (resp. (2)), it is easily seen that etale locally on X, there exists a chart $(P_X \to M, Q_Y \to N, h : Q \to P)$ of f satisfying the following condition (i) (resp. (i) and (ii)).

(i) For any $a \in P$, there exists $n \geqq 1$ (resp. $n \geqq 0$) such that $a^n \in h(Q)$. (resp. $a^{p^n} \in h(Q)$.

(ii) For any $a, b \in Q$ such that $h(a) = h(b)$, there exists $n \geqq 0$ such that $a^{p^n} = b^{p^n}$.)

Let $Q' = (h^{gp})^{-1}(P)$ where $h^{gp} : Q^{gp} \to P^{gp}$, let

$$Y' = Y \times_{\mathrm{Spec}(\mathbf{Z}[Q])} \mathrm{Spec}(\mathbf{Z}[Q']),$$

and endow Y' with the inverse image N' of the canonical log. str. of $\mathrm{Spec}(\mathbf{Z}[Q'])$. Then, $f' : (Y', N') \to (Y, N)$ is etale by (3.5). We prove that $f'' : (X, M) \to (Y', N')$ is exact. Note that $Q' \to P$ is exact. Let $x \in X$, and write the homomorphisms $P \to M_{\bar{x}}$, $Q' \to (f''^*N')_{\bar{x}}$, $Q' \to P$, $(f''^*N')_{\bar{x}} \to M_{\bar{x}}$ by s, t, h', g, respectively. Our task is to prove that if a, $b \in (f''^*N')_{\bar{x}}$ and $g(a) \in g(b)M_{\bar{x}}$, then $a \in b(f''^*N')_{\bar{x}}$. We may assume $a = t(a_0)$, $b = t(b_0)$ for some $a_0, b_0 \in Q'$. We have $h'(a_0)c = h'(b_0)d$ for some $c, d \in P$ such that the image of c in $M_{\bar{x}}$ belongs to $\mathcal{O}^*_{X,\bar{x}}$. Take $n \geqq 1$ such that $c^n = h'(e)$, $e \in Q'$. Then, $h'(a_0e) = h'(b_0)c^{n-1}d$. Since h' is exact, we have $a_0e \in b_0Q'$. Since $s(e) \in \mathcal{O}^*_{X,\bar{x}}$, we have $a \in b(f''^*N')_{\bar{x}}$. This completes the proof of (1). Furthermore, in the situation of (2), f'' is weakly purely inseparable as is easily seen, and we obtain the local existence of the factorization in (2). It remains to prove the uniqueness of the factorization in (2), from which the global existence follows from the local existence. Assume we have two factorizations $(X, M) \to (Y'_1, N'_1) \to (Y, N)$ and $(X, M) \to (Y'_2, N'_2) \to (Y, N)$ of f satisfying the condition stated in (2). Let $q_i : ((Y'_1, N'_1) \times_{(Y,N)} (Y'_2, N'_2))^{int} \to (Y'_i, N'_i)$ be the projections ($i = 1, 2$). Then, q_i is etale and purely inseparable. Hence q_i is an isomorphism by the following (4.11) (which we apply to the case where s and t are isomorphisms). This proves the uniqueness.

LEMMA (4.11). *Let p be a prime number and let*

$$\begin{array}{ccc} (T', L') & \xrightarrow{s} & (X, M) \\ {\scriptstyle i}\downarrow & & \downarrow{\scriptstyle f} \\ (T, L) & \xrightarrow{t} & (Y, N) \end{array}$$

be a commutative diagram of schemes with fine log. str.'s *over* $\mathbf{F}_p$ *such that* i *is purely inseparable, and such that* f *is etale. Then, there exists a unique morphism* $h : (T, L) \to (X, M)$ *such that* $hi = s$ *and* $fh = t$.

Proof. By taking a chart of f satisfying the condition (3.5.2), the proof proceeds just as the proof of (3.4).

THEOREM (4.12) (Cf. [DI], [Il$_2$].). *Let* p *be a prime number and let* $f : (X, M) \to (Y, N)$ *be a smooth morphism of schemes with fine* log. str.'s *over* $\mathbf{F}_p$. *Let* $f' : (X', M') \to (Y, N)$ *be the base change of* f *by the absolute frobenius* $F_{(Y,N)} : (Y, N) \to (Y, N)$, *let*

$$(X, M) \xrightarrow{F} (X', M') \to (X, M)$$

be the factorization of $F_{(X,M)}$ *characterized by the property* $f = f'F$, *and consider the factorization*

$$(X, M) \xrightarrow{g} (X'', M'') \xrightarrow{h} (X', M')^{int}$$

of $F^{int} : (X, M) \to (X', M')^{int}$ *given by* (4.10)(2).

(1) *Assume* f *is smooth. Let* s *be the composite morphism* $(X'', M'') \to (X', M') \to (X, M)$. *Then we have a canonical isomorphism of* $\mathcal{O}_{X''}$-modules

$$C^{-1} : \omega^q_{X''/Y} \to \mathcal{H}^q(\omega_{X/Y})$$

for any $q \in \mathbf{Z}$ characterized by

$$C^{-1}(ad \log(s^*(b_1)) \wedge \cdots \wedge d \log(s^*(b_q)))$$
$$= g^*(a) d \log(b_1) \wedge \cdots \wedge d \log(b_q)$$

$(a \in \mathcal{O}_{X''}, b_1, \ldots, b_q \in M)$.

(2) *Assume* f *is smooth and integral. Assume we are given a scheme with a fine* log. str. $(\tilde{Y}, \tilde{N})$ *such that* $\tilde{Y}$ *is flat over* $\mathbf{Z}/p^2\mathbf{Z}$, *and an isomorphism*

$$(Y, N) \cong (\tilde{Y}, \tilde{N}) \times_{\mathrm{Spec}(\mathbf{Z}/p^2\mathbf{Z})} \mathrm{Spec}(\mathbf{F}_p)$$

where Spec($\mathbf{Z}/p^2\mathbf{Z}$) *and* Spec($\mathbf{F}_p$) *are endowed with trivial* log. str.'s. *Then, there exists a canonical bijection between the set of all isomorphism classes of smooth liftings of* (X'', M'') *over* $(\tilde{Y}, \tilde{N})$ (3.14) *and the set of all splittings of* $\tau_{\leqq 1}\, \omega^{\cdot}_{X/Y}$ *in the derived category of the category of* $\mathcal{O}_{X''}$*-modules* (*cf.* [DI] *Section* 3). *If a smooth lifting of* (X, M) *over* $(\tilde{Y}, \tilde{N})$ *exists, there is an isomorphism*

$$\tau_{<p}\, \omega^{\cdot}_{X/Y} \cong \bigoplus_{0 \leqq i < p} \omega^{i}_{X''/Y}[-i]$$

in the derived category of the category of $\mathcal{O}''_X$*-modules.*

(3) *Assume the following* (i)–(iii).

(i) *f is smooth and of Cartier type.* (*Note that* $(X'', M'') = (X', M')$ *in this case.*)

(ii) *The underlying morphism* $X \to Y$ *is proper.*

(iii) *Etale locally on* Y, *there exist* $(\tilde{Y}, \tilde{N})$ *as in* (2) *and a smooth lifting of* (X', M') *over* $(\tilde{Y}, \tilde{N})$.

Then, the Hedge spectral sequence

$$E_1^{s,t} = R^t f_* \omega^s_{X/Y} \Rightarrow R^{s+t} f_* \omega^{\cdot}_{X/Y}$$

satisfies $E_1^{s,t} = E_\infty^{s,t}$ *for* s, t *such that* $s + t < p$. *Furthermore, the* $\mathcal{O}_Y$*-modules* $R^q f_* \omega^{\cdot}_{X/Y}$ *for* $q < p$ *are locally free and commute with any base changes.*

Proof. Since the proof is a simple modification of those given in [DI] (classically smooth case) and in [Il$_2$] (the case of morphisms of "semi-stable reduction type" between smooth schemes), we give here only the proof of the Cartier isomorphism (4.12)(1), and left the other part of the proof to the reader. (As in [Il$_2$], the other part is deduced from (1) and (3.14) by the arguments in [DI].) I just note that the assumption f is integral in (2) is used to have the flatness over $\mathbf{Z}/p^2\mathbf{Z}$ (4.5) of smooth liftings of (X'', M''). Now we prove (1). By a standard argument, we may assume that there is a cartesian diagram

$$\begin{array}{ccc} X & \longrightarrow & \mathrm{Spec}(\mathbf{F}_p[P]) \\ \downarrow & & \downarrow \\ Y & \longrightarrow & \mathrm{Spec}(\mathbf{F}_p[Q]) \end{array}$$

where P, Q are finitely generated integral monoids with a homomorphism $Q \to P$ such that $Q^{gp} \to P^{gp}$ is injective and the torsion part of P^{gp}/Q^{gp} is of order invertible on Y, and M and N are the inverse images of the canonical log. str.'s, respectively. Let H be the submonoid of P containing Q defined by

$$H = \{a \in P;\, a = b^p c \text{ in } P^{gp} \text{ for some } b \in P^{gp} \text{ and } c \in Q^{gp}\}.$$

Then we have

$$X'' = Y \times_{\mathrm{Spec}(\mathbf{F}_p[Q])} \mathrm{Spec}(\mathbf{F}_p[H]).$$

In this identification, $X'' \to X$ (resp. $X'' \to Y$, resp. $X \to X''$) is given by $F_{(Y,N)} \times (a \mapsto a^p;\, P \to H)$; $Y \times_{\mathrm{Spec}(\mathbf{F}_p[Q])} \mathrm{Spec}(\mathbf{F}_p[H]) \to Y \times_{\mathrm{Spec}(\mathbf{F}_p[Q])} \mathrm{Spec}(\mathbf{F}_p[P])$ (resp. $pr_1 : Y \times_{\mathrm{Spec}(\mathbf{F}_p[Q])} \mathrm{Spec}(\mathbf{F}_p[H] \to Y$, resp. id. $\times$ $(H \hookrightarrow P)$); $Y \times_{\mathrm{Spec}(\mathbf{F}_p[Q])} \mathrm{Spec}(\mathbf{F}_p[P]) \to Y \times_{\mathrm{Spec}(\mathbf{F}_p[Q])} \mathrm{Spec}(\mathbf{F}_p[H])$). For $v \in P^{gp}/Q^{gp} \otimes_{\mathbf{Z}} \mathbf{F}_p$, let E_v be the $\mathbf{F}_p[Q]$-submodule of $\mathbf{F}_p[P]$ generated by elements of P which belong to v, and define the complex $C_v^{\cdot}$ by

$$C_v^q = \mathcal{O}_Y \otimes_{\mathbf{F}_p[Q]} E_v \otimes_{\mathbf{F}_p} \wedge_{\mathbf{F}_p}^q((P^{gp}/Q^{gp}) \otimes \mathbf{F}_p)$$

with the differential $C_v^q \to C_v^{q+1}$ induced by

$$\wedge_{\mathbf{F}_p}^q((P^{gp}/Q^{gp}) \otimes \mathbf{F}_p) \to \wedge_{\mathbf{F}_p}^{q+1}((P^{gp}/Q^{gp}) \otimes \mathbf{F}_p); \qquad a \mapsto v \wedge a.$$

We have

$$\omega_{X/Y}^{\cdot} = \oplus_v C_v^{\cdot}.$$

The complex $C_v^{\cdot}$ is acyclic if $v \neq 0$, and so $\omega_{X/Y}^{\cdot}$ is quasi-isomorphic to $C_0^{\cdot}$. On the other hand, the differential of $C_0^{\cdot}$ is zero, $E_0 = \mathbf{F}_p[H]$, and

$$C_0^q = \mathcal{O}_Y \otimes_{\mathbf{F}_p[Q]} \mathbf{F}_p[H] \otimes_{\mathbf{F}_p} \wedge_{\mathbf{F}_p}^q((P^{gp}/Q^{gp}) \otimes \mathbf{F}_p) \cong \omega_{X''/Y}^q$$

(cf. (1.8)).

Remark (4.13). In (4.12)(1), if $Y = \mathrm{Spec}(k)$ for a field k and N is the trivial log. str., and if X is normal, X'' coincides with the normalization of X'.

5. Crystalline sites. A fact for log. str.'s which is different from the classical facts is that the crystalline cohomology theory is easier than the ℓ ($\neq$ char.)-adic etale cohomology theory. (I have not yet a good definition of the etale site for a log. str. A related problem is to define the K-group of a scheme with a log. str.)

(5.1) As a base, we take a 4-ple (S, L, I, γ) where S is a scheme such that $\mathcal{O}_S$ is killed by a nonzero integer, L is a fine log. str. on S, I is a quasi-coherent ideal on S, and γ is a PD (=divided power) structure on I.

(5.2) Let (S, L, I, γ) be as above, let (X, M) be a scheme with a fine log. str. over (S, L) such that γ extends to X. Then, we define the crystalline site $((X, M)/(S, L, I, \gamma))_{crys}$ (denoted also simply by $(X/S)^{\log}_{crys}$ if there is no risk of confusion) as follows. An object is a 5-ple (U, T, M_T, i, δ) where U is an etale scheme over X, (T, M_T) is a scheme with a fine log. str. over (S, L), i is an exact closed immersion (3.1) $(U, M) \to (T, M_T)$ over (S, L), and δ is a PD-structure on the ideal of $\mathcal{O}_T$ defining U which is compatible with γ. A morphism is defined in the evident way. A covering is a covering for the usual etale topology forgetting the log. str.'s.

The structure sheaf $\mathcal{O}_{X/S}$ on $(X/S)^{\log}_{crys}$ is defined by

$$\mathcal{O}_{X/S}(U, T, M_T, i, \delta) = \Gamma(T, \mathcal{O}_T).$$

We sometimes abbreviate (U, T, M_T, i, δ) simply as T. We sometimes denote $\gamma_n(a)$ and $\delta_n(a)$ as $a^{[n]}$.

We have the following fact by applying (1.4.1) to $U \to T$; If $g : T' \to T$ is a morphism in $(X/S)^{\log}_{crys}$, $g^*(M_T) \to M_{T'}$ is an isomorphism.

We have a logarithmic version of the PD-envelope:

PROPOSITION (5.3). *Let (S, I, γ) be as in* (5.1). (*We forget L here*). *Let $\mathcal{C}$ be the category of closed immersions* (3.1) $i : (X, M) \to (Y, N)$ *of schemes with* log. str.'s *over S such that M is fine and N is coherent*. (*By definition, a morphism $i' \to i$ is a commutative diagram*

$$\begin{array}{ccc} (X', M') & \xrightarrow{i'} & (Y', N') \\ \downarrow & & \downarrow \\ (X, M) & \xrightarrow{i} & (Y, N) \end{array}$$

over S.) Let $\mathcal{C}'$ be the category of pairs (i, δ) where i is an exact closed immersion (3.1) $(X, M) \to (Y, N)$ *of schemes with fine* log. str.'s *over S and δ is a PD-structure on the ideal of Y defining X which is compatible with γ. Then, the canonical functor $\mathcal{C}' \to \mathcal{C}$ has a right adjoint.*

Definition (5.4). In (5.3), let $i : (X, M) \to (Y, N)$ be an object of $\mathcal{C}$ and let $(\tilde{i} : (\tilde{X}, \tilde{M}) \to (\tilde{Y}, \tilde{N}), \delta)$ be the result of applying the right adjoint functor to i. We call $(\tilde{i}, \delta)$ (or sometimes $(\tilde{Y}, \tilde{N})$) the PD-envelope of (X, M) in (Y, N) with respect to γ, and denote it by $D_{(X,M)}((Y, N)/(S, I, \gamma))$ (or simply by $D_X^{\log}(Y)$).

(5.5). The construction of the PD-envelope given below shows the following facts:

(5.5.1). If i is an exact closed immersion, $D_X^{\log}(Y)$ coincides with the usual PD-envelope $D_X(Y)$ endowed with the inverse image of N.

(5.5.2). If γ extends to Y, $(\tilde{X}, \tilde{M}) \to (X, M)$ is an isomorphism.

(5.5.3). $\tilde{M}$ always coincides with the inverse image of M.

(5.6). *Proof of* (5.3). We construct $(\tilde{i}, \delta)$ of (5.4). We may assume N is fine, since $(\tilde{i}, \delta)$ for $(X, M) \to (Y, N)$ is the same thing with $(\tilde{i}, \delta)$ for $(X, M) \to (Y, N)^{int}$ (2.7). We may work etale locally, and hence we have a factorization $i = gi'$ with $i' : (X, M) \to (Z, M_Z)$ an exact closed immersion and g etale (4.10)(1). Let $(\tilde{i} : \tilde{X} \to D, \delta)$ be the PD-envelope of i' with respect to γ in the usual sense, and endow $\tilde{X}$ (resp. D) with the inverse image $\tilde{M}$ (resp. M_D) of M (resp. M_Z). It is not hard to see that $(\tilde{i} : (\tilde{X}, \tilde{M}) \to (D, M_D), \delta)$ has the desired universal property.

Example (5.7). Let k be a field of characteristic $p > 0$, let $X = \mathrm{Spec}(k[T])$, $Y = \mathrm{Spec}(k[T_1, T_2])$, $X \to Y$ by $T_i \mapsto T$ $(i = 1, 2)$. Endow X (resp. Y) with log. str. M (resp. N) corresponding to the divisor "$T = 0$" (resp. "$T_1 = 0$" $\cup$ "$T_2 = 0$"). Take the base $S = \mathrm{Spec}(k)$, $I = (0)$. Then the PD-envelope $D_X^{\log}(Y)$ is the usual PD-envelope of X in $Z = \mathrm{Spec}(k[T_1, T_2, T_1T_2^{-1}, T_2T_1^{-1}]) = \mathrm{Spec}(k[T_1, V, V^{-1}])$ $(V = T_1T_2^{-1}, V \mapsto 1$ on $X)$ endowed with the inverse image of the log. str. M_Z on Z corresponding to the divisor "$T_1 = 0$" (= "$T_2 = 0$"). Indeed, the closed immersion $(X, M) \to (Y, N)$ is not exact, but $(X, M) \to (Z, M_Z)$ is an exact closed immersion and $(Z, M_Z) \to (Y, N)$ is etale.

Remark (5.8). Let (S, L) be as in (5.1) (we forget here I and γ). We can define the n-th infinitesimal neighbourhood in the logarithmic sense as follows, similarly to the PD-envelopes. For $n \geqq 0$, let $\mathcal{C}_n$ be the category of exact closed immersions $(X, M) \to (Y, N)$ of schemes over (S, L) such that X is defined in Y by an ideal J with the property $J^{n+1} = 0$. Then the canonical functor $\mathcal{C}_n \to \mathcal{C}$ ($\mathcal{C}$ is as in (5.3)) has a right adjoint. Indeed, let (Z, M_Z) be as in (5.6), and let D be the n-th infinitesimal neighbourhood of X in Z in the usual sense endowed with the inverse image M_D of M_Z. Then, $(X, M) \to (D, M_D)$ is the desired universal object. In the case of the diagonal embedding $(X, M) \to (Y, N) = (X, M) \times_{(S,L)} (X, M)$ with $n = 1$, if we denote by J the ideal of X in D, we have

(5.8.1) $$\omega^1_{X/S} \cong J/J^2.$$

(5.9). We have the functoriality of the crystalline topoi. Let

$$\begin{array}{ccc} (X', M') & \xrightarrow{f} & (X, M) \\ \downarrow & & \downarrow \\ (S', L', I', \gamma') & \longrightarrow & (S, L, I, \gamma) \end{array}$$

be a commutative diagram where the assumptions of (5.1) (5.2) are satisfied by both $(X, M)/(S, L, I, \gamma)$ and $(X', M')/(S', L', I', \gamma')$. Then we have the morphism of topoi

$$f_{crys} : ((X'/S')^{\log}_{crys})^{\sim} \to ((X/S)^{\log}_{crys})^{\sim}$$

(the $\sim$ denote the topoi associated to sites) characterized by

$$f_{crys*}(\mathfrak{F})(U, T, M_T, i, \delta) = \mathrm{Mor}((U, T, M_T, i, \delta)^{\sim}, \mathfrak{F})$$

where $(U, T, M_T, i, \delta)^{\sim}$ is the sheaf on $(X'/S')^{\log}_{crys}$ whose value in $(U', T', M'_{T'}, i', \delta')$ is the set of all pairs (g, h) of morphisms $g : (U', M') \to (U, M)$, $h : (T', M'_{T'}) \to (M, T)$ for which the diagram

$$\begin{array}{ccccccc} (X', M') & \longrightarrow & (U', M') & \longrightarrow & (T', M'_{T'}) & \longrightarrow & (S', L') \\ {\scriptstyle f}\downarrow & & {\scriptstyle g}\downarrow & & {\scriptstyle h}\downarrow & & \downarrow \\ (X, M) & \longrightarrow & (U, M) & \longrightarrow & (T, M_T) & \longrightarrow & (S, L) \end{array}$$

commutes and such that h commutes with δ and δ'.

The proof of the fact f_{crys*} determines a morphism of topoi (i.e. f_{crys*} has a left adjoint which commutes with finite inverse limits) is proved by the same way as in the classical theory of crystalline topoi ([B] Chapter 3, Section 2), by using the notion (5.4) of PD-envelopes.

6. Crystals and crystalline cohomology. In this section, let (S, L, I, γ) be as in (5.1) and let $f : (X, M) \to (S, L)$ be a morphism of schemes such that M is fine and γ extends to X.

Definition (6.1). A crystal on $(X/S)^{\log}_{crys}$ is a sheaf of $\mathcal{O}_{X/S}$-modules $\mathfrak{F}$ on $(X/S)^{\log}_{crys}$ satisfying the following condition: For any morphism $g : T' \to T$ in $(X/S)^{\log}_{crys}$, if we denote by $\mathfrak{F}_T$ and $\mathfrak{F}_{T'}$ the sheaves on T_{et} and T'_{et} induced by $\mathfrak{F}$ respectively, $g^*(\mathfrak{F}_T) \to \mathfrak{F}_{T'}$ is an isomorphism.

THEOREM (6.2). *Let* (Y, N) *be a scheme with a fine* log. str. *which is smooth over* (S, L), *and let* $(X, M) \to (Y, N)$ *be a closed immersion* (3.1). *Denote the PD-envelope of* (X, M) *in* (Y, N) *as* (D, M_D). *Then, the following two categories* (a) (b) *are equivalent.*

(a) *The category of crystals on* $(X/S)^{\log}_{crys}$.

(b) *The category of* $\mathcal{O}_D$*-modules* $\mathfrak{M}$ *on* D_{et} *endowed with an additive map*

$$\nabla : \mathfrak{M} \to \mathfrak{M} \otimes_{\mathcal{O}_Y} \omega^1_{Y/S}$$

having the following properties (i)–(iii).

(i) $\nabla(ax) = a\nabla(x) + x \otimes da$ *for* $a \in \mathcal{O}_D$ *and* $x \in \mathfrak{M}$.

(ii) *The composite*

$$\mathfrak{M} \xrightarrow{\nabla} \mathfrak{M} \otimes_{\mathcal{O}_Y} \omega^1_{Y/S} \xrightarrow{\nabla} \mathfrak{M} \otimes_{\mathcal{O}_Y} \omega^2_{Y/S}$$

is zero, where we extend ∇ *to*

$$\mathfrak{M} \otimes_{\mathcal{O}_Y} \omega^q_{Y/S} \to \mathfrak{M} \otimes_{\mathcal{O}_Y} \omega^{q+1}_{Y/S};$$

$$x \otimes \omega \mapsto \nabla(x) \wedge \omega + x \otimes d\omega.$$

(iii) *Let* $x \in X$ *and let* t_i $(1 \leqq i \leqq r)$ *be elements of* $M_{\bar{x}}$ *such that* $(d\log(t_i))_{1 \leqq i \leqq r}$ *is a basis of* $\omega^1_{Y/S,\bar{x}}$. *Then, for any* i *and for any* $a \in \mathfrak{M}_{\bar{x}}$, *there exist* $m_1, \ldots, m_k, n_1, \ldots, n_k \in \mathbf{N}$ *such that*

$$\Big(\prod_{1 \leqq i \leqq r, 1 \leqq j \leqq k} (\nabla^{\log}_{t_i} - m_j)^{n_j}\Big)(a) = 0.$$

Here $\nabla^{\log}_{t_i}$ *is defined by: if* $\nabla(a) = \Sigma_{1 \leqq i \leqq r} a_i \otimes d\log(t_i)$, *then* $\nabla^{\log}_{t_i}(a) = a_i$.

(It is proved as in the classical case, that if the condition (iii) holds for one choice of $(t_i)_{1 \leqq i \leqq r}$, then it holds for any choice of $(t_i)_{1 \leqq i \leqq r}$.)

Remark (6.3). *If* $t_i \in \mathcal{O}^*_{X,\bar{x}}$ and ∇_{t_i} denotes $t_i^{-1}\nabla^{\log}_{t_i}$, then

$$(\nabla_{t_i})^n = t_i^{-n} \prod_{0 \leqq j \leqq n-1} (\nabla^{\log}_{t_i} - j).$$

Thus, the condition (b) (iii) in (6.2) is the natural logarithmic version of the classical notion of the "quasi-nilpotence" ([B] Chapter 2, Section 4.3).

THEOREM (6.4). *Let* (X, M), (Y, N) *and* D *be as in* (6.2), *let* $\mathfrak{F}$ *be a crystal on* $(X/S)^{\log}_{crys}$, *and let* $\mathfrak{M}$ *be the corresponding* $\mathcal{O}_D$*-module with* ∇. *Then,*

$$Ru^{\log}_{X/S*}(\mathfrak{F}) \cong \mathfrak{M} \otimes_{\mathcal{O}_Y} \omega^{\cdot}_{Y/S}.$$

Here $u^{\log}_{X/S}$ *is the canonical morphism* $((X/S)^{\log}_{crys})^{\sim} \to (X_{et})^{\sim}$ *characterized by*

$$(u^{\log}_{X/S})_*(\mathfrak{F})(U) = \textit{the global section of } \mathfrak{F} \textit{ on } (U/S)^{\log}_{crys}$$

for a sheaf $\mathfrak{F}$ *on* $(X/S)^{\log}_{crys}$.

For the proofs of (6.2) and (6.4), the following (6.5) is essential.

PROPOSITION (6.5). *Under the assumption of* (6.2), *let* $(D', M_{D'})$ *be the PD-envelope* (5.4) *of the diagonal morphism* $(X, M) \to (Y, N) \times_{(S,L)} (Y, N)$, *and let* $p_1, p_2 : (D', M_{D'}) \to (D, M_D)$ *be the first and the second projections, respectively. Let* $x \in X$, *take* $t_1, \ldots, t_r \in N_{\bar{x}}$ *such that* $(d\log(t_i))_{1 \leqq i \leqq r}$ *is a basis of* $\omega^1_{Y/S,\bar{x}}$, *and let* u_i $(1 \leqq i \leqq r)$ *be the elements of* $\mathrm{Ker}(\mathcal{O}^*_{D',\bar{x}} \to \mathcal{O}^*_{X,\bar{x}}) \subset (M_{D'})_{\bar{x}}$ *defined by* $p_2^*(t_i) = p_1^*(t_i)u_i$ *(the existence* *of* u_i *follows from* $(M_{D'})_{\bar{x}}/\mathcal{O}^*_{D',\bar{x}} \cong M_{\bar{x}}/\mathcal{O}^*_{X,\bar{x}})$. *Then we have the description of* D'

$$\mathcal{O}_{D,\bar{x}}\langle T_1, \ldots, T_r\rangle \cong \mathcal{O}_{D',\bar{x}}; \qquad T_i^{[n]} \mapsto (u_i - 1)^{[n]}$$

where $\mathcal{O}_{T,\bar{x}}\langle T_1, \ldots, T_r\rangle$ *denotes the PD-polynomial ring.*

(6.6). *Proof of* (6.5). By the construction of the PD-envelopes in (5.6), we may assume that $(X, M) \to (Y, N)$ is an exact closed immersion. Etale locally at x, take an exact closed immersion $(X, M) \to (Z, M_Z)$ where M_Z is fine and (Z, M_Z) is etale over $((Y, N) \times_{(S,L)} (Y, N))^{int}$ (4.10)(1). Let $q_i : Z \to Y\,(i = 1, 2)$ be the two projections. Then, the stalks at $\bar{x}$ of the sheaves M_Z and $q_i^*(N)\,(i = 1, 2)$ coincide. So, by replacing Z by an etale neighbourhood of $\bar{x} \to Z$, we may assume that $M_Z = q_i^*N\,(i = 1, 2)$. Then, q_i are smooth in the usual sense by (3.8). Since D (resp. D') is the usual PD-envelope of X in Y (resp. Z), and since $q_1^*(t_i)^{-1}q_2^*(t_i) - 1$ $(1 \leq i \leq r)$ form a smooth coordinate of Z over Y with respect to (say) q_1 and their restrictions to X are zero, the statement of (6.5) follows.

(6.7). *Proof of* (6.2). We follow the classical theory ([B] Chapter 4, Section 1). Let $(\mathfrak{M}, \nabla)$ be an object of (b). We show how to define the corresponding object of (a). Let $x \in X$, let $(t_i)_i$ be as in (b)(iii), and let D' and $(u_i)_i$ be as in (6.5). Then we have an isomorphism at $\bar{x}$

$$\eta : p_2^*\mathfrak{M} \cong p_1^*\mathfrak{M}; \tag{6.7.1}$$

$$1 \otimes a \mapsto \sum_{n \in \mathbf{N}^r} \left(\prod_{1 \leq i \leq r} (u_i - 1)^{[n_i]}\right) \otimes \left(\prod_{1 \leq i \leq r, 0 \leq j \leq n_i - 1} (\nabla_{t_i}^{\log} - j)\right)(a)$$

$(a \in \mathfrak{M})$, which satisfies the "transitivity condition"

$$p_{13}^*(\eta) = p_{23}^*(\eta)p_{12}^*(\eta)$$

([B] Chapter 2, 1.3.1) on the PD-envelope D'' of (X, M) in $(Y, N) \times_{(S,L)} (Y, N) \times_{(S,L)} (Y, N)$ where p_{12}, p_{13}, p_{23} are the projections $D'' \to D'$. Now we obtain an object $\mathfrak{F}$ of (a) from $(\mathfrak{M}, \nabla)$ as follows. Let (U, T, M_T, i, δ) be an object of $(X/S)^{\log}_{crys}$. Then, etale locally on T, $(X, M) \to (D, M_D)$ is extended to a morphism $h : (T, M_T) \to (D, M_D)$ over (S, L) by the smoothness of $(Y, N) \to (S, L)$. We define $\mathfrak{F}_T$ to be $h^*\mathfrak{M}$ etale locally on

T. If we have two such $h_i : (T, M_T) \to (D, M_D)$ $(i = 1, 2)$, they define $h' : (T, M_T) \to (D', M_{D'})$ such that $h_i = p_i h'$, and the isomorphism (6.7.1) induces $h_1^* \mathcal{M} \cong h_2^* \mathcal{M}$. Thus $\mathcal{F}$ is independent of h and defined globally on T.

Conversely if we have an object $\mathcal{F}$ of (a), let $\mathcal{M} = \mathcal{F}_D$. Then, the defining condition of the crystal gives an isomorphism $\eta : p_2^* \mathcal{M} = p_1^* \mathcal{M}$ satisfying the "transitivity condition". By writing η in the form

$$1 \otimes a \mapsto \sum_{n \in \mathbf{N}^r} \prod_{1 \leq i \leq r} (u_i - 1)^{[n_i]} \otimes \eta_n(a),$$

we define ∇ by

$$\nabla(a) = \sum_{1 \leq i \leq r} \eta_{e_i}(a) \otimes d \log(t_i)$$

where $(e_i)_{1 \leq i \leq r}$ is the canonical base of $\mathbf{N}$.

(6.9). *Proof of* (6.4). This is also a repetition of the classical argument. For an $\mathcal{O}_D$-module $\mathcal{N}$ on D_{et}, let $L(\mathcal{N})$ be the crystal on $(X/S)^{\log}_{crys}$ corresponding to the $\mathcal{O}_D$-module $p_1^* \mathcal{N}$ with

$$\nabla : p_1^* \mathcal{N} \to p_1^* \mathcal{N} \otimes_{\mathcal{O}_Y} \omega^1_{Y/S}; \qquad a \otimes v \mapsto v \otimes da$$

$(a \in \mathcal{O}_{D'}, v \in \mathcal{N}$, with D' as in (6.5)). Here d is the composite map

$$\mathcal{O}_{D'} \to \mathcal{O}_{D'} \otimes_{\mathcal{O}_Z} \omega^1_{Z/S} \cong \mathcal{O}_{D'} \otimes_{\mathcal{O}_Y} \omega^1_{Y/S}$$

with Z as in (6.6). The same argument as in the classical theory shows

$$R(u^{\log}_{X/S})_* L(\mathcal{N}) = \mathcal{N},$$

where we identified X_{et} with D_{et}. If $\mathcal{F}$ is a crystal corresponding to $(\mathcal{M}, \nabla)$, we have a resolution

$$\mathcal{F} \to L(\mathcal{M} \otimes_{\mathcal{O}_Y} \omega^{\cdot}_{Y/S}).$$

We obtain from this

$$R(u^{\log}_{X/S})_* \mathcal{F} \cong (u^{\log}_{X/S})_* L(\mathcal{M} \otimes_{\mathcal{O}_Y} \omega_{Y/S}) \cong \mathcal{M} \otimes_{\mathcal{O}_Y} \omega^{\cdot}_{Y/S}.$$

The following (6.10) (the base change theorem) and (6.11) (the Künneth formula) are proved by the same method in the classical theory ([B] Chapter 5, Sections 3 and 4).

THEOREM (6.10). *Assume we are given a commutative diagram*

$$\begin{array}{ccc} (X', M') & \xrightarrow{g'} & (X, M) \\ {\scriptstyle f'}\downarrow & & \downarrow{\scriptstyle f} \\ (Y', N') & \xrightarrow{g} & (Y, N) \\ \downarrow & & \downarrow \\ (S', L', I', \delta) & \xrightarrow{h} & (S, L, I, \delta) \end{array}$$

where all the log. str.'s *are fine, Y is quasi-compact, f is smooth and integral, the underlying morphism $X \to Y$ of f is quasi-separated, and the upper square is cartesian. Let $\mathcal{F}$ be a quasi-coherent crystal on $(X/S)^{\log}_{crys}$ which is flat over $\mathcal{O}_{X/S}$. Then,* we have a canonical isomorphism

$$Lg^{*}_{crys}(Rf_{crys*}(\mathcal{F})) \cong Rf'_{crys*}(g'^{*}_{crys}(\mathcal{F})).$$

THEOREM (6.12). *Let $f : (Y, M_Y) \to (X, M_X)$ and $g : (Z, M_Z) \to (X, M_X)$ be smooth integral morphisms between schemes with fine* log. str.'s *over (S, L), and assume that X is quasi-compact and the underlying morphisms $Y \to X$, $Z \to X$ are quasi-compact and quasi-separated. Let (V, M_V) be the fiber product of (Y, M_Y) and (Z, M_Z) over (X, M_X) with $p : (V, M_V) \to (Y, M_Y)$, $q : (V, M_V) \to (Z, M_Z)$, $h : (V, M_V) \to (X, M_X)$, and let $\mathcal{E}$ (resp. $\mathcal{F}$) be a quasi-coherent crystal on $(Y/S)^{\log}_{crys}$ (resp. $(Z/S)^{\log}_{crys}$) which is flat over $\mathcal{O}_{Y/S}$ (resp. $\mathcal{O}_{Z/S}$). Then, we have a canonical isomorphism*

$$Rf_{crys*}(\mathcal{E}) \otimes^{L}_{\mathcal{O}_{X/S}} Rg_{crys*}(\mathcal{F}) \cong Rh_{crys*}(p^{*}_{crys}(\mathcal{E}) \otimes_{\mathcal{O}_{V/S}} q^{*}_{crys}(\mathcal{F})).$$

Complement 1. We explain the relation between the log. str. of this note and that of Faltings in [Fa$_2$]. A definition of log. str. equivalent to that of Faltings was found by Deligne ([D]) independently.

The log. str. of Faltings in [Fa$_2$] on a scheme X is a family $(\mathcal{L}_i, x_i)_{1\leq i\leq r}$ of invertible sheaves $\mathcal{L}_i$ on X and global sections x_i of $\mathcal{L}_i$. An equivalent definition (take the dual) given in [D] is that a log. str. on X is a family $(\mathcal{L}_i, s_i)_{1\leq i\leq r}$ of invertible sheaves $\mathcal{L}_i$ on X and homomorphisms s_i :

$\mathcal{L}_i \to \mathcal{O}_X$ of $\mathcal{O}_X$-modules. To compare with our log. str., we adopt the latter definition of Deligne for it is nearer to our definition, and we call the log. str. $(\mathcal{L}_i, s_i)$ in the sense of Deligne the DF. log. str.

We claim that a DF. log. str. on X is equivalent to a pair (M, t) of a fine log. str. (in our sense) M on X and a homomorphism $t : \mathbf{N}^r_X \to M/\mathcal{O}^*_X$ which lifts etale locally on X to a chart $\mathbf{N}^r_X \to M$ of M. Indeed, for such a pair (M, t), we define $(\mathcal{L}_i, s_i)$ as follows. Let $(e_i)_{1 \leqq i \leqq r}$ be the cannical base of $\mathbf{N}^r$. Then, the inverse image of $t(e_i)$ under $M \to M/\mathcal{O}^*_X$ is a principal homogeneous space over $\mathcal{O}^*_X$ and corresponds to an invertible sheaf $\mathcal{L}_i$, and the homomorphism $M \to \mathcal{O}_X$ defines s_i. Conversely, we can reconstruct (M, t) from $(\mathcal{L}_i, s_i)$ as follows. Define first the pre-log. str. M' to be the sheaf of pairs (n, a) where $n \in \mathbf{N}^r$ and a is a local generator of $\otimes_i \mathcal{L}_i^{\otimes n_i}$, endowed with the homomorphism $M' \to \mathcal{O}_X$ induced by $\otimes_i s_i^{\otimes n_i} : \otimes_i \mathcal{L}_i^{\otimes n_i} \to \mathcal{O}_X$. Then, define M to be the log. str. associated to M', and t to be the composite of the inverse of the isomorphism

$$M'/\mathcal{O}^*_X \to \mathbf{N}^r_X; \qquad (n, a) \mapsto n$$

with the canonical homomorphism $M'/\mathcal{O}^*_X \to M/\mathcal{O}^*_X$.

The log. str. of Fontaine-Illusie is more general than the DF. log. str.: A fine log. str. M on a scheme X has a chart of the form $\mathbf{N}^r_X \to M$ with $r \geqq 0$ on an etale neighbourhood of $x \in X$ if and only if

$$M_{\bar{x}}/\mathcal{O}^*_{X,\bar{x}} \cong \mathbf{N}^s$$

for some $s \geqq 0$. One has also that for a finitely generated integral monoid P in which the unit element is the only invertible element, and for a non-zero ring R, the canonical log. str. on Spec$(R[P])$ comes from a DF. log. str. if and only if $P \cong \mathbf{N}^r$ for some r. Log. str.'s which do not come from DF. log. str.'s appear, for example, by taking a product of schemes with semi-stable reduction over a dvr., or by a ramified extension of the base dvr. of a scheme with semi-stable reduction.

Complement 2. The crystalline cohomology theory in this note is applied to the semi-stable reduction situation (3.7)(2) as follows. Let $X \to$ Spec(A) be as in (3.7)(2), let k be the residue field of A, and let Y be the special fiber $X \otimes_A k$ of X. Endow Y (resp. Spec(k)) with the inverse image $\bar{M}$ (resp. $\bar{N}$) of the log. str. M on X (resp. N on Spec(A)) in (3.7)(2). Assume k is perfect and char$(k) = p > 0$, and let $W_n(k)$ be the ring of Witt vectors of length n, and endow $\mathrm{Spec}(W_n(k))$ with the log. str. N_n associated to $\mathbf{N} \to W_n(k);\ 1 \mapsto 0$. By fixing a prime element π of A, we have morphisms

$$(Y, \bar{M}) \to (\mathrm{Spec}(k), \bar{N}) \to (\mathrm{Spec}(W_n(k)), N_n)$$

where the second arrow is induced from $\mathbf{N} \to A;\ 1 \mapsto \pi$. (Then, $N_1 \to \bar{N}$ is an isomorphism.) Consider the crystalline cohomology of (Y, M) over the base $(\mathrm{Spec}(W_n(k)), N_n)$ with the usual PD str. on the ideal $pW_n(k)$. Then this crystalline cohomology is very important, and serves as the mixed characteristic analogue of the limit Hodge str. [S]. For the details and the relation with the de Rham-Witt complex in [H_2], cf. [HK] and [K′].

UNIVERSITY OF TOKYO

REFERENCES

[B] Berthelot, P., Cohomologie crystalline des schémas de characteristic $p > 0$, Springer Lecture Notes **407** (1974).

[D] Deligne, P., letter to Illusie, L., June 1, 1988.

[DI] ______ and Illusie, P., Relèvements modulo p^2 et decomposition du complexe de de Rham, *Invent. Math.* **72** (1987) 247–270.

[Fa_1] Faltings, G., Crystalline cohomology and p-adic Galois-representations, preprint.

[Fa_2] ______, F-isocrystals on open varieties, results and conjectures, preprint.

[H_1] Hyodo, O., A cohomological construction of Artin representation over the Witt ring, preprint.

[H_2] ______, On the de Rham-Witt complex attached to semi-stable family, preprint.

[HK] ______ and Kato, K., Semistable reduction and crystalline cohomology with logarithmic poles, in preparation.

[I] Illusie, L., Réduction semi-stable et dégénérescence de suites spectrales de Hodge, preprint.

[K′] Kato, K., Semistable reduction and p-adic etale cohomology, in preparation.

[K] Katz, N., Nilpotent connections and the monodromy theorem: Applications and a result of Turrittin, *Publ. Math. IHES.*, **39** (1976) 229–257.

[KKMS] Kempf, G., Knudsen, F., Mumford, D., Saint-Donat, B., Toroidal embeddings, I, Springer Lecture Notes, **339** (1973).

[S] Steenbrink, J., Limits of Hodge structures, *Invent. Math.* **31** (1976) 229–257.

[SGAI] Grothendieck, A., Revêtement étale et groupe fondamental, Springer Lecture Notes **224** (1971).

PROJECTIVE COORDINATE RINGS OF ABELIAN VARIETIES

By George R. Kempf

Let $A = \oplus_{n \in \mathbf{N}} A_n$ be a graded commutative ring. Then $k = A / \oplus_{n>0} A_n$ is a graded A-module. The ring A is called wonderful if for all $i \geq 0$ the graded A-module $\mathrm{Tor}_i^A(k, k)$ is purely of degree i.

Let $\mathfrak{N}$ be an ample invertible sheaf on an abelian variety X of dimension g over an algebraically closed field k. We will prove

Theorem 1. *If $\mathfrak{N}$ is an p-power where $p \geq 4$ then the ring $\oplus_{n \in \mathbf{N}} \Gamma(X, \mathfrak{N}^{\otimes n})$ is wonderful.*

If $p \geq 3$ we may represent the coordinate ring as a quotient of the polynomial ring $k[\Gamma(X, \mathfrak{N})]$ modulo an ideal I [4, 1, 5]. We will also prove

Theorem 2. a) *If $p \geq 4$ then I is generated by its terms of degree* 2.

b) *If $p = 3$ then I is generated by its terms of degree* 2 *and* 3.

In case a) this implies that if X is embedded by the corresponding complete linear system then X is the scheme-theoretic intersection of quadrics. This was originally proven by Mumford in [2, 4]. In case b) the similar statement is due to Sekiguchi [5]. Theorem 2a) follows formally from Theorem 1 but all the above results are special cases of a more technical result, Theorem 6, to be found in the paper proper. For reasons of exposition we will prove the theorem about relations explicitly before we embark on the generalizations to syzygies. I should remark that Theorem 2a) was conjectured by R. Lazarfeld as part of a more general conjecture about the syzygies for the coordinate ring as a module over the polynomial ring.

In this paper we will almost prove Lazarfeld's conjecture. Among other results we will prove the most desirable case of Mumford's Theorem on the equations of the moduli of abelian varieties.

1. The relations. Let $\mathfrak{M}$ be a fixed ample invertible sheaf on our abelian variety. We will be working with invertible sheaves $\mathfrak{L}_i$ on X which

---

Manuscript received 20 May 1988.

are algebraically equivalent to $\mathcal{M}^{\otimes \ell_i}$ for positive integers ℓ_i. Let $\mathcal{P}$ be the Poincaré sheaf on $X \times X^\wedge$ where $X^\wedge$ is the dual abelian variety. For any point α of $X^\wedge$ we denote $\mathcal{P}|_{X\times\{\alpha\}}$ by $\mathcal{P}_\alpha$.

We begin with the fundamental lemma of Mumford [4]

$$\sum_{\alpha\in X^\wedge} \Gamma(X, \mathcal{L}_1 \otimes \mathcal{P}_\alpha)\cdot\Gamma(X, \mathcal{L}_2 \otimes \mathcal{P}_{-\alpha}) = \Gamma(X, \mathcal{L}_1 \otimes \mathcal{L}_2).$$

We will use the following interpretation of this lemma. Let λ be a linear functional on $\Gamma(X, \mathcal{L}_1 \otimes \mathcal{L}_2)$ then we have an associated family $\{\lambda_\alpha\}$ of linear functionals on $\Gamma(X, \mathcal{L}_1 \otimes \mathcal{P}_\alpha) \otimes \Gamma(X, \mathcal{L}_2 \otimes \mathcal{P}_{-\alpha})$ defined by composition with multiplication. Mumford's lemma means that λ is determined by the family $\{\lambda_\alpha\}$.

The interesting question is what families $\{\lambda_\alpha\}$ can be constructed this way. Clearly $\{\lambda_\alpha\}$ has to depend regularly on α in $X^\wedge$. Our first result is that this is the only condition. To state it in standard form we will use sheaf theory.

For any $\mathcal{L}$ on X be $\mathcal{W}^\pm(\mathcal{L}) = \pi_{X^\wedge *}(\pi_X^*\mathcal{L} \otimes \mathcal{P}^{\otimes \pm 1})$. Then we have multiplication

$$\mathcal{W}^+(\mathcal{L}_1) \otimes \mathcal{W}^-(\mathcal{L}_2) \to \Gamma(X, \mathcal{L}_1 \otimes \mathcal{L}_2) \otimes_k \mathcal{O}_{X^\wedge}.$$

Thus we have a reverse homomorphism

$$M : \Gamma(X, \mathcal{L}_1 \otimes \mathcal{L}_2)^\wedge \to \Gamma(X, (\mathcal{W}^+(\mathcal{L}_1) \otimes \mathcal{W}^-(\mathcal{L}_2))^\wedge).$$

The formal statement is

PROPOSITION 3. a) *M is an isomorphism and* b) *the higher cohomology groups of* $(\mathcal{W}^+(\mathcal{L}_1) \otimes \mathcal{W}^-(\mathcal{L}_2))^\wedge$ *are zero.*

Proof. For one proof see the remark in section 10 of [2]. I will give a different proof here. For a) as M is injective by Mumford's lemma it suffices to check that the two spaces are isomorphic. Now $\mathcal{K} \equiv (\mathcal{W}^+(\mathcal{L}_1) \otimes \mathcal{W}^-(\mathcal{L}_2))^\wedge = (\mathcal{W}^+(\mathcal{L}_1))^\wedge \otimes (\mathcal{W}^-(\mathcal{L}_2))^\wedge$ is isomorphic to the only nonzero direct image $R^{2g}\pi_{X^\wedge *}\mathcal{F}$ by $\pi_{X^\wedge}$ of a sheaf $\mathcal{F}$ on $X \times X \times X^\wedge$, where $\mathcal{F} = \pi_1^*(\mathcal{L}_1^{\otimes -1} \otimes \pi_{1,3}^*\mathcal{P}^{\otimes -1}) \otimes \pi_2^*(\mathcal{L}_2^{\otimes -1} \otimes \pi_{2,3}^*\mathcal{P})$. This follows from Serre duality and the Künneth formula. Thus by the Leray Spectral sequence

$$H^i(X, \mathcal{K}) \approx H^{2g+i}(X \times X \times X^\wedge, \mathcal{F}).$$

We next compute via the projection onto $X \times X$.

Claim. $R^g \pi_{12*}\mathfrak{F} \approx \pi_1^* \mathcal{L}_1^{\otimes -1} \otimes \pi_2^* \mathcal{L}_2^{\otimes -1}|_\Delta \approx (\mathcal{L}_1 \otimes \mathcal{L}_2)^{\otimes -1}$ is the only nonzero higher direct image of $\mathfrak{F}$ via π_{12}. First we note that this claim implies the Proposition. By the Leray Spectral sequence $H^{2g+i}(X \times X \times X^\wedge, \mathfrak{F}) \approx H^{g+i}(X, (\mathcal{L}_1 \otimes \mathcal{L}_2)^{-1})$ which is $H^{-i}(X, \mathcal{L}_1 \otimes \mathcal{L}_2)^\wedge$. Thus the result follows because $\mathcal{L}_1 \otimes \mathcal{L}_2$ is ample.

To prove the claim recall from Mumford's book that ⊛ $R^g \pi_{X^*} \mathcal{P} \approx \mathcal{O}_0$ is the only nonzero direct image of $\mathcal{P}$ via π_{X^*}. On the other hand $\mathfrak{F} = \pi_{12}^*(\pi_1^* \mathcal{L}_1^{\otimes -1} \otimes \pi_2^* \mathcal{L}_2^{\otimes -1}) \otimes (\pi_2 - \pi_1, \pi_3)^* \mathcal{P}$. Thus the claim follows from ⊛ by flat base extension and the projection formula. Q.E.D.

If $\mathfrak{F}$ and $\mathcal{G}$ are two sheaves on X we denote the kernel of the multiplication

$$\Gamma(X, \mathfrak{F}) \otimes_k \Gamma(X, \mathcal{G}) \to \Gamma(X, \mathfrak{F} \otimes \mathcal{G})$$

by $R(\mathfrak{F}, \mathcal{G})$.

Proposition 4. *If $\ell_1 + \ell_2 \geq 5$ and $\ell_1, \ell_2 \geq 2$ then*

$$\sum_{\alpha \in X} R(\mathcal{L}_1, \mathcal{L}_2 \otimes \mathcal{P}_\alpha)\Gamma(X, \mathcal{L}_3 \otimes \mathcal{P}_{-\alpha}) = R(\mathcal{L}_1, \mathcal{L}_2 \otimes \mathcal{L}_3).$$

Proof. By [1 or 5] the multiplication

$$\Gamma(X, \mathcal{L}_1) \otimes_k \Gamma(X, \mathcal{L}_2 \otimes \mathcal{L}_3) \to \Gamma(X, \mathcal{L}_1 \otimes \mathcal{L}_2 \otimes \mathcal{L}_3)$$

is surjective. Hence the dual K of the second space is contained in the dual of the tensor product. Clearly K is contained in the perpendicular to the sum and we want to show that they are equal. Consider a linear functional λ on the tensor product which vanishes on the sum. By Mumford's lemma λ is determined by the family $\{\lambda_\alpha\}$ of linear functionals on $\Gamma(X, \mathcal{L}_1) \otimes \Gamma(X, \mathcal{L}_2 \otimes \mathcal{P}_\alpha) \otimes \Gamma(X, \mathcal{L}_3 \otimes \mathcal{P}_{-\alpha})$. By [1] again the multiplication $\Gamma(X, \mathcal{L}_1) \otimes \Gamma(X, \mathcal{L}_2 \otimes \mathcal{P}_\alpha) \to \Gamma(X, \mathcal{L}_1 \otimes \mathcal{L}_2 \otimes \mathcal{P}_\alpha)$ is surjective. Hence $R(\mathcal{L}_1, \mathcal{L}_2 \otimes \mathcal{P}_\alpha)$ is a vector bundle over X and the family $\{\lambda_\alpha\}$ is induced by a regular family $\{\mu_\alpha\}$ of linear functionals on $\Gamma(X, \mathcal{L}_1 \otimes \mathcal{L}_2 \otimes \mathcal{P}_\alpha) \otimes \Gamma(X, \mathcal{L}_3 \otimes \mathcal{P}_{-\alpha}\}$. By Proposition 3 $\{\mu_\alpha\}$ is given by a linear function on $\Gamma(X, \mathcal{L}_1 \otimes \mathcal{L}_2 \otimes \mathcal{L}_3)$. This proves that $(\Sigma)^\perp \subset K$. Q.E.D.

Next we get a definitive result.

THEOREM 5. *If* $\ell_1 \geq 3$, $\ell_2 \geq 4$ *(or* $\ell_1 \geq 2$ *and* $\ell_2 \geq 5$*) and* $\ell_3 \geq 2$ *then*

$$R(\mathcal{L}_1, \mathcal{L}_2 \otimes \mathcal{L}_3) = R(\mathcal{L}_1, \mathcal{L}_2)\Gamma(X, \mathcal{L}_3).$$

Proof. Write $\mathcal{L}_2 = \mathcal{L}_4 \otimes \mathcal{L}_5$ where $\ell_5 = 2$. Thus by Proposition 4 $R(\mathcal{L}_1, \mathcal{L}_2 \otimes \mathcal{L}_3) = \Sigma R(\mathcal{L}_1, \mathcal{L}_4 \otimes \mathcal{P}_\alpha)\Gamma(X, \mathcal{L}_5 \otimes \mathcal{L}_3 \otimes \mathcal{P}_{-\alpha})$ but by [1] for general α, $\Gamma(X, \mathcal{L}_5 \otimes \mathcal{P}_{-\alpha})\Gamma(X, \mathcal{L}_3) = \Gamma(X, \mathcal{L}_5 \otimes \mathcal{L}_3 \otimes \mathcal{P}_{-\alpha})$. Thus we get

$$R(\mathcal{L}_1, \mathcal{L}_2 \otimes \mathcal{L}_3) = (\Sigma R(\mathcal{L}_1, \mathcal{L}_4 \otimes \mathcal{P}_\alpha)\Gamma(X, \mathcal{L}_5 \otimes \mathcal{P}_{-\alpha}))\Gamma(X, \mathcal{L}_3).$$

Hence the result is true. Q.E.D.

We will next see how Theorem 5 implies Theorem 2. For part a) let $\ell_1 \geq 4$. Then we have $R(\mathcal{L}_1, \mathcal{L}^{\otimes}{}_1{}^{n+1}) = R(\mathcal{L}_1, \mathcal{L}_1)\Gamma(X, \mathcal{L}^{\otimes}{}_1{}^{n})$ if $n \geq 0$. Thus the quadratic relations $R(\mathcal{L}_1, \mathcal{L}_2)$ generate all relations. For part b) let $\ell_1 = 3$. Then $R(\mathcal{L}_1, \mathcal{L}_1^{\otimes n+2}) = R(\mathcal{L}_1, \mathcal{L}_2^{\otimes 2})\Gamma(X, \mathcal{L}_1^{\otimes n})$ if $n \geq 0$. Thus the relations are generated by $R(\mathcal{L}_1, \mathcal{L}_1)$ and $R(\mathcal{L}_1, \mathcal{L}_1^{\otimes 2})$. This proves Theorem 2.

To prove Theorem 1 requires a great generalization of the above arguments.

2. The syzygies. Let $\mathfrak{F}_1, \ldots, \mathfrak{F}_n$ be sheaves on X. Then we define $S(\mathfrak{F}_1, \ldots, \mathfrak{F}_n)$ to be the kernel of the multiplication

$$S(\mathfrak{F}_1, \ldots, \mathfrak{F}_{n-1}) \otimes \Gamma(X, \mathfrak{F}_n) \to S(\mathfrak{F}_1, \ldots, \mathfrak{F}_{n-1} \otimes \mathfrak{F}_n)$$

where $S(\mathfrak{F}) = \Gamma(X, \mathfrak{F})$. Thus $S(\mathfrak{F}_1, \ldots, \mathfrak{F}_n)$ is a subspace of $\otimes_i \Gamma(X, \mathfrak{F}_i)$ and the above multiplication involve only the last tensor factor. Clearly $S(\mathfrak{F}_1, \mathfrak{F}_2) = R(\mathfrak{F}_1, \mathfrak{F}_2)$ and the S stands for syzygies in general.

The main objective is

THEOREM 6. *If* $\ell_1 \geq 3, \ell_2 \geq 4, \ldots, \ell_{n-1} \geq 4$ *(or* $\ell_1 \geq 2, \ell_2 \geq 5, \ell_3 \geq 4, \ldots, \ell_{n-1} \geq 4$*) and* $\ell_n \geq 2$ *then*

$$S(\mathcal{L}_1, \ldots, \mathcal{L}_{n-2}, \mathcal{L}_{n-1} \otimes \mathcal{L}_n) = S(\mathcal{L}_1, \ldots, \mathcal{L}_{n-1})\Gamma(X, \mathcal{L}_n).$$

We need some auxilary statement to prove this.

PROPOSITION 7. *In the situation of Theorem* 6 *then* $S(\mathcal{L}_1, \ldots, \mathcal{L}_n)$ *form a vector bundle on the appropriate component of* $\mathrm{Pic}(X)^n$.

The first point is obvious.

a) Theorem $6(n)$ + Proposition $7(n-1) \Rightarrow$ Propositon $7n$). The next statement is

PROPOSITION 8. *If* $\ell_1 \geq 3, \ell_2 \geq 4, \ldots, \ell_{n-2} \geq 4$, (*or* $\ell_1 \geq 2, \ell_2 \geq 5, \ell_3 \geq 4, \ldots, \ell_{n-2} \geq 4$) *and* $\ell_{n-1} \geq 2$ *then*

$$S(\mathcal{L}_1, \ldots, \mathcal{L}_{n-2}, \mathcal{L}_{n-1} \otimes \mathcal{L}_n)$$
$$= \sum_{x \in X^\wedge} S(\mathcal{L}_1, \ldots, \mathcal{L}_{n-1} \otimes \mathcal{P}_\alpha)\Gamma(X, \mathcal{L}_n \otimes \mathcal{P}_{-\alpha}).$$

The idea of the proof of Theorem 5 shows b) Proposition $8(n) \Rightarrow$ Theorem $6(n)$. Thus the crux of the problem is c) Theorem $6(n-1)$ + Proposition $7(n-1) \Rightarrow$ Proposition $8(n)$. This step is accomplished by sheaf-theory.

By Proposition $7(n-1)$ the family $S(\mathcal{L}_1, \ldots, \mathcal{L}_{n-1} \otimes \mathcal{P}_\alpha)$ of vector spaces form a vector bundle on $X^\wedge$. Let $\mathcal{F}_{n-1}$ be the corresponding local free sheaf on $X^\wedge$. The strengthening of Proposition 8 is

PROPOSITION 9. *In the situation of Proposition* 8, a) *we have an isomorphism* $S(\mathcal{L}_1, \mathcal{L}_{n-1} \otimes \mathcal{L}_n)^\wedge \to \Gamma(X^\wedge, (\mathcal{F}_{n-1} \otimes \mathcal{W}^{-1}(\mathcal{L}_n))^\wedge)$ *and* b) *the higher cohomology groups of* $(\mathcal{F}_{n-1} \otimes \mathcal{W}^{-1}(\mathcal{L}_n))^\wedge$ *are zero.*

Proof. We have exact sequence by Theorem $6(n)$ $0 \to S(\mathcal{L}_1, \ldots, \mathcal{L}_{n-1} \otimes \mathcal{P}_\alpha) \otimes \Gamma(X, \mathcal{L}_n \otimes \mathcal{P}_{-\alpha}) \to S(\mathcal{L}_1, \ldots, \mathcal{L}_{n-2}) \otimes \Gamma(X, \mathcal{L}_{n-1} \otimes \mathcal{P}_{+\alpha}) \otimes \Gamma(X, \mathcal{L}_n \otimes \mathcal{P}_{-\alpha}) \to S(\mathcal{L}_1, \ldots, \mathcal{L}_{n-2} \otimes \mathcal{L}_{n-1} \otimes \mathcal{P}_\alpha) \otimes \Gamma(X, \mathcal{L}_n \otimes \mathcal{P}_{-\alpha}) \to 0$. In terms of sheaves this gives $0 \to (\mathcal{F}_{n-2} \otimes \mathcal{W}^{-1}(\mathcal{L}_n)^\wedge \to (S(\mathcal{L}_1, \ldots, \mathcal{L}_{n-1}) \otimes \mathcal{W}^{+}(\mathcal{L}_{n-1}) \otimes \mathcal{W}^{-}(\mathcal{L}_n))^\wedge \to (\mathcal{F}_{n-1} \otimes \mathcal{W}^{-1}(\mathcal{L}_n))^\wedge \to 0$ for an appropriate $\mathcal{F}_{n-2}$. The long exact sequence of cohomology gives the statement b) by induction on n and Proposition 3b). Thus the last arrow is surjective on sections. This means that any regular family $\{\lambda_\alpha\}$ of linear functionals on $S(\mathcal{L}_1, \ldots, \mathcal{L}_{n-1} \otimes \mathcal{P}_\alpha) \otimes \Gamma(X, \mathcal{L}_n \otimes \mathcal{P}_{-\alpha})$ comes from a regular family $\{\mu_\alpha\}$ of linear functionals on $S(\mathcal{L}_1, \ldots, \mathcal{L}_{n-2}) \otimes \Gamma(X, \mathcal{L}_{n-1} \otimes \mathcal{P}_\alpha) \otimes \Gamma(X, \mathcal{L}_n \otimes \mathcal{P}_{-\alpha})$. By Proposition 3a) $\{\mu_\alpha\}$ is induced by a linear functional μ on $S(\mathcal{L}_1, \ldots, \mathcal{L}_{n-2}) \otimes \Gamma(X, \mathcal{L}_{n-1} \otimes \mathcal{L}_n)$. Thus we have a surjection

$$A : (S(\mathcal{L}_1, \ldots, \mathcal{L}_{n-2}) \otimes \Gamma(X, \mathcal{L}_{n-1} \otimes \mathcal{L}_n))^\wedge$$

$$\to \Gamma(X^\wedge, (\mathcal{F}_{n-1} \otimes \mathcal{W}^{-1}(\mathcal{L}_n))^\wedge),$$

but the kernel of A is $(S(\mathcal{L}_1, \ldots, \mathcal{L}_{n-2} \otimes \mathcal{L}_{n-1} \otimes \mathcal{L}_n))^\wedge$ by induction. This statement follows from the obvious exact sequence.

3. Applications. We first give the proof of Theorem 1. One simply builds a projective resolution of k over the ring $A = \oplus_{n \in \mathbf{N}} \Gamma(X, \mathfrak{N}^{\otimes n})$ as follows;

$$\to S(\mathfrak{N}(n \text{ times})) \otimes_k A \cdots \to S(\mathfrak{N}, \mathfrak{N}) \otimes_k A$$

$$\to \Gamma(\mathfrak{N}) \otimes_k A \to A \to k \to 0.$$

This is exact because the kernel of

$$S(\mathfrak{N}(n\text{-times})) \otimes_k A \to S(\mathfrak{N}(n-1 \text{ times})) \otimes_k A$$

is $\oplus_{q \geq 0} S(\mathfrak{N}, \ldots, \mathfrak{N}^{\otimes 1+q})$ which is generated by the term with $q = 0$ by Theorem 6.

Now let $\mathfrak{N} = \mathcal{L}^{\otimes p}$ and let $\mathfrak{M}$ be algebraically equivalent to $\mathcal{L}^{\otimes m}$.

THEOREM 10. *If $P \geq 4$ and $m \geq 3$ then the A-module $M = \oplus_{n \geq 0} \Gamma(X, \mathfrak{M} \otimes \mathfrak{N}^{\otimes n})$ is generated by $\Gamma(X, \mathfrak{M})$ and the relations as generated by $R(\mathfrak{M}, \mathfrak{N})$.*

This follows from Theorem 5 but we could equally state a result from [7], which is

COROLLARY 11. *In the situation of Theorem 10 $\mathfrak{M}$ is strongly normally generated for $\mathfrak{N}$.*

We will next work out the general properties of a finitely generated wonderful ring A over a field k. The precise result is

THEOREM 12. *If $P = \mathrm{Sym}_k A_1$ then A is quotient ring of P. The kernel I of the surjection $P \to A$ is generated by the quadric relations I_2. The kernel of the surjection $I_2 \otimes_k P \to I$ is generated by linear syzygies of the form $\Sigma\, q_i \otimes a_i$ where q's are I_2 and the a's are in A_1 together the banal syzygies of the form $q_1 \otimes q_2 - q_2 \otimes q_1$ where the q's are in I_2.*

Proof. Let m be the relevant ideal $\oplus_{n \geq 0} A_n$ of A. Then $\mathrm{Tor}_1^A(k, k)$ is isomorphic to m/m^2. As the tor has degree one, m/m^2 and hence m are generated by A_1 as an A-ideal. As A is generated it follows that A is generated by A_1 as a k-algebra. As $\mathrm{Tor}_2^A(k, k)$ is of degree the kernel of the surjection $A_1 \otimes A \to \mathfrak{M}$ is generated by elements of the form $\Sigma\, \alpha_i \otimes \beta_i$ where the α's and the β's are in A_1. It follows that any polynomial relation between a basis of A_1 is generated by the $\Sigma\, \alpha_i\beta_i = 0$. This proves the second statement.

For the third statement we need some machinery. Consider the change of rings spectral sequence

$$\mathrm{Tor}_p^A(k, \mathrm{Tor}_q^P(A, k)) \Rightarrow \mathrm{Tor}_{p+q}^P(k, k).$$

This is spectral sequence of rings; i.e. E_r for $r \geq 2$ are graded rings, $d_r : E_r \to E_r$ is a derivation and $E_{r+1} = H(E_r)$ has the quotient ring structure. Now $\mathrm{Tor}_n^P(k, k)$ has pure degree n as P is a polynomial ring on linear generators. As $\mathrm{Tor}_p^A(k, \mathrm{Tor}_q^P(A, k) \cong \mathrm{Tor}_p^A(k, k) \otimes \mathrm{Tor}_q^P(A, k)$, we can determine its degree from those of $\mathrm{Tor}_q^P(A, k)$ as $\mathrm{Tor}_p^A(k, k)$ has pure degree P. The differential d_r increases q by $r - 1$ and lower p by r. By computing degree we must have $\mathrm{Tor}_0^P(A, k) = k$ and $\mathrm{Tor}_1^P(A, k)$ is a pure degree two. This reproves the last results. In fact $d_2 : \mathrm{Tor}_2^A(k, k) \otimes k \to k \otimes \mathrm{Tor}_1^P(A, k)$ as we already know. Proceeding further $k \otimes \mathrm{Tor}_2^P(A, k)$ must have degree >2. Thus in the sequence it must die but there are only two differentials coming into $k \otimes \mathrm{Tor}_2^P(A, k)$, d_3 from a quotient of $\mathrm{Tor}_3^A(k, k) \otimes k$ (of degree 3) and d_2 from $\mathrm{Tor}_2^A(k, k) \otimes \mathrm{Tor}_1^P(A, k)$ (of degree 4). Therefore $\mathrm{Tor}_2^P(A, k)$ has degree 3 and 4 and its degree four part comes from $d_2(b \cdot q)$ where b is linear relation in $\mathrm{Tor}_2^A(k, k)$ and q is a quadratic relation in $\mathrm{Tor}_1^P(A, k)$. Now $d_2(b \cdot q) = d_2(b) \cdot q$ which is the product in $\mathrm{Tor}_*^P(A, k)$ of two quadratic relations. A simple calculation shows that the product is represented by the banal syzygies $d_2(b) \otimes q - q \otimes d_2(b)$. Q.E.D.

Therefore the theorem applies to the ring of Theorem 1 coming from an abelian variety embedded by a fourth or better powers. This reproves a result Theorem 24 of [2].

4. The moduli problem. We will use the basic set up of section 11 of [2]. There we have an embedding i : moduli $\subset$ M-R where M-R is the Mumford-Riemann scheme when n is a n even integer ≥ 4. The result is

THEOREM 13. *i is an open immersion.*

Thus the new result is the case $n = 4$. The proof is almost the same. We need to show that the family $\mathfrak{X} \to M\text{-}R$ is flat at the point q_X of $M\text{-}R$ corresponding to an abelian variety. By the last section we need only lift the degree zero and degree one syzygies because the degree two relations are banal and hence automatically lift.

The argument for lifting these syzygies is that the cubic relations between $\Gamma(\mathfrak{M})$ and $\Gamma(\mathfrak{M}^{\otimes 2})$ are generated by the quadrics by Theorem 2.

Then knowing these we can do the lifting using the injectivity Theorem 27 for $\Gamma(X, \mathfrak{M}^{\otimes 3})$ instead of $\Gamma(X, \mathfrak{M}^{\otimes 4})$ in the same way. This is the proof.

5. The obvious relations. We return to the notations of Section 1. We will now assume that is excellent and gives a separable polarization of X and the characteristic of the ground field is not 2. Assume that ℓ is an even integer ≥ 2. Then $\mathfrak{L}$ is totally symmetric. Let $\mathfrak{L}'$ be $T_x^*\mathfrak{L}$ for some point x of X.

By the discussion before Theorem 14 of [2]. The obvious relations between $\Gamma(X, \mathfrak{L})$ and $\Gamma(X, \mathfrak{L}')$ are generated by relations of the form

$$\sum_{(c_1,c_2)\in B(\mathfrak{L})^2} (r_{c_1,c_2}(1, 2) - r_{c_1,c_2}(2, 1) \sum_{\substack{k_2\in B(\mathfrak{L}^{\otimes 2}) \\ -k_1+k_2=c, \\ k_1+k_2=0c_2}} \delta_{k_2}|y$$

where k_1 runs through $B(\mathfrak{L}^{\otimes 2})$, $2y = x$ and $r_{c_1,c_2}(u, v) = \Sigma_{(d_1,d_2)\in B(\mathfrak{L})^2}\, \delta_{d_1} \otimes \delta_{d_2}$ such that $d_1 - c_1 = p$, $d_1 + c_1 = \eta_u$, $d_2 - c_2 = \sigma$ and $d_2 + c_2 = \eta_v$ where $\rho, \eta_1, \eta_2, \sigma$ are in $B(\mathfrak{N})$.

Continuing with section 7 of [2] we want to study the equivariance properties of the multiplication $m : \Gamma(X, \mathfrak{L}) \otimes \Gamma(X, \mathfrak{L}') \to \Gamma(X, \mathfrak{L} \otimes \mathfrak{L}')$ under the abelian group C which acts on $\Gamma(X, \mathfrak{L})$ via $H(\mathfrak{L})$ and on $\Gamma(X, \mathfrak{L}')$ by T_x^* of the action on $\Gamma(X, \mathfrak{L})$. Consider the isogeny $\xi : X \times X \to X \times X$ given by $\xi(u, v) = (u + v, u - v)$. Thus $\xi(y, -y) = (0, x)$. Hence as $\xi^*(\pi_1^*\mathfrak{L} \otimes \pi_2^*\mathfrak{L}) \approx \pi_1^*\mathfrak{L}^{\otimes 2} \otimes \pi_2^*\mathfrak{L}^{\otimes 2}$ we have $\mathfrak{L} \otimes \mathfrak{L}' \approx T_y^*(\mathfrak{L}^{\otimes 2})$ and an action of C on $\Gamma(X, \mathfrak{L} \otimes \mathfrak{L}')$ via T_y^* the action on $\Gamma(X, \mathfrak{L}^{\otimes 2})$.

Let $Y'(c, \chi) = T_y^*$ of the previous $Y(c, \chi)$ in $\Gamma(X, \mathfrak{L}^{\otimes 2})$ and $X(a, b, \chi) = (1 \times T_x)^*$ of the previous $X(a, b, \chi)$. Then we have a form of Corollary 16; i.e.

$$m(X(a, b, \chi)) = Y'(u, \chi) \cdot Y(v, \chi)|_{-y}$$

where $a = u + v$ and $b = u - v$ in $B(\mathcal{L}^{\otimes 2})$. Furthermore in Lemma 17 and Proposition 20 we may replace $|_0$ by $|_{-y}$. With the replacement $|_0$ by $|_y$ the proof of Theorem 20 gives the desired Theorem B as $-y$ is an arbitrary element of X in Lemma 17.

Thus by these results and Theorem 5 we may conclude.

THEOREM 14. *If ℓ_1 and $\ell_2 \geq 4$ then $R(\mathcal{L}_1, \mathcal{L}_2)$ is generated by obvious relations.*

This is because the above discussion gives the case $\ell_1 = \ell_2 = 4$ and a multiple of an obvious relation is obvious. Thus this follows from Proposition 4.

6. Syzygies over the polynomial ring. We return to the situation of Section 2. Let $\mathcal{L}$ be a fixed invertible sheaf on X. Let $P = \mathrm{Sym}_k[\Gamma(X, \mathcal{L})]$ and A be the graded ring $\bigoplus_{n \geq 0} \Gamma(X, \mathcal{L}^{\otimes n})$.

We will prove the following special case of Lazarfeld's conjecture.

THEOREM 15. *$\mathrm{Tor}_i^P(A, k)$ is purely of degree $i + 1$ if $i \leq \ell/2$ when $\ell \geq 5$ and $i > 0$.*

We first make a trivial reduction of the problem by Theorem 2. $\mathrm{Tor}_2^P(A, k)$ is purely of degree 2. Now min degree of $\mathrm{Tor}_i^P(A, k)$ is strictly increasing. Therefore by induction we need only prove

Claim. $\mathrm{Tor}_i^P(A, k)$ is zero in degree $>i + 1$ in the required cases.

Next we will present an idea for computing $\mathrm{Tor}_i^P(M, k)$ where M is a finite generated grated P-module. Now M is generated by M_d for $d \leq$ some minimal degree, say $d(M)$. Here $d(M)$ is the smallest number such that $M \times_p k$ is zero in degree $>d(M)$. Consider the exact sequence

$$0 \to T'(M) \to M[1] \otimes_k P_1 \xrightarrow{\alpha} M$$

where $M[1]$ is M with an increase of degree by one. The image M' of α is a submodule of M such that the artin module M/M' is zero in degree $>d(M)$. Therefore by an obvious filtration of M/M' by $k[e]$'s we have

$$\mathrm{Tor}_i^P(M/M', k) \text{ is zero in degree } >i + d(M).$$

Consequently by the long exact sequence $\mathrm{Tor}_i^P(M', k) \to \mathrm{Tor}_i(M, k)$ is surjective in degrees $>i + d(M)$.

We also have the long exact sequence

$$\mathrm{Tor}_i^P(T'(M, k) \longrightarrow \mathrm{Tor}_i^P(M[1] \otimes_k P_1, k) \xrightarrow{\alpha *} \mathrm{Tor}_i^P(M', k)$$

$$\xrightarrow{\delta} \mathrm{Tor}_i^P(T'(M), k) \longrightarrow \ldots .$$

In degree $>i + d(M) + \mathrm{Tor}_i^P(M[1] \otimes P_k, k) \overset{\alpha_*}{\to} \mathrm{Tor}_i^P(M', k)$ $\mathrm{Tor}_i^P(M, k) \otimes P_1 \overset{m}{\to} \mathrm{Tor}_i^P(M, k)$ where m is multiplication by P_1 in the first variable. As m is also multiplication by P_1 is the second variable it is zero. Therefore *) In degrees $>i + d(M)$ we have an surjective $\mathrm{Im}\,\delta \to$ $>\mathrm{Tor}_i^P(M, k)\ \mathrm{Tor}_{i-1}^P(T'(M, k)$.

We may repeat this procedure i times. Then $\mathrm{Tor}_{-1}^P(T^{i+1}(M), k)$ is zero where $T^jM = T^{j-1}(T'(M))$ and $T^0M = M$. Hence we get

Lemma 16. *In degrees* $>i - j + d(T^j(M))$ for all $0 \leq j \leq i$ we have $\mathrm{Tor}_i^P(M, k)$ is zero.

We first compute $T^j(M)$ where $M = \oplus_n \Gamma(X, \mathcal{L}' \otimes \mathcal{L}^{\otimes n})$ for some invertible sheaves. This will require some new functors.

Let $K(\mathcal{L}_1) = \Gamma(X, \mathcal{L}_1)$ and when $n > 1$ define $K(\mathcal{L}_1, \ldots, \mathcal{L}_n)$ be the exact sequence

$$0 \to K(\mathcal{L}_1, \ldots, \mathcal{L}_n) \to K(\mathcal{L}_1, \mathcal{L}_3, \ldots, \mathcal{L}_n) \otimes \Gamma(X, \mathcal{L}_2)$$

$$\to K(\mathcal{L}_1 \otimes \mathcal{L}_2, \mathcal{L}_3, \ldots, \mathcal{L}_n).$$

Repeating the same ideas as in Section 2 we prove

Theorem 17. *If* $n \geq 3$, $K(\mathcal{L}_1, \mathcal{L}_3, \ldots, \mathcal{L}_n) \otimes \Gamma(X, \mathcal{L}_2) \to$ $K(\mathcal{L}_1, \otimes \mathcal{L}_2, \mathcal{L}_3, \ldots, \mathcal{L}_n)$ *is surjective if* $\ell_2 \geq 2$, *either* $\ell_1 \geq 2n - 2$, $\ell_3 \geq 2$, $\ell_{n-1} \geq 2$, $\ell_n \geq 3$ *or* $\ell_1 \geq 2n - 1$, $\ell_3 \geq 2$, $\ell_n \geq 2$.

In terms of the K's $T^j(M) = \oplus_n K(\mathcal{L}' \otimes \mathcal{L}^{\otimes n-j+1}, \mathcal{L}, \ldots, \mathcal{L})$ where $\mathcal{L}$ repeats $j - 1$ times. If $\ell \geq 5$ the theorem states that $d^j(M) =$ $\min n$ such that $\ell' + (n - j + 1)\ell \geq 2j$ or, rather, $-1 + j + \{1/\ell[2j - \ell']\}$. For Theorem 16 by Lemma 16 we need $i \geq i - j + d(T^j(M))$ for all $0 \leq j \leq i$. Here $\mathcal{L}' = 0_X$. We we want

$$0 \geq -j + j - 1 + \left\{\frac{1}{\ell}[2j]\right\} \quad \text{or} \quad 2j \leq \ell \quad \text{if} \quad 0 \leq j \leq i.$$

More simply we want $2i \leq \ell$ or $i \leq \ell/2$. Q.E.D.

JOHNS HOPKINS UNIVERSITY

REFERENCES

[1] G. Kempf, Multiplication over abelian varieties, *Amer. Jour. of Math.* **110** (1988), 765-774.
[2] ______, Linear systems over abelian varieties, to appear *Amer. Jour. of Math.*
[3] D. Mumford, On the equations defining abelian varieties I, *Inven. Math.* **1** (1966), 287-354.
[4] ______, Varieties defined by quadratic equations, in "Questions on algebraic varieties", Centro Inter. Mate., Estrivo, Roma (1970), 31-100.
[5] T. Sekiguchi, On the normal generation by a line bundle on an abelian variety, *Proc. of Japan Aca.* **54** (1978), 185-188.
[6] ______, On the cubics defining abelian varieties, *J. Math. Soc. Japan*, **30** (1978), 703-721.
[7] G. Kempf, Notes on abelian integrals, I, to appear.

CLASSICAL PROJECTIVE GEOMETRY AND MODULAR VARIETIES

By Robert MacPherson[1] and Mark McConnell[2]

CONTENTS

Introduction. This paper has two purposes. The first is to announce the construction of cell complexes which are equivalent to certain locally symmetric spaces and their compactifications. The second is to explore

---

[1]Partially supported by NSF grant DMS-8803083.
[2]Partially supported by NSF grant DMS-8715305.
Manuscript received November 22, 1988.

some projective geometry in the style of Desargues and Steiner which is related to the construction of these cell complexes.

1. The simplest example. We begin with an example involving geometry in the projective line. This is less geometrically rich than the higher dimensional examples treated later, but it illustrates how projective geometry is related to the construction of cell complexes.

If $\mathbb{P}^1$ is a projective line over a field of characteristic $\neq 2$, we will be interested in a class of configurations of points in $\mathbb{P}^1$ which we call $\mathcal{C}$-*configurations*.

Definition. A $\mathcal{C}$-*configuration* in a projective line $\mathbb{P}^1$ is either

(1) three distinct points of $\mathbb{P}^1$, all of which are called *primary points*; or

(2) four distinct points of $\mathbb{P}^1$, two called *primary points* and two called *secondary points*, such that the two secondary points are harmonic conjugates with respect to the two primary points.

(Recall that points C and D are *harmonic conjugates* with respect to points A and B if the cross ratio of $ABCD$ is -1, or equivalently if affine coordinates can be introduced in $\mathbb{P}^1$ such that A, B, C, and D have coordinates 0, ∞, -1, and 1 respectively.)

In drawing pictures of $\mathcal{C}$-configurations, we will symbolize primary points by solid black dots •, and secondary points by small circles ∘. Thus an example of a $\mathcal{C}$-configuration of the first type in the real projective line is

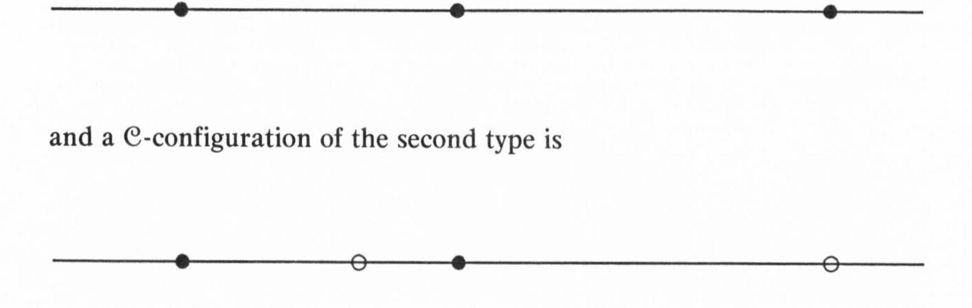

and a $\mathcal{C}$-configuration of the second type is

We put an order relation $<$ on the set of all $\mathcal{C}$-configurations in $\mathbb{P}^1$ in the following way:

Definition. Let $\mathbf{X}$ and $\mathbf{X}'$ be two $\mathcal{C}$-configurations in $\mathbb{P}^1$. We say that $\mathbf{X} < \mathbf{X}'$ if and only if $\mathbf{X}'$ is a harmonic quadruple, and $\mathbf{X}$ is a triple obtained from $\mathbf{X}'$ by omitting one of the secondary points and changing the other to a primary point.

For example, $\mathbf{X} < \mathbf{X}'$ holds for the real configurations $\mathbf{X}$ and $\mathbf{X}'$ drawn above.

Now we are ready to build a cell complex.

Definition. The *$\mathcal{C}$-complex* associated to the projective line $\mathbb{P}^1$ is the graph with a vertex for every $\mathcal{C}$-configuration $\mathbf{X}$, and an edge joining the vertex for $\mathbf{X}$ and the vertex for $\mathbf{X}'$ whenever $\mathbf{X} < \mathbf{X}'$.

(This is called the *geometric realization* of the set of $\mathcal{C}$-configurations.)

Example. Let's examine the case of the projective line $\mathbb{P}^1(\mathbb{F}_3)$ over the field of three elements. This projective line has four points: 0, 1, 2, and ∞ with respect to affine coordinates.

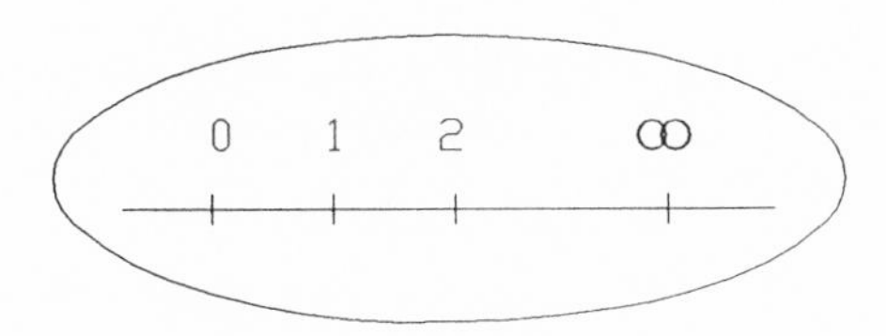

There are four $\mathcal{C}$-configurations in $\mathbb{P}^1(\mathbb{F}_3)$ of the first type, one for each three-element subset of $\mathbb{P}^1(\mathbb{F}_3)$. There are six $\mathcal{C}$-configurations of the second type, one for each two element subset S: we let S be the primary points, and the remaining two points are automatically harmonically conjugate with respect to the primary pair. (This is a special property of the three-element field.)

We now draw the $\mathcal{C}$-complex associated to $\mathbb{P}^1(\mathbb{F}_3)$, labelling each vertex with the corresponding projective configuration:

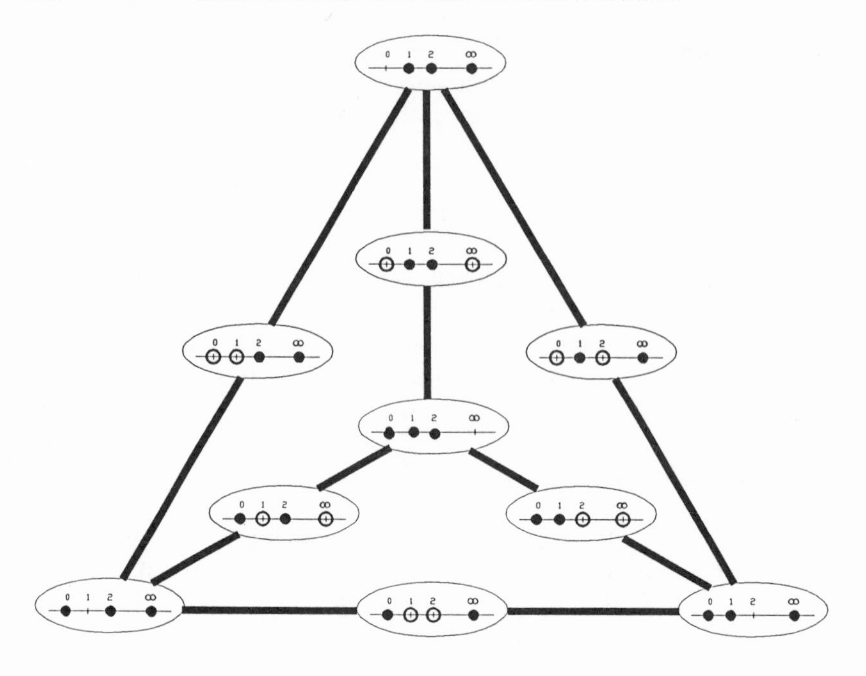

So the cell complex associated to $\mathbb{P}^1(\mathbb{F}_3)$ is this graph:

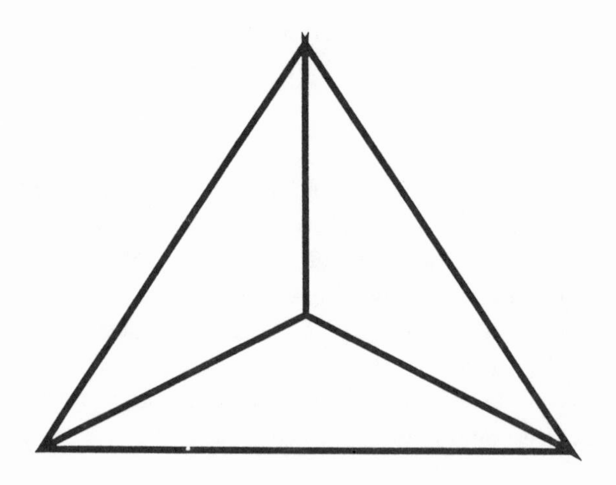

Connection with Modular Varieties. We now explain the question about modular varieties to which this graph provides the answer. Consider the group SL(2, $\mathbb{Z}$) of two-by-two matrices with integral entries and determinant 1, acting on the upper half plane $\mathbb{H}$ in the usual way. For any integer N, $\Gamma(N)$ is the subgroup of SL(2, $\mathbb{Z}$) consisting of matrices that are congruent to the identity modulo N. Then the quotient space $\Gamma(N)\backslash\mathbb{H}$ is an important modular curve. The question is how to calculate its topological type.

THEOREM. *If p is a prime which is congruent to* 3 (mod 4), *then the* $\mathcal{C}$*-complex associated to the projective line* $\mathbb{P}^1(\mathbb{F}_p)$ *over the field with p elements is a strong deformation retract of* $\Gamma(p)\backslash\mathbb{H}$.

In the case of $\mathbb{F}_3$ examined above, it is well known that $\Gamma(3)\backslash\mathbb{H}$ is homeomorphic to the sphere minus four points. This is homeomorphic to an open disk minus three points, which strong deformation retracts to the graph calculated above.

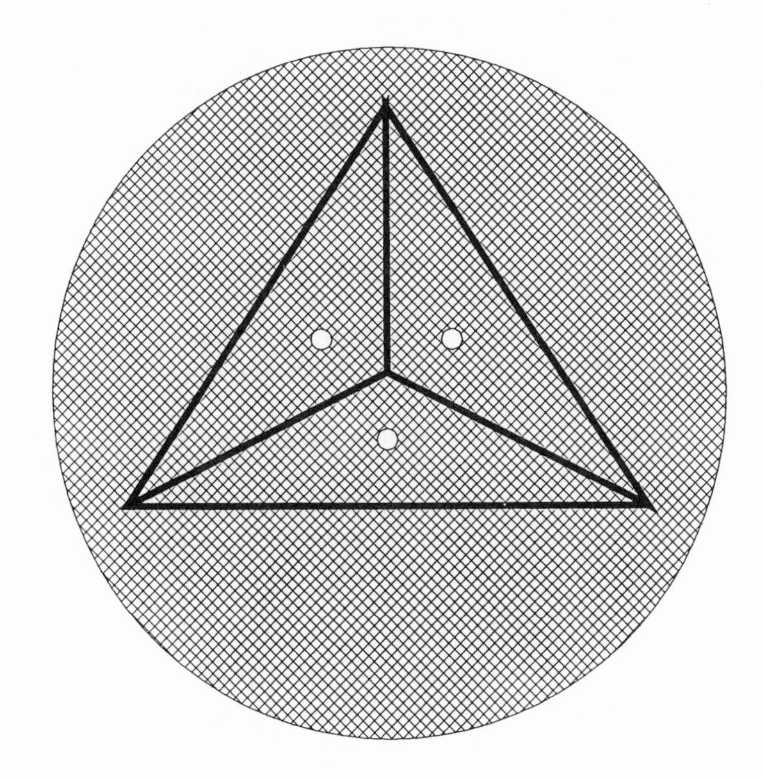

2. Contents of this paper.

Projective Geometry. In addition to the $\mathcal{C}$-configurations in the projective line which we described above, we introduce $\mathcal{C}$-configurations in the projective plane and in a symplectic projective three-space (defined in Chapter 4). In each case, the rough outlines of the theory are the same as above. We work in $\mathbb{P}^m(k)$, k a field. The $\mathcal{C}$-configurations consist of primary and secondary points. There is a partial order relation $\leq$ on the set of $\mathcal{C}$-configurations in $\mathbb{P}^m$. The geometric realization of this partially ordered set is called the *$\mathcal{C}$-complex* associated to $\mathbb{P}^m$. For certain k the $\mathcal{C}$-complex is a strong deformation retract of an appropriate modular variety.

The idea of assembling cells representing geometric figures into a cell complex is, of course, inspired by the beautiful Bruhat-Tits theory of buildings [B]. However, individual $\mathcal{C}$-configurations are more intricate than the figures of Bruhat-Tits. Many were studied by geometers of the seventeenth through the nineteenth centuries, who regarded these configurations as jewels of their subject.

Classical projective geometry was a beautiful field of mathematics. It died, in our opinion, not because it ran out of theorems to prove, but because it lacked organizing principles by which to select theorems that were important. Also, it was isolated from the rest of mathematics. Much of what we do may be regarded as a direct continuation of nineteenth century synthetic geometry. In fact, we hope the new motivation of studying $\mathcal{C}$-complexes will provide projective geometry with one organizational principle, and with one relation tying it to "mainstream" mathematics. (It also has other ties with current research mathematics. We note, in particular, the importance of representable matroids [G-G-M-S], arrangements of hyperplanes [O-S] [G-M, Part III], and motivic cohomology [B-M-S, Section 2.3] [B-G-S-V].)

A large part of this paper's exposition is motivated by this dream of continuing classical projective geometry. We include results about individual $\mathcal{C}$-configurations (Chapters 2, 3, and 5), an extension of the old idea of nets (6.6), and the study of a new geometry with a very classical flavor—symplectic projective geometry (Chapter 4).

$\mathcal{C}$-complexes and Modular Varieties. The main interest in $\mathcal{C}$-complexes stems from the fact that, for certain finite projective spaces, they are deformation retracts of locally symmetric spaces. The latter are among the most important spaces known, because of their relations to modular forms, discrete groups, number theory, and other fields.

For projective spaces over the rationals, the $\mathcal{C}$-complexes are related to unions of symmetric spaces with Voronoi or Voronoi-like cellulations [V] [M-McC1].

If $\mathbb{P}^1$ is the projective line, then the $\mathcal{C}$-complexes are deformation retracts of locally symmetric spaces for $SL(2, \mathbb{R})$ as explained above. If $\mathbb{P}^2$ is the projective plane, the $\mathcal{C}$-complexes are deformation retracts of locally symmetric spaces for $SL(3, \mathbb{R})$. If $\mathbb{P}^3$ is symplectic projective three-space, the $\mathcal{C}$-complexes are deformation retracts of locally symmetric spaces for $Sp(4, \mathbb{R})$. The cell decompositions for $SL(2, \mathbb{R})$ and $SL(3, \mathbb{R})$ are classical [V] [Sou1] [Sou2]; what is new in these cases is the construction of these cell complexes from projective configurations. The cell decomposition for $Sp(4, \mathbb{R})$ is new [M-McC1]. (This is the first explicit cell decomposition for a Hermitian locally symmetric space other than the classical modular curves.)

Another new aspect is the extension of these cell decompositions to the Satake compactifications of the locally symmetric spaces. This is important because the intersection homology of the Satake-Baily-Borel compactification of a Hermitian locally symmetric space has recently become interesting to number theorists. We discuss this topic in Chapter 7, which may be skipped by more geometrically-minded readers.

Directions for further work. We hope there will eventually be a higher-dimensional generalization of the theory we present here. Can the topology of higher-dimensional locally symmetric spaces be related to projective configurations in higher-dimensional projective spaces? We do not even know what to conjecture, since we do not have a conceptual definition of a $\mathcal{C}$-configuration even in the present cases, but have only a list (Chapters 2, 3, and 5). Preliminary indications are that for the higher cases one may need m-ary points as well as primary and secondary ones.

We have definitions of $\mathcal{C}$-configurations and the $\mathcal{C}$-complex for $G = SL(4, \mathbb{R})$. This involves configurations in $\mathbb{P}^3$. The results are analogous to those in Chapters 2, 3, 6 and 7. For simplicity we have not described them here. For $SL(n, \mathbb{R})$, $n \geq 5$, the situation is more complicated.

3. Acknowledgments. For cellulations of $\Gamma \backslash SL(n, \mathbb{R})/K$, previous work was done in [Mi] [V] [V2] [V3] [Sou1] [Sou2] [AMRT] [Ash] [Što]

[L-S]. Siegel [Sie] found a fundamental domain for the upper half-spaces $\Gamma\backslash\mathrm{Sp}(2n, \mathbb{R})/K$ for all n in a nonconstructive way. Our cellulation of $\Gamma\backslash\mathrm{Sp}(4, \mathbb{R})/K$ will be described in [M-McC1].

We are grateful to A. Ash, A. Borel, W. Casselman, H. S. M. Coxeter, W. Fulton, M. Goresky, A. Nicas, and C. Soulé for helpful conversations. We thank the Institute for Advanced Study and the IHES for their hospitality while many of the ideas in this paper were being worked out.

The first author would like to point out that those parts of this paper relating to $\mathrm{SL}(n, \mathbb{R})$ symmetric spaces are not joint work, but are rather an exposition of some of the second author's thesis.

Chapter 1—Generalities on $\mathcal{C}$-configurations

The actual definition of a $\mathcal{C}$-configuration will be given in a case-by-case way in Chapters 2, 3, and 5. However, certain general ideas apply in all three of these cases. We collect these ideas here.

(1.1). A $\mathcal{C}$-configuration $\mathbf{X}$ in a projective space $\mathbb{P}^m$ will always be a finite set of points in $\mathbb{P}^m$ decomposed into two disjoint subsets, one called the set of *primary points* of $\mathbf{X}$ and the other the set of *secondary points* of $\mathbf{X}$. Such a finite set will be a $\mathcal{C}$-configuration if and only if its projective equivalence class belongs to a certain finite list (to be found in (2.1), (3.1), (5.1)).

(1.2). *Remark.* As mentioned in the introduction, we do not have a collection of projective properties which characterize the $\mathcal{C}$-configurations. However, there are some general conditions which all the $\mathcal{C}$-configurations satisfy.

(1) $\mathcal{C}$-configurations fix the ambient projective space. That is, the only automorphism of $\mathbb{P}^m$ which fixes a $\mathcal{C}$-configuration pointwise is the identity automorphism.

(2) $\mathcal{C}$-configurations have no continuous moduli. That is, there are only finitely many $\mathcal{C}$-configurations up to projective equivalence, even if $\mathbb{P}^m$ is infinite.

(3) Any configuration in $\mathbb{P}^m$ that is projectively equivalent to a $\mathcal{C}$-configuration is a $\mathcal{C}$-configuration (by definition).

(4) If $\mathbf{X}$ is a $\mathcal{C}$-configuration, any line meets at most four points of $\mathbf{X}$. If a line meets exactly four points, two of which are primary and two of

which are secondary, then the secondary points are harmonically conjugate with respect to the primary pair.

(1.3). *Definition*. If $\mathbf{X}$ and $\mathbf{X}'$ are two $\mathcal{C}$-configurations in the same projective space, we say $\mathbf{X} \leq \mathbf{X}'$ if the set of points of $\mathbf{X}'$ contains the set of points of $\mathbf{X}$, and the set of primary points of $\mathbf{X}$ contains the set of primary points of $\mathbf{X}'$.

In other words, the $\mathcal{C}$-configurations which are $\leq \mathbf{X}$ are those that can be obtained from $\mathbf{X}$ by deleting some secondary points and by changing some others to primary points.

For a fixed projective space $\mathbb{P}^m$, the set of all $\mathcal{C}$-configurations in $\mathbb{P}^m$ forms a partially ordered set under $\leq$, which we denote by $\mathcal{C}$.

$\mathbf{X} < \mathbf{X}'$ means, of course, that $\mathbf{X} \leq \mathbf{X}'$ but $\mathbf{X} \neq \mathbf{X}'$.

In our pictures, primary points are drawn as solid dots •; secondary points, as small circles ∘.

(1.4). With respect to $\leq$, $\mathcal{C}$ is a ranked poset. More precisely, there is a *rank function r* on $\mathcal{C}$ characterized by the following properties:

(a) any minimal element of $\mathcal{C}$ has rank 0;

(b) for any $\mathbf{X}, \mathbf{Y} \in \mathcal{C}$, $r(\mathbf{Y}) - r(\mathbf{X}) = k$ whenever there is a maximal chain $\mathbf{X} = \mathbf{X}_0 < \mathbf{X}_1 < \mathbf{X}_2 < \cdots < \mathbf{X}_k = \mathbf{Y}$.

In the case of $\mathbb{P}^1$ which we treated in the Introduction, triples of points have rank 0 and harmonic quadruples have rank 1.

(1.5). We recall the definition of the *geometric realization* (or *order complex* [Al] [G-M, p. 237]) of $\mathcal{C}$. This is a simplicial complex with one vertex for every element $\mathbf{X} \in \mathcal{C}$ and one k-simplex for every chain $\mathbf{X}_0 < \mathbf{X}_1 < \mathbf{X}_2 < \cdots < \mathbf{X}_k$ of elements of $\mathcal{C}$.

If $\mathcal{S}$ is any partially ordered set, let $K(\mathcal{S})$ denote its geometric realization.

Given $\mathbf{X} \in \mathcal{C}$, define $\mathcal{B}_{\leq \mathbf{X}}$ to be the partially ordered set of all $\mathbf{Y} \in \mathcal{C}$ such that $\mathbf{Y} \leq \mathbf{X}$.

Proposition. *For any $\mathcal{C}$-configuration $\mathbf{X}$, $K(\mathcal{B}_{\leq \mathbf{X}})$ is homeomorphic to a closed r-ball, where r is the rank of* $\mathbf{X}$.

This proposition may with enjoyment be verified case by case. Some of these verifications will be included later in this paper.

COROLLARY. *The set of cells* $\{K(\mathcal{B}_{\leq \mathbf{X}}) \mid \mathbf{X} \in \mathcal{C}\}$ *gives* $K(\mathcal{C})$ *the structure of a regular cell complex with one cell for each* $\mathbf{X} \in \mathcal{C}$.

The proof is left as an exercise.

Definition. The $\mathcal{C}$*-complex* $\mathcal{P}$ *associated to* $\mathbb{P}^m$ is the geometric realization of $(\mathcal{C}, \leq)$, endowed with the regular cell complex structure which comes from the Corollary.

(1.6). *Remark*. All the projective spaces we consider are over fields of characteristic $\neq 2$. From some points of view, this is a very restricted class of spaces. For instance, a projective plane is of this form if and only if it satisfies Desargues', Pappus', and Fano's axioms [Ha]. We do not know how much of this paper remains true in more general spaces.

CHAPTER 2—THE $\mathcal{C}$-CONFIGURATIONS IN THE PROJECTIVE LINE

(2.1). *Definition of the* $\mathcal{C}$*-configurations.* Let k be a field of characteristic $\neq 2$. A $\mathcal{C}$*-configuration in* $\mathbb{P}^1(k)$ is any set $\mathbf{X}$ of primary and secondary points of $\mathbb{P}^1(k)$ such that $\mathbf{X}$ is projectively isomorphic to one of the following two pictures.

Configurations of rank 0
The Projective Basis

Configurations of rank 1
The Harmonic Quadruple

In the second picture the points (•, •; ∘, ∘) are required to form a harmonic quadruple. As stated in the Introduction, four points $(A, B; C, D)$ form a harmonic quadruple if and only if the cross-ratio of $ABCD$ is -1. If this holds, $(B, A; C, D)$, $(A, B; D, C)$, and $(C, D; A, B)$ are also harmonic; in particular, the statement that (•, •; ∘, ∘) is harmonic is unambiguous. [V-Y, Sections 31, 56] [Ha, p. 57 ff.]

(2.2). *Remarks on the Geometry of the $\mathcal{C}$-configurations.* The projective basis is the configuration that must be fixed in order to introduce projective coordinates in $\mathbb{P}^1(k)$ [V-Y, Section 35]. In the practice of nineteenth century geometry, the harmonic quadruple was the most important configuration on the line.

(2.3). *Remarks on the Topology of the $\mathcal{C}$-complex.*

PROPOSITION. *The $\mathcal{C}$-complex $\mathcal{P}$ associated to $\mathbb{P}^1(k)$ (1.5) is a graph. Each vertex meets precisely three edges.*

In the next sections we restate this result, splitting it into two parts and proving each part. The proofs are important as examples of how $\mathcal{C}$-configurations are manipulated.

(2.3.1). PROPOSITION. *Each rank-1 configuration is $>$ exactly two rank-0 configurations.*

Proof. Erase either secondary point, and make the other secondary point primary. □

Note. This is the verification of Proposition 1.5 in the $SL(2, \mathbb{R})$ case.

(2.3.2). PROPOSITION. *Each vertex of $\mathcal{P}$ meets precisely three edges.*

Proof. We want to prove that any rank-0 configuration **X** is $<$ precisely three rank-1 configurations. Choose a point A of **X**, make it secondary, and add the harmonic conjugate of this point (with respect to the other two points of **X**) as a new secondary point B.

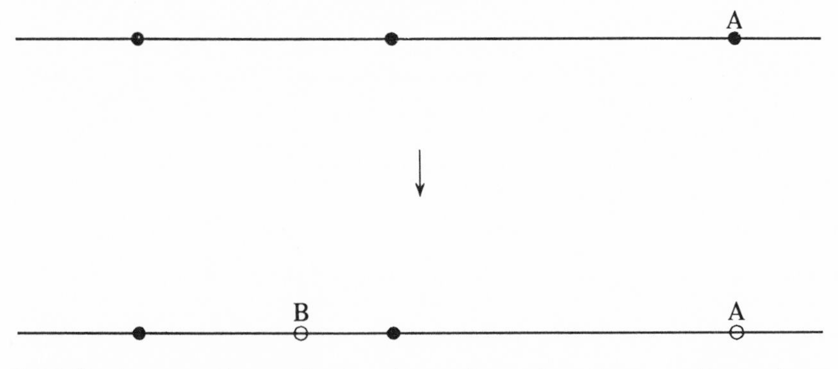

Initially there are three points to choose from; thus there are three edges which meet the given vertex. □

CHAPTER 3—THE $\mathcal{C}$-CONFIGURATIONS IN THE PROJECTIVE PLANE

(3.1). *Definition of the $\mathcal{C}$-configurations.* Let k be a field of characteristic $\neq 2$. A $\mathcal{C}$-configuration in $\mathbb{P}^2(k)$ is any set $\mathbf{X}$ of primary and secondary points in $\mathbb{P}^2(k)$ such that $\mathbf{X}$ is projectively isomorphic to one of the following five pictures.

Configurations of rank 0

The Complete Quadrilateral

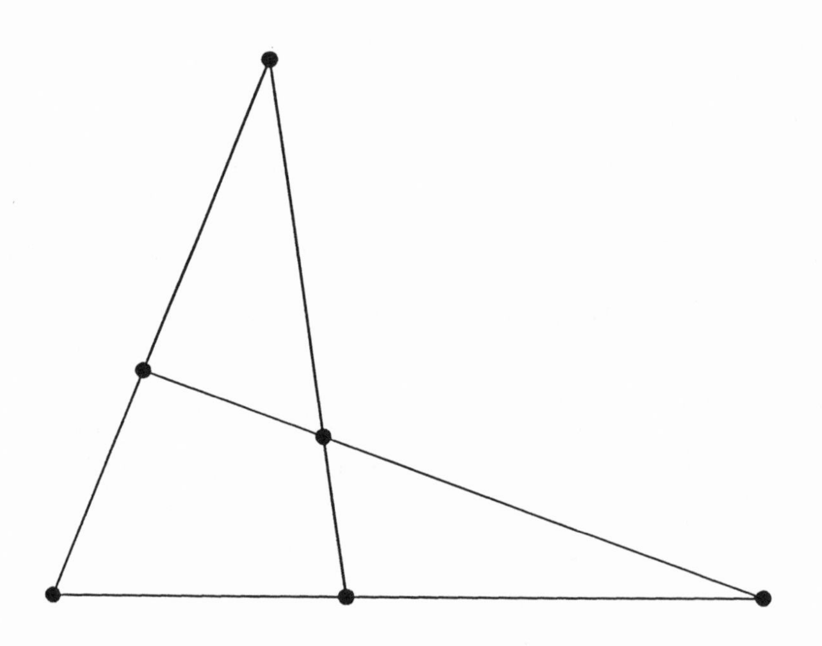

Configurations of rank 1

The Carnot Configuration

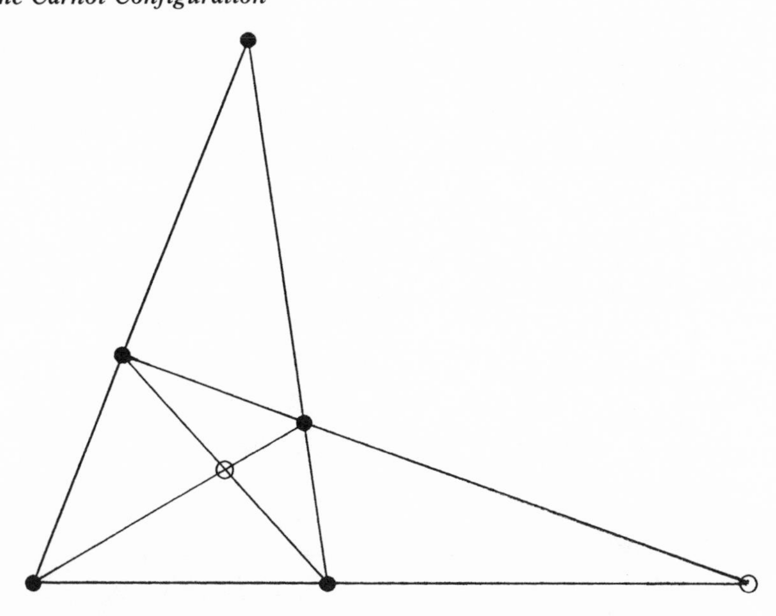

Configurations of rank 2

The Complete Quadrangle

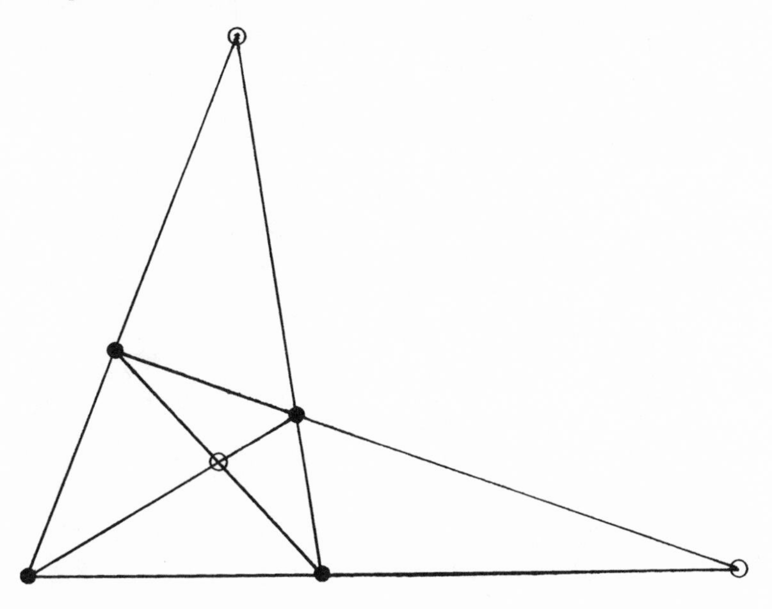

The Pascal-Brianchon Configuration

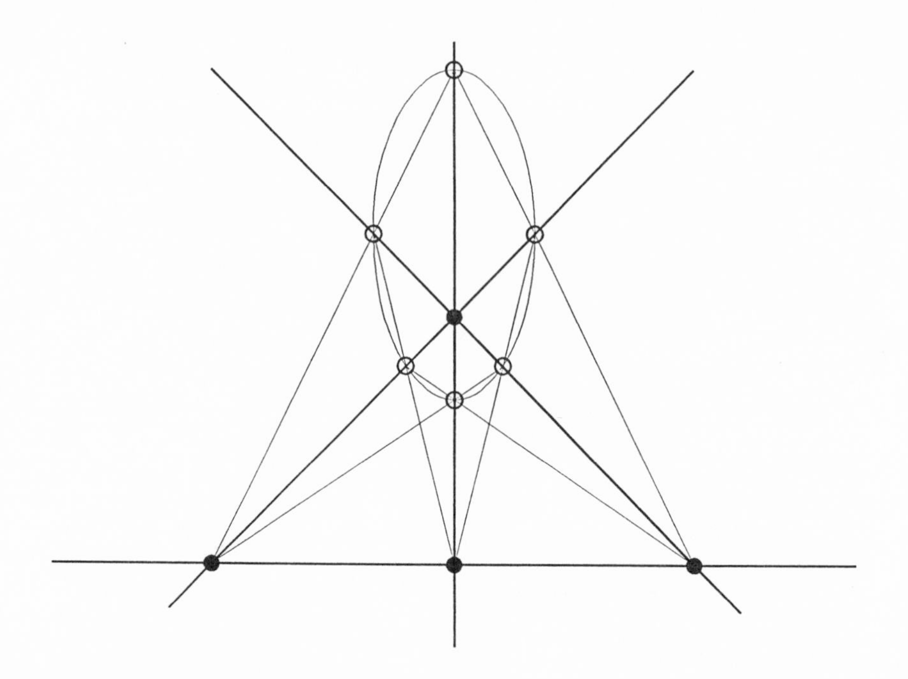

Configurations of rank 3

The Cremona Configuration

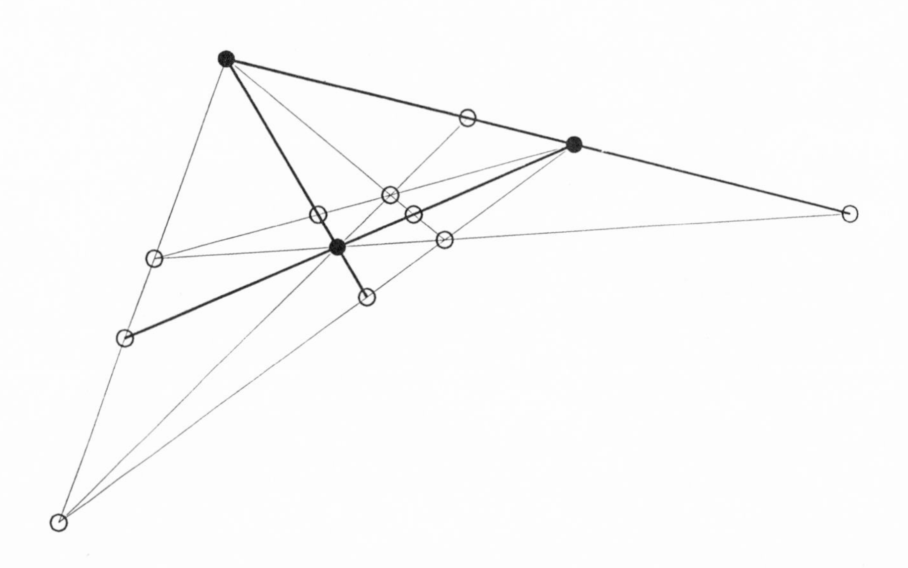

(3.2). *Remarks on the Geometry of the* $\mathcal{C}$*-configurations.*

(3.2.1). *The Complete Quadrilateral configuration.* Classically, a set of four lines in general position was called a complete quadrilateral. This is the dual of a projective basis.

By the definition in (3.1), a complete quadrilateral $\mathcal{C}$-configuration is any set of six (primary) points which are the pairwise intersection points of four lines in general position.

(3.2.2). *The Carnot configuration.* In many classical books on projective geometry, the first nontrivial theorem is that, in the following figure, $(F, G; P, Q)$ is a harmonic quadruple.

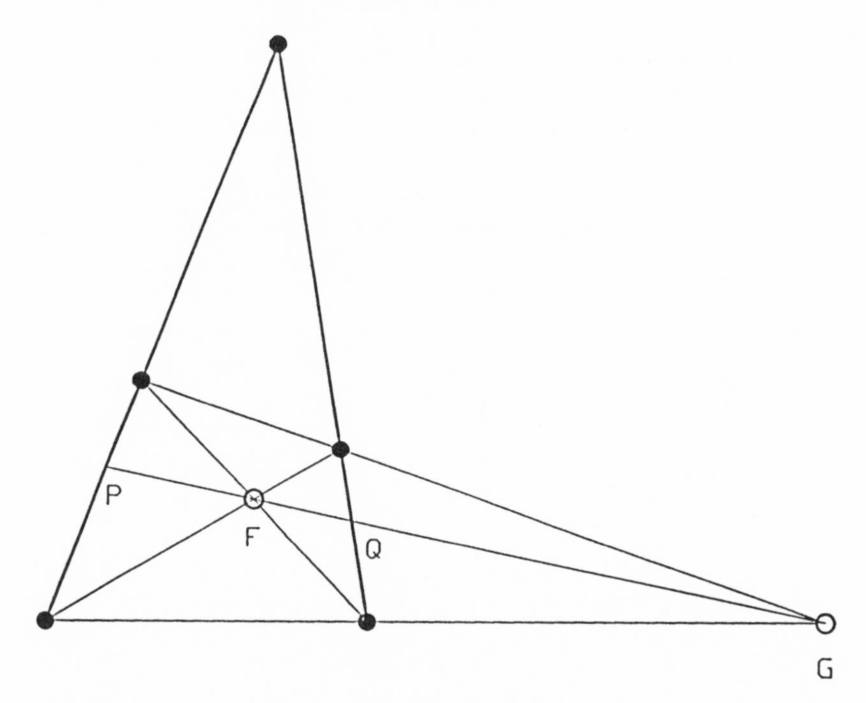

Desargues, who first defined the notion of harmonic quadruple, would surely have known this theorem by 1639, the year he published his major work on geometry [K, pp. 288–289]. We would have liked to name this configuration after him, but his name was already taken (5.2.1).

Carnot was the first person who resurrected the work of Desargues and Pascal after it went through a dry spell during the eighteenth century. His *Geometry of Position* (1803) gave a completely synthetic treatment of projective geometry, and contained (as far as we know) the first proof of this theorem which had been given in print since 1700. [C, p. 46] [K, p. 841].

(3.2.3). *The Complete Quadrangle configuration.* Classically, a set of four points in general position was called a complete quadrangle. This is a projective basis [V-Y, Sections 35, 63] [Ha, p. 93].

By the definition in (3.1), a complete quadrangle $\mathcal{C}$-configuration is any quadrangle of four primary points in general position, together with the quadrangle's diagonal points as secondary points.

(3.2.4). *The Pascal-Brianchon configuration.* From the point of view of projective geometry, this configuration may be the most interesting of the five on our list.

Construction of the configuration. All Pascal-Brianchon configurations may be constructed by the following procedure. Choose a line l, a point S not on l, and three points X_1, X_2, X_3 on l. Choose A_1 on X_1S. (X_1, X_2, X_3, S are primary; A_1 is secondary.) We construct secondary points A_2, A_3, . . . in a cyclic pattern by the following rule: A_{i+1} is the intersection of the line connecting X_{i+1} and S with the line connecting X_{i-1} and A_i. (The subscripts on the X's are interpreted mod 3.)

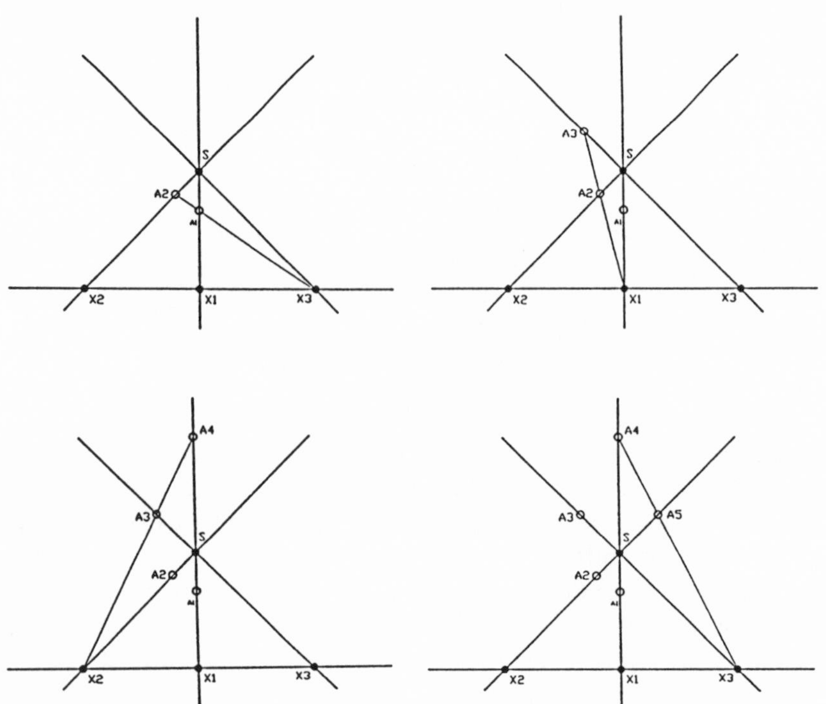

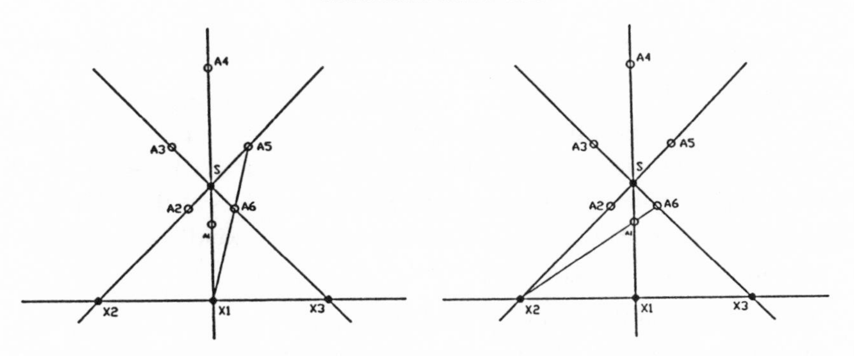

Proposition 1. $A_7 = A_1$. *(Thus $A_i = A_{i+6}$ for all i.)*

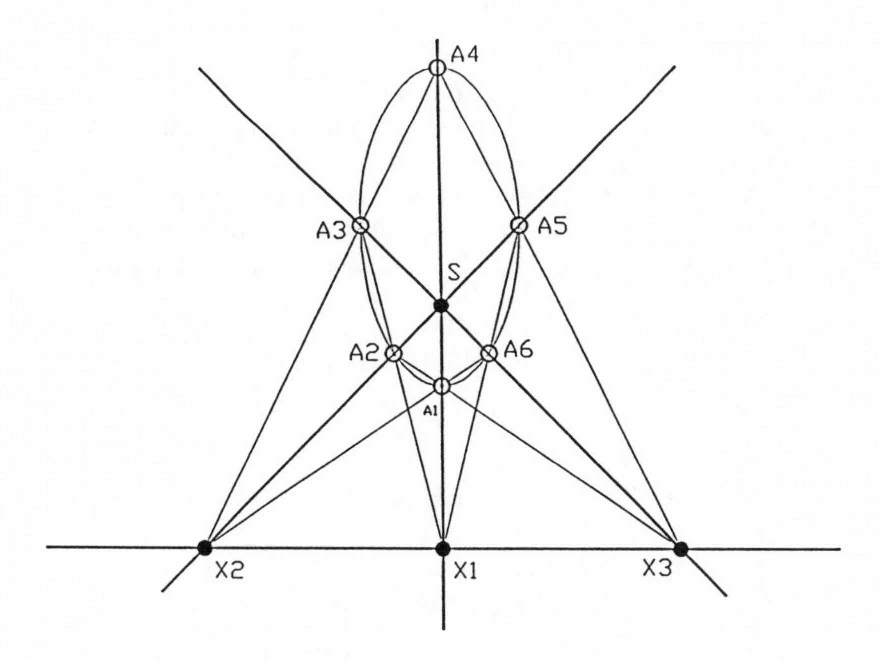

This can be proved in many ways. One can prove it analytically, or one can send l to infinity and use Euclidean geometry. Here we give a proof in the style of classical projective geometry. References for the theory of pole and polar are [V-Y, Section 44] [K, p. 294].

Proof. Let W be the conic through $A_2, A_3, \ldots, A_6$. Let s be the polar of S with respect to W.

LEMMA. $s = l$.

Proof of Lemma. $A_2A_3A_5A_6$ is a quadrilateral inscribed in W whose diagonals meet in S. So the intersection $X_1 = A_2A_3.A_5A_6$ of the pair of opposite sides must lie on the polar of S. Secondly, $(A_2, A_5; S, X_2)$ is a harmonic quadruple (to see this, examine the sides of the complete quadrangle $X_1X_3A_3A_4$). Since A_2 and A_5 are on the conic, X_2 is on the polar of S. Thus $s = X_1X_2$, or $s = l$. □

We continue the proof of the proposition. Define A_1' as the point (other than A_4) where W meets X_1S. $A_4A_5A_1'A_2$ is a quadrilateral inscribed in W whose diagonals meet at S, so $A_4A_5.A_1'A_2$ is on l. $A_4A_5.l = X_3$, so X_3 is on $A_1'A_2$. A similar proof shows X_2 is on $A_1'A_6$. Thus A_2X_3, A_6X_2, and X_1S concur at A_1'. However, by the definition of A_2, $A_1 = A_2X_3.X_1S$; so $A_1 = A_1'$. And by definition $A_7 = A_6X_2.X_1S$, meaning $A_7 = A_1'$. □

Connection with Pascal's and Brianchon's Theorems. Since the six secondary points have a cyclic pattern, we form them into a hexagon consisting of the six lines $A_1A_2, A_2A_3, \ldots, A_6A_1$ in that order. The next two propositions show that this hexagon has two remarkable properties: (1) it can be inscribed in a conic; (2) it circumscribes a conic. In other words, the configuration realizes Pascal's and Brianchon's theorems simultaneously.

PROPOSITION 2. *The points $A_1, A_2, \ldots, A_6$ lie on a conic.*

Proof. This is exactly Pascal's theorem.

PROPOSITION 2′. *The sides $A_1A_2, A_2A_3, \ldots, A_6A_1$ are tangent to a conic.*

Proof. This is exactly Brianchon's theorem.

(Alternatively, both propositions follow from the proof of Proposition 1.)

Here is another way to state the content of Propositions 2 and 2′. Assume $A_1, \ldots, A_6$ are distinct points lying on a conic such that

(1) the lines A_jA_{j+3} $(j = 1, 2, 3)$ concur at some point S;

(2) the points X_j defined by $X_j = A_{j-2}A_{j-1}.A_{j+1}A_{j+2}$ $(j = 1, 2, 3)$ are collinear.

(The subscripts are interpreted mod 6.) Then the A's are the secondary points of a Pascal-Brianchon $\mathcal{C}$-configuration whose primary points are the X's and S.

Other properties. The configuration is self-dual. Duality interchanges l and S; the points X_i and the lines X_iS; the vertices and the sides of the hexagon; and finally, it carries the inscribed conic to a circumscribed conic, and vice versa.

(3.2.5). *The Cremona configuration*. Veblen and Young [V-Y, Section 18] called this the *quadrangle-quadrilateral* configuration. Cremona considered it in [C, Section 272]. It can be constructed in two ways, which are essentially dual. We describe the two methods by stating the two dual propositions which the configuration realizes.

PROPOSITION.

Choose a complete quadrangle $ABCD$. Its six sides meet in pairs to form the three vertices E, F, G of the diagonal triangle. For each side p of the diagonal triangle, exactly two of the six sides of the quadrangle meet p in new points (as opposed to the old points E, F, G). Call these new points $H_1, \ldots, H_6$.
Then $H_1, \ldots, H_6$ are the vertices of a complete quadrilateral $wxyz$.

Choose a complete quadrilateral $abcd$. Connect its six vertices in pairs to form the three sides e, f, g of the diagonal triangle. For each vertex P of the diagonal triangle, exactly two of the six vertices of the quadrilateral can be joined to P by new lines (as opposed to the old lines e, f, g). Call these new lines $h_1, \ldots, h_6$.
Then $h_1, \ldots, h_6$ are the sides of a complete quadrangle $WXYZ$.

Furthermore, if $wxyz = abcd$ then $WXYZ = ABCD$, and conversely. □

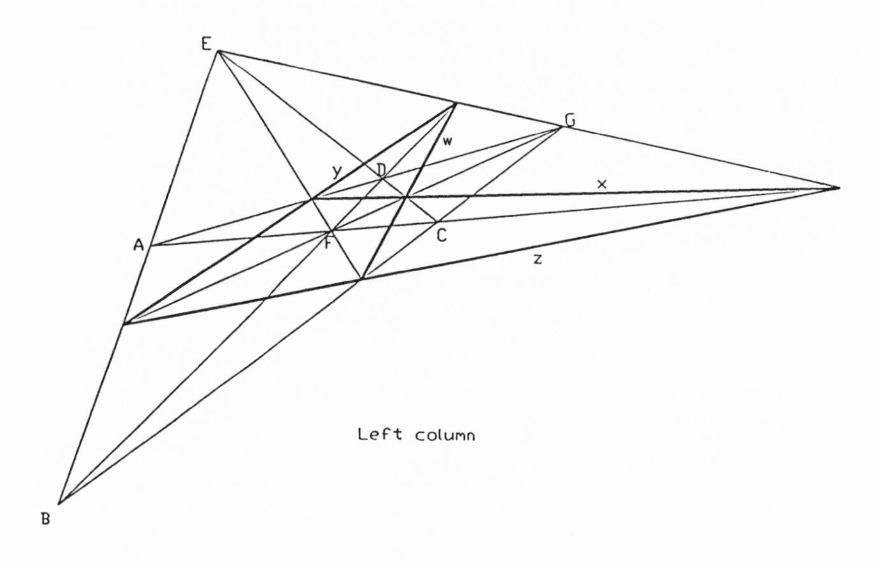

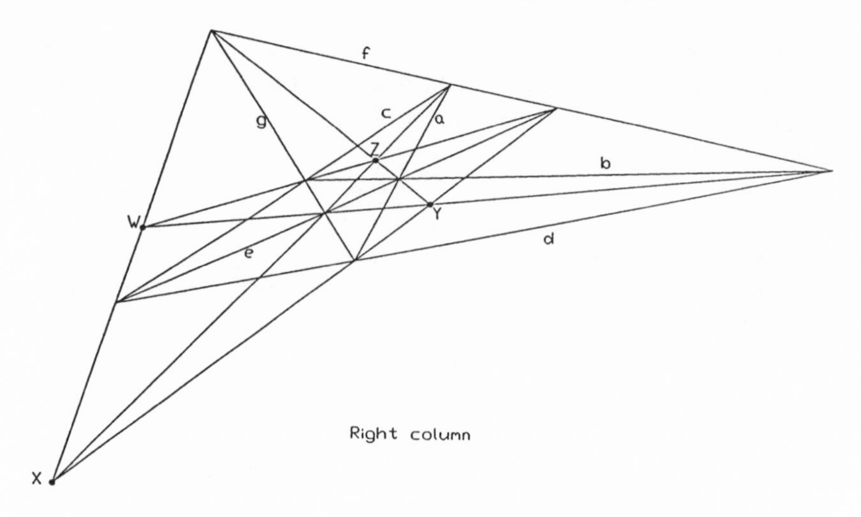

Every Cremona configuration arises in this way, where (using the notation of the left column, for example) the primary points are E, F, G and the secondary points are $A, \ldots, D$ and $H_1, \ldots, H_6$.

Closer study of the construction reveals that it is based on the following very elegant properties [V-Y, Section 18].

The diagonal triangle EFG is in perspective with each of the four triangles formed by a set of three of the vertices $ABCD$, the center of perspectivity being in each case the fourth vertex. By Desargues' theorem, this gives rise to four axes of perspectivity, one corresponding to each of the points $ABCD$. These lines are $wxyz$.

The diagonal triangle efg is in perspective with each of the four triangles formed by a set of three of the lines $abcd$, the line of perspectivity being in each case the fourth line. By Desargues' theorem, this gives rise to four centers of perspectivity, one corresponding to each of the lines $abcd$. These points are $WXYZ$.

(3.3). *Remarks on the Topology of the $\mathcal{C}$-complex.*

PROPOSITION. *The $\mathcal{C}$-complex $\mathcal{P}$ associated to $\mathbb{P}^2(k)$ (1.5) is a regular cell complex of dimension three.* □

The complex has five types of cells, corresponding to the five types of configurations; these are listed in the following table. The rank of the configuration equals the dimension of the cell.

$\mathcal{C}$-config.	quadrilateral	Carnot	quadrangle	Pascal-Bri.	Cremona
Cell in $\mathcal{P}$	vertex	edge	triangle	hexagon	Soulé cube
Rank	0	1	2	2	3

For example, a "triangle" is a cell of dimension two whose boundary consists of three edges and three vertices. The Soulé cube is a cell of dimension three whose boundary contains four triangles and six hexagons in the following arrangement:

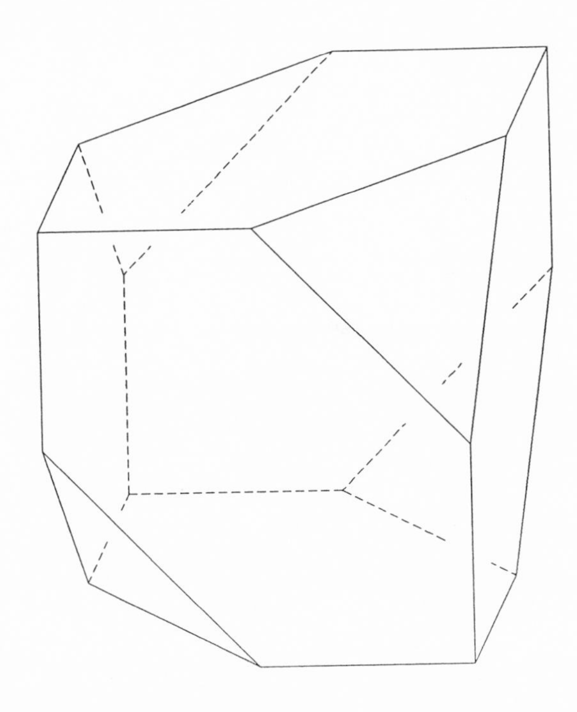

Note. When $k = \mathbb{Q}$, this complex was first constructed by Soulé and Lannes [Sou1] [Sou2].

(3.3.1). *Examples of how the $\mathcal{C}$-configurations determine the cells' combinatorics.* We now give a few of the steps which go into the full proof of Proposition 3.3. As in (2.3), the main reason we do this is to illustrate how $\mathcal{C}$-configurations are used.

(1) *The cell corresponding to a Carnot configuration is an edge.*

Choose a Carnot configuration $\mathbf{X}_1$. We want to check that there are exactly two complete quadrilateral $\mathcal{C}$-configurations $\mathbf{X}$ such that $\mathbf{X} < \mathbf{X}_1$. But this is clear—if $\mathbf{X}_1$ looks like this:

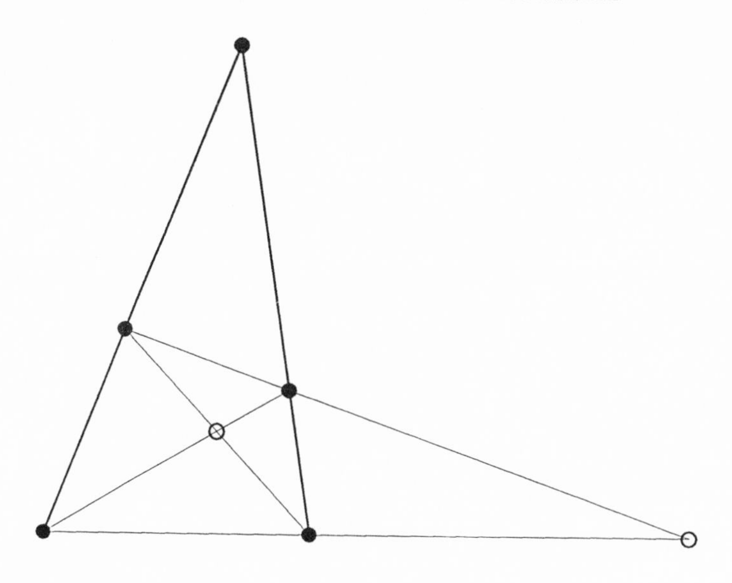

the possible **X** are these:

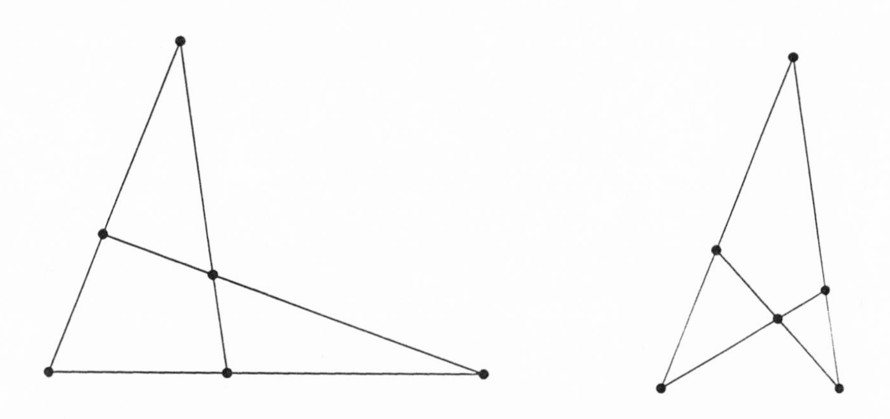

Note. This example and the next one are verifications of cases of Proposition 1.5 in the SL(3, ℝ) case.

(2) *The cell corresponding to a complete quadrangle configuration is a triangle*.

Let **X** be this complete quadrangle configuration:

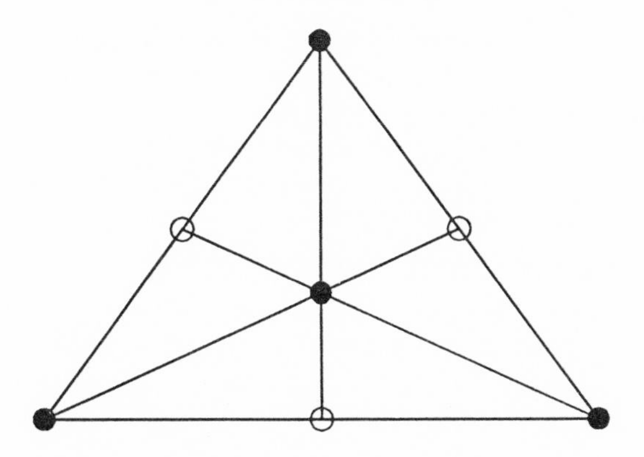

The cell associated with **X** is the following, where the vertices, the edges, and the 2-cell are labelled with the corresponding $\mathcal{C}$-configuration.

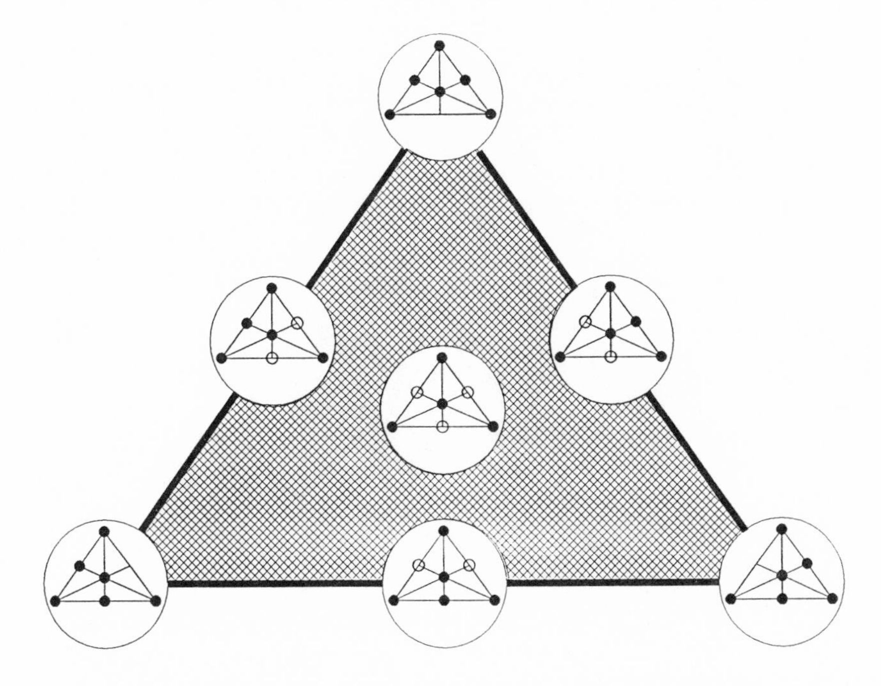

(3) *Each Soulé cube meets six hexagons and four triangles.*

Let **X** be a Cremona configuration (3.2.5), corresponding to a cell $C_3 \in \mathcal{P}$ of dimension three. Let's find all the configurations $\mathbf{X}' < \mathbf{X}$ of rank two. Let S be the set of primary points of **X**. Choose any secondary point

A, and let $S' = S \cup \{A\}$. There are two possibilities. First, S' might consist of four points in general position. Then there is a unique $\mathbf{X}' < \mathbf{X}$ for which the set of primary points is S':

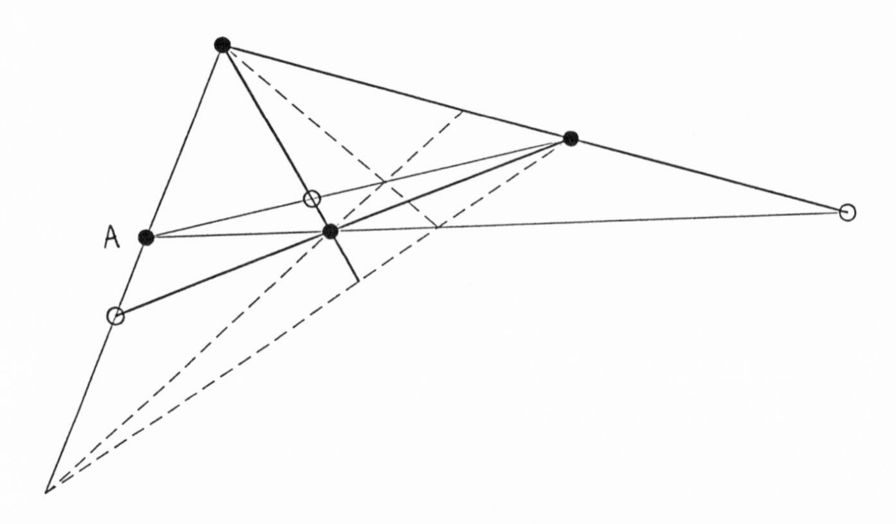

Second, S' might consist of three collinear points and a fourth point off the line of the first three. Again, there is a unique $\mathbf{X}' < \mathbf{X}$:

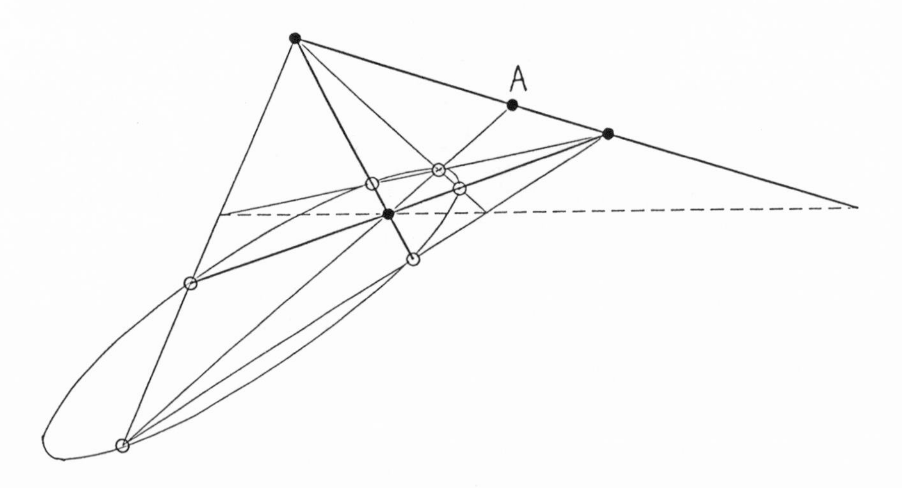

Since there are four secondary points of the first type and six of the second, we are done.

(3.3.2). *Table of Incidences in* $\mathcal{P}$. In the following table, a given cell R of the type in row i is on a_{ij} cells of the type in column j.

More precisely, choose numbers i, j between 1 and 5. Let a_{ij} be the entry in the ith row and jth column of the table below. If $i > j$ (i.e. you're below the diagonal), a given cell R of row i's type will contain a_{ij} cells of column j's type in its boundary. If $i < j$ (i.e. you're above the diagonal), a given cell R of row i's type will be contained in the boundary of a_{ij} cells of column j's type.

	vertex	edge	triangle	hexagon	Soulé cube
vertex	—	6	3	12	12
edge	2	—	1	4	8
triangle	3	3	—	—	4
hexagon	6	6	—	—	3
Soulé cube	12	24	4	6	—

Example. Each vertex meets six edges.

Choose a vertex C_0 in $\mathcal{P}$, represented by the following complete quadrilateral **X**:

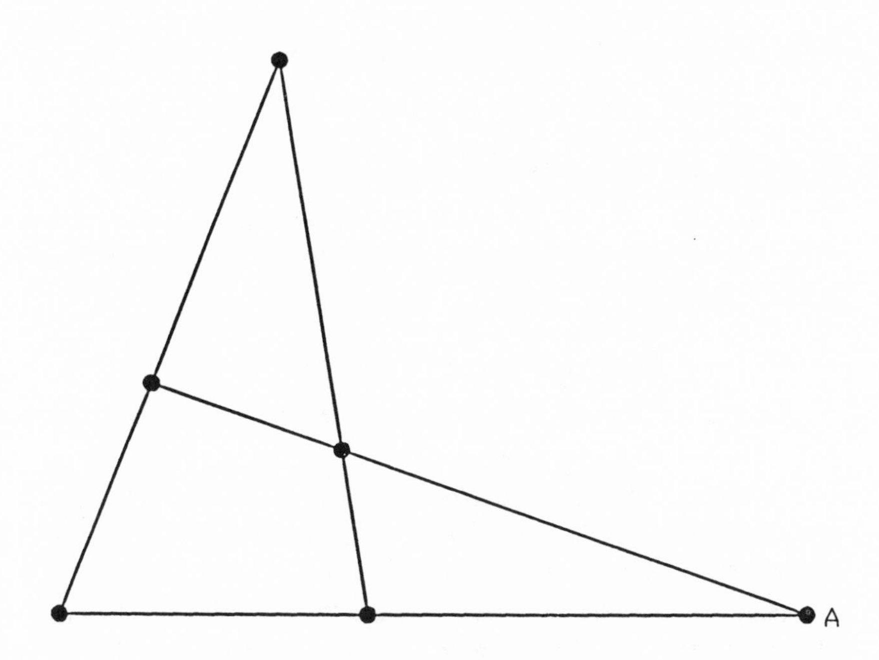

Choose a point $A \in \mathbf{X}$. There is a unique Carnot configuration $\mathbf{X}'$ such that $\mathbf{X}' > \mathbf{X}$ and A is a secondary point of $\mathbf{X}$:

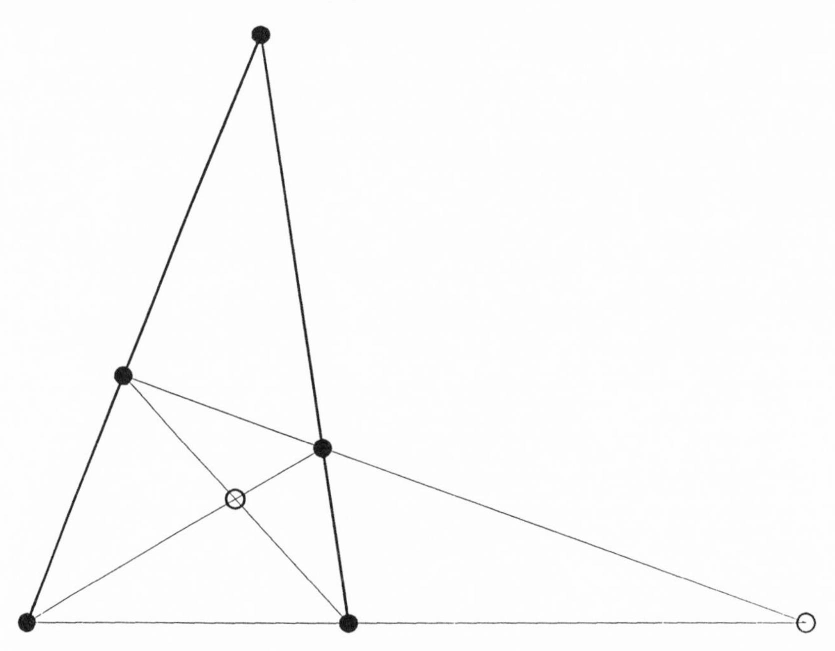

Since there are six ways to choose A, there are six edges meeting C_0.

CHAPTER 4—ELEMENTS OF SYNTHETIC SYMPLECTIC GEOMETRY

(4.1). Let k be a field of characteristic $\neq 2$.

Definition. The *symplectic projective three-space* over k is the space $\mathbb{P}^3(k)$ together with a distinguished projectivity $\phi : \mathbb{P}^3(k) \to \mathbb{P}^3(k)^*$ satisfying the axiom:

(4.1.1) for any point P in $\mathbb{P}^3(k)$, P is on $\phi(P)$.

Here $\mathbb{P}^3(k)^*$ is the dual of $\mathbb{P}^3(k)$, viewed as the space of hyperplanes in $\mathbb{P}^3(k)$.

Definition. An *automorphism* of a symplectic projective three-space is any linear automorphism of $\mathbb{P}^3$ which commutes with ϕ. The group of such automorphisms is denoted $\mathrm{Aut}_{\mathrm{Sp}}(\mathbb{P}^3(k))$.

(4.2). Recall that any projectivity $\psi : \mathbb{P}^3(k) \to \mathbb{P}^3(k)^*$ gives rise to a projectivity $\psi^* : \mathbb{P}^3(k)^* \to \mathbb{P}^3(k)$ in the following way. For any plane π, $\{\phi(P) \mid P \in \pi\}$ is a set of planes which concur at a point Q; define $\psi^*(\pi) = Q$.

PROPOSITION. *In a symplectic* $\mathbb{P}^3$, $\phi^* \circ \phi =$ *the identity.*

Proof. Choose a basis in $\mathbb{P}^3(k)$; write points as column vectors v, and planes as row vectors w. Then ϕ will have the form $v \mapsto (\Omega \cdot v)^t$, where Ω is an invertible matrix and $(\quad)^t$ denotes the transpose. Axiom 4.1.1 is equivalent to:

$$(\forall v) \quad v^t \cdot \Omega^t \cdot v = 0.$$

This implies Ω is skew-symmetric.

ϕ^* will send $w \mapsto (w \cdot \Omega^{-1})^t$. Thus

$$\begin{aligned} \phi^* \circ \phi : v \mapsto & ((\Omega \cdot v)^t \cdot \Omega^{-1})^t \\ & = \Omega^{-1t} \cdot \Omega \cdot v \\ & = -v. \end{aligned}$$

So modulo scalars, $\phi^* \circ \phi$ is the identity. □

Remarks.

(1) The previous proof shows why this is called symplectic geometry.

(2) Classically, duality morphisms $\mathbb{P}^3 \to \mathbb{P}^{3*}$ were defined with respect to a fixed nondegenerate quadric surface $\mathfrak{Q}$. Such maps were called *polarities* [V-Y, p. 89]. In this case the points satisfying 4.1.1 are exactly the points of $\mathfrak{Q}$, and Ω is symmetric. We have not found "symplectic" duality morphisms in the literature of classical projective geometry.

(3) We would like to have a synthetic proof of Proposition 4.2.

(4) The results of this chapter can be generalized to all odd-dimensional projective spaces.

Notation. $\phi(P)$ is denoted $P^\perp$, and $\phi^*(\pi)$ is denoted $\pi^\perp$.

COROLLARY.

(a) *For all points P in a symplectic* $\mathbb{P}^3$, $P = P^{\perp\perp}$.

(b) *For all planes π in a symplectic* $\mathbb{P}^3$, $\pi = \pi^{\perp\perp}$. □

(4.3). *Definition.* A line l in a symplectic $\mathbb{P}^3$ is *Lagrangian* if and only if there is a point P on l such that the plane $P^\perp$ is on l.

PROPOSITION. *If l is Lagrangian, then*

(1) *for any point Q on l, $Q^\perp$ is on l;*
(2) *for any plane ρ on l, $\rho^\perp$ is on l.*

Proof. (1) Q is on $P^\perp$, so $Q^\perp$ is on $P^{\perp\perp}$. By Corollary 4.2, $Q^\perp$ is on P. Also, $Q^\perp$ is on Q by Axiom 4.1.1. So $Q^\perp$ contains the line $PQ = l$.
The proof of (2) is (by Corollary 4.2) dual to that of (1). □

Remarks.

(1) In particular, in the definition of Lagrangian line, the choice of P was arbitrary.
(2) Fix a Lagrangian line l. Intuitively, as P moves forward along l, the plane $P^\perp$ spins around l, completing one 180° revolution when P completes one circuit.

(4.4). Let l be a line on a given point P but which does not lie on $P^\perp$. Then Proposition 4.3 implies l is *non-Lagrangian*.

Definition. For any line l, the *symplectic dual* of l is the line of intersection of the pencil of planes $\{\pi \mid \pi^\perp \text{ is on } l\}$. This line is denoted $l^\perp$.

PROPOSITION.

(1) *$l^\perp = l$ if and only if l is Lagrangian.*
(2) *If l is non-Lagrangian, $l^\perp$ and l are skew.*
(3) *For any line l, $l^{\perp\perp} = l$.*

Proof. (1) is clear from Proposition 4.3, and (3) is clear from Corollary 4.2. To prove (2), assume l and $l^\perp$ meet at P. That P is on l implies $l^\perp$ is on $P^\perp$. So $l^\perp$ is on both P and $P^\perp$, meaning $l^\perp$ (and hence l) was Lagrangian after all—a contradiction. □

Remark. We have seen that the choice of projectivity ϕ determines the Lagrangian lines. Conversely, the set of Lagrangian lines determines ϕ. For given any point P, the Lagrangian lines through P span a plane which must be $P^\perp$.

(4.5). We state a few more propositions of symplectic geometry. Their proofs are left as exercises.

PROPOSITION 1. *Given a point P and a line l on P. Then l is Lagrangian if and only if l is on $P^\perp$.*

PROPOSITION 2. *Given P, and Q not on $P^\perp$. Then the line PQ is the symplectic dual of the line $P^\perp \cap Q^\perp$.*

PROPOSITION 3. *Given P, choose Lagrangian lines l, m on P. Then $P^\perp$ is the plane spanned by l and m. Conversely, if Lagrangian lines l, m lie in a plane π, then they meet in $\pi^\perp$.*

PROPOSITION 4. *Let l be a non-Lagrangian line. For any points P on l and Q on $l^{\perp}$, the line PQ is Lagrangian.*

PROPOSITION 5. *Let A_1, A_2, A_3, B_1, B_2 be points such that all the lines A_iB_j ($i = 1, 2, 3; j = 1, 2$) are Lagrangian. Then A_1, A_2, and A_3 are collinear by a non-Lagrangian line whose dual is the line B_1B_2.*

PROPOSITION 6. *Let l be a Lagrangian line, and P a point not on l. Then there is a unique Lagrangian line m through P which meets l.*

This is called *dropping a Lagrangian perpendicular* from P to l.

COROLLARY. *Let l, m be two skew Lagrangian lines. For any P on l, drop a Lagrangian perpendicular from P to m, and say it meets m at Q. Then the Lagrangian perpendicular dropped from Q to l meets l at P.*

Thus there is a distinguished projectivity between any two skew Lagrangian lines. This is called the ladder structure of the two lines.

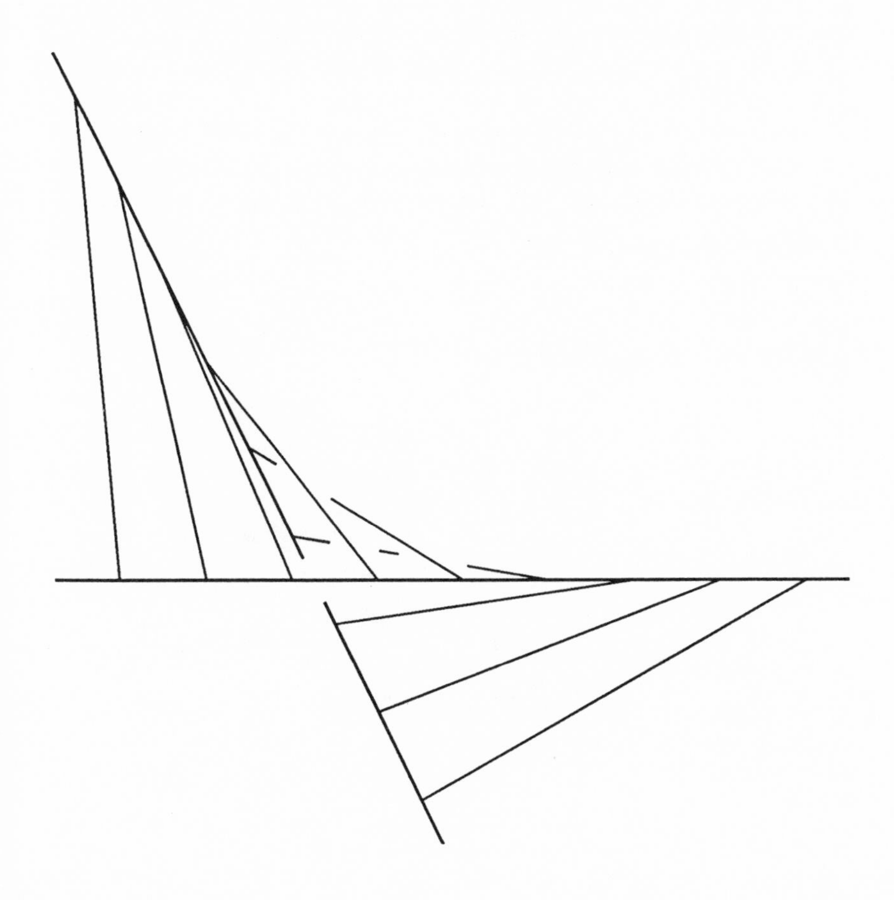

Remark. The set $\{n \mid n$ meets l and $m\}$ is a quadric hypersurface. (Over $k = \mathbb{R}$, it is the hyperboloid of one sheet.)

(4.6). *Definition.* A *symplectic complete quadrilateral* is a plane complete quadrilateral in which two of the lines are Lagrangian and two are not.

By Proposition 4.5.3, the point of intersection of the Lagrangian lines is the dual of the plane.

CHAPTER 5—THE $\mathcal{C}$-CONFIGURATIONS IN SYMPLECTIC PROJECTIVE THREE-SPACE

(5.1). *Definition of the $\mathcal{C}$-configurations.* We will use a symplectic three-space over k, where k is any field of characteristic $\neq 2$. The action of ϕ and ϕ^* will always be denoted by $(\quad)^\perp$ (see (4.2)).

In our pictures, Lagrangian lines are drawn heavy and solid:

and non-Lagrangian lines are drawn in a dotted pattern:

A *$\mathcal{C}$-configuration in symplectic* $\mathbb{P}^3(k)$ is any set $\mathbf{X}$ of primary and secondary points in the symplectic three-space such that $\mathbf{X}$ is isomorphic (by an element of $\mathrm{Aut}_{\mathrm{Sp}}(\mathbb{P}^3(k))$, defined in (4.1)) to one of the following.

Configurations of rank 0

Desargues' configuration

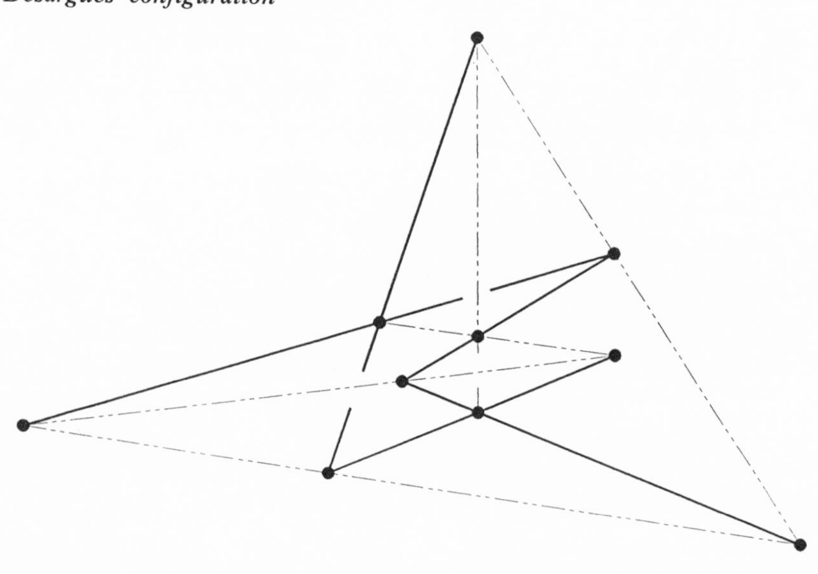

Reye's configuration

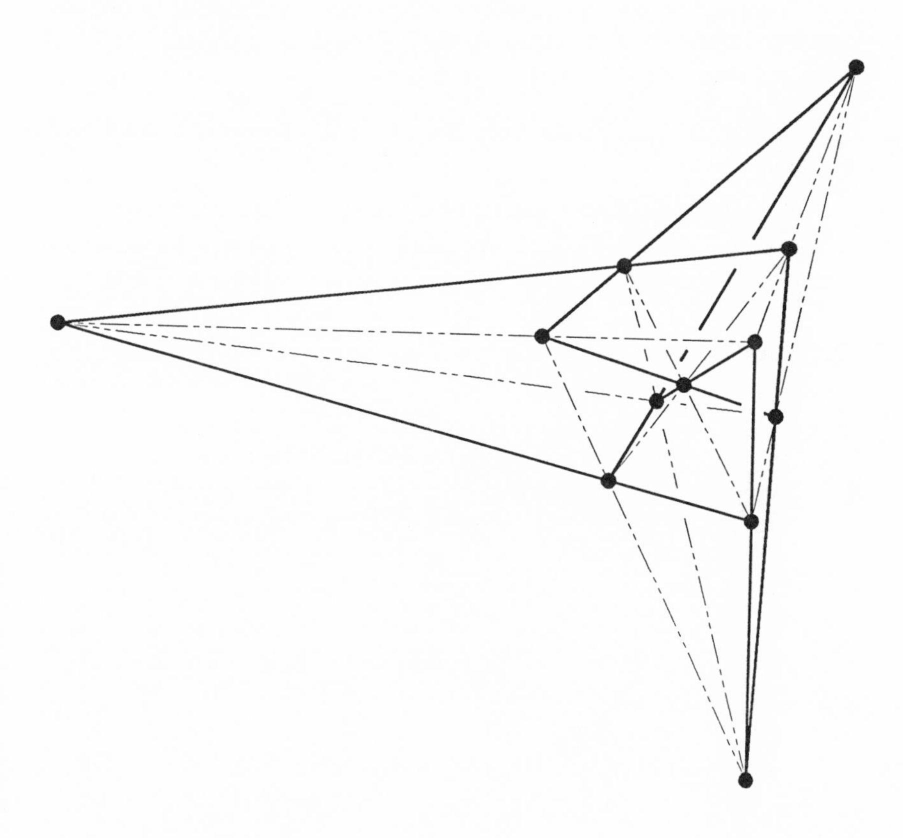

Configurations of rank 1

The DR configuration

The primary points of a DR configuration are the nine points remaining after a minor point has been removed from a Desargues' configuration. (The term "minor point" is defined in (5.2.1).)

PROPOSITION. *These nine points are contained in a unique Desargues' configuration and in a unique Reye's configuration.* □

The set of points (both primary and secondary) of the DR configuration is the union of the points of the Desargues' and Reye's configurations of the proposition.

The RR configuration

The primary points of an RR configuration are the nine points remaining after three Lagrangian-collinear points have been removed from a Reye's configuration.

PROPOSITION. *These nine points are contained in exactly two Reye's configurations.* □

The set of points (both primary and secondary) of the RR configuration is the union of the points of the two Reye's configurations of the proposition.

Configurations of rank 2

The Three Utilities configuration

For any primary point A of this configuration, the plane $A^{\perp}$ contains a Pascal-Brianchon configuration.

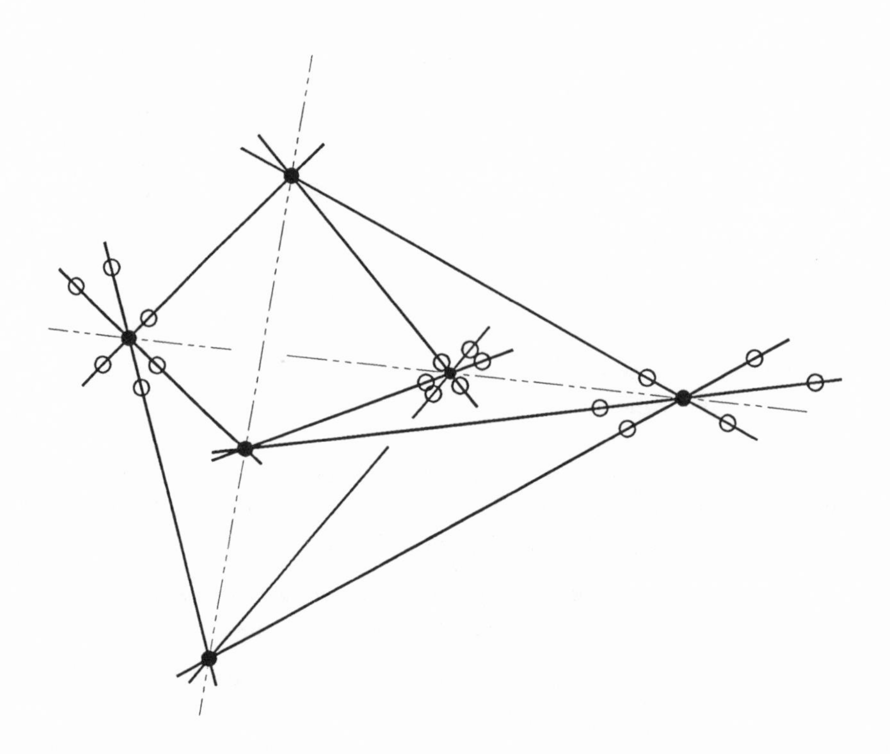

Type 2b

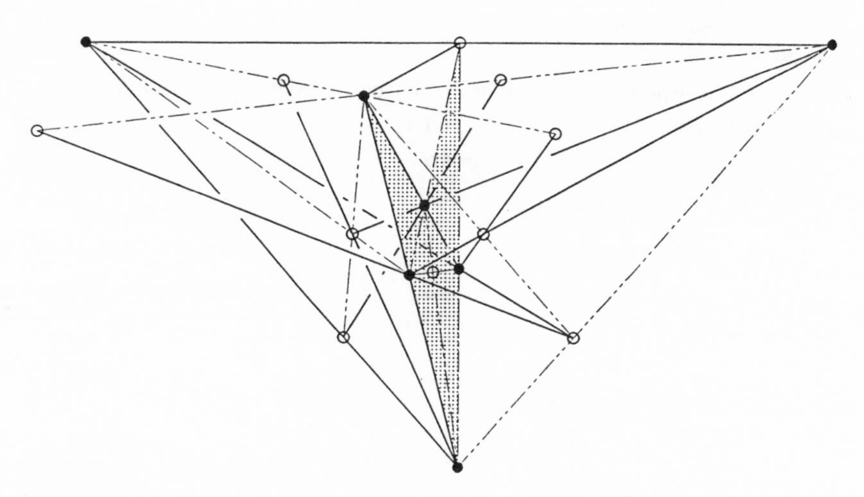

Type 2c

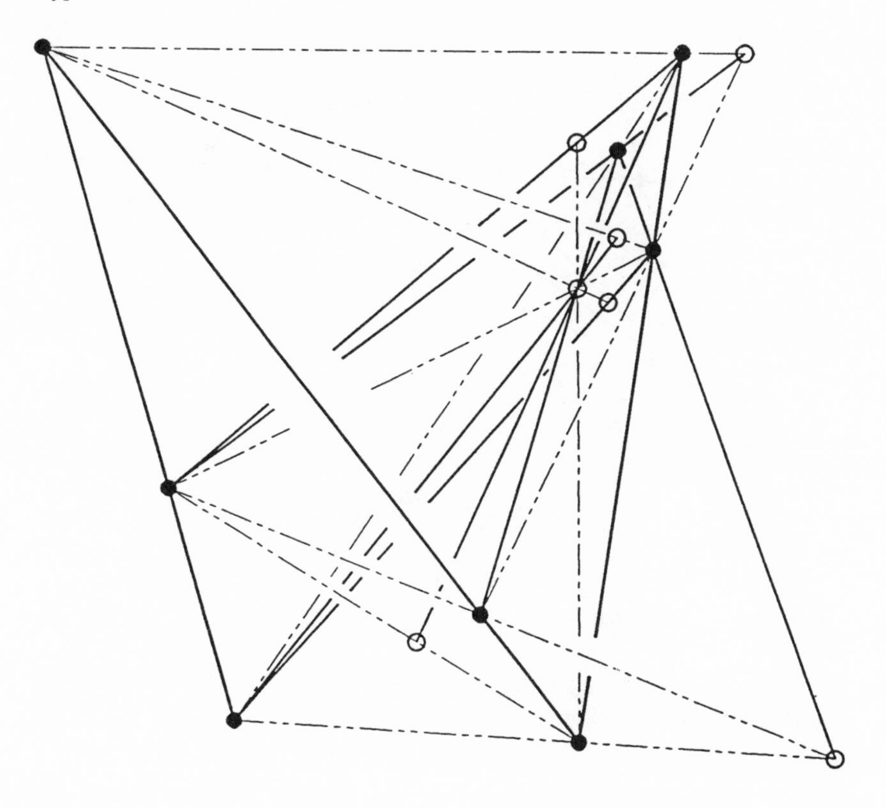

Configurations of rank 3

The Two Utilities configuration

This is a sub-configuration of the Three Utilities configuration. A is a former primary point, and B is the harmonic conjugate of A with respect to the other two points on the line AB.

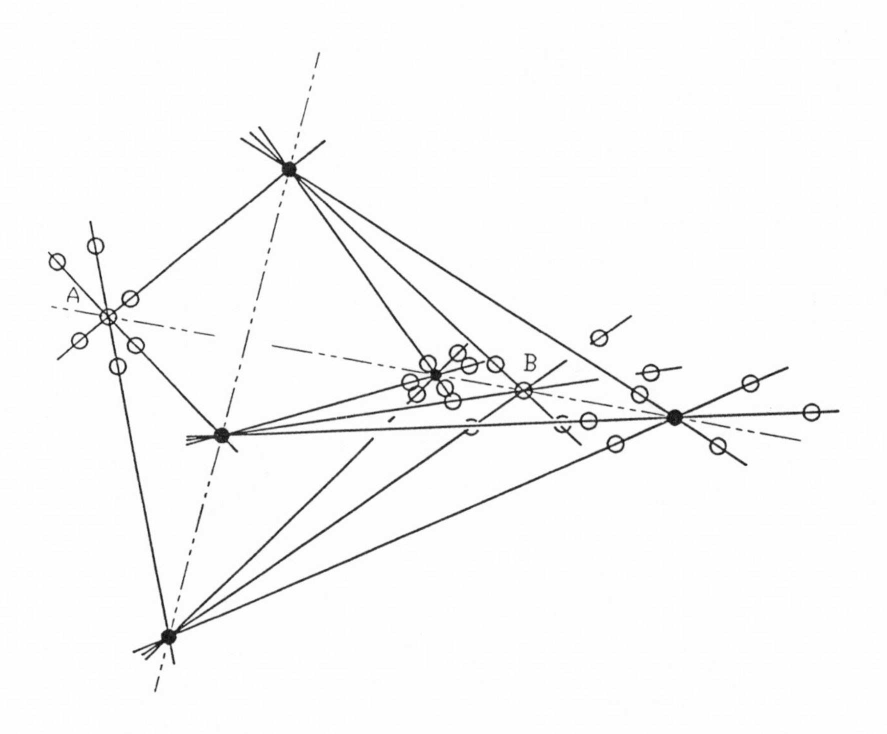

The Ladder configuration

For each primary point A of this configuration, the plane $A^{\perp}$ contains a Pascal-Brianchon configuration.

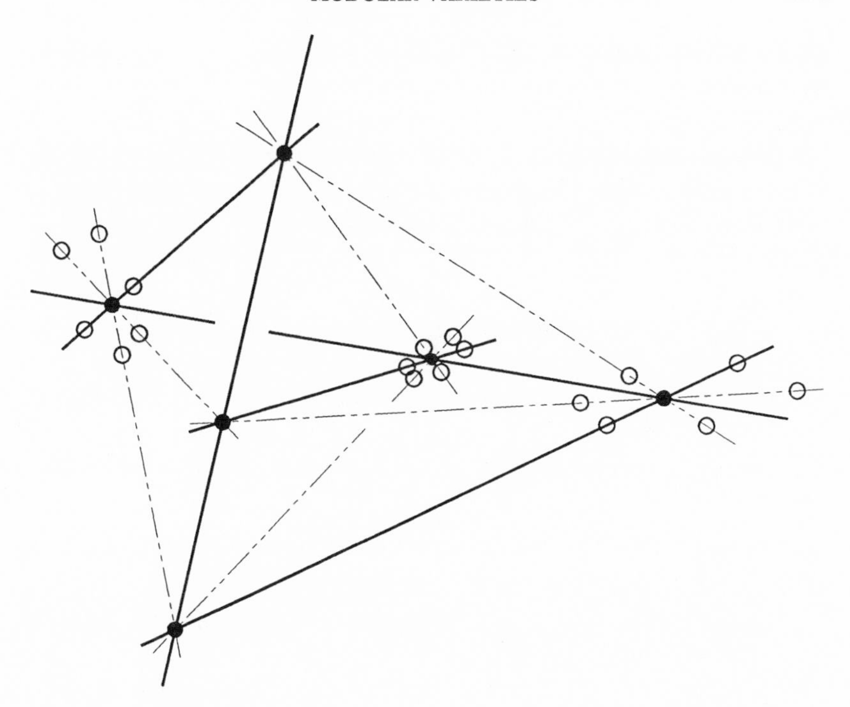

Type 3c

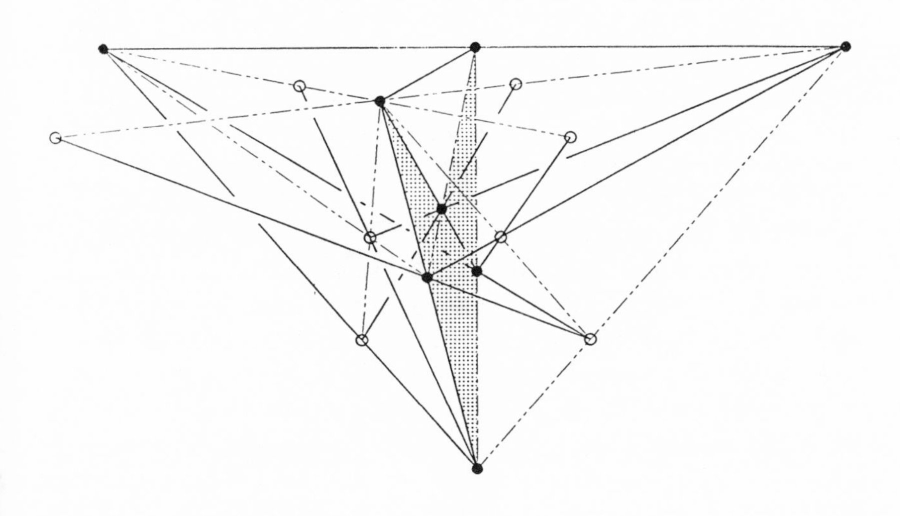

Configurations of rank 4

Type 4

The configuration consists of all points $[a_0 : a_1 : a_2 : a_3] \in \mathbb{P}^3(k)$ with $a_i \in \{-1, 0, 1\}$. The primary points are those with only one coordinate nonzero. (Here we are assuming the symplectic form is $a_0b_3 + a_1b_2 - a_2b_1 - a_3b_0$.)

(5.2). *Remarks on the Geometry of the $\mathcal{C}$-configurations.*

(5.2.0). *Preliminary Remarks.* We said above ((4.4), Remark 1) that the Lagrangian lines in $\mathbb{P}^3$ determine the symplectic projectivity ϕ on $\mathbb{P}^3$. The next proposition makes the analogous statement for the $\mathcal{C}$-configurations.

PROPOSITION. *A $\mathcal{C}$-configuration is determined (up to an element of the group* $\mathrm{Aut}_{\mathrm{Sp}}(\mathbb{P}^3(k)))$ *by its points and Lagrangian lines.*

More precisely, let $\mathbf{X}$ be a $\mathcal{C}$-configuration, and let S be an arbitrary set of primary and secondary points in $\mathbb{P}^3$. Assume there is a one-to-one correspondence between $\mathbf{X}$ and S, such that Lagrangian-collinear subsets of $\mathbf{X}$ correspond to Lagrangian-collinear subsets of S and vice versa. Then there is an element of $\mathrm{Aut}_{\mathrm{Sp}}(\mathbb{P}^3(k))$ which carries S to $\mathbf{X}$.

Proof. Check this for a Desargues' $\mathcal{C}$-configuration directly. It can then be shown (using $<$) for any $\mathcal{C}$-configuration containing a Desargues', and then (using $>$) for all the other $\mathcal{C}$-configurations. □

Remark. A general configuration is not determined by its Lagrangian lines. For example, take more than five points in general position such that every line connecting two of them is non-Lagrangian.

(5.2.1). *Desargues' configuration.* Classically, Desargues' configuration was the set of ten points obtained by taking the ten three-fold intersections of five planes in general position. The ten points lie on ten lines, with three points on each line and with three lines through each point. (There are no secondary points.) It was not noticed, of course, that it could be positioned so that five of its lines were Lagrangian. But it was noticed that its lines naturally broke into two sets of five lines each (in several different ways). It was said that the configuration consisted of *two pentagons inscribed in each other* [H-CV, p. 125 ff.] (compare [V-Y, Section 36, Theorem 22]).

To make Desargues' configuration into a symplectic configuration, we capitalize on this last idea. Choose any *Lagrangian pentagon*—any pentagon *ABCDE* (in space, not a plane) whose sides *AB*, *BC*, . . . , *EA* are Lagrangian (its diagonals are necessarily non-Lagrangian). Then the ten three-fold-intersection points of the planes $A^{\perp}$, $B^{\perp}$, . . . , $E^{\perp}$ form a Desargues' $\mathcal{C}$-configuration. Conversely, any Desargues' $\mathcal{C}$-configuration can be obtained in this way.

The ten points are of two types: *A*, *B*, *C*, *D*, *E*, called the *major points*, and *F*, *G*, *H*, *I*, *J*, the *minor points*. Each minor point is the foot of the Lagrangian perpendicular dropped from a vertex of the Lagrangian pentagon down to its opposite side (4.5.6). The intersection pattern of the configuration's Lagrangian lines looks like this.

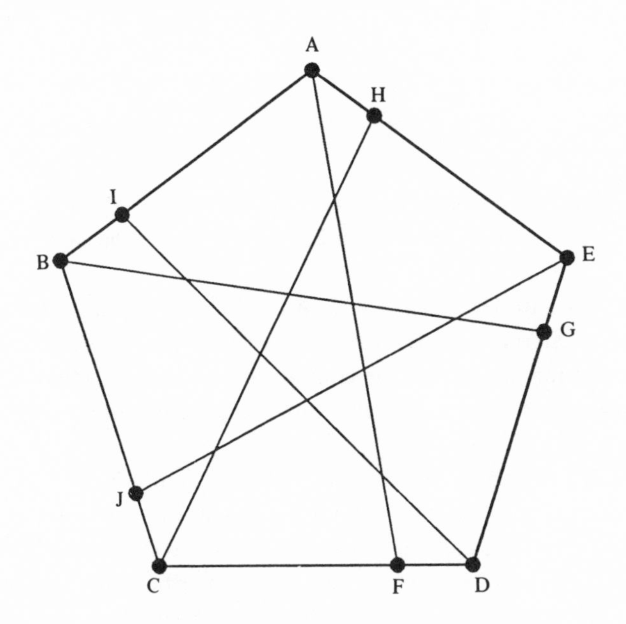

PROPOSITION. *The following triples of points are collinear by non-Lagrangian lines*: *AJG*, *BFH*, *CGI*, *DHJ*, *and EIF*.

Proof. To prove this for *AJG*, note that all three of these points are connected by Lagrangian lines to both *B* and *E*. Apply Proposition 4.5.5. The other cases are similar. □

These five non-Lagrangian lines are schematically arranged in this way:

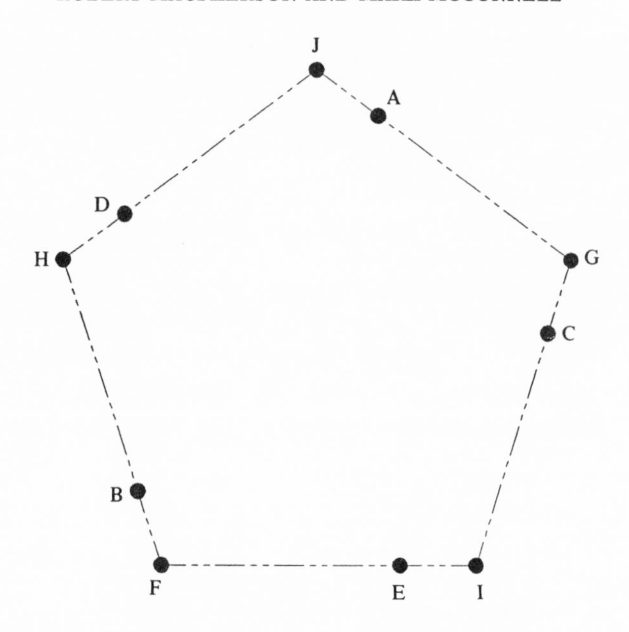

Thus the symplectic Desargues' configuration is indeed two pentagons inscribed in each other: one Lagrangian, and one non-Lagrangian, pentagon.

The configuration is self-dual in the following symplectic sense: the operator $\perp$ interchanges the five planes which cut out the configuration with the five major points. The configuration is not self-dual in the ordinary sense (the dual of a complete 5-plane is a complete 5-point).

(5.2.2). *Reye's configuration.* Reye's configuration is less well known today than Desargues', though it was studied seriously in the nineteenth century. Classically, it was defined in $\mathbb{P}^3(\mathbb{R})$ in several ways [H-CV, Section 22]. First, if we take a regular 24-sided polytope centered at the origin in $\mathbb{R}^4$ and projectivize, then the body's twenty-four vertices become the twelve points of a Reye's configuration in $\mathbb{P}^3(\mathbb{R})$. Second, given four disjoint spheres in general position in $\mathbb{R}^3$, then there are twelve cones with the property of being tangent to two of the spheres, and these cones' twelve vertices form a Reye's configuration. The definitions are extended from $\mathbb{P}^3(\mathbb{R})$ to $\mathbb{P}^3(k)$.

Even if we ignore its symplectic structure, the configuration embodies several theorems of projective geometry. It is self-dual. Also, a given projective 3-space contains a Reye's configuration if and only if the space is $\mathbb{P}^3(k)$ for some field k of characteristic $\neq 2$ [Cox, pers.].

In a symplectic Reye's configuration $\mathbf{X}$, all the points $P \in \mathbf{X}$ are equivalent. (There are no secondary points.) Each point is on two Lagrangian lines and two non-Lagrangian lines. There are twelve planes π which contain a symplectic complete quadrilateral (4.6). (Let Π be the set of these planes.) The configuration is Lagrangianly self-dual: each $\pi \in \Pi$ equals $P^{\perp}$ for some $P \in \mathbf{X}$, and conversely. If l is any non-Lagrangian configuration line (i.e. l meets three points of $\mathbf{X}$), then $l^{\perp}$ also meets three points of $\mathbf{X}$.

(5.2.3). *The Three Utilities configuration.* All Three Utilities configurations are constructed in the following way. Choose two dual non-Lagrangian lines l_1, l_2 and choose three primary points A_{i1}, A_{i2}, A_{i3} ($i =$ 1, 2) on l_i. Let m_{jk} be the line $A_{1j}A_{2k}$. By Proposition 4.5.4, the m_{jk} are Lagrangian. The eighteen secondary points of the configuration are determined by the following facts:

(1) in any plane $A_{ij}^{\perp}$ there are exactly four primary and six secondary points, arranged in a Pascal-Brianchon configuration;
(2) once one secondary point is chosen, the position of all the others is determined (by (3.2.4)).

The eighteen secondary points lie on a unique quadric surface. l_1 and l_2 are dual with respect to this quadric (in the sense of (4.2), Remark 2). The quadric is doubly ruled. In each ruling, the secondary points lie on six lines of the ruling, three points on each line. In one ruling, all the lines are Lagrangian; in the other, all are non-Lagrangian.

(5.2.4). *Type 2c.* The primary points of this configuration can be described in two equivalent ways:

(1) Take a Desargues' configuration and delete two consecutive points from the pentagon of minor points.
(2) Start with the primary points $\mathbf{X}$ of an RR configuration. Say that a line is a *configuration line of* $\mathbf{X}$ if it meets at least three points of $\mathbf{X}$. There is a unique Lagrangian configuration line l which meets every Lagrangian configuration line of $\mathbf{X}$. Choose any point of $\mathbf{X}$ which doesn't lie on l, and delete it.

(5.2.5). *The Two Utilities configuration.* Let $\mathbf{X}$ be a Three Utilities configuration. Take one of the base lines l_i (say l_1) and delete one of its three primary points (say A_{11}). Put secondary points $P (= A_{11})$ and P' on l_1 such that P and P' are harmonically conjugate with respect to A_{12} and A_{13}. There is a unique Three Utilities configuration $\mathbf{X}'$ whose primary points are $\{A_{2i}\} \cup \{P', A_{12}, A_{13}\}$ and such that $\mathbf{X}$ and $\mathbf{X}'$ have twelve secondary points in common. If S, S' are the sets of secondary points for $\mathbf{X}$, $\mathbf{X}'$ respectively, then the Two Utilities configuration has twenty-six secondary points, $S \cup S' \cup \{P, P'\}$. Each Two Utilities configuration can be obtained in this manner.

(5.2.6). *The Ladder configuration.* Choose two skew Lagrangian lines l_1, l_2 and choose three primary points A_1, A_2, A_3 on l_1. By Corollary 4.5.6 these determine three primary points B_1, B_2, B_3 on l_2 such that $A_j B_j$ is Lagrangian. Let m_{jk} $(j, k = 1, 2, 3)$ join A_j to B_k; m_{jk} is Lagrangian if and only if $j = k$. The eighteen secondary points of the Ladder configuration are determined by the following facts:

(1) in any plane $A_{ij}^{\perp}$ there are exactly four primary and six secondary points, arranged in a Pascal-Brianchon configuration;
(2) once one secondary point is chosen, the position of all the others is determined (by (3.2.4)).

The secondary points lie on a unique quadric surface; one of its double rulings is Lagrangian, and the other is not.

(5.2.7). *Type 4.* The primary points form a "Lagrangian square"—four points $ABCD$ in space (not coplanar, no three collinear) such that the sides AB, BC, CD, DA of the "square" are Lagrangian (while the diagonals AC, BD are necessarily non-Lagrangian).

(5.3). *The Topology of the $\mathcal{C}$-complex.*

PROPOSITION. *The $\mathcal{C}$-complex $\mathcal{P}$ associated to the symplectic $\mathbb{P}^3(k)$ is a regular cell complex of dimension four.* □

The complex has eleven types of cells, corresponding to the eleven types of configurations; these are listed in the following table. In the left column, the rank of the configuration is the dimension of the cell.

Rank	$\mathcal{C}$-configuration	Cell
0	Desargues'	vertex
0	Reye's	vertex
1	DR	edge
1	RR	edge
2	Three Utilities	hexagon
2	2b	square
2	2c	triangle
3	Two Utilities	vertebra
3	Ladder	crystal
3	3c	square pyramid
4	4	top-dimensional cell

(5.3.1). *Description of the cells and their closures.* As the names suggest, the boundary of a DR edge is one Desargues and one Reye vertex; the boundary of an RR edge is two Reye vertices. The cells of dimension two contain edges and vertices in the following patterns, where D and R stand for Desargues and Reye vertices respectively.

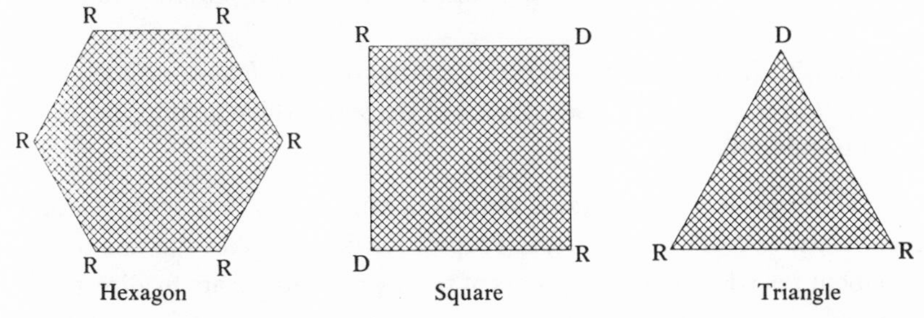

Hexagon Square Triangle

Here are schematic diagrams (Schlegel diagrams) of the boundaries of the cells of dimension three, flattened out into the plane by stereographic projection.

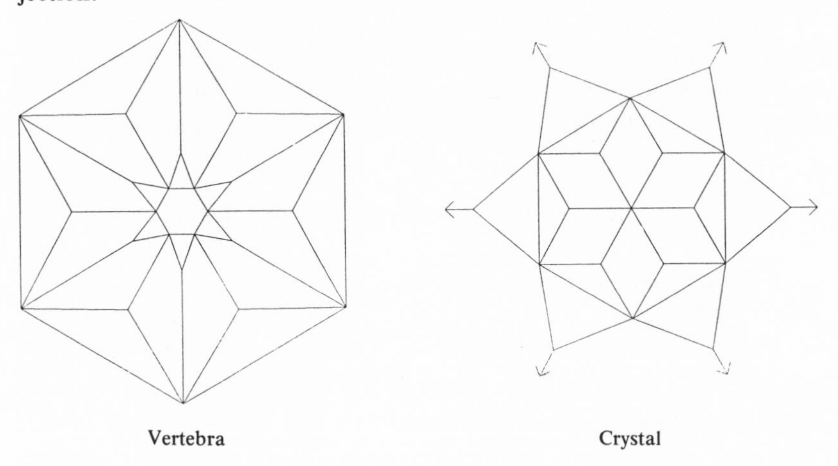
Vertebra Crystal

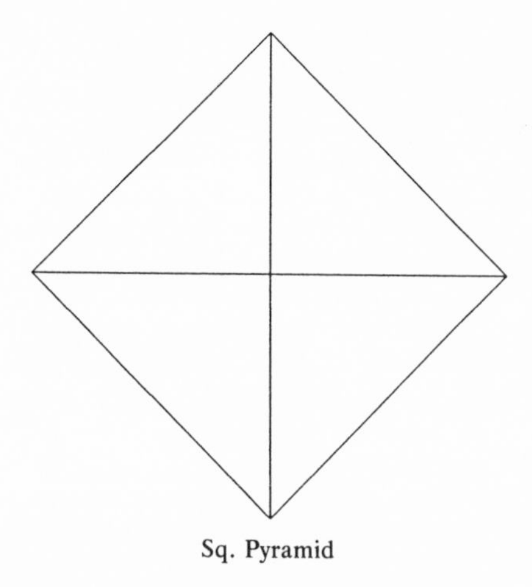

Sq. Pyramid

The top-dimensional cell is a polytope of dimension four whose boundary consists of four vertebrae, four crystals, and thirty-two square pyramids.

(5.3.2). *Examples of how the $\mathcal{C}$-configurations determine the cells' combinatorics.* We now discuss a few of the steps which go into the proof of Proposition 5.3.

(1) Proposition 5.1.3 stated that a DR edge meets exactly one Desargues vertex and one Reye vertex. Proposition 5.1.4 makes the analogous statement for RR edges. The proofs of these propositions are by direct calculation.

(2) *Exactly two Three Utilities configurations are $<$ a given Two Utilities configuration.*

Let **Y** be a Two Utilities configuration (with notation as in (5.2.5)). One Three Utilities configuration is obtained by making P primary, deleting P', and deleting the secondary points on the lines $P'A_{2i}$ ($i = 1, 2, 3$). The other Three Utilities configuration arises in the same way after reversing the roles of P and P'.

Remark. S' lies on a two-parameter family of quadric surfaces.

(5.3.3). *Table of Incidences in $\mathcal{P}$.* In the following table, a given cell R of the type in row i is on a_{ij} cells of the type in column j.

More precisely, choose numbers i, j between 1 and 11. Let a_{ij} be the entry in the ith row and jth column of the table below. If $i > j$ (i.e. you're below the diagonal), a given cell R of row i's type will contain a_{ij} cells of column j's type in its boundary. If $i < j$ (i.e. you're above the diagonal), a given cell R of row i's type will be contained in the boundary of a_{ij} cells of column j's type.

	D	R	DR	RR	hex	sq	tri	vrtb	crys	pyr	top
Desargues	—	—	5	0	0	5	5	5	5	5	15
Reye	—	—	24	8	4	24	48	24	16	36	42
DR	1	1	—	—	0	2	2	3	3	3	11
RR	0	2	—	—	1	0	6	6	3	6	15
hexagon	0	6	0	6	—	—	—	6	0	0	9
square	2	2	4	0	—	—	—	2	2	1	8
triangle	1	2	2	1	—	—	—	1	1	2	7
vertebra	12	12	36	12	2	12	12	—	—	—	3
crystal	12	8	36	6	0	12	12	—	—	—	3
sq. pyramid	2	3	6	2	0	1	4	—	—	—	4
top cell	48	28	176	40	4	64	112	4	4	32	—

Chapter 6—The $\mathcal{C}$-complex and Modular Varieties

In Chapters 2, 3, and 5 we constructed regular cell complexes $\mathcal{P}$ called $\mathcal{C}$-*complexes*. In this chapter we relate $\mathcal{P}$ to the locally symmetric space $\Gamma(p)\backslash M$, where M is the symmetric space associated to $G = \mathrm{SL}(2, \mathbb{R})$, $\mathrm{SL}(3, \mathbb{R})$, or $\mathrm{Sp}(4, \mathbb{R})$.

In (6.1) we state the main theorems. In (6.2)–(6.4) we discuss some of the ingredients of the proof in the $\mathrm{SL}(n, \mathbb{R})$ case ($n = 2, 3$). In (6.5) we mention the corresponding topics in the symplectic case. In (6.6) we discuss and extend the classical notion of a net.

(6.1). *Statement of the Main Theorems.* Let G be $\mathrm{SL}(2, \mathbb{R})$, $\mathrm{SL}(3, \mathbb{R})$, or $\mathrm{Sp}(4, \mathbb{R})$, with K a maximal compact subgroup. $M = G/K$ is the symmetric space associated to G. $G(\mathbb{Z})$ is $\mathrm{SL}(n, \mathbb{Z})$ or $\mathrm{Sp}(4, \mathbb{Z})$. Let $\Gamma(N)$ be the subgroup of $G(\mathbb{Z})$ consisting of those matrices congruent to the identity mod N.

Let $\mathcal{C}$ be the set of all $\mathcal{C}$-configurations, as defined in Chapters 2, 3, and 5 in the case $k = \mathbb{F}_p$. Let $\mathcal{P}$ denote the $\mathcal{C}$-complex (1.5).

Theorem. (Main Theorem in the SL case.) [McC1]. *Let* $G = \mathrm{SL}(n, \mathbb{R})$ ($n = 2, 3$). *Let p be an odd prime satisfying*

$$p \not\equiv 1 \pmod 4 \quad \textit{when} \quad n = 2$$

(6.1.1)

$$p \not\equiv 1 \pmod 6 \quad \textit{when} \quad n = 3.$$

Then, with the above notation, $\Gamma(p)\backslash M$ *strongly deformation-retracts onto a subspace which is homeomorphic to* $\mathcal{P}$. □

Remark. When p does not satisfy (6.1.1), $\mathcal{P}$ has n connected components which are homeomorphic to each other. Let $\mathcal{P}_0$ be one such component. Then $\Gamma(p)\backslash M$ is an n-fold covering space of a space Y which strongly deformation-retracts onto $\mathcal{P}_0$.

THEOREM. (Main Theorem in the Sp case.) [M-McC1]. *Let* $G = \mathrm{Sp}(4, \mathbb{R})$. *Let p be a prime satisfying*

$$p \not\equiv 1 \pmod 4. \tag{6.1.2}$$

Then, with the above notation, $\Gamma(p)\backslash M$ *strongly deformation-retracts onto a subspace which is homeomorphic to* $\mathcal{P}$. □

As before, there is a slightly more complicated statement when $p \equiv 1 \pmod 4$.

(6.2). In this and the next two sections we discuss parts of the proof of Theorem 6.1 in the SL case.

We begin by considering the symmetric space M upstairs, rather than $\Gamma(p)\backslash M$ downstairs. We will describe a certain cell decomposition of M due to Voronoi [V]. In (6.3) we will say how Voronoi's decomposition is related to the $\mathcal{C}$-complex.

Voronoi decomposed M as a disjoint union of subsets $R \subset M$. Denote the set of these subsets by $\mathcal{R}$. Each element of $\mathcal{R}$ is analytically isomorphic to an open ball and is called a *cell* in M. $\mathcal{R}$ is equivariant under $\mathrm{SL}(n, \mathbb{Z})$. For any subgroup Γ of finite index in $G(\mathbb{Z})$, the projection $M \to \Gamma\backslash M$ induces a cell decomposition of $\Gamma\backslash M$ with only finitely many cells, and in such a way that the map is cellular. The proofs of these facts are either in [V] or follow easily from it.

(6.3). Let $\mathcal{P}$ be the $\mathcal{C}$-complex associated to $\mathrm{SL}(n, \mathbb{R})$ ($n = 2, 3$) in the case $k = \mathbb{Q}$. All the connected components of $\mathcal{P}$ are homeomorphic to each other; choose one connected component and call it $\mathcal{P}_0$.

PROPOSITION. [McC1]. $\mathcal{P}_0$ *and* $\mathcal{R}$ *are canonically dual to each other as cell complexes. That is, there is a canonical bijection between* $\mathcal{P}_0$ *and* $\mathcal{R}$ *which reverses the face relations among the cells.*

Example. Here is a picture of $M = \mathrm{SL}(2, \mathbb{R})/\mathrm{SO}(2, \mathbb{R})$ showing $\mathcal{P}_0$ and $\mathcal{R}$. The Voronoi cells are the triangles, together with the line segments which form the triangles' borders. (The vertices of the triangles are not Voronoi cells, and do not even belong to M: they lie at infinity.) Because of space limitations we are only able to draw the first few triangles. $\mathcal{P}_0$ is a graph, which we draw using dotted lines; each vertex of the graph is in the center of a triangle, and each edge of the graph crosses a boundary cell between two triangles.

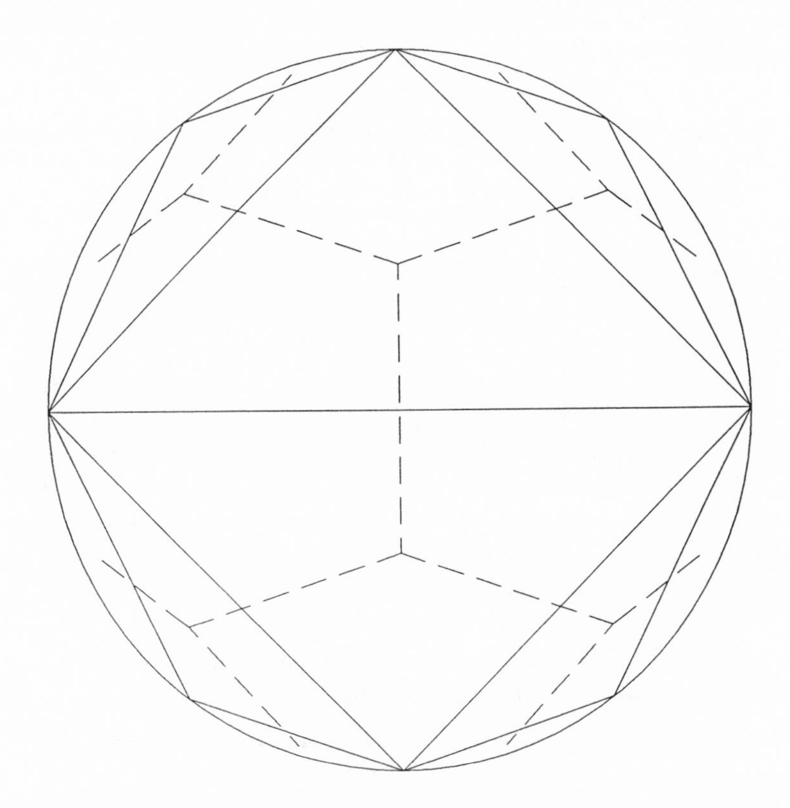

(6.4). We have not yet described how one retracts $\Gamma(p)\backslash M$ onto $\mathcal{P}_0$. Soulé and Lannes [Sou1] [Sou2] defined $\mathfrak{N}$, a $G(\mathbb{Z})$-equivariant strong deformation-retract of M. $\Gamma\backslash\mathfrak{N}$ is compact for all subgroups $\Gamma \subseteq G(\mathbb{Z})$ of finite index. $\mathfrak{N}$ is a locally finite regular cell complex. The cell structure is $G(\mathbb{Z})$-equivariant, and it induces a finite cell complex structure on $\Gamma\backslash\mathfrak{N}$.

PROPOSITION. *$\mathcal{R}$ and $\mathcal{N}$ are duals of each other as cell complexes.*

In the cases we need ($n = 2, 3$), this follows easily from [V] and [Sou2]. (It can be proved for all n.)

COROLLARY. *With notation as in* (6.3), *$\mathcal{N}$ and $\mathcal{P}_0$ are isomorphic as cell complexes and thus homeomorphic as spaces.*

(6.5). When $G = \mathrm{Sp}(4, \mathbb{R})$, results analogous to those in (6.2)–(6.4) are true. We have constructed [M-McC1] an $\mathrm{Sp}(4, \mathbb{Z})$-equivariant cell decomposition of M which enjoys most of the formal properties of Voronoi's complex $\mathcal{R}$. If $\mathcal{P}$ is the $\mathcal{C}$-complex in the symplectic case and for $k = \mathbb{Q}$, and $\mathcal{P}_0$ is a connected component of $\mathcal{P}$, then our cell decomposition of M is dual to $\mathcal{P}_0$. The existence of a retraction from $\Gamma(p)\backslash M$ onto a subspace homeomorphic to $\mathcal{P}_0$ will follow from [M-McC1] and [M-McC2].

(6.6). *Remarks on Nets.* Classical projective geometers used the term *net* [V-Y, p. 36] to describe the following two phenomena:

(1) Let $\mathbb{P}$ be a projective plane over $\mathbb{Q}$. Take any four points A, B, C, D in $\mathbb{P}$ in general position. Repeatedly add new points $l_1 \cap l_2$ for lines l_i joining two old points. You will eventually get every point of $\mathbb{P}$ in this way.

(2) The unique projectivity from $\mathbb{P}$ to $\mathbb{P}^2(\mathbb{Q})$ taking A, B, C, D to $[1:0:0]$, $[0:1:0]$, $[0:0:1]$, $[1:1:1] \in \mathbb{P}^2(\mathbb{Q})$ may be realized by sending each point of $\mathbb{P}$ constructed from A, B, C, D as in (1) to the point correspondingly constructed from $[1:0:0]$, $[0:1:0]$, $[0:0:1]$, $[1:1:1]$.

These facts allowed von Staudt to introduce coordinates in abstract projective planes $\mathbb{P}$ by purely projective methods. [Ha] [K, p. 850 ff.]

With the next definition we refine these ideas.

Definition. A net for $\mathbb{P}$ is a connected component of the $\mathcal{C}$-complex of $\mathbb{P}$.

Statements (1) and (2) can then be sharpened:

PROPOSITION 1. *Let S be a net for $\mathbb{P}$, where $\mathbb{P}$ is a projective space over a prime field. For any point P in $\mathbb{P}$, there is a cell in $\mathcal{S}$ such that P is one of the points in the $\mathcal{C}$-configuration corresponding to that cell.*

PROPOSITION 2. *Let $\mathcal{S}_1$ and $\mathcal{S}_2$ be two nets for $\mathbb{P}$. Let $\mathbf{X}_1$, $\mathbf{X}_2$ be $\mathcal{C}$-configurations corresponding to cells in $\mathcal{S}_1$, $\mathcal{S}_2$ respectively. Assume there is an isomorphism $\beta : \mathbb{P} \to \mathbb{P}$ sending $\mathbf{X}_1$ to $\mathbf{X}_2$. Then β induces a unique isomorphism $\tilde{\beta}$ of cell complexes from $\mathcal{S}_1$ to $\mathcal{S}_2$.*

(By Proposition 1, if $\mathbb{P}$ is defined over a prime field then $\tilde{\beta}$ determines β.)

CHAPTER 7—CELLULATIONS OF SATAKE COMPACTIFICATIONS OF MODULAR VARIETIES

In this chapter we use the $\mathcal{C}$-configurations to define a cellulation of the Satake compactifications of the locally symmetric spaces $\Gamma(p)\backslash M$.

For simplicity we discuss only the SL(3, $\mathbb{R}$) case. Analogous results hold for SL(2, $\mathbb{R}$) and Sp(4, $\mathbb{R}$). (We make some remarks on Sp(4, $\mathbb{R}$) in (7.7).)

For simplicity, again, we discuss only the maximal Satake compactification. General references on Satake compactifications and boundary components are [Sa1] [Sa2] [Z].

We follow the general plan of the paper so far: we define a set $\mathcal{C}'$ of configurations (7.1), put a partial ordering on $\mathcal{C} \cup \mathcal{C}'$ (7.3)-(7.5), form the geometric realization, and relate this space to the Satake compactification of a certain modular variety. The main theorem is stated in (7.2).

(7.1). Fix an odd prime $p \not\equiv 1 \pmod 6$, and let $k = \mathbb{F}_p$. Define the set $\mathcal{C}$ of all $\mathcal{C}$-configurations in $\mathbb{P}^2(k)$ as in (3.1).

Definition. By a *flag* Φ in $\mathbb{P}^2(k)$ we mean either a point, a line, or a pair $P \in l$ consisting of a point on a line. $P \in l$ is called a *full flag*.

Definition. A *boundary $\mathcal{C}$-configuration in* $\mathbb{P}^2(k)$ is a pair $(\mathbf{Y}, \Phi)$ where $\mathbf{Y}$ is a set either of primary and secondary points or of primary and secondary lines, Φ is a flag, and $\mathbf{Y} \cup \Phi$ is projectively isomorphic to one of the following five pictures:

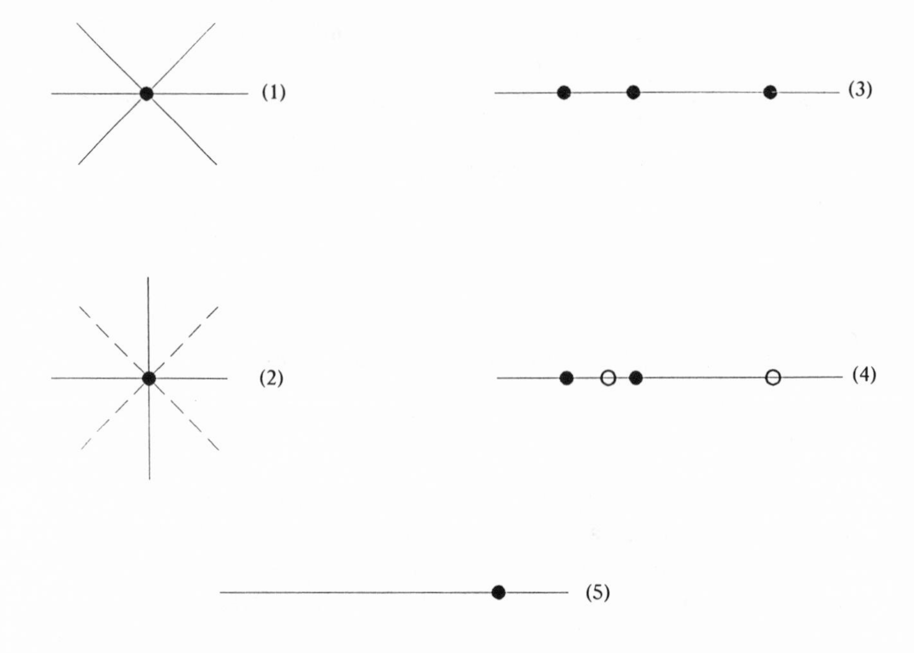

In (1) and (2), Φ is a point; primary lines are drawn as solid lines, and secondary lines are drawn as dotted lines. In (1), $\mathbf{Y}$ contains three primary lines. In (2), $\mathbf{Y}$ contains two primary and two secondary lines such that the secondary lines are harmonically conjugate with respect to the primary lines. In (3) and (4), Φ is a line; (3) and (4) are the usual $\mathcal{C}$-configurations (2.1) on Φ. In (5), Φ is a full flag and $\mathbf{Y} = \Phi$.

Note. (1) and (2) are the duals of (3) and (4).

Notation.

(a) Let $\mathcal{C}'$ denote the set of all boundary $\mathcal{C}$-configurations.

(b) When $(\mathbf{Y}, \Phi)$ is a boundary $\mathcal{C}$-configuration, we will express this fact by saying $\mathbf{Y}$ is a boundary $\mathcal{C}$-configuration on Φ.

(7.2). In (7.5) we will define a partial ordering $\sqsubseteq$ on the set $\mathcal{C} \cup \mathcal{C}'$. If we take $\sqsubseteq$ for granted for a moment, we can state our main theorem on compactifications. A reference for the definition of *normalization* is [G-M1, p. 151].

THEOREM. [McC2]. *Let* $p \not\equiv 1 \pmod 6$. *Then the normalization of the geometric realization of the partially ordered set* $(\mathcal{C} \cup \mathcal{C}', \sqsubseteq)$ *is ho-*

meomorphic to the maximal Satake compactification $\overline{\Gamma(p)\backslash M}$ *of* $\Gamma(p)\backslash M$.

Remarks. $\mathcal{R}$ (6.2) is a decomposition of M into noncompact cells. We will compactify M by adjoining new cells to $\mathcal{R}$. The new cells are in one-to-one correspondence with the boundary $\mathcal{C}$-configurations.

There is a finite-to-one map α from the set of all boundary components of the maximal Satake compactification $\overline{\Gamma(p)\backslash M}$, onto the set of flags in $\mathbb{P}^2(\mathbb{F}_p)$. For a given line l in $\mathbb{P}^2$, $\alpha^{-1}(l)$ is $1/2(p-1)$ different rank-one boundary components of $\overline{\Gamma(p)\backslash M}$. For a point P, $\alpha^{-1}(P)$ is again $1/2(p-1)$ rank-one boundary components. (They are of a different type from the components in $\alpha^{-1}(l)$. This happens because there are two types of maximal rational parabolic subgroups in G.) For a flag $P \in l$, $\alpha^{-1}(P \in l)$ is $(1/2(p-1))^2$ rank-zero boundary components of $\overline{\Gamma(p)\backslash M}$.

A boundary $\mathcal{C}$-configuration $\mathbf{Y}$ on l corresponds to a single cell in the geometric realization of $(\mathcal{C} \cup \mathcal{C}', \sqsubseteq)$. But after taking the normalization of this complex, $\mathbf{Y}$ gives rise to $1/2(p-1)$ cells in the maximal Satake. Similar results hold for flags P and $P \in l$.

In (7.3) and (7.4) we give some definitions which are preliminary to our discussion of $\sqsubseteq$.

(7.3). Let $\mathbf{X}$ be a $\mathcal{C}$-configuration. We say that $\mathbf{X}$ *respects the line* l if and only if at least two primary points of $\mathbf{X}$ lie on l. $\mathbf{X}$ *respects the point* P if and only if

(1) P is a primary point of $\mathbf{X}$;
(2) there are at least two distinct lines which pass through P and another primary point of $\mathbf{X}$.

$\mathbf{X}$ *respects the flag* $P \in l$ if and only if it respects both P and l.

Note. Respecting a point and respecting a line are dual conditions.

(7.4). Whenever a configuration respects a line flag l or a point flag P, we get an induced boundary $\mathcal{C}$-configuration. The next two definitions describe how this comes about. (We will use this material in parts (5), (6) of (7.5).)

PROPOSITION-DEFINITION. *Let* $\Phi = l$ *be a flag consisting of a line in* $\mathbb{P}^2(k)$, *and let* $\mathbf{X}$ *be a* $\mathcal{C}$*-configuration respecting* l. *If* $J = \mathbf{X} \cap l$ *is a boundary* $\mathcal{C}$*-configuration on* l, *then the* induced configuration on l *is de-*

fined to be J. Otherwise, there is an isomorphism of $\mathbb{P}^2$ *carrying J to one of the three pictures in the left-hand column below; the* induced configuration on *l is a harmonic quadruple constructed as shown in the right-hand column.*

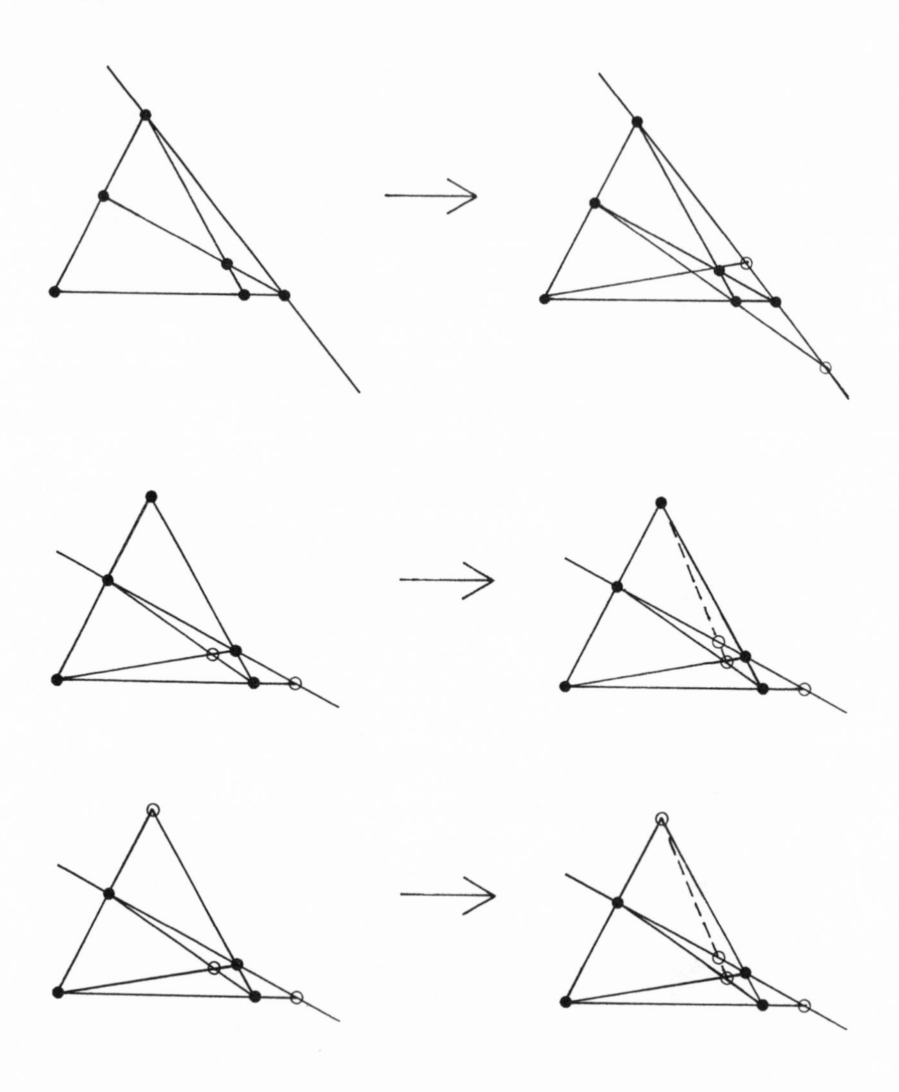

Proposition-Definition. *Let* $\Phi = P$ *be a flag consisting of a point in* $\mathbb{P}^2(k)$*, and let* $\mathbf{X}$ *be a* $\mathfrak{C}$*-configuration respecting P. Let J be the set of all lines which pass through P and at least one other point of* $\mathbf{X}$*. (A line in J is called* primary *if it passes through at least one primary point*

other than P; other lines of J are secondary.) *Then J is a boundary $\mathcal{C}$-configuration on P, and the* induced configuration on *P is defined to be J.*

Notation. We will not be using induced configurations on full flags. For the sake of consistency, though, we make the following definition. If Ψ is a full flag, Ψ itself will be called a boundary $\mathcal{C}$-configuration on Ψ (and will sometimes be denoted by $\mathbf{X}$).

(7.5). We can now define $\sqsubseteq$.

Definition. For $\mathbf{X}_1, \mathbf{X}_2 \in \mathcal{C} \cup \mathcal{C}'$, we write $\mathbf{X}_1 \sqsubseteq \mathbf{X}_2$ if and only if one of the following holds:

(1) $\mathbf{X}_1, \mathbf{X}_2 \in \mathcal{C}$ and $\mathbf{X}_1 \geq \mathbf{X}_2$;
(2) $\mathbf{X}_1, \mathbf{X}_2$ are boundary $\mathcal{C}$-configurations on a common line l, and $\mathbf{X}_1 \geq \mathbf{X}_2$;
(3) $\mathbf{X}_1, \mathbf{X}_2$ are boundary $\mathcal{C}$-configurations on a common point P, and the plane dual of $\mathbf{X}_1$ is $\geq$ the plane dual of $\mathbf{X}_2$;
(4) $\mathbf{X}_1 = \mathbf{X}_2 =$ a full flag;
(5) $\mathbf{X}_1$ is a boundary $\mathcal{C}$-configuration on a flag l, $\mathbf{X}_2 \in \mathcal{C}$, and $\mathbf{X}_2$ induces $\mathbf{X}_1$ on l (7.4);
(6) $\mathbf{X}_1$ is a boundary $\mathcal{C}$-configuration on a flag P, $\mathbf{X}_2 \in \mathcal{C}$, and $\mathbf{X}_2$ induces $\mathbf{X}_1$ on P (7.4);
(7) $\mathbf{X}_1$ is a full flag, $\mathbf{X}_2 \in \mathcal{C}$, and $\mathbf{X}_1$ respects $\mathbf{X}_2$;
(8) $\mathbf{X}_1$ is a full flag $P \in l$, and $\mathbf{X}_2$ is a boundary $\mathcal{C}$-configuration on l.
(9) $\mathbf{X}_1$ is a full flag $P \in l$, and $\mathbf{X}_2$ is a boundary $\mathcal{C}$-configuration on P.

Explanation. (a) This definition breaks up into cases because a cell in $\overline{\Gamma(p)\backslash M}$ can lie on one of several boundary components. Let C_1, C_2 be cells corresponding to $\mathbf{X}_1$, $\mathbf{X}_2$ respectively. $\mathbf{X}_1 \sqsubseteq \mathbf{X}_2$ just means that C_1 is in the closure of C_2. The nine cases can be summarized in the following diagram:

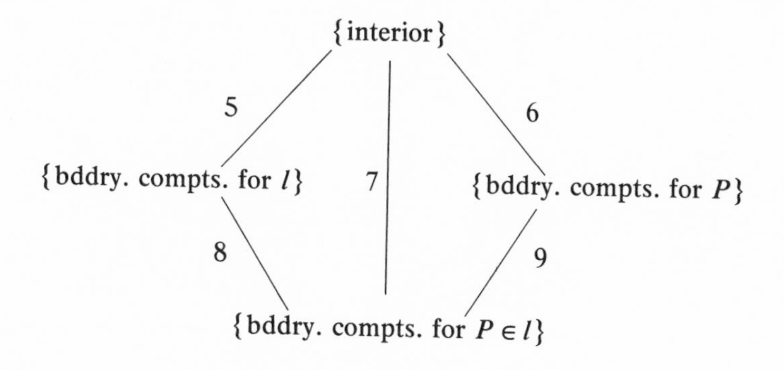

For instance, the segment labelled 6 represents case (6) of the definition. In this situation C_1 lies on a boundary component associated to a point flag P, C_2 lies in the interior $\Gamma(p)\backslash M$, and $C_1 \subset \partial(C_2)$.

In cases (1)-(4), C_1 and C_2 both lie in the same part of the diagram, so there is no segment between them.

(b) We have inverted the order relation on $\mathcal{C}$. In case (1), for example, $\sqsubseteq$ corresponds to $\geq$ and vice versa. To see why, recall the discussion of $\mathcal{R}$ and $\mathcal{N}$ in (6.2)-(6.4). It was convenient to work with $\mathcal{N}$ in Chapter 6 because $\Gamma(p)\backslash\mathcal{N}$ was compact. We are now going to compactify M, though, so we no longer want a retract of M; we prefer to work with all of M. $\mathcal{R}$ is a decomposition of all of M, so it is suitable for this purpose. But $\mathcal{N}$ is the nerve of $\mathcal{R}$—boundary relations in $\mathcal{R}$ are reversed and become coboundary relations in $\mathcal{N}$. This is why $\sqsubseteq$ coincides with $\geq$ wherever both are defined.

(7.6). *Remark.* Imagine we removed all the cells lying on the boundary of the compactification. We would have a decomposition of $\Gamma(p)\backslash M$ into open cells, and the face relations among these cells would be given by the restriction of $\sqsubseteq$ to $\mathcal{C}$. In other words, with the $\mathcal{C}$-configurations we can build a space which equals $\Gamma(p)\backslash M$ *up to homeomorphism*, not merely up to homotopy.

(7.7). *Remark on the Sp case.* The results of this chapter can be generalized to modular varieties associated to $\mathrm{Sp}(4, \mathbb{R})$. The main difference concerns the definition of a flag. By a *symplectic flag* we mean one of the following:

(1) a pair $P \in \pi$ where $\pi = P^{\perp}$;
(2) a Lagrangian line l;
(3) a full flag $P \subset l \subset P^{\perp}$.

(These correspond to the different types of parabolic subgroups of $\mathrm{Sp}(4, \mathbb{R})$.) The rest of the construction goes through in essentially the same way as for the SL(3) case [M-McC2].

MASSACHUSETTS INSTITUTE OF TECHNOLOGY
HARVARD UNIVERSITY

REFERENCES

[Al] P. S. Alexandrov, *Diskrete Raüme*, Mat. Sb. (N. S.) **2** (1937), 501-518.

[AMRT] A. Ash, D. Mumford, M. Rapoport, and Y. Tai, *Smooth Compactifications of Locally Symmetric Varieties*, in "Lie Groups: History, Frontiers, and Applications, vol. 4," Math. Sci. Press, 1975.

[Ash] A. Ash, *Small-Dimensional Classifying Spaces for Arithmetic Subgroups*, Duke Math J. **51** (1984), 459-468.

[B-G-S-V] A. Beilinson, A. Goncharov, V. Schectman, and A. Varchenko, *Aomoto Dilogarithms, Mixed Hodge Structures, and Motivic Cohomology of Pairs of Triangles on the Plane*, (to appear in volume honoring A. Grothendieck's sixtieth birthday).

[B-M-S] A. Beilinson, R. MacPherson, and V. Schectman, *Notes on Motivic Cohomology*, Duke J. **54** (1987), 679-710.

[B] N. Bourbaki, "Groupes et Algèbres de Lie," chs. 4-6, Actu. Sci. Ind. no. 1337, Hermann, Paris, 1968.

[Cox] H. S. M. Coxeter, personal communication.

[C] L. Cremona, "Elements of Projective Geometry," trans. C. Leudesdorf, 3rd ed., Oxford Univ. Press, London, 1913.

[G-G-M-S] I. M. Gelfand, R. M. Goresky, R. D. MacPherson, and V. V. Serganova, *Combinatorial Geometries, Convex Polyhedra, and Schubert Cells*, Advances in Math. **63** (1987), 301-316.

[G-M1] M. Goresky and R. MacPherson, *Intersection Homology Theory*, Topology **19** (1980), 135-162.

[G-M] ______ and ______, "Stratified Morse Theory," (The Book), Springer-Verlag, Berlin, 1988.

[Ha] R. Hartshorne, "Foundations of Projective Geometry," Benjamin/Cummings, Reading, Mass., 1967.

[H-CV] D. Hilbert, S. Cohn-Vossen, "Geometry and the Imagination," trans. P. Nemenyi, Chelsea Publ. Co., New York, 1952.

[K] M. Kline, "Mathematical Thought from Ancient to Modern Times," Oxford Univ. Press, New York, 1972.

[L-S] R. Lee and R. H. Szczarba, *On the Torsion in* $K_4(\mathbb{Z})$ *and* $K_5(\mathbb{Z})$, Duke Math. J. **45** (1978), 101-129.

[M-McC1] R. MacPherson and M. McConnell, *Explicit Reduction Theory for* $\mathrm{Sp}(4, \mathbb{Z})$, to appear.

[M-McC2] ______ and ______, *Cellulations of Satake Compactifications of Modular Varieties associated to* $\mathrm{Sp}(4, \mathbb{R})$, to appear.

[McC1] M. McConnell, *Projective Geometry and Cellulations of Locally Symmetric Spaces associated to* $\mathrm{SL}(n, \mathbb{R})$, to appear.

[McC2] ______, *Cellulations of Satake Compactifications of Locally Symmetric Spaces associated to* $\mathrm{SL}(n, \mathbb{R})$, to appear.

[Mi] H. Minkowski, *Diskontinuitätsbereich für Arithmetische Äquivalenz*, Crelle **129**.

[O-S] P. Orlik, L. Solomon, *Combinatorics and Topology of Complements of Hyperplanes*, Inv. Math. **56** (1980), 167-189.

[Sa1] I. Satake, *On representations and compactifications of symmetric Riemannian spaces*, Annals of Math. **71** (1960), 77-110.

[Sa2] ______, *On compactifications of the quotient spaces for arithmetically defined discontinuous groups*, Annals of Math. **72** (1960), 555-580.

[Sie] C. L. Siegel, *Symplectic Geometry*, Amer. J. **65** (1943), 1-86.

[Sou1] C. Soulé, "Thèse," Univ. Paris VII.

[Sou2] ______, *The Cohomology of* SL(3, $\mathbb{Z}$), Topology **17** (1978), 1-22.

[Što] Štogrin, *Locally quasi-densest lattice packings of spheres*, Soviet Math. Doklady **15** (1974), 1288-1292.

[V-Y] O. Veblen, J. W. Young, "Projective Geometry," vol. 1, Ginn and Co., Boston, 1910.

[V] G. Voronoi, *Sur quelques propriétés des formes quadratiques positives parfaites*, Crelle **133** (1908), 97-178.

[V2] ______, *Recherches sur les paralléloèdres primitifs*, Crelle **134** (1908), 198-287.

[V3] ______, *Domaines de formes quadratiques correspondant aux différents types de paralléloèdres primitifs*, Crelle **136** (1909), 67-181.

[Z] S. Zucker, *Satake Compactifications*, Comm. Math. Helvetici **58** (1983), 312-343.

STABLE HOMOTOPY AND LOCAL NUMBER THEORY

By JACK MORAVA

Dedicated to the memory of Frank Adams

---

1. The technique of classifying objects of mathematical interest by representing their moduli functors is one of the great contributions of algebraic topology to the general mathematical culture of our day; by now we are not only accustomed to encountering such objects as the classifying spaces for the symmetric groups, the braid groups, or the general linear groups, but we are also used to the notion that objects of such central significance can be expected to appear in unexpected contexts precisely because of their universality. In this talk I want to explain how groups of units in division algebras over the p-adic field have become important in stable homotopy theory, and how their representations play the role of local factors for some kind of adelization of the stable homotopy category.

It will probably be easier to reverse historical order and work from local to global, and to begin (for expository purposes) with objects from local number theory.† We will be concerned with a division algebra D, say with center C; and a good deal of work in the first part of this century was denoted to the study and classification of such things, especially in the case in which C is a locally compact field. By now there is a very general theory of Azumaya algebras (which are noncommutative algebras, separable and of finite rank over the center C) under the relation of Morita equivalence (which identifies A with any algebra of finite matrices over it, cf. [6; 10, Chapter 8]): tensor product (over the center) defines a law of composition for such algebras, and the set of equivalence classes is the Brauer group of C. If C is connected, the only candidate for a nontrivial element is the quaternion algebra; in other words the Brauer group of $\mathbf{R}$ is the integers

---

Manuscript received 15 December 1988.

†In this talk I have slighted the recent amazing applications of global number theory and modular forms to algebraic topology motivated by the discovery, by quantum field theorists, of elliptic cohomology [cf. 25, 43]; that story is still unfolding.

mod two, and the Brauer group of $\mathbf{C}$ is trivial; but the Brauer group of the example of interest here, i.e. a totally disconnected nontrivial locally compact field such as $\mathbf{Q}_p$, is the much larger group $\mathbf{Q}/\mathbf{Z}$.

Thus over a p-adic field there are many inequivalent division algebras. If m and n are coprime integers we can construct an example with invariant $m/n \in Br(\mathbf{Q}_p)$ as the quotient

$$K\langle F\rangle/(F^n - p^m)$$

where $K = W(\mathbf{F}_q) \otimes \mathbf{Q}$ is the quotient field of the ring $W(\mathbf{F}_q)$ of Witt vectors over the finite field with $q = p^n$ elements (alternatively K is the extension obtained from $\mathbf{Q}_p$ by adjoining a primitive $(q - 1)^{\text{th}}$ root of unity) and the pointed brackets signify that

$$Fa = a^\sigma F$$

for $a \in K$, with $\sigma \in \mathrm{Gal}(K/\mathbf{Q}_p) \cong \mathrm{Gal}(\mathbf{F}_q/\mathbf{F}_p)$ the Frobenius endomorphism lifted to the Witt vectors. The case of most interest in homotopy theory (and in local class-field theory, as it turns out) has $m = 1$; in what follows I will write D for such an algebra, and $S(D)$ for the maximal compact subgroup of strict units defined by the exact sequence

$$1 \longrightarrow S(D) \longrightarrow D^\times = S(D)\cdot \mathbf{Z} \overset{p-\mathrm{ord}}{\longrightarrow} \mathbf{Z} \longrightarrow 0$$

where p-ord(F) is normalized to be one. Note that a generator of $\mathbf{Z}$ acts nontrivially (if $n > 1$) on $S(D)$, sending α to $F^{-1}\alpha F$.

It is the groups $D^\times$, for $[D] = 1/n \in Br(\mathbf{Q}_p)$, which play the role of *motivic groups* for the stable homotopy category, in a sense which I will try to make precise below. To a topologist it is the cohomology of the groups $S(D)$ which is most immediately useful; these are profinite groups, and it is the cohomology based on continuous cochains which is of interest. This gives the whole subject a flavor very much like Galois cohomology, and I hope to be able to make convincing the claim that the groups $H_c^*(S(D))$ are objects which ought to be of interest to number-theorists.

Before discussing these groups in any detail I should make some general remarks about them. In the first place the groups $S(D)$ have virtual p-cohomological dimension n^2, and have finite cohomological dimension whenever $(p - 1) \nmid n$; if $(p - 1)$ does divide n then the cohomology groups are periodic in high dimension, so the $H_c^*(S(D))$ are in a certain sense virtually finite-dimensional. Furthermore, when $(p - 1) \nmid n$ they are p-Poincaré duality groups.

Now the standard way of approaching the cohomology of such groups is by use of the cohomology of simpler subgroups, but unfortunately cohomology does not behave under induction as nicely as representation theory. Nevertheless there are some observations to be made. In the first place, finite subgroups of the units of a (skew) field are extremely restricted; for example they are necessarily cyclic if the field is commutative. In general such groups have cyclic Sylow subgroups, and they have been completely classified, cf. [3]. Their study is intimately connected with the study of the free actions of finite groups on spheres, which began in antiquity and which profoundly affected the early development of algebraic topology. (This influence can be seen in the development of the Tate cohomology of finite groups as presented, for example, in the book of Cartan and Eilenberg. In fact number theorists had calculated the cohomology of the finite subgroups of division algebras by the late fifties, but their interest seems to have then turned in other directions, cf. [44].)

Now even in the simplest case, i.e. when $n = 1$ and $S(D)$ is commutative, it is easy to see that the study of $H_c^*(S(D))$ has deep number-theoretic roots. To simplify matters let us assume that p is odd, so $(p - 1) \nmid n$! Then $S(D)$, which is just the group $\mathbf{Z}_p^\times$ of p-adic units, has cohomological dimension one, and its cohomology with coefficients in the module $\mathbf{F}_p$ with trivial action is an exterior algebra—i.e. $\mathbf{Z}_p^\times$ is a kind of p-adic analogue of the circle, with cohomology satisfying Poincaré duality. However, to topologists it was the cohomology of $\mathbf{Z}_p^\times$ with coefficients in the "Tate module" which first manifested itself, cf. [5]: in particular if $\mathbf{Z}_p^{\otimes n}$ denotes the p-adic integers with the action of $\alpha \in \mathbf{Z}_p^\times$ as multiplication by α^n, then

$$\begin{aligned} H_c^1(\mathbf{Z}_p^\times, \mathbf{Z}_p^{\otimes n}) &= \text{cyclic of order } p^{\nu+1}, \\ &\quad \text{if } n = (p-1)p^\nu \cdot (\text{rel. prime to } p) \\ &= 0, \qquad \text{otherwise} \end{aligned}$$

is the subgroup of $\mathbf{Q}/\mathbf{Z}$ generated by the Bernoulli quotient B_n/n, according to the theorem of von Staudt, cf. [9]. Topologists, on the other hand, have known this group as the image of Whitehead's J-homomorphism

$$\pi_*(\mathbf{O}) \to \pi_*^S(S^0)$$

since the work of Adams [1].

Thus the p-component of the image of the J-homomorphism is (at least for p odd) the cohomology of $\mathbf{Z}_p^\times$ with coefficients in the 'Tate algebra' $\mathbf{Z}_p[t]$ where $\alpha, t \mapsto \alpha t$. The next level of approximation to homotopy theory involves p-adic quaternion groups, which we take up now.

[It is worth remarking here that most of the complexity of the group $S(D)$ in concentrated in the pro-Sylow p-subgroup $\mathbf{Sl}(D)$ of the special linear group of D, defined as the kernel of the determinant from $D^\times$ to $\mathbf{Q}_p^\times$; indeed (analogous to the filtration of local Galois groups by ramification groups) we have a decomposition

$$D^\times = \mathbf{Z} \cdot \mu_{q-1} \cdot \mathbf{S}(D)$$

where $\mathbf{S}(D)$ is a pro-p-group, and the sequence

$$1 \to \mathbf{Sl}(D) \to \mathbf{S}(D) \overset{\det}{\to} (1 + p\mathbf{Z})^\times \to 1$$

is exact. The action of the group μ_{q-1} of roots of unity (by conjugation) on $H_c^*(\mathbf{Sl}(D))$ defines a $(q-1)$-periodic grading on the cohomology, invariant under the action of $\mathbf{Z}$ (by "Frobenius"), and it is this bigraded cohomology $H_c^{*,*}(\mathbf{Sl}(D))$ which occurs in the topological applications.]

Thus when $n = 2$ and $p > 3$, the (Poincaré duality) group $\mathbf{Sl}(D)$ has p-cohomological dimension 3, and calculating its mod p cohomology reduces to the identification of its abelianization. In fact one finds that

$$H_c^*(\mathbf{Sl}(D), \mathbf{F}_p) = \begin{array}{ll} \mathbf{F}_p & \text{in dimension } 3 \\ \mathbf{F}_{p^2} & \text{in dimension } 2 \\ \mathbf{F}_{p^2} & \text{in dimension } 1 \\ \mathbf{F}_p & \text{in dimension } 0. \end{array}$$

In homotopy theory this calculation is reflected in the structure of the stable homotopy of a space of type $V(2)$, (introduced by Larry Smith [49] and studied extensively by H. Toda (cf. [50]) and his school (cf. eg. [46])).

[It is startling how little attention has been paid to the cohomology of these groups. Ravenel [39] calculated the cohomology when $n = 3$ for large primes, and Yamagachi did the case $n = 4$ [51]; but they seem not to have received the attention they deserve from number-theorists. (For example, Ravenel shows that the mod p cohomology of $S(D)$ when $n = 2$ has Poincaré series

$$\frac{(1 + t)^2(1 + t^2)}{1 - t}$$

when $p = 3$, and

$$\frac{(1 + t)^2(1 - t^5)}{(1 - t)^2(1 + t^2)}$$

when $p = 2$.) Some beautiful things have been proved about the groups themselves by Serre [45], and Deninger and Singhoff [16] have studied related Lie algebra cohomology groups.]

Now when n is two, an element of $\mathbf{Sl}(D)$ can be written as a p-adic quaternion $a + b\sqrt{p}$ with $a, b \in W(\mathbf{F}_{p^2})$, or as a 2×2 matrix $\left[\begin{smallmatrix} a & pb \\ b^\sigma & a^\sigma \end{smallmatrix}\right]$, with $|a|^2 - p|b|^2 = 1$ (where $|a|^2 = aa^\sigma$). Thus $a \in W(\mathbf{F}_{p^2})$ is a unit, and (following Mike Hopkins [20]) we can define the analogue of the Tate algebra (in the case $n = 2$) as the algebra $W(\mathbf{F}_{p^2})[[t]]$ with the (continuous) fractional linear action

$$[a + b\sqrt{p}](t) = \frac{at + pb}{b^\sigma t + a^\sigma} \in W(\mathbf{F}_{p^2})[[t]]$$

of $\mathbf{Sl}(D)$. (Note that the "action" with p in the *lower* left hand corner is not well-defined!)

In contrast to our almost complete understanding of the cohomology of such groups with connected coefficients (cf. [8]), it seems to be difficult to say very much about $H_c^*(\mathbf{Sl}(D), W(\mathbf{F}_{p^2})[[t]])$. Indeed, while it is rather elementary to calculate the Lie algebra cohomology $H^*(\mathfrak{sl}(D), K[[t]])$, there is no very obvious way to relate it to the group cohomology with the same coefficients; and at the time of the conference, a great deal of effort was going on trying to understand the groups

$$H_c^*(\mathbf{Sl}(D), W(\mathbf{F}_{p^2})[[t]] \otimes \mathbf{Q});$$

in August, following up some hints in work of Shimomura, they were conjecturally calculated by Hopkins and Ravenel in a remarkable piece of work [22].

2. At this point it becomes impossible to proceed any further without 'digressing' into homotopy theory. One place to begin is with a question raised by Adams, at the 1973 Manifold conference [2] in Tokyo. Algebraic topologists are by now familiar with an adelization technique by which homotopy-theoretic problems are treated one prime at a time; but the existence (and great usefulness) of K-theory makes us aware that there are other ways to 'see' the homotopy category. Adams asks how much of the homotopy category is 'seen' by the system

$$E(n)_* = Z_{(p)}[v_1, \ldots, v_{n-1}, v_n, v_n^{-1}]$$

of localizations of the Brown-Peterson homology functors BP_*, cf. [23]. Since Adams's question was posed, Bousfield [11] has developed a powerful machine for the study of such questions; in particular we can now talk of the localization $L_n S^0$ of the stable sphere with respect to the homology theory $E(n)_*$ in a way which includes the classical p-adic localization (as the case $n = \infty$) and (the p-adic localization of) the part of the sphere seen by K-theory as the case $n = 1$.

Recently Devinatz and Hopkins have used Bousfield's localizations to imbed the J-homomorphism into a beautiful systematic context suggested by Ravenel [41]. There are natural maps from the nth localization of the sphere spectrum to the $(n - 1)^{\text{th}}$, and they interpret the cofibre of this map as an analogue $J(n)S^0$ of the spectrum classically called Im J, when n is one. This procedure replaces with geometry some fairly vague algebraic intuitions based on the cohomological degree filtration of the Adams-Novikov spectral sequence [[28], [32], [36], [52]]. Moreover, they construct a spectral sequence with E_2-term

$$H_c^{*,*}(\mathbf{Sl}(D), \hat{E}(n))$$

which conjecturally converges to the Pontrajagin dual of the homotopy of $J(n)S^0$; here $\hat{E}(n)$ is a completion

$$\mathbf{Z}_p[v_1, \ldots, v_{n-1}]((v_n^{-1}))$$

of the localization $E(n)_*$ of BP_* mentioned above when $n > 1$; when $n = 1$ we recover the cohomology

$$H_c^{*,*}(\mathbf{Z}_p^\times, \mathbf{Z}_p[t, t^{-1}])$$

of (a localization of) the Tate algebra, as above, and when $n = 2$ we get the cohomology of the group of quaternion units with coefficients in Hopkins's algebra (under the correspondence $v_2 \mapsto t$, $v_1 \mapsto 1$; we are suppressing here an action of $\mathbf{Z}$, which induces the Frobenius automorphism on $W(\mathbf{F}_{p^2})$ and gets us down to coefficients in $W(\mathbf{F}_p) = \mathbf{Z}_p$).

To get some idea of how this works, I recall that L_1S^0 has a (somewhat mystifying) copy of $\mathbf{Q}_p/\mathbf{Z}_p$ in dimension -2 [41]; otherwise it has $\mathbf{Z}_p$ in dimension zero, and in positive dimensions its homotopy is the p-component of Im J. (There is another copy of Im J, upside down, in negative dimensions, which pairs under multiplication into the $\mathbf{Q}_p/\mathbf{Z}_p$, which thus provides L_1S^0 with a kind of Poincaré duality). On the other hand, L_0S^0 is just a copy of $\mathbf{Q}_p$, concentrated in dimension 0; and the Devinatz-Hopkins construction for small values of n gives us an exact sequence

$$\cdots \to \pi_*L_1S^0 \to \pi_*L_0S^0 \to \pi_*J(1)S^0 \to \cdots$$

which can be conveniently displayed, modulo a finite amount of torsion in any dimension, as the table

dim 0	$\mathbf{Z}_p$	$\mathbf{Q}_p$	$\mathbf{Q}_p/\mathbf{Z}_p$
-1	0	0	$\mathbf{Q}_p/\mathbf{Z}_p$.
-2	$\mathbf{Q}_p/\mathbf{Z}_p$	0	0

That is,

$$(\pi_{-*}J(1)S^0)^\wedge \equiv \mathbf{Z}_p \text{ (mod finite torsion)}$$

if $* = 0, 1$; but this is just a reformulation of the fact that $\mathbf{Z}_p^\times$ is a p-adic circle, with cohomology an exterior algebra on a single one-dimensional generator. (Here $A^\wedge = \mathrm{Hom}(A, \mathbf{Q}/\mathbf{Z})$, for a torsion group A.)

More generally, the calculations of Hopkins and Ravenel (modulo the conjectured convergence of the spectral sequence) suggest that

$$\pi_* L_n S^0 \equiv \mathbf{Z}_p \qquad \text{in dimension } 0$$

$$\equiv \mathbf{Q}_p/\mathbf{Z}_p \qquad \text{in dimension } -2n$$

modulo a finite amount of torsion in each degree. However we at present do not know very much about this torsion; and in particular we do not know how to generalize the duality pairing in the case $n = 1$.

Nevertheless there is some reason to hope for a 'geometric' understanding of these questions. The E_2 term of the Devinatz-Hopkins spectral sequence is the cohomology of a very complicated, but by now moderately well-understood, object which is the subject of a considerable literature in arithmetic algebraic geometry: the Lubin-Tate moduli space for the liftings of one-dimensional formal groups. In fact it was known to Quillen, back in 1971, that the E_2-term of Novikov's Adams spectral sequence for the stable homotopy of spheres in terms of complex cobordism is, in the language of algebraic geometry, the (equivariant) cohomology of the modular scheme of one-dimensional formal group laws, with coefficients in the coherent structure sheaf of algebraic functions, cf. [37]. Bousfield's localizations furnish us with a technique for studying the sphere spectrum over various interesting subschemes of this modular scheme; for example the bordism theory $E(n)_*$ is just the restriction of complex cobordism, as a sheaf over the space of all one-dimensional formal group laws, to the locally closed sub-space of group laws of height $\leq n$ at p.

It is worth the trouble to state this precisely. Quillen showed that the scheme representing the functor which assigns to a commutative ring A, the set of formal one-dimensional group laws in A, can be identified with the spectrum of the complex bordism ring MU_*; this ring being coherent [47] it is also possible to interpret the complex bordism of a finite complex as a coherent sheaf over this modular scheme. We can thus think of complex bordism as a homology theory, taking values in the category of sheaves over a nice, although rather big, algebraic variety. This functor has many good properties, including a Künneth theorem for products.

Now arithmetic algebraic geometers have devoted a good deal of effort to the study of what are now called Tannakian categories, (although in the best of all possible universes they would be named after Grothendieck): these are *very roughly* [42] abelian categories with an internal hom-functor and $\otimes$, together with a faithful functor to vector spaces over a field k. A by now standard approach to the study of a cohomology functor ω (defined for example on some interesting class of algebraic varieties) is to consider the functor which assigns to the (commutative) k-algebra A, the group $\mathrm{Aut}(\omega \otimes_k A)$ of automorphism of the cohomology functor $\omega \otimes_k A$; in favorable cases it can be shown that the functor $A \mapsto \mathrm{Aut}(\omega \otimes_k A)$ is representable, i.e. by a groupscheme $\mathbf{Aut}(\omega)$. This enables us to "enrich" the cohomology functor ω, and to regard it as taking values in some category of representations of the groupscheme $\mathbf{Aut}(\omega)$. Sometimes it is even conjectured that this enriched functor maps some 'motivic' category of (pieces of) varieties fully faithfully to some category of group representations.

Now this in broad outline is quite like what is done in modern homotopy theory. For example, it is by now standard to try to understand a (co)homology functor as taking values in a category of modules or comodules over some Hopf algebra. In particular, the automorphism functor of mod 2 cohomology was identified by Atiyah and Hirzebruch in 1961 as the functor which assigns to $A \in (\mathbf{F}_2\text{-algebras})$, the group $(A\langle\langle F\rangle\rangle)^\times$ of units in the Hilbert ring $A\langle\langle F\rangle\rangle$ of formal power series in an indeterminate F satisfying $Fa = a^pF$ (with $p = 2$), cf. [4]; the affine algebra of functions on the groupscheme so defined is Milnor's dual to the Steenrod algebra. Similarly, the Hopf algebra of Landweber-Novikov cooperations in complex cobordism is the affine algebra of functions on the groupscheme which assigns to the commutative ring A, the group (under composition) of invertible formal power series over A. Thus when ω is the complex cobordism functor, we can understand $\mathbf{Aut}(\omega)$ to be the group of formal diffeomorphisms f of the line, fixing the origin, acting on the modular variety of formal group laws, sending $F(X, Y)$ to $f^{-1}(F(f(X), f(Y)))$, i.e. acting on formal group laws by changing the coordinate parameters. We can then interpret the complex cobordism functor as having its values in a Tannakian category of equivariant sheaves over this modular stack of formal groups; there is a big open orbit above each (classical) prime, and the fiber above a point of this open set is the p-completion of ordinary K-theory, while the isotropy group is the multiplicative group $\mathbf{Z}_p^\times$ of Adams operations.

There are of course some differences between the situations in algebraic geometry and homotopy theory. In homotopy theory we lack the faithful forgetful functor, but as a kind of replacement we have the Adams spectral sequence. That is, when ω is sufficiently well-behaved, e.g. in the cases of complex cobordism or ordinary (mod p) cohomology, we have a spectral sequence

$$H^*(\mathbf{Aut}(\omega); \omega(X^D \wedge Y)) \Rightarrow \pi_*(X^D \wedge Y)$$

(where X and Y are (for simplicity) finite complexes, and X^D is the S-dual of X) which calculates the stable morphisms from X to Y by what algebraic geometers might call descent, from the Tannakian category of values taken by the homology functor ω. When ω is the complex cobordism functor, we know a good deal about this value category—e.g. its stratification by orbits—and we are able, now, to decompose the homotopy category itself along lines suggested by the (essentially number-theoretic) classification of one-dimensional formal group laws.

[It is perhaps not inappropriate to remark here that while the construction of l-adic cohomology was quickly 'completed' by the construction of an l-adic homotopy theory, the construction of a p-adic homotopy theory has not followed so quickly after the development of p-adic cohomology theories for algebraic varieties. In fact at this conference, in Faltings's talk, we have begun to hear about something like a p-adic homotopy theory; and I would like to complement his talk with the suggestion that the p-adic Hodge theory of algebraic varieties may eventually be understood as a specialization of deep p-adic properties of finite complexes.]

The most convincing piece of evidence for the homotopy-theoretic significance of these 'number-theoretic' constructions comes from the recent construction by Devinatz, Hopkins, and Smith [18], of a prime ideal spectrum for the stable homotopy category. There is a very general construction of the prime ideal spectrum of a commutative ring in terms of the abelian category of its modules, by Gabriel [19]; and Hopkins has pointed out that one can still recover the prime spectrum from the derived category. Thus open sets of the prime spectrum correspond to classes of objects in the derived category which are closed under cofibrations and retractions. Constructions of Steve Mitchell [33] and Jeff Smith [47] show that the prime ideals of the homotopy category correspond precisely to the 'prime ideals' in the moduli scheme for formal groups. [It is classical that over an algebraically closed field of finite characteristic (i.e. at a geometric point of the moduli space) a one-dimensional formal group law is classified up to isomorphism by its height, which can be any positive integer. One reason that the stable homotopy category is so difficult to work with is that the primes corresponding to these geometric points are *embedded*, each in the next, in the language of algebraic geometry.] The Devinatz-Hopkins spectral sequence can be interpreted as the contribution to the stable homotopy of spheres at the nth chromatic prime above p. Its E_2-term is the cohomology of the isotropy group of the corresponding geometric point, with coefficients in the functions on the space of formal deformations of that point. A more down-to-earth description of the finite complexes supported by a single chromatic prime comes from the proof (cf. [18]) of Ravenel's nilpotence conjecture [that a stable endomorphism of a finite complex is nilpotent if and only if it induces a nilpotent map on complex cobordism]; Devinatz, Hopkins, and Smith show in fact that the algebra of stable endomorphisms of a finite complex has Krull dimension one, so that every finite complex possesses a nonnilpotent self-map. Moreover, these self-maps are in some sense in the *center* of the stable endomorphism ring, and induce a map on cobordism which is roughly multiplication by some power of v_n. Thus complexes of type $V(n)$, which were once thought to be very special, are now seen to be generic. A similar paradoxical fact is that the induction which proves the existence of a central nonnilpotent endomorphism starts with Nishida's theorem [34], which characterizes the ideal of positive-dimensional classes in the stable homotopy of spheres as a nil ideal. [One consequence is that the relevant structure space for the stable homotopy category is very different from the prime ideal spectrum of the stable homotopy ring itself.]

Because we haven't had much practical experience with the higher chromatic primes they currently appear to be rather mysterious, but there is no doubt that they are important. For example, some of the smoothings of a homotopy sphere can be constructed geometrically (from singularities, following Brieskorn); these smoothings are closely related to the homotopy of Im J. The remaining smoothings are classified by elements of coker J, and are not at all well-understood geometrically.

Fortunately we can see the chromatic primes very clearly in the cohomology of finite groups. Atiyah showed that the K-theory of a finite group is closely related to the set of its conjugacy classes, and Quillen showed that (modulo nilpotent elements) its cohomology can be constructed out of the category of its elementary abelian subgroups. We know now, from work of Hopkins, Kuhn, and Ravenel [21], that the localization of a p-group at the nth chromatic prime is determined by its conjugacy classes of commuting n-tuples of elements. Thus the filtration on the spectrum of the cohomology algebra defined by rank of the supporting elementary abelian subgroup corresponds to the chromatic filtration on homotopy. [One of the first places this filtration has appeared in 'nature' is in the role played by conjugacy classes of commuting pairs of elements in the character theory of the Fischer-Griess simple group [35], which has important connections with elliptic cohomology.] Now that Segal's conjectures have been proven, it would be very interesting to understand their relation to the chromatic filtration, along the lines suggested by Davis et al. [15].

3. After this long modulation into chromatic homotopy theory I will return to the local key in which we began, with a brief account of some properties of $D^\times$ considered as a Lie group. It is important to realize that $D^\times$ is not very connected, and that it can have several distinct sorts of maximal torus. Indeed, it is fundamental to Weil's approach [53] to local classfield theory that any extension field L of $\mathbf{Q}_p$, of degree n, can be imbedded as a maximal commutative subfield of D (or indeed in any division algebra of degree n with center $\mathbf{Q}_p$). Thus the maximal toruses $L^\times$ of $D^\times$ correspond naturally with the set of isomorphism classes of such extensions; moreover if L is a normal extension of $\mathbf{Q}_p$, its Galois group of automorphisms is the Weyl group of $L^\times$ in $D^\times$. However, when the class $[D] = 1/n \in Br(\mathbf{Q}_p)$, we can say a good deal more. The normalizer of $L^\times$ in $D^\times$ is an extension

$$1 \to L^\times \to N(L) \to \mathrm{Gal}(L/\mathbf{Q}_p) \to 1$$

classified by an element in the cohomology group $H^2(\mathrm{Gal}(L/\mathbf{Q}_p), L^\times)$, which is cyclic of order n; when $[D] = 1/n$ this class is always a generator, and is the Artin-Takagi 'fundamental class' of local classfield theory. Moreover, the (p-adic crystallographic) group $N(L)$ possesses a striking interpretation in Galois-theoretic terms: the theorem of Weil and Shafarevich identifies it as the Weil group of the maximal abelian extension L_{ab} of L, over $\mathbf{Q}_p$, cf. [27, 29]. Thus its profinite completion is $\mathrm{Gal}(L_{ab}/\mathbf{Q}_p)$. In spite of this, however, we do not seem to know anything about its cohomology (except that it is of virtual cohomological dimension n, and so forth). Even if we did, we would not be in a position to make any deductions about the cohomology of $D^\times$.

[In the classical case, one of the quickest ways to get from the cohomology of a maximal torus in a compact connected simply-connected Lie group to the cohomology of the group itself, is to use a Becker-Gottlieb transfer argument [7] based on the fact that the Euler characteristic of the normalizer of a maximal torus in such a group is necessarily one. In the p-adic case we currently have no such transfer (though recent results of Carlsson point in that direction), and I at least have no idea what the Euler characteristic of such a normalizer ought to be (other than that it might be expected to reflect the arithmetic of L in some way, if it isn't 1). This is in great contrast to the situation in representation theory, where (because of their relevance to Langland's conjectures) such groups as $D^\times$ have been extensively studied (cf. eg. [12]).]

Perhaps it would be most natural to think of the stable category as having compatible L-adic completions (with homotopy groups taking values in compact modules eg. over the valuation ring of L) indexed by nontrivially locally compact fields of characteristic zero, so that Im J is the part of QS^0 'defined over Q_p', cf. [24], [30].

Finally, in view of the interest of the set of commutative subfields of D, it may be worth noting that there is naturally associated to D a combinatorial object vaguely resembling the buildings studied lately by group theorists, i.e. the simplicial object having for its d-simplices the (totally disconnected topological) space of chains $L_0 \subset L_1 \subset \cdots \subset L_{d-1}$ of commutative proper extensions of $\mathbf{Q}_p$ in D, cf. [38]. I will stop with the hope that many mathematicians will find something interesting in the various compartments of this somewhat rambling edifice; but I should not end without thanking and acknowledging the support of the many mathematicians whose work has been expropriated here in summary form, including in particular Frank Adams, Andy Baker, Mike Boardman, Pete Bousfield, Don Davis, Ethan Devinatz, Mike Hopkins, Dave Johnson, Keith Johnson, Nick Kuhn, Peter Landweber, Goro Nishida, Haynes Miller, Larry Smith, Jeff Smith, Hirosi Toda, Doug Ravenel, Steve Wilson, Friedhelm Waldhausen, Urs Würgler, and Nobuaki Yagita. For personal support and inspiration I would like to add Michael Barratt, William Browder, Mark Mahowald, Jun-Ichi Igusa, Michio Kuga, and Takashi Ono to this list of people to whom I owe special debts.

THE JOHNS HOPKINS UNIVERSITY

REFERENCES

[1] J. F. Adams, On the groups $J(X)$, IV, *Topology*, **5** (1966), 21–71.

[2] ______, in *International Conference on Manifolds and Related Topics*, U. Tokyo Press (1975).

[3] S. A. Amitsur, Finite subgroups of division rings, *Trans. AMS*, **80** (1955), 361–386.

[4] M. F. Atiyah and F. Hirzebruch, Kohomologie Operationen und Characteristische Klassen, *Math. Zeit.*, **77** (1961), 149–187.

[5] ______ and D. O. Tall, Group representations, lambda-rings, and the *J*-homomorphism, *Topology*, **8** (1969), 253–297.

[6] H. Bass, *Lectures on Topics in Algebraic K-Theory*, notes by A. K. Roy, Tata Institute for Fundamental Research, Lectures in Mathematics, **41** (1975).

[7] J. C. Becker and D. H. Gottlieb, The transfer map and fiber bundles, *Topology*, **14** (1975), 1–12.

[8] A. Borel and N. Wallach, Continuous cohomology, discrete subgroups, and representations of reductive groups, *Ann. Math. Studies*, **94**, Princeton (1980).

[9] Z. I. Borevich and I. R. Shafarevich, *Number Theory*, Academic Press (1960).

[10] N. Bourbaki, *Algébre, Livre* II, Hermann Act. Sci. et Ind., **1291** (1962).

[11] A. K. Bousfield, The localization of spectra with respect to homology, *Topology*, **18** (1979), 257–281.

[12] C. Bushnell and A. Fröhlich, *Gauss Sums and p-Adic Division Algebras*, Lecture Notes 987, Springer (1983).

[13] H. Cartan and S. Eilenberg, *Homological Algebra*, Princeton University Press, 1956.

[14] P. Cartier, Relèvements des groupes formels commutatifs, Seminaire Bourbaki, **359** (1968–69).

[15] D. Davis, D. Johnson, J. Klippenstein, M. Mahowald, and S. Wegmann, The spectrum $(P \wedge BP\langle 2\rangle)_{-\infty}$, *Trans. AMS*, **296** (1987), 95–110.

[16] Ch. Deninger and W. Singhof, On the cohomology of nilpotent Lie algebras, *Bull. Soc. Math. France*, **116** (1988), 3–14.

[17] E. Devinatz and M. Hopkins, in preparation.

[18] ______, ______, and J. H. Smith, Nilpotence in stable homotopy I, *Ann. Math.*, **128** (1988), 207–241.

[19] P. Gabriel, Des categories abeliennes, *Bull. Soc. Math. France*, **90** (1962), 323–448.

[20] M. Hopkins, in Proceedings of the Barratt-Whitehead conference. Durham (N.H.).

[21] ______, N. Kuhn, and D. Ravenel, Complex oriented cohomology of classifying spaces, preprint (1988).

[22] ______ and D. Ravenel, in preparation.

[23] D. Johnson and W. S. Wilson, Projective dimension and Brown-Peterson homology, *Topology*, **12** (1973), 327–353.

[24] K. Johnson, The Conner-Floyd map for formal *A*-modules, *Trans AMS*, **302** (1987), 319–332.

[25] P. S. Landweber (ed.), *Elliptic Curves and Modular Forms in Algebraic Topology*, Lecture Notes 1326, Springer (1988).

[26] J. Lubin and J. Tate, Formal moduli for one-parameter formal Lie groups, *Bull. Soc. Math. France*, **94** (1966), 49–60.

[27] J. Morava, The Weil group as automorphisms of the Lubin-Tate group, *Asterisque*, **63** (1979), 169–178.

[28] ______, Noetherian localizations of categories of cobordism comodules, *Ann. Math.*, **121** (1965), 1–39.

[29] ______, Some Weil group representations motivated by algebraic topology, in [25].

[30] ______, Browder-Fröhlich symbols, in *Proc. Arcata Conference on Algebraic Topology.* Lecture Notes in Math., Springer, to appear.

[31] ______, On the complex cobordism ring as a Fock representation, *Proc. Kinosaki Conference in honor of H. Toda*, Lecture Notes, to appear.

[32] H. Miller, D. Ravenel, and W. S. Wilson, Periodic phenomena in the Adams-Novikov spectral sequence, *Ann. Math.*, **106** (1977), 469–516.

[33] S. Mitchell, Finite complexes with $A(n)$-free cohomology, *Topology*, **24** (1985), 227–246.

[34] G. Nishida, The nilpotency of elements of the stable homotopy groups of spheres, *J. Math. Soc. Japan*, **25** (1973), 707–732.

[35] S. Norton, Generalized Moonshine, (appendix to a paper of G. Mason), *Proc. Symposia in Pure Math.*, **47** (1987), 208–210.

[36] S. P. Novikov, The methods of algebraic topology from the point of view of cobordism theories, *Math. USSR (Izvestija)*, **1** (1967), 827–913.

[37] D. Quillen, Elementary proofs of some results of cobordism theory using Steenrod operations, *Adv. in Math.*, **7** (1971), 29–51.

[38] ______, Homotopy properties of posets of non-trivial p-subgroups of a group, *Adv. in Math.*, **28** (1978), 101–128.

[39] D. Ravenel, The cohomology of Morava stabilizer algebras, *Math. Zeits.*, **152** (1977), 287–297.

[40] ______, *Complex Cobordism and Stable Homotopy Groups of Spheres*, Academic Press, Orlando, Florida, 1986.

[41] ______, Localization with respect to certain periodic homology theories, *Amer. J. Math.*, **106** (1984), 351–414.

[42] N. Saavedra Rivano, *Categories Tannakiennes*, Lecture Notes 265, Springer (1972).

[43] G. Segal, Elliptic cohomology, Seminaire Bourbaki, **695**, February 1988.

[44] J. P. Serre, Groupes finis á cohomologie périodique, Seminaire Bourbaki, **209** (1960–61).

[45] ______, Un 'formule de masse' pour les extensions totalment ramifiées de degré dominé d'un corps local, Comptes Rendus, Adad. Sci. Paris, **286** (1978), 1031–1036.

[46] K. Shimomura, Nontriviality of some products of β-elements in the stable homotopy of spheres, *Hiroshima Math. J.*, **17** (1987), 349–353.

[47] J. H. Smith, Stable splitting derived from the symmetric group, to appear.

[48] L. Smith, On the finite generation of $\Omega_*^U(X)$, *J. Math. Mech.*, **18** (1968/69), 1017–1023.

[49] ______, On realizing complex bordism modules I–IV, *Amer. J. Math.*, **92** (1970), 793–856; **93** (1971), 226–263; **94** (1972), 875–890; **99** (1977), 418–436.

[50] H. Toda, On realizing exterior parts of the Steenrod algebra, *Topology*, **10** (1971), 53–65.

[51] A. Yamaguchi, thesis, Kyoto University.

[52] F. Waldhausen, Algebraic K-theory of spaces, localization, and the chromatic filtration of spaces, in *Algebraic Topology, Aarhus 1982*, Lecture Notes 1051 (1984).

[53] A. Weil, *Basic Number Theory*, Grundlehren 144, Springer (1967).

BIRATIONAL CLASSIFICATION OF ALGEBRAIC 3-FOLDS

By Shigefumi Mori

This is an announcement of my joint work with János Kollár.

1. Review of results for 3-folds. Let X be a nonsingular projective variety over $\mathbf{C}$. Then the plurigenera

$$P_\nu(X) = \dim H^0(X, \mathcal{O}(\nu K_X)) \quad (\nu > 0)$$

are birational invariants of X and hence so is the *Kodaira dimension* $\kappa(X)$ which is defined as

$$\kappa(X) = \begin{cases} -\infty & \text{if} \quad P_\nu(X) = 0 \quad \text{for all} \quad \nu > 0 \\ \kappa \text{ s.t. } \exists a, b > 0 \quad \forall \nu > 0 \quad a\nu^\kappa < P_{\nu e}(X) < b\nu^\kappa \\ \qquad \text{if} \quad P_e(X) > 0 \quad \text{for some} \quad e > 0. \end{cases}$$

This κ introduced by Iitaka and Moishezon is the most fundamental birational invariant in the birational classification of algebraic varieties. It takes the values $-\infty, 0, \ldots, \dim X$, and the cases $\dim X, 0, -\infty$ generalize curves of genus ≥ 2, 1, 0, respectively. If $\kappa(X) = \dim X$, then X is called of *general type*.

Recent results of extremal ray theory (or minimal model theory) by Benveniste, Kawamata, Kollár, Mori, Reid and Shokurov imply two important results for 3-folds

Theorem 1.1. (via Benveniste and Kawamata cf. (1.9)) *The canonical ring*

$$R(X) = \bigoplus_{\nu \geq 0} H^0(X, \mathcal{O}(\nu K_X))$$

is finitely generated if X is a 3-fold of general type.

Manuscript received August 24, 1988.

THEOREM 1.2. (via Miyaoka cf. (1.7)) *A 3-fold X has $\kappa(X) = -\infty$ iff X is uniruled, i.e. there is a surface Y and a dominating rational map from $\mathbf{P}^1 \times Y$ to X.*

These results are related to canonical models and minimal models which are defined as follows.

Definition 1.3. ([R]) Let (P, X) be a germ of a normal algebraic variety. Let $f: Y \to (P, X)$ be a resolution. We say that (P, X) is a *canonical* (resp. *terminal*) singularity if

(i) there is an integer $r > 0$ such that rK_X is a Cartier divisor (the smallest such r is called the *index* of (P, X)), and
(ii) let $f: Y \to (P, X)$ be an arbitrary resolution, and let $E_1, \ldots, E_n$ be all the exceptional divisors. Then one has $rK_Y = f^*(K_X) + \Sigma_i a_i E_i$ with all $a_i \geq 0$ (resp. $a_i > 0$).

We say that an algebraic variety X is a *canonical* (resp. *minimal*) *model* if X has only canonical (resp. terminal) singularities and K_X is ample (resp. *nef*, i.e. $(K_X \cdot C) \geq 0$ for an arbitrary irreducible curve C.) X is said to have *only* **Q**-*factorial* singularities if every (global) Weil divisor is **Q**-Cartier.

The minimal model theory says

THEOREM 1.4. *A nonsingular projective 3-fold X is birational to a projective 3-fold Y with only* **Q**-*factorial terminal singularities such that either*

(a) *K_X is nef (minimal model case), or*
(b) *Y has a morphism to a normal projective variety Z such that* $\dim Z < \dim Y$ *and $-K_Y$ is relatively ample over Z.*

In the case (b), X is uniruled by the following.

THEOREM 1.5. (Miyaoka-Mori [MM]) *Let V be a nonsingular projective variety and $C \subset V$ an irreducible curve such that $(C \cdot K_V) < 0$. Then through any point of C there is a rational curve on V.*

In the case (a), $\kappa(X) \geq 0$ by the following.

THEOREM 1.6. (Miyaoka [My1, My2, My3]) *If a 3-fold X is a minimal model, then $\kappa(X) \geq 0$.*

Putting these together, one has

THEOREM 1.7. *For a nonsingular projective 3-fold X, the following are equivalent*

(a) $\kappa(X) \geq 0$,
(b) *X is birational to a minimal model*,
(c) *X is not uniruled*.

The application to varieties of general type is due to

THEOREM 1.8. (Benveniste [B], Kawamata [Ka1, KMM]) *If a variety X of general type is birational to a minimal model, then X is birational to a canonical model, or equivalently the canonical ring of X is finitely generated.*

Thus follows

THEOREM 1.9. *A 3-fold of general type has a finitely generated canonical ring.*

2. Main results and corollaries. We are concerned with two basic problems. The approach via $\kappa(X)$ necessarily raises the question whether $\kappa(X)$ is deformation invariant. Because of the complexity of 3-dimensional birational geometry, the existence of the coarse moduli space for

$$\{\text{variety of general type}\}/\text{birational equivalence}$$

is not clear at all. It is pointed out by Nakayama [N] that the minimal model theory for 4-folds settles the first problem.

In short, our main result is the minimal model theory not for 4-folds but for a 1-parameter family of 3-folds with only $\mathbf{Q}$-factorial terminal singularities. However this settles the above problems reasonably well.

Let $f : X \to \bar{\Delta}$ be a flat projective morphism of an analytic space X to a closed 1-dimensional unit disk

$$\bar{\Delta} = \{z \in \mathbf{C} \mid |z| \leq 1\}$$

(to be precise, to $\{z \in \mathbf{C} \mid |z| < 1 + \epsilon\}$ for some $\epsilon > 0$) such that each fiber X_t is a 3-fold with only $\mathbf{Q}$-factorial terminal singularities.

Then we prove that the minimal model theory for such $X/\bar{\Delta}$ can be performed and as a result we have

THEOREM 2.1. *There exists another flat projective morphism of an analytic space Y to $\bar{\Delta}$ such that each fiber Y_t is a 3-fold with only $\mathbf{Q}$-factorial terminal singularities with properties*:

(a) *there exists a $\bar{\Delta}$-birational mappings from Y to X inducing birational mappings from Y_t to X_t for all $t \in \bar{\Delta}$,*

(b) *either all Y_t are minimal models or all Y_t are uniruled.*

This means, in short, that minimal models vary continuously. We use a slight modification of [L] for deformation invariance of $\kappa(X)$ and $P_\nu(X)$. Since the abundance conjecture [KMM] is not completely proved for 3-folds, certain cases (cf. (2.3)) are left unsettled.

THEOREM 2.2. *Under the above notation, $\kappa(X_t) = \kappa(X_0)$ for all $t \in \bar{\Delta}$.*

THEOREM 2.3. *If $\kappa(X_0) \neq 0$, then $P_\nu(X_t) = P_\nu(X_0)$ for all $\nu > 0$ and all $t \in \bar{\Delta}$.*

To state the application to birational moduli, we need to consider deformations of restricted type for canonical singularities.

PROPOSITION-DEFINITION 2.4. *Let X_0 be a projective 3-fold with only canonical singularities and let $f_0 : Y_0 \to X_0$ be a projective $\mathbf{Q}$-factorial terminal modification such that*

$$K_{Y_0} = f_0{}^* K_{X_0}.$$

Then there is a natural map of infinitesimal deformation spaces

$$\mathrm{Def}(Y_0) \to \mathrm{Def}(X_0)$$

and it is finite. Furthermore the subspace

$$\mathrm{Im}(\mathrm{Def}(Y_0) \to \mathrm{Def}(X_0))$$

is independent of choice of Y_0. It is called the SFCT subspace of $\mathrm{Def}(X_0)$. *A flat family of 3-folds X/S satisfies SFCT if for every $s \in S$ the image of the natural map $(s, S) \to$ $\mathrm{Def}(X_s)$ lies in the SFCT subspace of* $\mathrm{Def}(X_s)$.

Now birational moduli should represent the following functor.

Definition 2.5. Fix a function $P(k)$ for $k > 0$ (the Hilbert function). Let $\mathfrak{M}_P$ be the functor

$$\mathfrak{M}_P(S) = \left\{ \begin{matrix} \text{proper flat family } X/S \quad \text{s.t.} \quad X_s \text{ is a canonical} \\ \text{3-fold and } P_k(X_s) = P(k) \forall s, \text{ satisfying } SFCT \end{matrix} \right\}.$$

Here is the application to the birational moduli.

THEOREM 2.6. (i) *For every $P(k)$ the functor $\mathfrak{M}_P$ is coarsely represented by a separated algebraic space M_P of finite type.*

(ii) *Let Y/S be a flat projective family of algebraic 3-folds of general type with only $\mathbf{Q}$-factorial terminal singularities and assume that S is connected. For some $s \in S$, let $P(k) = P_k(Y_s)$. Then there is a unique morphism $f: S \to M_P$ such that for every $s \in S$, $f(s)$ is the moduli point of the canonical model of Y_s.*

NAGOYA UNIVERSITY

REFERENCES

[B] X. Benveniste, Sur l'anneau canonique de certaines variétés de dimension 3, *Inv. Math.* **73** (1983), 157–164.

[Ka1] Y. Kawamata, On the finiteness of generators of a pluri-canonical ring for a 3-fold of general type, *Am. J. Math.* **106** (1984), 1503–1512.

[Ka2] ______, The crepant blowing-ups of 3-dimensional canonical singularities and its application to degenerations of surfaces, *Ann. of Math.* **127** (1988) 93–163.

[KMM] ______, K. Matsuda and K. Matsuki, Introduction to the minimal model problem, Proc. Sympos. Algebraic Geom., Sendai 1985, *Adv. in Pure Math.* **10** (1987), 293–344.

[L] M. Levine, Pluri-canonical divisors on Kähler manifolds, *Invent. Math.* **74** (1983) 293–303.

[My1] Y. Miyaoka, Deformations of a morphism along a foliation and a birational criterion of conic bundles, *Proc. Sympos. Pure Math.* (Summer Research Inst., Bowdoin, 1985) **46** (1987) 245–268.

[My2] ______, The pseudo-effectivity of $3c_2 - c_1^2$ for threefolds with numerically effective canonical classes, *Proc. Sympos. Algebraic Geom.*, Sendai 1985, *Adv. in Pure Math.* **10** (1987), 449–476.

[My3] ______, On the Kodaira dimension of Minimal Threefolds, Preprint.

[MM] Y. Miyaoka and S. Mori, A numerical criterion of uniruledness, *Ann. Math.* **124** (1986) 65–69.

[Mo] S. Mori, Flip theorem and the existence of minimal models for 3-folds, *J. of AMS*, **1** (1988) 117–253.

[N] N. Nakayama, Invariance of the plurigenera of algebraic varieties under minimal model conjectures, *Topology* **25** (1986) 237–251.

[R] M. Reid, Minimal models of canonical 3-folds, in *Algebraic Varieties and Analytic Varieties*, S. Iitaka ed., Advanced Studies in Pure Math. **1** (1983), Kinokuniya Tokyo, and North-Holland Amsterdam, 131–180.

[Sh] V. V. Shokurov, The nonvanishing theorem, *Izv. Akad. Nauk, SSSR, Ser. Matem.* **49** (1985), 635–651, *Math. USSR, Izv.*, **19** (1985).

ASYMPTOTIC BEHAVIOR OF FLENSTED-JENSEN'S SPHERICAL TRACE FUNCTIONS WITH RESPECT TO SPECTRAL PARAMETERS

By Toshio Oshima

---

0. Introduction. Let G be a connected real semisimple Lie group with finite center, σ be an involutive automorphism of G and H be an open subgroup of the fixed point group of σ. Then the homogeneous space G/H is called a semisimple symmetric space. Fix a σ-stable maximal compact subgroup K of G. Then under the assumption

$$\text{rank}(G/H) = \text{rank}(K/K \cap H), \tag{0.1}$$

Flensted-Jensen proves in [FJ] that there exist countably many discrete series for G/H, which are, by definition, equivalence classes of irreducible unitary representations of G realized in $L^2(G/H)$.

In [FJ] a certain function ψ_λ is introduced by the analytic continuation of the Poisson integral of a certain measure on a boundary of the noncompact riemannian form G^d/K^d of G/H, which is a K-finite simultaneous eigenfunction of the invariant differential operators on G/H. Here the parameter λ corresponds to the eigenvalue and also to the infinitesimal character of the representation generated by the function ψ_λ on G/H. Then [FJ] proves that if λ is sufficiently regular, ψ_λ is square integrable on G/H under the condition (0.1) and thus we have discrete series for G/H.

In this paper we will proves an estimate for ψ_λ when λ tends to infinity. Denoting by C the normalizer of $K \cap H$ in K, Theorem 2.1 says that if x does not belong to the set CH/H, $\psi_\lambda(x)$ converges to 0 when λ tends to infinity under a certain condition. Here we remark that $|\psi_\lambda(x)| = 1$ if $x \in CH/H$.

Suppose Γ is a torsion free cocompact discrete subgroup of G such that $\Gamma \cap H$ is also cocompact subgroup of H. In [T-W] it is shown that the left translations of a certain cohomology class in $H^R(K\backslash G/\Gamma,\ E)$ cor-

---

Manuscript received 22 November 1988.

responds to a discrete series for G/H. Here E is a locally constant bundle over $K\backslash G/\Gamma$ defined through a certain harmonic form on $K\backslash G/\Gamma$. In [T-W], it is essential to consider the infinite sum

$$\sum_{\gamma\in\Gamma/\Gamma\cap H} \psi_\lambda(x\gamma), \tag{0.2}$$

which converges if $\psi_\lambda \in L^1(G/H)$. In fact the sum (0.2) converges when λ is sufficiently regular. On the other hand, the estimate obtained in this paper assures that the sum does not vanish for a sufficiently regular λ. This implies that the corresponding representation of G belonging to the discrete series for G/H is realized in $L^2(G/\Gamma)$. The estimate also gives a better condition for Γ in the main result in [T-W, Section 6].

Suppose G is a compact semisimple Lie group. Then Flensted-Jensen's spherical trace function ψ_λ is nothing but a zonal spherical function on the compact symmetric space G/H. We first consider the estimate in this case.

1. Zonal spherical functions. Let G_c be a simply connected and connected complex semisimple Lie group, θ be a holomorphic involutive automorphism of G_c and U be a θ-stable compact real form of G_c. We denote by $\mathfrak{g}_c$ and $\mathfrak{u}$ the Lie algebras of G_c and U, respectively. Let $\mathfrak{u} = \mathfrak{k} + \sqrt{-1}\mathfrak{p}$ be the decomposition of $\mathfrak{u}$ into $+1$ or -1 eigenspaces for the induced involution, which will be also denoted by θ. Let define a real form $\mathfrak{g}$ of $\mathfrak{g}_c$ by $\mathfrak{g} = \mathfrak{k} + \mathfrak{p}$. Let $\mathfrak{a}_\mathfrak{p}$ be a maximal abelian subspace of $\mathfrak{p}$ and $\mathfrak{t}$ be a maximal abelian subspace of the centralizer of $\mathfrak{a}$ in $\mathfrak{k}$. Then $\mathfrak{j} = \mathfrak{t} + \mathfrak{a}_\mathfrak{p}$ is a Cartan subalgebra of $\mathfrak{g}$. Let Σ and $\Sigma(\mathfrak{j})$ be the root systems for the pairs $(\mathfrak{g}, \mathfrak{a})$ and $(\mathfrak{g}_c, \mathfrak{j}_c)$, respectively. Here $\mathfrak{j}_c$ denotes the complexification of $\mathfrak{j}$ in $\mathfrak{g}_c$. We fix their compatible positive systems Σ^+ and $\Sigma(\mathfrak{j})^+$. Let G, K and $A_\mathfrak{p}$ be the analytic subgroups of G_c with the Lie algebras $\mathfrak{g}$, $\mathfrak{k}$ and $\mathfrak{a}_\mathfrak{p}$, respectively, $G = KA_\mathfrak{p}N$ be the Iwasawa decomposition of G corresponding to Σ^+, $\mathfrak{m}$ be the centralizer of $\mathfrak{a}_\mathfrak{p}$ in $\mathfrak{k}$ and $\mathfrak{n}$ be the Lie algebra of N. Let $\mathfrak{j}_c^*$ and $(\mathfrak{a}_\mathfrak{p})_c^*$ be the complexifications of the duals $\mathfrak{j}^*$ and $\mathfrak{a}_\mathfrak{p}^*$ of $\mathfrak{j}$ and $\mathfrak{a}_\mathfrak{p}$, respectively. We identify $\mathfrak{j}_c^*$ with $\mathfrak{j}_c$ by the Killing form $\langle\ ,\ \rangle$ of $\mathfrak{g}_c$ and therefore we identify $(\mathfrak{a}_\mathfrak{p})_c^*$ and $\mathfrak{a}_\mathfrak{p}^*$ with subspaces of $\mathfrak{j}_c^*$.

Let $(U/K)^\wedge$ be the set of equivalence classes of the irreducible unitary representations of U with nontrivial K-fixed vectors. Then $(U/K)^\wedge$ is naturally identified with the set of equivalence classes of the finite dimensional irreducible holomorphic representations of G_c with nontrivial K-fixed vectors. Moreover owing to Cartan and Helgason [Wa, Theorem 3.3.1.1], the set of highest weights of the representations belonging to $(U/K)^\wedge$ is given by

$$L = \left\{ \Lambda \in \mathfrak{a}_{\mathfrak{p}}^*; \ \frac{\langle \Lambda, \alpha \rangle}{\langle \alpha, \alpha \rangle} \in \{0, 1, 2, 3, \ldots\} \text{ for any } \alpha \in \Sigma^+ \right\}.$$

Let $\mathfrak{k}_c$ be the complexification of $\mathfrak{k}$ and K_c be the corresponding analytic subgroup of G_c. We denote by $\mathcal{O}(K_c \backslash G_c)$ the space of holomorphic functions on $K_c \backslash G_c$. Then $\mathcal{O}(K_c \backslash G_c)$ is a G_c-module through the right regular representation and it is known by a theorem due to Cartan that any element Λ in $(U/K)^\wedge$ can be realized in a unique subspace V_Λ of $\mathcal{O}(K_c \backslash G_c)$. We denote by (π_Λ, V_Λ) the corresponding holomorphic representation of G_c in V_Λ and by f_Λ a highest weight vector of the representation with respect to the positive system $\Sigma(\mathfrak{j})^+$. This implies

$$f_\Lambda(kg(\exp X)n) = f_\Lambda(g) \exp \Lambda(X) \tag{1.1}$$

for any $k \in K$, $g \in G_c$, $X \in \mathfrak{a}_{\mathfrak{p}}$ and $n \in N$. Since $\mathfrak{g} = \mathfrak{k} + \mathfrak{a}_{\mathfrak{p}} + \mathfrak{n}$ and f_Λ is holomorphic, we can normalize f_Λ by

$$f_\Lambda(e) = 1. \tag{1.2}$$

Let $N_U(K)$ be the normalizer of K in U. Note that $K \backslash N_U(K)$ is a finite group. Then we have the following

LEMMA 1.1.

$$|f_\Lambda(u)| \leq 1 \quad \textit{for any} \quad u \in U \tag{1.3}$$

and

$$|f_\Lambda(u)| = 1 \quad \textit{for any} \quad u \in N_U(K). \tag{1.4}$$

Moreover if Λ *is regular (i.e.* $\langle \Lambda, \alpha \rangle \neq 0$ *for any* $\alpha \in \Sigma$*), then for any* $u \in U - N_U(K)$ *satisfying* $|f_\Lambda(u)| = 1$*, there exists an element* $\tilde{X} \in \mathfrak{k}$ *with*

$$\left. \frac{\partial^2}{\partial t^2} |f_\Lambda(u \exp(t\tilde{X}))|^2 \right|_{t=0} < 0. \tag{1.5}$$

Proof. Fix a Hermitian inner product (,) on V_Λ so that $\pi_{\Lambda|U}$ is unitary and that it satisfies $(f_\Lambda, f_\Lambda) = 1$. Then $\pi_\Lambda(Z)$ is skew Hermitian for any $Z \in \mathfrak{u}$ and hence for $X \in \mathfrak{k}$, $Y \in \mathfrak{p}$, $u \in V_\Lambda$ and $v \in V_\Lambda$, we have

$$(u, \pi_\Lambda(X + Y)v) = (u, \pi_\Lambda(X)v - \sqrt{-1}\pi_\Lambda(\sqrt{-1}Y)v)$$

$$= (\pi_\Lambda(-X + Y)u, v) = (\pi_\Lambda(-\theta(X + Y))u, v),$$

which implies $(u, \pi_\Lambda(g)v) = (\pi_\Lambda(\theta(g^{-1}))u, v)$ for $g \in G$. Thus for $g \in G$, $f_\Lambda^2(g) = (\pi_\Lambda(g)f_\Lambda, \pi_\Lambda(g)f_\Lambda) = (\pi_\Lambda(\theta(g^{-1})g)f_\Lambda, f_\Lambda)$. Since f_Λ is a holomorphic function, we have

$$f_\Lambda^2(g) = (\pi_\Lambda(\theta(g^{-1})g)f_\Lambda, f_\Lambda) \quad \text{for any} \quad g \in G_c. \tag{1.6}$$

This equation proves (1.3) because $\theta(u^{-1})u \in U$ for $u \in U$. Moreover (1.4) follows from the fact that the map $N_U(K) \ni u \mapsto f_\Lambda(u)$ induces a character of the finite group $K \backslash N_U(K)$.

Let u be an element in $U - N_U(K)$ with $|f_\Lambda(u)| = 1$. Then

$$\pi_\Lambda(\theta(u^{-1})u)f_\Lambda = Cf_\Lambda \tag{1.7}$$

with a complex number C satisfying $|C| = 1$. Since the space $\{X + \theta(X);\ X \in \mathfrak{n}\}$ generates $\mathfrak{k}$ as a Lie algebra, there exists $X \in \mathfrak{n}$ which satisfies $\mathrm{Ad}(u)(X + \theta(X)) \notin \mathfrak{k}$. Put

$$u' = \theta(u^{-1})u \quad \text{and} \quad \tilde{X} = X + \theta(X).$$

Then $\theta(\mathrm{Ad}(u)\tilde{X}) \neq \mathrm{Ad}(u)\tilde{X}$ and hence

$$\mathrm{Ad}(u')\tilde{X} \neq \tilde{X}. \tag{1.8}$$

Now we put

$$\pi_\Lambda(\mathrm{Ad}(u')\tilde{X} - \tilde{X})f_\Lambda = C'f_\Lambda + v$$

with suitable $C' \in \mathbf{C}$ and $v \in V_\Lambda$ satisfying $(f_\Lambda, v) = 0$. Suppose $v = 0$. Since Λ is regular, the element $\mathrm{Ad}(u')\tilde{X} - \tilde{X}$ of $\mathfrak{k}$ belongs to $\mathfrak{m} + \sqrt{-1}\mathfrak{a}_\mathfrak{p}$ and $\langle \mathrm{Ad}(u')\tilde{X} - \tilde{X}, \tilde{X} \rangle = 0$ and therefore

$$\langle \mathrm{Ad}(u')\tilde{X} - \tilde{X}, \mathrm{Ad}(u')\tilde{X} - \tilde{X} \rangle = \langle \mathrm{Ad}(u')\tilde{X}, \mathrm{Ad}(u')\tilde{X} \rangle - \langle \tilde{X}, \tilde{X} \rangle$$

$$= 0,$$

which contradicts (1.8).

Thus we can conclude that $v \neq 0$. For $t \in \mathbf{R}$, put

$$\pi_\Lambda(\exp(-t\tilde{X})\exp(t\,\mathrm{Ad}(u')\tilde{X}))f_\Lambda = c(t)f_\Lambda + tv(t).$$

Here $c(t) \in \mathbf{C}$ and $v(t) \in V_\Lambda$ with $(v(t), f_\Lambda) = 0$ and $v(0) = v$. Since $|f_\Lambda(u' \exp t\tilde{X})|^4 = |c(t)|^2$ and since $|c(t)|^2 + t^2(v(t), v(t)) = 1$, we have the last claim in the lemma. ∎

The zonal spherical functions ϕ_Λ on the simply connected compact symmetric space U/K are parametrized by the elements Λ of L and are given by

$$\phi_\Lambda(g) = \int_K f_\Lambda(gk)dk \tag{1.9}$$

with the normalized Haar measure dk on K. They have the following asymptotic for Λ:

PROPOSITION 1.2.

$$|\phi_\Lambda(u)| = 1 \quad \textit{for any} \quad u \in N_U(K). \tag{1.10}$$

Conversely for any compact subset V of $U - N_U(K)$, there exists a positive number C_V such that

$$|\phi_\Lambda(u)| \leq (1 + C_V \min_{\alpha\in\Sigma^+} \langle \Lambda, \alpha \rangle^{1/2})^{-1} \quad \textit{for any} \quad u \in V. \tag{1.11}$$

Proof. Since K is a normal subgroup of $N_U(K)$, $f_\Lambda(gk) = f_\Lambda(g)$ for any $g \in N_U(K)$ and $k \in K$ and therefore we have (1.10) from (1.9) and Lemma 1.1.

Fix a regular $\Lambda_0 \in L$ and put $m(\Lambda) = \min_{\alpha\in\Sigma^+} \langle \Lambda, \alpha \rangle$. Since

$$f_{\Lambda+\Lambda'} = f_\Lambda f_{\Lambda'} \quad \text{for} \quad \Lambda, \Lambda' \in L,$$

we have a positive constant C with $|f_\Lambda(g)| \leq |f_{\Lambda_0}(g)|^{Cm(\Lambda)}$. Here we remark the existence of a positive number C' which satisfies $m(\Lambda) \geq C'$ for any regular $\Lambda \in L$. Then dividing K into a finite number of sufficiently small disjoint open subsets V_i with $\int_{K-\cup V_i} dk = 0$ and applying the following lemma to the function $|f_{\Lambda_0}(g)|^2$, we have the proposition by Lemma 1.1. ■

LEMMA 1.3. *Let $f(x)$ be a real valued C^∞-function defined on a neighborhood of the origin of $\mathbf{R}^n$ which satisfies $0 \leq f(x) \leq 1$. Suppose $f(0) < 1$ or $\frac{\partial^2 f}{\partial x_1^2}(0) < 0$. Then for any sufficiently small connected open neighborhood V of the origin, there exists a positive number ϵ such that*

$$\int_V f(x)^\lambda dx \leq \left(1 + \epsilon \frac{\lambda}{1 + \lambda^{1/2}}\right)^{-1} \int_V dx$$

for any positive real number λ.

Proof. We may assume $f(0) = 1$ because the lemma is clear if $f(0) < 1$. Denoting $x' = (x_2, \ldots, x_n)$, there exist C^∞-functions $a(x)$, $b(x')$ and $c(x')$ such that

$$f(x)^{-1} = 1 + a(x)(x_1^2 + 2b(x')x_1 + c(x')).$$

in a neighborhood of the origin. Here the assumption implies $a(0) > 0$ and $c(x') \geq 0$. Replacing x_1 by $x_1 - b(x')$, we may moreover assume $b(x') = 0$. For a small connected open neighborhood V of the origin, we can choose a positive number C which satisfies $a(x) \geq C$ if $x \in V$.

Let $v(x_1)$ be the volume of the open set $\{x' \in \mathbf{R}^{n-1}; (x_1, x') \in V\}$ in $\mathbf{R}^{n-1}$. Put $t(x_1) = \int_0^{x_1} v(x_1)dx_1$, $t_+ = t(\infty)$ and $t_- = t(-\infty)$. Then $t_+ > 0$, $t_- < 0$, $\int_V dx = t_+ - t_-$ and there exists a positive number C' with $Cx_1^2 \geq C't(x_1)^2$. Thus we have

$$\begin{aligned}\int_V f(x)^\lambda &\leq \int_V (1 + Cx_1^2)^{-\lambda} dx \\ &= \int v(x_1)(1 + Cx_1^2)^{-\lambda} dx_1 \\ &\leq \int_{t_-}^{t_+} (1 + C'\lambda t^2)^{-1} dt\end{aligned}$$

$$= \frac{1}{\sqrt{C'\lambda}} (\arctan \sqrt{C'\lambda} t_+ - \arctan \sqrt{C'\lambda} t_-).$$

Since $s^{-1} \arctan s \leq (1 + C'' \frac{s^2}{1+s})^{-1}$ for all $s > 0$ with a suitable $C'' > 0$, the lemma is clear. ■

2. Flensted-Jensen's spherical trace functions. Retain the notation in Section 1. Let σ be an involutive automorphism of G which commutes with θ, $\mathfrak{g} = \mathfrak{h} + \mathfrak{q}$ be the decomposition of $\mathfrak{g}$ into $+1$ and -1 eigenspaces of the induced involution denoted also by σ and H be the analytic subgroup of G with the Lie algebra $\mathfrak{h}$. Fix a maximal abelian subspace $\mathfrak{a}$ of $\mathfrak{p} \cap \mathfrak{q}$. In this section we assume

$$\text{rank}(G/K) = \text{rank}(H/H \cap K), \tag{2.1}$$

which assures that we can choose $\mathfrak{a}_\mathfrak{p} \subset \mathfrak{h} \cap \mathfrak{p}$. For an element $\lambda \in (\mathfrak{a}_\mathfrak{p})_c^*$ Flensted-Jensen [FJ] defined a function

$$\psi_\lambda(x) = \int_{K \cap H} e^{\langle -\lambda - \rho, H(xk) \rangle} dk \tag{2.2}$$

on G. Here dk is the normalized Haar measure on $K \cap H$, ρ is half the sum of roots in Σ^+ counted with multiplicity and the element $H(x)$ in $\mathfrak{a}_\mathfrak{p}$ is defined by the Iwasawa projection with $\exp H(x) \in KxN$. Let $\Sigma(\mathfrak{h}, \mathfrak{a}_\mathfrak{p})$ be the root system for the pair $(\mathfrak{h}, \mathfrak{a}_\mathfrak{p})$, $\Sigma(\mathfrak{h}, \mathfrak{a}_\mathfrak{p})^+$ be its positive system with $\Sigma(\mathfrak{h}, \mathfrak{a}_\mathfrak{p})^+ \subset \Sigma^+$ and ρ_t be half the sum of roots in $\Sigma(\mathfrak{h}, \mathfrak{a}_\mathfrak{p})^+$ counted with multiplicity in $\mathfrak{h}$. Then [FJ, (3.13)] says

$$\psi_\lambda(\ell \exp(X)) = \int_{K \cap H} e^{\langle -\lambda - \rho, H(\exp(-X)k) \rangle} e^{\langle \lambda + \rho - 2\rho_t, H(\ell k) \rangle} dk \tag{2.3}$$

for $\ell \in H$ and $X \in \mathfrak{p} \cap \mathfrak{q}$.

Put $\mathfrak{k}^d = \mathfrak{k} \cap \mathfrak{h} + \sqrt{-1}(\mathfrak{p} \cap \mathfrak{h})$, $\mathfrak{h}^d = \mathfrak{k} \cap \mathfrak{h} + \sqrt{-1}(\mathfrak{k} \cap \mathfrak{q})$ and $\mathfrak{g}^d = \mathfrak{k}^d + \sqrt{-1}(\mathfrak{k} \cap \mathfrak{q}) + \mathfrak{p} \cap \mathfrak{q}$. Let G^d be the simply connected Lie group with the Lie algebra $\mathfrak{g}^d$ and let K^d and H^d be the analytic subgroups of G^d with the Lie algebras $\mathfrak{k}^d$ and $\mathfrak{h}^d$, respectively. Then K^d is a maximal compact subgroup of G^d modulo center of G^d. Denoting

(2.4) $L' = \{\Lambda \in \mathfrak{a}_{\mathfrak{p}}^*;\ \Lambda$ coincides with the highest weight of an irreducible unitary representation of K^d with a nontrivial $K^d \cap H^d$-fixed vector$\}$

under the ordering defined by $\Sigma(\mathfrak{h}, \mathfrak{a}_{\mathfrak{p}})^+$, we assume that $\lambda \in (\mathfrak{a}_{\mathfrak{p}})_c^*$ satisfies

$$\langle \lambda + \rho, \alpha \rangle \geq 0 \quad \text{for} \quad \alpha \in \Sigma^+ \tag{2.5}$$

and

$$\lambda + \rho - 2\rho_t \in L'. \tag{2.6}$$

The condition (2.6) assures that the function

$$\mathfrak{h} \ni Y \mapsto e^{\langle \lambda + \rho - 2\rho_t, H(\exp(Y)) \rangle}$$

can be holomorphically extended on the complexification of $\mathfrak{h}$ and defines a function on K^d. Since $G^d = K^d \exp(\mathfrak{a}) H^d$, we define a function $\psi_\lambda(gH^d)$ on the semisimple symmetric space G^d/H^d by the right hand of (2.3) with $\ell \in K^d$ and $X \in \mathfrak{a}$.

Then [FJ] proves that $\psi_\lambda(gH^d)$ is a simultaneous eigenfunction of the invariant differential operators on G^d/H^d and moreover that it is square integrable modulo the center of G^d with respect to the invariant measure on G^d/H^d if λ is sufficiently regular. Moreover [MO] proves that the condition $\langle \lambda, \alpha \rangle > 0$ for $\alpha \in \Sigma^+$ with (2.6) is equivalent to the square integrability. To have an estimate of this function when $\langle \lambda, \alpha \rangle$ tends to infinity for all $\alpha \in \Sigma^+$, we review the argument in [FJ].

Let $\{\alpha_1, \ldots, \alpha_n\}$ be the fundamental system for Σ^+. Then it follows from [FJ, Lemma 4.6 and Lemma 4.7] that there exist finite number of real analytic functions $b_{ij}(k)$ on $K \cap H$ and elements μ_{ij} of $\mathfrak{a}_{\mathfrak{p}}^*$ ($i = 1, \ldots, d_i$ and $j = 1, \ldots, n$) such that for $X \in \mathfrak{a}$ and $k \in K \cap H$

$$e^{\langle \lambda + \rho, H(\exp(X)k) \rangle} = \prod_{j=1}^{n} \left(\sum_{i=1}^{d_j} |b_{ij}(k)|^2 \cosh \langle \mu_{ij}, X \rangle \right)^{\langle \lambda + \rho, \alpha_j \rangle / 2\langle \alpha_j, \alpha_j \rangle}, \tag{2.7}$$

$$\sum_{i=1}^{d_j} |b_{ij}(k)|^2 = 1 \quad \text{for} \quad j = 1, \ldots, n$$

and

$$(2.8)\quad e^{\langle\lambda+\rho, H(\exp(X)k)\rangle} > 1 \quad \text{if} \quad X \neq 0$$

$$\text{and} \quad \langle\lambda + \rho, \alpha\rangle > 0 \quad \text{for} \quad \alpha \in \Sigma^+.$$

Since $K \cap H$ is compact, we can conclude the existence of a positive number ϵ satisfying

$$(2.9)\qquad e^{\langle-\lambda-\rho, H(\exp(-X)k)\rangle} \leq (\cosh(\epsilon \min_{\alpha\in\Sigma^+}\langle\lambda + \rho, \alpha\rangle \|X\|))^{-1}$$

for $X \in \mathfrak{a}$ and $k \in K \cap H$ by denoting $\|X\| = \langle X, X\rangle^{1/2}$.

Let $\mathfrak{k}_0^d$ be the center of $\mathfrak{k}^d$ and put $\mathfrak{k}_r^d = [\mathfrak{k}^d, \mathfrak{k}^d]$. Let K_0^d and K_r^d be the analytic subgroups of K^d with the Lie algebras $\mathfrak{k}_0^r$ and $\mathfrak{k}_0^d$, respectively. Then $K^d = K_0^d K_r^d$ and for $k_0 \in K_0^d$ and $k_r \in K_r^d$

$$e^{\langle\lambda+\rho-2\rho_t, H(k_0k_r)\rangle} = e^{\langle\lambda+\rho-2\rho_t, H(k_0)\rangle}e^{\langle\lambda+\rho-2\rho_t, H(k_r)\rangle}.$$

Since $|e^{\langle\lambda+\rho-2\rho_t, H(k_0)\rangle}| = 1$, by applying the argument of the proof of Proposition 1.2 to the function

$$K_r^d \ni k_r \mapsto e^{\langle\lambda+\rho-2\rho_t, H(k_r)\rangle}$$

we have the following

THEOREM 2.1. *Under the above notation we can find a positive number ϵ which implies the following:*

Let λ be an element satisfying (2.5) *and* (2.6). *Then*

$$|\psi_\lambda(k \exp(X)H^d)| \leq (\cosh(\epsilon\|X\| \min_{\alpha\in\Sigma^+}\langle\lambda + \rho, \alpha\rangle))^{-1}$$

for $k \in K^d$ and $X \in \mathfrak{a}$.

Let A be the normalizer of $K^d \cap H^d$ in K^d. Then for any compact subset V of $K^d - A$ there exists a positive number C_V such that

$$(2.10)\quad |\psi_\lambda(k_0k \exp(X)H^d)| \leq (\cosh(\epsilon\|X\| \min_{\alpha\in\Sigma^+}\langle\lambda + \rho, \alpha\rangle))^{-1}$$

$$\cdot(1 + C_V \min_{\alpha\in\Sigma(\mathfrak{h},\mathfrak{a}_\mathfrak{p})^+}\langle\lambda + \rho - 2\rho_t, \alpha\rangle^{1/2})^{-1}$$

for $k_0 \in K_0^d$, $k \in V$ and $X \in \mathfrak{a}$.

We will give an application of the above theorem. Fix a subgroup Z of G^d contained in the center of G^d so that K^d/Z is compact and put $G' = G^d/Z$, $K' = K^d/Z$ and $H' = H^d Z/Z$. Suppose $\lambda + \rho - 2\rho_t$ equals the highest weight of a certain irreducible unitary representation of K' with a nontrivial $K' \cap H'$-fixed vector. Then the function ψ_λ defines a square integrable function on G'/H' if $\langle \lambda, \alpha \rangle > 0$ for $\alpha \in \Sigma^+$.

Let Γ be a discrete subgroup of G' so that

(2.11)

We denote by $Z(\mathfrak{g})$ the center of the universal enveloping algebra of $\mathfrak{g}_c$ and by $N_{K'}(K' \cap H')$ the normalizer of $K' \cap H'$ in K'. Then [T-W, Lemma 3.7] says that if a function $\psi \in L^1(G'/H')$ is $Z(\mathfrak{g})$-finite and left K'-finite, the series

$$\sum_{\gamma \in \Gamma/\Gamma \cap H'} \psi(g\gamma) \tag{2.12}$$

converges absolutely and uniformly on compact subsets of G' and defines a function on G'/Γ.

Applying this to the function ψ_λ, we can conclude the existence of a positive number M such that the series

$$\phi_\lambda(g) = \sum_{\gamma \in \Gamma/\Gamma \cap H'} \psi_\lambda(g\gamma) \tag{2.13}$$

converges absolutely and uniformly on every compact subset of G' and defines a nonzero function ϕ_λ on G'/Γ if

$$\Gamma \cap N_{K'}(K' \cap H')H' \subset H' \tag{2.14}$$

and

$$\langle \lambda, \alpha \rangle \geq M \quad \text{for} \quad \alpha \in \Sigma^+. \tag{2.15}$$

In fact, if λ satisfies (2.15) for a sufficiently large M, then it follows from Theorem 2.1 that $\phi_\lambda \in L^1(G'/H')$ (cf. [O, Corollary 1.3] or [T-W, Proposition 2.4]) and that $\phi_\lambda(e) \neq 0$. Combining this fact with the main result in [M-O], we have

THEOREM 2.2. *Let G be a connected real semisimple Lie group with finite center, σ be an involutive automorphism of G and H be an open subgroup of the fixed point group of σ. Assume* (0.1) *in place of* (2.1). *Let Γ be a torsion free discontinuous subgroup of G so that the normalizer of H in Γ is contained in H, G/Γ is compact and that $H/\Gamma \cap H$ has finite volume. Then every discrete series for G/H with a sufficiently regular infinitesimal character can be realized in $L^2(G/\Gamma)$.*

UNIVERSITY OF TOKYO

REFERENCES

[FJ] M. Flensted-Jensen, Discrete series for semisimple symmetric spaces, *Ann. of Math.* **111** (1980), 253–311.

[M-O] T. Matsuki and T. Oshima, A description of discrete series for semisimple symmetric spaces, *Advanced Studies in Pure Math.* **4** (1984), 331–390.

[O] T. Oshima, Asymptotic behavior of spherical functions on semisimple symmetric spaces, *Advanced Studies in Pure Math.* **14** (1988), 561–601.

[T-W] Y. L. Tong and S. P. Wang, Geometric realization of discrete series for semisimple symmetric spaces, to appear in *Invent. Math.* (1989).

[W] G. Warner, *Harmonic Analysis on Semisimple Lie Groups, I, II*, Springer-Verlag, Berlin-Heidelberg-New York, 1972.

𝔇-MODULES AND NONLINEAR INTEGRABLE SYSTEMS

MIKIO SATO

LECTURE GIVEN ON MAY 17, 1988

NOTED BY T. SHIOTA

---

1. Introduction. Algebraic analysis is the study of differential equations and functions using algebraic tools. Although its original program included both linear and nonlinear equations, it was initially developed only for linear equations as a general theory of functions and pseudodifferential or differential equations ($\mathcal{E}$- or $\mathfrak{D}$-modules) [18, 9]. This was used in the study of algebraic as well as analytic aspects of linear equations and of functions. Linear equations, however, serve to study too limited a range of functions. It was therefore expected that algebraic analysis include nonlinear equations, infinite order equations and infinite dimensional cases in this framework as well. For example, infinite order operators are not only of theoretical importance but also are applied to study theta zerovalues [7], and algebraic aspects of soliton equations are essential in the study of the theta function of a Jacobian variety [10, 13, 21].

In contrast to its linear counterpart nonlinear algebraic analysis was first developed with holonomic quantum field theory and soliton theory, study of very nice special classes of equations described by monodromy- or spectrum-preserving deformations of linear differential operators [17, 19, 11, 15]. It revealed rich algebraic and geometric structures behind them through the deformation problems and described their solutions explicitly. However, as an algebraic variety or a commutative ring is more intrinsic than the equations defining it, a system of *general* algebraic nonlinear differential equations can be described more intrinsically when it is put in a differential algebraic formalism, and this extends the philosophy of linear algebraic analysis. Abstractly, this can be done in the framework of $\mathfrak{D}$-modules, although the ring $\mathfrak{D}$ here will not be the usual ring of differential

---

Manuscript received November 2, 1988.

operators on a base manifold. As linear algebraic objects (modules) are naturally attached to a variety (a commutative ring), linear equations, or $\mathfrak{D}$-modules, are naturally attached to a system of nonlinear differential equations, thus showing a strong connection between the linear theory and the nonlinear theory. For example, in the linear approximation problem the solution manifold of a system of nonlinear equations becomes the parameter space for a family of linear equations. Somewhat remarkably, this is the same geometric situation as in the case of monodromy- or spectrum-preserving deformation problems for linear equations, although the linear equations play different roles in those two cases.

We start this lecture with a brief summary of the linear theory in Section 2, which will serve as a prototype when we try to extend the framework to general (algebraic) nonlinear equations in Section 3. We will then come back to the more concrete situation and discuss special cases related to deformations of linear differential operators in Section 4.

The lecture was delivered on May 17, 1988 at the JAMI Conference. Based on the notes taken by Y. Shimizu, it was written, using Harvard Mathematics Department computers, by T. Shiota, who is solely responsible for its contents. Sections 2 and 3 follow the actual talk. Section 4 presents the theory of integrable systems and its higher dimensional analogue due to Sato and his collaborators, A. Nakayashiki, Y. Oyama and K. Takasaki. In the talk this was discussed only briefly due to time constraints, and a little more material is added here. T. S. thanks Atsushi Nakayashiki and Yosuke Oyama for their prompt, informative answers to his inquiries, from which the last few paragraphs of Section 4 greatly benefit.

2. Linear differential equations. Let $\mathfrak{D} = \mathfrak{D}_X$ be the sheaf of rings of differential operators on a manifold X. A system of linear differential equations $Pu = 0$ on X, $P \in H^0(\mathrm{Mat}(n \times m, \mathfrak{D}))$, $u = {}^t(u_0, \ldots, u_{m-1})$, is described intrinsically as a coherent left $\mathfrak{D}$-module $\mathfrak{M}$, where

$$\mathfrak{D}^n \xrightarrow{\cdot P} \mathfrak{D}^m \to \mathfrak{M} \to 0$$

is exact. The space of its solutions in a known $\mathfrak{D}$-module $\mathfrak{N}$, e.g., $\mathfrak{N} = \mathcal{O}_X$, the structure sheaf of X, is $\mathrm{Hom}_{\mathfrak{D}}(\mathfrak{M}, \mathfrak{N})$, and the obstruction to solving the inhomogeneous equation $Pu = f$ is $\mathrm{Ext}^1_{\mathfrak{D}}(\mathfrak{M}, \mathfrak{N})$.

$\mathfrak{D}$ has a natural filtration by the order of a differential operator, and the associated graded ring is commutative, i.e., noncommutativity of $\mathfrak{D}$ is rather tame. An element of $\mathrm{gr}_k\mathfrak{D}$ is a function on the cotangent bundle $T^*X \xrightarrow{\pi} X$ homogeneous of degree k. If $P \in \mathfrak{D}$ is of order m, its image $\sigma(P)$ in $\mathrm{gr}_m\mathfrak{D}$ is called the principal symbol of P. $P \in \mathfrak{D}_{X,x}$ is *microlocally elliptic* at $(x, \xi) \in T^*X$ if $\sigma(P)(x, \xi) \neq 0$. Such P's form a multiplicatively closed subset

$$S = S_{x,\xi} = \{P \in \mathfrak{D}_{X,x} \mid \sigma(P)(x, \xi) \neq 0\}$$

of $\mathfrak{D}_{X,x}$ and may be inverted to localize $\mathfrak{D}_X$ to get the sheaf $\mathcal{E} = \mathcal{E}_{T^*X} \supset \pi^{-1}\mathfrak{D}_X$ of *microdifferential* (or *pseudodifferential*) *operators* on T^*X or on $\mathbf{P}(T^*X) = \mathrm{Proj}(\mathrm{gr}\mathfrak{D})$: The commutativity of $\mathrm{gr}\mathfrak{D}$ implies that the pair $(S, \mathfrak{D}_{X,x})$ satisfies the *Ore condition*: For any $(P, Q) \in S \times \mathfrak{D}_{X,x}$ there exists $(P', Q') \in S \times \mathfrak{D}_{X,x}$ such that $P'Q = Q'P$. Thus one can construct the ring of fractions

$$S^{-1}\mathfrak{D}_{X,x} = \{P^{-1}Q \mid (P, Q) \in S \times \mathfrak{D}_{X,x}\}$$

as in the commutative case (Ore construction). The stalk of $\mathcal{E}_{T^*X}$ at $(x, \xi) \in T^*X$ is $S^{-1}\mathfrak{D}_{X,x}$ or some completion of it. In Section 4 we will use a special case of this with $\dim X = 1$ and with completion in the topology such that $P_n \to 0$ if and only if $\mathrm{ord}\, P_n \to -\infty$. Then (locally) $\mathcal{E} = \mathcal{O}_X((\partial^{-1}))$ with ∂ a nonvanishing vector field, and is called the sheaf of *formal pseudodifferential operators*.

Introducing the notion of a quantized contact transformation, one can transform a generic linear differential equation to the de Rham system. Taking a good filtration F on $\mathfrak{M}$, the characteristic variety $\mathrm{ch}(\mathfrak{M})$ of $\mathfrak{M}$ is defined as the support of $\mathrm{gr}^F\mathfrak{M}$ as an $\mathcal{O}_{T^*X}$-module. This definition is independent of F and $\mathrm{ch}(\mathfrak{M})$ is involutive with respect to the standard symplectic structure on T^*X. See [6] and references therein. $\mathfrak{M}$ is said to be *holonomic* if $\mathrm{ch}(\mathfrak{M})$ is Lagrangian, i.e., of the lowest dimension possible. In this case the space of solutions of $\mathfrak{M}$ is finite dimensional. This suggests the possibility of studying a function by the holonomic system it satisfies. Among its successful applications is the study of b-functions. See [8] and references therein. At any point where $\mathrm{ch}(\mathfrak{M}) \to X$ is flat, $\mathfrak{M}$ is $\mathcal{O}$-flat, so that $\mathfrak{M} \otimes_{\mathcal{O}} \mathfrak{M}'$, $\mathfrak{M}^{\otimes n}$, or $\mathfrak{M}^Y \subset \mathfrak{M}^{\otimes n}$ for any Young diagram Y with $|Y| = n$, behave nicely.

3. Nonlinear differential equations. In the general theory of $\mathfrak{D}$-modules, the existence of a nice filtration on $\mathfrak{D}$ led us to the notion of microlocalization. We will observe that this idea can be introduced to the Ritt school's theory of differential algebras to develop a new framework for general nonlinear differential equations.

A general system of algebraic nonlinear differential equations can be described in terms of a differential algebra. Let $\mathcal{A}$ be a commutative algebra. Let Θ be a left $\mathcal{A}$-module and a Lie algebra which acts on $\mathcal{A}$ as derivations. Let $\mathfrak{D}$ be the enveloping algebra of Θ over $\mathcal{A}$, filtered by the degree. Then $\mathcal{A}$ and Θ can be recovered from $\mathfrak{D}$ as $\mathcal{A} = \mathfrak{D}^{(0)}$ and $\Theta \cong \mathfrak{D}^{(1)}/\mathfrak{D}^{(0)}$. Although this $\mathfrak{D}$ is much larger than the usual ring $\mathfrak{D}_X$ of differential operators, they have similar properties. In particular, $\mathrm{gr}\mathfrak{D}$ is commutative, so $\mathfrak{D}$ can be microlocalized as in Section 2. It might be interesting to apply this idea to nonlinear analysis. $\mathcal{A}$ is a left $\mathfrak{D}$-module and the pair $(\mathcal{A}, \mathfrak{D})$ represents a system of differential equations: $\mathcal{A}$ contains the independent and dependent variables, and the differential ideal of the relations (or a set of generators thereof) is the system of differential equations. A solution to the system is identified with a homomorphism of the pair to a known one, say $(\mathcal{O}_X, \mathfrak{D}_X)$.

Let $\mathcal{C} \subset \mathcal{A}$ be the constants. Assume for simplicity that $\mathcal{A}$ is finitely generated over $\mathcal{C}$ and that $\mathcal{A}$ has no zero divisors. We denote by $\mathcal{K}$ the field of fractions of $\mathcal{A}$. Let $\Theta_{\mathcal{K}} = \Theta \otimes \mathcal{K}$. We have $\mathcal{K}\Theta_{\mathcal{K}} \subset \Theta_{\mathcal{K}}$. Let $(\delta_i)_{0 \le i < r}$ be a basis for the $\mathcal{K}$-vector space $\Theta_{\mathcal{K}}$. Assuming that Θ acts faithfully on $\mathcal{A}$, there exists $(x_i)_{0 \le i < r}$ in $\mathcal{A}$ such that $\det(\delta_i(x_j)) \neq 0$. By replacing δ_i if necessary, we may assume $\delta_i(x_j) = \delta_{ij}$, i.e., $\delta_i = \partial/\partial x_i$. Let $\mathcal{A}' \subset \mathcal{A}$ be the subring generated over $\mathcal{C}$ by the x_j's. Suppose $\mathcal{A}$ is generated over the subring $\mathcal{A}'[\delta_0, \ldots, \delta_{r-1}]$ of $\mathfrak{D}$ by y_j, $0 \le j < s$. Then

$$\mathcal{A} = \mathcal{C}[x_i, y_j^{(\alpha)}]/(\text{relations}),$$

where $y_j^{(\alpha)} = \delta^\alpha(y_j)$ for $0 \le j < s$, $\alpha \in \mathbf{N}^r$, and the generators of the differential ideal of relations

$$F_k(x_i, y_j^{(\alpha)}) = 0, \qquad k = 0, 1, \ldots \tag{1}$$

are the differential equations associated to the pair. Unlike the usual Ritt school formulation, we do not specify the choice of independent variables x_j, or the abelian Lie subalgebra $\mathcal{C}\delta_0 \oplus \cdots \oplus \mathcal{C}\delta_{r-1}$ of $\Theta_{\mathcal{K}}$, since it is not natural to do so from the viewpoint of transformation theory for general differential equations.

Expanding the y_j in (1) in power series in a parameter ϵ, we get

(2) $\quad F_k(x_i, y_j^{(\alpha)} + \epsilon u_j^{(\alpha)} + \epsilon^2 v_j^{(\alpha)} + \cdots) = 0, \qquad k = 0, 1, \ldots,$

where $u_j^{(\alpha)} = \delta^\alpha u_j$, etc. At any point (y_j) of the solution manifold of (1), Taylor expanding (2) in ϵ gives an infinite sequence of systems of linear differential equations, or $\mathfrak{D}$-modules $\mathfrak{M}^{(n)}$, $n = 1, 2, \ldots$, for $u_j^{(\alpha)}$, $v_j^{(\alpha)}$, etc:

Let $\Omega^{(1)} = \Omega^{(1)}_{\mathfrak{A}/\mathfrak{C}} = \mathfrak{I}/\mathfrak{I}^2$, where $\mathfrak{I}$ is the ideal in $\mathfrak{A} \otimes_{\mathfrak{C}} \mathfrak{A}$ generated by $f \otimes 1 - 1 \otimes f, f \in \mathfrak{A}$. $\Omega^{(1)}$ is an $\mathfrak{A}$-module. Let $\Omega = \Sigma_n^{\oplus} \Omega^{(n)}$ be the *symmetric* algebra of $\Omega^{(1)}$ over $\Omega^{(0)} = \mathfrak{A}$. Let $d : \Omega^{(n)} \to \Omega^{(n+1)}$ be the derivation over $\mathfrak{C}$ defined by

$$d : \Omega^{(0)} = \mathfrak{A} \to \Omega^{(1)}$$

$$f \mapsto f \otimes 1 - 1 \otimes f \bmod \mathfrak{I}^2$$

and $d(\omega\omega') = d(\omega)\omega' + \omega d(\omega')$. For any $\delta \in \Theta$ let $C_\delta^{(i)} : \Omega^{(n)} \to \Omega^{(n-i)}$ be the derivation such that for any $f \in \mathfrak{A}$, $C_\delta^{(i)}(d^n f) = (n!/(n-i)!)d^{n-i}\delta(f)$, and $C_\delta^{(i)}(\Omega^{(n)}) = 0$ if $n < i$. Then $C_\delta^{(0)}$ is the Lie derivation and

$$C_{f\delta}^{(0)} = fC_\delta^{(0)} + \frac{df}{1!} C_\delta^{(1)} + \frac{d^2 f}{2!} C_\delta^{(2)} + \cdots.$$

Then for $n = 1, 2, \ldots$, we define

$$\mathfrak{M}^{(n)} = \{\omega \in \Omega^{(n)} \mid C_{f\delta}^{(0)}(\omega) = fC_\delta^{(0)}(\omega) \text{ for any } f \in \mathfrak{A},\ \delta \in \Theta\}$$

$$= \{\omega \in \Omega^{(n)} \mid C_\delta^{(i)}(\omega) = 0 \text{ for any } 0 < i \le n,\ \delta \in \Theta\}.$$

Θ acts on $\mathfrak{M}^{(n)}$ by Lie derivative, making it a Θ-module, hence a $\mathfrak{D}$-module. $\mathfrak{M}^{(n)}$ describes the nth approximation of (1). In particular, $\mathfrak{M}^{(1)}$, which is by definition the annihilator of Θ with respect to the interior product of Θ and $\Omega^{(1)}$, can be considered as the conormal bundle of the solution manifold of (1).

4. The universal Grassmann manifold. Certain problems in mathematical physics are described by deformation problems for linear differential operators. Solutions to these problems are then constructed by solving the corresponding linear equations. By associating to the linear equations certain subspaces of a vector space, one often sees that the solution manifold of the original problem is canonically isomorphic to a Grassmann manifold or a flag manifold [15, 16, 20]. Linear equations thus play more basic roles here than in the general case with the abstract $\mathfrak{D}$ that we saw above, and the ring $\mathfrak{D}$ here will be the usual ring of differential operators again. We start this section with one of those examples, the Kadomtsev-Petviashvili (KP) hierarchy, and then discuss some generalizations. In this example, we identify the solution manifold with a Grassmannian. The τ-function (Hirota's dependent variable) then becomes a canonical section of the determinant line bundle, and the $gl(\infty)$-symmetry behind the Grassmannian is explicitly realized as the $gl(\infty)$-commutation relations satisfied by the infinitesimal Bäcklund transformations. Hence this is the right approach to the KP hierarchy.

4.1. The KP hierarchy. In what follows a formal pseudodifferential operator is simply called a pseudodifferential operator. Let $\tilde{\mathfrak{B}} = \mathbf{Q}[u_i^{(j)}; 2 \leq i < \infty, 0 \leq j < \infty]$, and let ∂ be the derivation on $\tilde{\mathfrak{B}}$ defined by $\partial u_i^{(j)} = u_i^{(j+1)}$. Define a pseudodifferential operator $\tilde{L}$ by $\tilde{L} = \partial + u_2\partial^{-1} + \cdots + u_i\partial^{-i+1} + \cdots$, where $u_i = u_i^{(0)}$. Let Θ be the Lie algebra of derivations on $\tilde{\mathfrak{B}}$ generated over $\tilde{\mathfrak{B}}$ by ∂_m, $m \in \mathbf{N}$, where the action of ∂_m on $\tilde{\mathfrak{B}}$ is given by $[\partial_m, \partial] = 0$ and $\partial_m\tilde{L} = [(\tilde{L}^m)_+, \tilde{L}]$. Here and in what follows for any pseudodifferential operator P, P_+ denotes the differential operator part of P and $P_- = P - P_+$. Then we have $[\partial_m, \partial_n] = 0$, and $\Theta = \sum_{m=1}^{\infty} \tilde{\mathfrak{B}}\partial_m$. Denoting by $\mathfrak{U}(\Theta)$ the universal enveloping algebra of Θ, the pair $(\tilde{\mathfrak{B}}, \mathfrak{U}(\Theta))$ represents the KP hierarchy in the sense of Section 3.

Let $(\mathfrak{B}, \partial)$ be a differential algebra over a ring $\mathfrak{C} \supset \mathbf{Q}$, say $(\mathfrak{C}[[x]], d/dx)$. We assume for simplicity that $\mathfrak{C}$ is a field (or a local ring). A (formal) solution of the KP hierarchy in $\mathfrak{B}$ is a pseudodifferential operator

$$L \in \partial + \mathfrak{B}[[t]][[\partial^{-1}]]\partial^{-1}, \tag{3}$$

where $t = (t_1, t_2, t_3, \ldots)$, such that

$$\frac{\partial L}{\partial t_m} = [(L^m)_+, L], \qquad m = 1, 2, \ldots \tag{4}$$

holds. Given any $L_0 \in \partial + \mathfrak{B}[[\partial^{-1}]]\partial^{-1}$ there exists a unique formal solution L of (4) with initial value $L|_{t=0} = L_0$. Let L be a solution. Since $[\partial_m, \partial_n] = 0$, assuming the functions can be integrated ($\partial : \mathfrak{B} \to \mathfrak{B}$ surjective) we can formally solve the system of linear differential equations

$$Lw = \frac{1}{z} w, \tag{5}$$

$$\frac{\partial w}{\partial t_m} = (L^m)_+ w, \qquad m = 1, 2, \ldots, \tag{6}$$

where z is a formal scalar variable 'near' zero and w is of the form

$$w = w(t, z) = \left(1 + \sum_{n=1}^{\infty} w_n(t) z^n\right) \exp\left(x/z + \sum_{m=1}^{\infty} t_m / z^m\right) \tag{7}$$

with $x \in \partial^{-1}1 \subset \mathcal{B}$ and $w_n(t) \in \mathcal{B}[[t]]$. Setting

$$W = 1 + \sum_{n=1}^{\infty} w_n(t) \partial^{-n} \in 1 + \mathcal{B}[[t]][[\partial^{-1}]]\partial^{-1},$$

(7) becomes

$$w = W \exp\left(x/z + \sum_{m=1}^{\infty} t_m / z^m\right).$$

Conversely if L is of the form (3) and w is of the form (7) satisfying (5), (6), then L satisfies (4). L is uniquely determined by w, and w, called the wave function, is determined by L up to multiplication by an element of $1 + z\mathbb{C}[[z]]$.

Suppose $\mathcal{B} = \mathbb{C}[[x]]$. Then

$$w \in \mathbb{C}[[x, t, z]] \exp\left(x/z + \sum_{m=1}^{\infty} t_m / z^m\right) \subset (\mathbb{C}((z)))[[x, t]]. \tag{8}$$

Let V be the rank 1 free $\mathbb{C}((z))$-module $\mathbb{C}((z))\sqrt{d(1/z)}$ considered as a big $\mathbb{C}$-vector space with z-adic topology, and let $V^{(0)} \subset V$ be its linearly compact subspace $z\mathbb{C}[[z]]\sqrt{d(1/z)}$. Let $w_0 = w|_{t=0}$, i.e., $w_0(z) = w(0, z)$. Taylor expanding $w_0(z)\sqrt{d(1/z)}$ in x, we get a sequence in V. Let V_{w_0} be the $\mathbb{C}$-vector subspace of V spanned by the sequence.

The space V_{w_0} satisfies the condition:

$$\pi|_{V_{w_0}} : V_{w_0} \to V/V^{(0)} \text{ is an isomorphism,}$$

where π is the natural projection $V \to V/V^{(0)}$. Thus V_{w0} is in the 'big cell' $\mathrm{UGM}^{\emptyset}$ of the *universal Grassmann manifold* $\mathrm{UGM} = \mathrm{UGM}(V, V^{(0)})$, which is by definition the set of all the subspaces U of V which are comparable to $V^{\emptyset} \overset{\mathrm{def}}{=} \mathbb{C}[z^{-1}]\sqrt{d(1/z)}$ of relative dimension zero, i.e., $\pi|_U : U \to V/V^{(0)} \cong V^{\emptyset}$ has kernel and cokernel of the same finite dimension. Conversely, given any $U \in \mathrm{UGM}^{\emptyset}$, there exists a unique $w_0 = w_{0,U} \in (1 + z\mathbb{C}[[x, z]])\exp(x/z)$ with $U = V_{w0}$. Indeed, if we define

$$U^* = \exp(-x/z)U[[x]] \subset V[[x]],$$

then $\pi|_{U^*} : U^* \to (V/V^{(0)})[[x]]$ is still an isomorphism, so by the formula

$$w_0\sqrt{d(1/z)} = \exp(x/z)(\pi|_{U^*})^{-1}\sqrt{d(1/z)} \in U[[x]] \tag{9}$$

w_0 is well defined and satisfies the requirements. To restore the time evolutions, we observe:

(6) implies that the coefficients of the expansion for $w(t, z)\sqrt{d(1/z)}$ in all the (x, t)-variables still belong to V_{w0}. Conversely, if a w of the form (7) satisfies this property, then there exists a unique L satisfying (5) and (6) (and hence (4)).

So we define

$$U_t = \exp\left(-\sum_{m=1}^{\infty} t_m/z^m\right)U[[t]] \subset V[[t]]. \tag{10}$$

Then $\pi|_{U_t} : U_t \to (V/V^{(0)})[[t]]$ is still an isomorphism, and w is defined by $w = w_{0,U_t}\exp(\Sigma_{m=1}^{\infty} t_m/z^m)$, or by the formula analogous to (9):

$$w\sqrt{d(1/z)} = \exp\left(x/z + \sum_{m=1}^{\infty} t_m/z^m\right)(\pi|_{U_t^*})^{-1}\sqrt{d(1/z)} \in U[[x, t]],$$

$$U_t^* = U_{t(x)}, \qquad t(x) = (x + t_1, t_2, \ldots, t_n, \ldots).$$

Hence $\mathrm{UGM}^{\emptyset}$ parametrizes the regular formal solutions of the KP hierarchy, and the KP hierarchy can be identified with the (formal) dynamical system on $\mathrm{UGM}^{\emptyset}$ defined by (10). The restriction to $\mathrm{UGM}^{\emptyset} \subset \mathrm{UGM}$ comes from the assumption $\mathcal{B} = \mathbb{C}[[x]]$, i.e., that there are no negative powers of x allowed. We introduce the notion of *regularizable operators* in order to treat the general case: $W_0 \in 1 + \mathcal{C}((x))[[\partial^{-1}]]\partial^{-1}$ is regularizable if there exists $m \in \mathbf{N}$ such that $x^m W_0$, $x^m(W_0^*)^{-1}$ are regular, i.e., have the coefficients in $\mathcal{C}[[x]]$. Here W_0^* is the formal adjoint of W_0. We have seen that through the Taylor expansion of $W_0 \exp(x/z)\sqrt{d(1/z)}$ in x the set of the regular monic W_0's is isomorphic to $\mathrm{UGM}^{\emptyset}$. This isomorphism is extended to an isomorphism γ between the space $\mathcal{W}$ of all the regularizable W_0's and the whole UGM [15, p. 268, Theorem 5], [16, Section 9], so that the KP hierarchy becomes the dynamical system defined on the whole Grassmann manifold by (10).

Let $\mathcal{E} = \mathcal{E}_{\mathcal{C}[[x]]} = \mathcal{C}[[x]](((d/dx)^{-1}))$ and $\mathcal{D} = \mathcal{D}_{\mathcal{C}[[x]]} = \mathcal{C}[[x]][d/dx]$. Consider V as a cyclic right $\mathcal{E}$-module by the isomorphism

$$V \cong \mathcal{E}/x\mathcal{E}$$

$$(P \exp(x/z)\sqrt{d(1/z)})|_{x=0} \cong P \bmod x\mathcal{E}, \qquad P \in \mathcal{E}.$$

Let $W_0 \in \mathcal{W}$, let $m \in \mathbf{N}$ be such that $x^m W_0$, $W_0^{-1}x^m \in \mathcal{E}$, and let $w_0 = W_0 \exp(x/z)$. Then this isomorphism induces

$$\gamma(W_0) = \{(Px^m w_0(z)\sqrt{d(1/z)})|_{x=0} \in V \,|\, P \in \mathcal{D}\}$$

$$\cong \{Px^m W_0 \bmod x\mathcal{E} \in \mathcal{E}/x\mathcal{E} \,|\, P \in \mathcal{D}\}.$$

The action of $f(z) \in \mathcal{C}((z))$ on V is given by the right multiplication of the constant coefficient pseudodifferential operator $f(\partial^{-1}) \in \mathcal{C}((\partial^{-1}))$ on $\mathcal{E}/x\mathcal{E}$. $\mathcal{C}((\partial^{-1}))$ acts faithfully on $\mathcal{E}/x\mathcal{E}$. Let us denote by q the map $f(z) \mapsto f(\partial^{-1})$. Although a proper subspace U of V is no longer an $\mathcal{E}$-module, it is often invariant under a nontrivial subring of $\mathcal{C}((z))$ or $\mathcal{C}((\partial^{-1}))$. If $U = \gamma(W_0)$ comes from the Krichever construction (see below), the corresponding point on UGM has a natural $H^0(X - p, \mathcal{O})$-module structure, where we embed $H^0(X - p, \mathcal{O})$ in $\mathcal{C}((z))$ by the expansion in the local coordinate z at p. Then we observe through the second picture here and the Krichever theory that $W_0 q(H^0(X - p, \mathcal{O}))W_0^{-1}$ is a commutative subring of the ring of ordinary differential operators.

The subspaces $V^{\emptyset}$ and $V^{(0)}$ of V are maximal isotropic and dual to each other with respect to the residue pairing

$$\langle f\sqrt{d(1/z)}, g\sqrt{d(1/z)}\rangle = \mathrm{res}_{z=0} f(z)g(z)d(1/z). \tag{11}$$

They can be replaced by another pair of vector spaces with this property in order to obtain the KP solution space with different analytic properties [15, p. 263], [20].

For $W_0 \in \mathcal{W}$ let $U = \gamma(W_0)$ and let $U' \in \mathrm{UGM}$ be defined similarly by Laurent expanding $(W_0^*)^{-1} \exp(-x/z)\sqrt{d(1/z)}$ in x. Then U and U' are the orthogonal complement of each other with respect to the residue pairing (11) [3, Lemma 1.1]. Combining this with a previous observation, we get another definition of the KP hierarchy:

A function $w = w(t, z)$ of the form (7) is a wave function of the KP hierarchy if and only if there exists another function $w^*(t, z)$ of the form

$$w^*(t, z) = \left(1 + \sum_{n=1}^{\infty} v_n(t) z^n\right) \exp\left(-x/z - \sum_{m=1}^{\infty} t_m/z^m\right),$$

where $v_n(t) \in \mathbb{C}((x))[[t]]$, such that for any t, t'

$$\operatorname{res}_{z=0} w(t, z) w^*(t', z) d(1/z) = 0 \tag{12}$$

holds. Then w^*, called the adjoint wave function, is related to w by

$$w = W \exp\left(x/z + \sum_{m=1}^{\infty} t_m/z^m\right),$$

$$w^* = (W^*)^{-1} \exp\left(-x/z - \sum_{m=1}^{\infty} t_m/z^m\right).$$

This was first observed by Cherednik [2] for the solutions constructed by Krichever's method (see below): If w is obtained from a sheaf $\mathcal{F}$ and w^* is obtained from a sheaf $\mathcal{G}$, then (12) holds if and only if $\mathcal{G} = \mathcal{H}om(\mathcal{F}, \omega)$, where ω is the dualizing sheaf of X. Writing w and w^* in terms of the τ-function, (12) gives Hirota's bilinear equations for τ, which can be identified with the Plücker relations for the projective embedding of UGM, and deduce many nontrivial bilinear equations for Riemann's theta function of the Jacobian of a curve.

4.2. Krichever's theory. Krichever [10] constructed a large class of KP solutions from geometric data:

1. X a complete reduced irreducible curve over $\mathbb{C}$, and $p \in X$ a smooth point,

2. $\mathcal{F}$ a torsion-free rank 1 sheaf on X such that $\chi(\mathcal{F}) = 0$,
3. $z \in \hat{\mathfrak{m}}_p - \hat{\mathfrak{m}}_p^2$ a (formal) local coordinate at p,
4. $\rho : \hat{\mathcal{F}}_p(p) \cong \hat{\mathcal{O}}_p\sqrt{d(1/z)}$ a trivialization.

In the UGM set-up the corresponding solution is given by $H^0(X - p, \mathcal{F})$, identified with a subspace U of V by z and ρ in the data. Thus $\chi(\mathcal{F}) = 0$ implies that U has relative dimension 0, and $U \in \mathrm{UGM}^\emptyset$ if and only if $h^0(\mathcal{F}) = h^1(\mathcal{F}) = 0$. Changing ρ results in multiplying the wave function by an element of $1 + z\mathbb{C}[[z]]$, so that to describe the KP solution in L we do not need ρ. Our description of the KP time evolutions on the Grassmannian gives the following description of them in terms of deformations of $\mathcal{F}$: Let $D \cong \mathrm{Spec}\, \mathbb{C}[[z]]$ and $D^* \cong \mathrm{Spec}\, \mathbb{C}((z))$ be the infinitesimal disc and the infinitesimal punctured disc at p, respectively. Define the homomorphism $\phi : \mathrm{Spec}(\mathbb{C}[[t]]) \to \mathrm{Pic}^0(X)$ by $\phi(t) = \mathcal{O}$ on D and on $X - p$, but glued to itself on D^* by the transition function $\exp(\Sigma_{n=1}^{\infty} t_n z^{-n})$. Denote by σ_t the standard identification $\widehat{\phi(t)}_p \cong \hat{\mathcal{O}}_p$. Then the KP time evolution in the Krichever data $(X, p, \mathcal{F}, z, \rho)$ is given by

$$X_t = X, \qquad p_t = p, \qquad \mathcal{F}_t = \phi(t) \otimes \mathcal{F}, \qquad z_t = z,$$

$$\rho_t = \sigma_t \otimes \rho.$$

That the KP flows deform $\mathcal{F}$ with X fixed is an analogue of isospectral deformation of L. We will see it more clearly after interpreting $X - p$ as the spectrum of a commutative ring of ordinary differential operators. Since ϕ has a finite dimensional image, the KP orbit in L is finite dimensional. We will see that this property characterizes these solutions. This means, by (4), that there are sufficiently many ordinary differential operators which commute with L [1, 12, 14]: Let

$$\mathcal{A} = \{P \in \mathbb{C}((x))[[t]][d/dx] \,|\, [P, L] = 0\}.$$

Then $\mathcal{A}$ is a commutative ring and for any $n \gg 0$ in $\mathbf{N}$, there is $P \in \mathcal{A}$ of order n. If, moreover, $\mathcal{F}$ satisfies

$$\mathcal{H}om_{\mathcal{O}}(\mathcal{F}, \mathcal{F}) \cong \mathcal{O}, \tag{13}$$

i.e., if $\mathcal{F}$ is not the direct image of a sheaf on a less singular curve, then the geometric data can be recovered from $\mathcal{A}$, hence from finite dimensionality of the orbit: We have $\mathrm{Spec}\, \mathcal{A} = (X - p) \times \mathrm{Spec}\, \mathbb{C}[[t]]$, etc.

In the case $\mathcal{C} = \mathbf{C}$ the solution extends globally in t, and the KP orbit in L is isomorphic to $\mathrm{Jac}(X)$. If moreover X is smooth, the solution is said to be quasiperiodic, and the corresponding τ-function becomes Riemann's theta function evaluated at $\mathfrak{F}_t$, up to some exponential factor. Generalizations of this construction to higher rank cases are discussed in [4, 13].

Finally let us discuss a generalization of this to higher dimensional cases. In the above construction, the inclusion $\bar{\mathfrak{A}} \stackrel{\text{def}}{=} H^0(X - p, \mathcal{O}) \subset \mathcal{C}((z))$ via z makes $\bar{\mathfrak{A}}$ act on V, hence on UGM. If $U \in$ UGM is a fixed point of this action, then U has a torsion-free $\bar{\mathfrak{A}}$-module structure which respects the filtrations on $\bar{\mathfrak{A}}$ and on U, hence it has rank 1 and comes from Krichever data: $U \cong H^0(X - p, \mathfrak{F})$ for some torsion-free rank 1 sheaf $\mathfrak{F}$ on X. Thus the fixed point set $\mathrm{UGM}^{\bar{\mathfrak{A}}}$ of this action consists of all the U's obtained from the above geometric data with given (X, p, z), and its image in $\mathrm{UGM}/(1 + z\mathcal{C}[[z]])$ ($\cong$ the space of L's) is isomorphic to $\overline{\mathrm{Pic}^{g-1}(X)}$, or an open dense subset therein if one requires (13). Here g is the arithmetic genus of X and $\overline{\mathrm{Pic}^{g-1}(X)}$ is the moduli space of torsion-free rank 1 sheaves of degree $g - 1$ on X [5].

Now we replace the triple (X, p, z) by $(X; H; z_0, \ldots, z_{r-1})$, where X is an irreducible projective variety with $\dim X = r$, $H \subset X$ is an ample irreducible divisor and $z_0, \ldots, z_{r-1} \in \hat{\mathfrak{m}}_p$ is a (formal) local coordinate system on X at $p \in H$, such that H and X are smooth at p and $H = \{z_0 = 0\}$ near p. Let $\mathfrak{F}$ be a torsion-free rank 1 sheaf on X and ρ a trivialization of $\mathfrak{F}$ at p. Then via $z_0, \ldots, z_{r-1}$, $H^0(X - H, \mathfrak{F})$ is embedded into $V_r = \mathcal{C}[[z_{r-1}, \ldots, z_2, z_1]]((z_0))$ and the image is preserved by the natural action of $\bar{\mathfrak{A}} \stackrel{\text{def}}{=} H^0(X - H, \mathcal{O}) \subset \mathcal{C}[[z_{r-1}, \ldots, z_2, z_1]]((z_0))$ on V_r. Therefore, if we replace $V = \mathcal{C}((z))$ by V_r and make a Grassmann manifold out of this big vector space with a suitable choice of the reference space $V_r^{(0)} \subset V_r$, then $H^0(X - H, \mathcal{O})$ acts on it, and a fixed point of this action is given by $H^0(X - H, \mathfrak{F})$ for some torsion-free rank 1 sheaf $\mathfrak{F}$ on X. The choice of $V_r^{(0)}$ may depend on $\mathfrak{F}$, but the fixed point set of this action on the family of Grassmann manifolds over the space of the equivalence classes of $V_r^{(0)}$ becomes the union of certain components of $\overline{\mathrm{Pic}(X)}$ as in the 1 dimensional case. We can also introduce the KP-like time evolutions to this fixed point set and observe that the orbit of a point is isomorphic to $\mathrm{Pic}^0(X)$ if $\mathfrak{F}$ satisfies the condition parallel to (13). Unfortunately, this does not quite lead us to a higher dimensional analogue of the KP theory on the *whole* Grassmann manifold or the family of Grassmann manifolds. Hence we still do not know the right definition of the higher dimensional analogue of UGM, which, besides the fact that there is no longer a standard choice of the reference space $V_r^{(0)}$ like $V^{(0)}$ in the 1 dimensional case, may not even be a Grassmann manifold.

The r dimensional analogue of the affine open cell $\mathrm{UGM}^\emptyset$ of UGM has been relatively well-understood. Let us denote it by $\mathrm{UGM}_r^\emptyset$, although at the moment we do not have the right definition of UGM_r yet. For any $U \subset V_r$ and $n \in \mathbf{N}$ let $U(n) = U \cap z_0^{-n}\mathcal{C}[[z]]$. Suppose $V_r^{(0)} \subset V_r$ satisfies the condition $\dim V_r(n)/V_r^{(0)}(n) < \infty$ for any n. Then we define

$$\mathrm{UGM}_r^\emptyset = \mathrm{UGM}_r^\emptyset(V_r, V_r^{(0)})$$

$$= \{U \subset V_r \mid U(n) + V_r^{(0)}(n) = V_r(n),\ U(n) \cap V_r^{(0)}(n) = \{0\} \text{ for any } n\}.$$

Note that if $r = 1$ and $V_r^{(0)} = V^{(0)}$, this coincides with $\mathrm{UGM}^\emptyset$. In the higher dimensional case, the condition that $U(n) \oplus V_r^{(0)}(n) = V_r(n)$ for any n seems to play a crucial role in the theory (e.g., as a natural condition to guarantee nondivergence in the time evolution of a general point on $\mathrm{UGM}_r^\emptyset$). Note that if $U \in \mathrm{UGM}_r^\emptyset$ comes from the above geometric construction, i.e., if U is in the fixed point set $(\mathrm{UGM}_r^\emptyset)^{\bar{\mathcal{A}}}$ of the $\bar{\mathcal{A}}$-action, then we have

$$\dim H^0(X, \mathfrak{F}(nH)) = \dim V_r(n)/V_r^{(0)}(n)$$

for any n. In particular, the Hilbert polynomial puts a strong condition for $V_r^{(0)}$ in order to have $(\mathrm{UGM}_r^\emptyset)^{\bar{\mathcal{A}}} \neq \emptyset$.

In some special cases this geometric construction can be more explicit. Let $\alpha : X \to \mathrm{Alb}(X)$ be the Albanese map. Suppose that $\mathfrak{F}$ is the sheaf of sections of the pull back $\alpha^*\mathcal{L}$ of a line bundle $\mathcal{L}$ on $\mathrm{Alb}(X)$, rather than a general torsion-free rank 1 sheaf on X, and that $\mathcal{L}$ satisfies some nice conditions, e.g., that $X \subset \mathrm{Alb}(X)$ via α and that any element of $H^0(X - H, \alpha^*\mathcal{L})$ is a pullback of a meromorphic section of $\mathcal{L}$ on $\mathrm{Alb}(X)$. Then the corresponding point $V_{\mathcal{L}}$ on $\mathrm{UGM}_r^\emptyset$ can be described in terms of theta functions on $\mathrm{Alb}(X)$, analogously to the 1 dimensional case where the wave function can be expressed by the quotient of Riemann's theta function of the Jacobian variety (or rather, the τ-function) and a translate of it. For example, $V_{\mathcal{L}}$ has been studied in details using identities for theta functions when X is a principally polarized abelian variety.

KYOTO UNIVERSITY

REFERENCES

[1] J. L. Burchnall, T. W. Chaundy: Commutative ordinary differential operators. Proc. London Math. Soc. **21**, 420–440 (1922); Proc. R. Soc. London (A) **118**, 557–583 (1928); Ibid. **134**, 471–485 (1931).
[2] I. V. Cherednik: Differential equations for the Baker-Akhiezer functions of algebraic curves. Funct. Anal. Appl. **12** 195–203 (1978).
[3] E. Date, M. Jimbo, M. Kashiwara, T. Miwa: Transformation groups for soliton equations. In: Proceedings RIMS Symp. Nonlinear integrable systems—classical theory and quantum theory (Kyoto 1981), pp. 39–119. World Scientific 1983.
[4] V. G. Drinfeld: Commutative subrings of certain noncommutative rings. Funct. Anal. Appl. **11** 9–12 (1977).
[5] C. D'Souza: Compactification of generalized Jacobians. Proc. Indian Acad. Sci. (A) 88, (5) 419–457 (1979).
[6] O. Gabber: The integrability of characteristic variety. Amer. J. Math. **103** 445–468 (1981).
[7] T. Kawai: Systems of linear differential equations of infinite order—an aspect of infinite analysis. RIMS preprint 599, 1987.
[8] M. Kashiwara, T. Kawai: Microlocal analysis. Publ. RIMS, Kyoto Univ. **19** 1003–1032 (1983).
[9] ______, ______, T. Kimura: Foundations of algebraic analysis. Princeton University Press 1986.
[10] I. M. Krichever: Methods of algebraic geometry in the theory of nonlinear equations. Russ. Math. Surv. **32**, (6) 185–213 (1977).
[11] P. D. Lax, Integrals of nonlinear equations of evolution and solitary waves. Commun. Pure Appl. Math. **21** 467–490 (1968).
[12] ______, Periodic solutions of the K-dV equation. Commun. Pure Appl. Math. **28** 141–188 (1975).
[13] D. Mumford: An algebro-geometric construction of commuting operators and of solutions to the Toda lattice equation, Korteweg-de Vries equation and related non-linear equations. In: Proceedings Int. Symp. Algebraic Geometry, Kyoto (1977), pp. 115–153. Kinokuniya Book Store, Tokyo 1978.
[14] S. P. Novikov, The periodic problem for the Korteweg-de Vries equation. Funct. Anal. Appl. **8** 236–246 (1974).
[15] M. Sato, Y. Sato, Soliton equations as dynamical systems on infinite dimensional Grassmann manifold. In: Nonlinear partial differential equations in applied science (Proc. U.S.-Japan Seminar, Tokyo 1982) pp. 259–271. North Holland 1983.
[16] ______, Soliton equations and universal Grassmann manifold (written by Noumi in Japanese). Math. Lect. Note Ser. No 18. Sophia University, Tokyo 1984.
[17] ______, T. Miwa, M. Jimbo, Holonomic quantum fields II—the Riemann-Hilbert problem—. Publ. RIMS, Kyoto Univ. **15** 201–278 (1979).
[18] ______, T. Kawai, M. Kashiwara: Microfunctions and pseudo-differential equations. In: Hyperfunctions and pseudo-differential equations (Proc. Conference at Katata, 1971), Lect. Notes Math., vol. 287, pp. 265–529. Springer 1973.

[19] L. Schlesinger: Über eine Klasse von Differentialsystemen beliebiger Ordnung mit festen kritischen Punkten. J Reine u. Angew. Math. **141** 96–145 (1912).

[20] G. Segal, G. Wilson: Loop groups and equations of KdV type. IHES Publ. Math. **61**, 5–65 (1985).

[21] T. Shiota: Characterization of Jacobian varieties in terms of soliton equations. Invent. Math. **83** 333-382 (1986).

L-FUNCTIONS AND EIGENVALUE PROBLEMS

By Goro Shimura

Introduction. The eigenvalue problems we consider in this paper concern an inhomogeneous equation

$$(1) \qquad (L_m - \lambda)f(z) = p(z).$$

Here f and p are functions, possibly with singularities, on the complex upper half plane

$$(2) \qquad H = \{z \in \mathbf{C} \mid \mathrm{Im}(z) > 0\},$$

and L_m is the differential operator given by

$$L_m = -y^{-2}(\partial^2/\partial x^2 + \partial^2/\partial y^2) + miy\partial/\partial x,$$

where $z = x + iy$ as usual and m is an integer. We fix a congruence subgroup Γ of $\mathrm{SL}_2(\mathbf{Z})$, and assume that both f and p are Γ-automorphic forms of weight m in the sense that they are invariant under

$$(3) \qquad f \mapsto f(\gamma z)(cz + d)^{-m}|cz + d|^m$$

for every $\gamma = \begin{pmatrix} * & * \\ c & d \end{pmatrix} \in \Gamma$. It should be noted that L_m commutes with the map of (3). The homogeneous equation

$$(4) \qquad (L_m - \lambda)\varphi = 0$$

has been extensively studied. One of the most noteworthy facts about such λ and φ is the existence of a certain zeta function, defined by Selberg, whose set of zeros coincides essentially with the set $S(\Gamma, m)$ of complex numbers s such that $\lambda = s(1 - s)$ occurs as an eigenvalue of L_m as in (4)

Manuscript received September 1, 1988.

with a cusp form φ. However, no connection of such λ or s with the zeros of classical zeta functions has been discovered.

Now the fundamental principle governing the present paper can be condensed as follows:

A zero of a zeta function is a complex number s such that $s(1-s)$ occurs as a value λ in the inhomogeneous equation (1) *with a fixed p determined by the zeta function and a rapidly decreasing f.*

We shall present two types of zeta functions and the corresponding f and p. To describe the first type, we let $\Gamma = \mathrm{SL}_2(\mathbf{Z})$ and take two holomorphic cusp forms $g(z) = \Sigma_{n=1}^{\infty} a_n e^{2\pi inz}$ of weight k and $h(z) = \Sigma_{n=1}^{\infty} b_n e^{2\pi inz}$ of weight l with respect to Γ, and put $\kappa = (k+l)/2$, $m = l - k$, and $R(s) = R_0(s + \kappa - 1)$, where

$$R_0(s) = (4\pi)^{-s}\Gamma(s) \sum_{n=1}^{\infty} \bar{a}_n b_n n^{-s}.$$

It is well known that $s(1-s)\pi^{-s}\Gamma(s + |m/2|)\zeta(2s)R(s)$ can be continued to an entire function invariant under $s \mapsto 1 - s$. Now we shall show that there is a function $f(z, s)$ meromorphic on the whole s-plane with the following properties (Theorem 3.1):

(I-0) *$f(z, s)$ is finite at s if $s \notin \{1/2\} \cup S(\Gamma, m)$ and $\Gamma(s + |m/2|)\zeta(2s) \neq 0$. Put $f_s(z) = f(z, s)$ for such an s.*

(I-1) *f_s is a C^∞ Γ-automorphic form of weight m.*

(I-2) $[L_m - s(1-s)]f_s = y^\kappa \overline{g(z)} h(z)$.

(I-3) $f_s(z) = (2s-1)^{-1}R(s)y^{1-s} + O(e^{-\pi y})$ *as* $y \to \infty$.

The significance of the last fact is that f_s is rapidly decreasing at every cusp of Γ if and only if $R(s) = 0$, which is an example of the above principle.

Such a function f can be obtained as an integral

$$f(z, s) = -\int_{\Gamma\backslash H} G_s(z, w)\overline{g(w)}h(w)\mathrm{Im}(w)^\kappa d_H w, \tag{5}$$

where $d_H w$ is the standard invariant measure on H and G_s is the resolvent kernel of equation (1), which was introduced by Roelcke [9] and subsequently studied by several authors. (See Fadeev [3], Neunhöffer [8], Elstrodt [1], Fay [4], Hejhal [7]. In this paper, we employ the notation and results of [7], which is the most comprehensive treatment of the subject.)

As its name suggests, relation (I-2) is an immediate consequence of (5). Likewise, (I-0) and (I-1) follow from the corresponding properties of G_s. Thus the essential point is (I-3), which is also closely connected with the following fact: for a fixed w, $G_s(z, w)$ as a function of z has a Fourier expansion for sufficiently large $\mathrm{Im}(z)$, and the expansion shows that

$$G_s(z, w) = (1 - 2s)^{-1} y^{1-s} E_{-m}(w, s) + O(e^{-\pi y}) \quad \text{as} \quad y \to \infty$$

with a certain Eisenstein series E_{-m} of weight $-m$. Now the above R can be obtained as an integral

$$R(s) = \int_{\Gamma\backslash H} E_{-m}(w, s)\mathrm{Im}(w)^{\kappa}\overline{g(w)}h(w)d_H w, \tag{6}$$

and therefore termwise integration, if valid, would give (I-3), establishing $(2s - 1)^{-1}R(s)y^{1-s}$ as the constant term of $f(z, s)$. This is not so, however. Our proof of (I-3), which is rather involved, will show that the supposed constant term is actually the dominant part of the true constant term of f, the remaining part being rapidly decreasing.

Let us next note a few consequences of this fact. First of all, the self-adjointness of L_m combined with (I-2) and (I-3) shows that if $R(s) = 0$, then

(I-4)

$$\mathrm{Im}[s(1 - s)]\langle f_s, f_s\rangle = \mathrm{Im}\langle p, f_s\rangle$$

$$= -\mathrm{Im}\left[\int_{\Gamma\backslash H}\int_{\Gamma\backslash H} G_s(z, w)\overline{p(z)}p(w)d_H z d_H w\right],$$

where $p(z) = y^{\kappa}\overline{g(z)}h(z)$ and $\langle\ ,\ \rangle$ denotes the Petersson inner product. Thus for such a zero s of R, one has $s(1 - s) \in \mathbf{R}$ if and only if the last double integral is real. We shall also prove (Theorems 3.1 and 3.4):

(I-5) $4t \cdot \mathrm{Im}\langle f_s, p\rangle = |R(s)|^2$ *if* $s = (1/2) + it$ *with* $0 \neq t \in \mathbf{R}$.
(I-6) $(1 - 2s)\langle E_m(z, \bar{s}), f_s\rangle = dR/ds$ *if* $R(s) = 0$.

There is a significant aspect of these relations beyond the fact that they hold: namely, all of them have their counterparts in the second case, in which R is replaced by the L-function $L(s, \chi)$ of a Hecke character χ of an imaginary quadratic field K. We take the archimedean factor of χ is of the type $a \mapsto a^m |a|^{-m}$, and assume for simplicity that the restriction of χ to $\mathbf{Q}$ is trivial. In this case f of (1) is given by

$$g(z, s) = \mu(A/\pi)^s \Gamma(s + |m/2|)\zeta(2s) \sum_{t \in T} \chi(\mathbf{Z}\tau + \mathbf{Z}) G_s(z, \tau).$$

Here T is a finite subset of $K \cap H$ such that the lattices $\mathbf{Z}\tau + \mathbf{Z}$ for $\tau \in T$ represent certain ideal classes in K; μ, A, and T are chosen so that

$$\mathcal{R}(s, \chi) = \mu(A/\pi)^s \Gamma(s + |m/2|)\zeta(2s) \sum_{\tau \in T} \chi(\mathbf{Z}\tau + \mathbf{Z}) E_{-m}(\tau, s), \tag{7}$$

where $\mathcal{R}(s, \chi)$ is $L(s, \chi)$ times its standard gamma factors. Then we have (Theorem 5.2):

(II-0) *If $z \notin \Gamma T$, $g(z, s)$ is meromorphic in s with poles only in $\{1/2\} \cup S(\Gamma, m)$; put $g_s(z) = g(z, s)$ for $s \notin \{1/2\} \cup S(\Gamma, m)$.*

(II-1) *g_s is a Γ-automorphic form of weight m with singularities in ΓT.*

(II-2) *$[L_m - s(1 - s)]g_s = 0$ outside ΓT.*

(II-3) *$g_s(z) = (1 - 2s)^{-1}\mathcal{R}(s, \chi)y^{1-s} + O(e^{-\pi y})$ as $y \to \infty$.*

We again observe that g_s is rapidly decreasing if and only if $\mathcal{R}(s, \chi) = 0$. This time (II-2) shows that g_s is an eigenfunction of L_m, but in order to gain a better comprehension of g_s, we have to look more closely at its singularities. Since g_s is a linear combination $\Sigma_{\tau \in T} c_s(\tau) G_s(z, \tau)$ with complex numbers $c_s(\tau)$, it can be obtained as an integral

$$\int_{\Gamma \backslash H} G_s(z, w) \sum_{\tau \in T} c_s(\tau)\delta_\tau(w) d_H w \tag{8}$$

with Dirac's point-distribution δ_τ. Thus g_s should be viewed as a solution of (1) with $\Sigma_\tau c_s(\tau)\delta_\tau$ as p. In fact, for each $\tau \in T$, we have

$$g_s(z) = (2\pi)^{-1} e_\tau c_s(\tau) \log|z - \tau| + q_{\tau,s}(z)$$

in a neighborhood of τ, where $q_{\tau,s}$ is continuous at τ and e_τ is a certain integer. Also, (7) may be written in the form

$$(9) \qquad \mathcal{R}(s, \chi) = \int_{\Gamma\backslash H} E_{-m}(w, s) \sum_{\tau} c_s(\tau)\delta_\tau(w) d_H w.$$

Now the analogues of (I-4, 5, 6) can be given as follows (Theorem 5.5):

(II-4) $\operatorname{Im}[s(1-s)]\langle g_s, g_s\rangle = -\Sigma_\tau \operatorname{Im}[c_s(\tau)q_{\tau,s}(\tau)]$ *if* $\mathcal{R}(s, \chi) = 0$.

(II-5) $4t\,\Sigma_\tau \operatorname{Im}[c_s(\tau)q_{\tau,s}(\tau)] = |\mathcal{R}(s, \chi)|^2$ *if* $s = (1/2) + it$ *with* $0 \neq t \in \mathbf{R}$.

(II-6) $(2s-1)\langle E_m(z, \bar{s}), g_s\rangle = d\mathcal{R}(s, \chi)/ds$ *if* $\mathcal{R}(s, \chi) = 0$.

Thus there is an unmistakable parallelism between the two cases. It is this parallelism that sustains our interpretation of the zeros of a zeta function in terms of the inhomogeneous equation (1). Though we treat only these two types, it is expected that many more cases can be comprehended according to this principle. Such probabilities are enhanced by the fact or expectations that many other kinds of zeta functions are given by integrals similar to (6) or (9). The reader is referred to Section 6 for more comments on this. It should be noted that the cases with congruence subgroups present some nontrivial difficulties because of their plural inequivalent cusps. We touch little on this, except that our treatment of the second type include groups of prime power level.

Another related problem concerns a *canonical* expression of a Hecke L-function as a linear combination of special values of Eisenstein series, of which (7) is an example, and which is far from obvious for an arbitrary Hecke character of K. Since this is a subject of independent interest, we devote two full sections (4 and 7) to the study of such an expression for a Hecke L-function of an arbitrary CM-field, though we shall not consider equation (1) in the higher-dimensional case.

The introduction would be utterly defective if a work [6] of Hejhal were not mentioned. Indeed, he observed in the paper that the constant term of $G_s(z, \omega)$ for $\omega = e^{2\pi i/3}$ is essentially $\zeta(2s)^{-1}$ times the zeta function of $\mathbf{Q}(\omega)$ and therefore $G_s(z, \omega)$ is rapidly decreasing if the zeta function vanishes at s. The author wishes to acknowledge his indebtedness to this and also to thank heartily Dennis Hejhal for clarifying some technical aspects of G_s on several occasions and communicating some of his observations, though they are not incorporated in the present paper.

1. Automorphic forms with logarithmic singularities. Throughout the paper, we denote by H the upper half plane as in (2) of the introduction. For every 2×2-matrix α, we denote by a_α, b_α, c_α, and d_α the entries of α in the standard order. We let every α of $SL_2(\mathbf{R})$ act on H as usual by $\alpha(z) = \alpha z = (a_\alpha z + b_\alpha)/(c_\alpha z + d_\alpha)$, and define two factors of automorphy $j_\alpha(z)$ and $J_\alpha(z)$ by

$$j_\alpha(z) = j(\alpha, z) = c_\alpha z + d_\alpha, \tag{1.1a}$$

$$J_\alpha(z) = J(\alpha, z) = j_\alpha(z)/|j_\alpha(z)| \qquad (\alpha \in SL_2(\mathbf{R}),\ z \in H). \tag{1.1b}$$

Given an integer m, a function f on H, and $\alpha \in SL_2(\mathbf{R})$, we define a function $f\|_m\alpha$ on H by

$$(f\|_m\alpha)(z) = J_\alpha(z)^{-m}f(\alpha z) \qquad (z \in H). \tag{1.2}$$

We also define differential operators δ_m, ϵ_m, and L_m acting on C^∞ functions f on H or on its open subsets by

$$\delta_m f = y^{1-(m/2)}(\partial/\partial z)(y^{m/2}f) = y\partial f/\partial z - (mi/4)f, \tag{1.3a}$$

$$\epsilon_m f = -y^{1+(m/2)}(\partial/\partial \bar{z})(y^{-m/2}f) = -y\partial f/\partial \bar{z} + (mi/4)f, \tag{1.3b}$$

$$\begin{aligned} L_m &= -y^2(\partial^2/\partial x^2 + \partial^2/\partial y^2) + miy\partial/\partial x \\ &= 4\delta_{m-2}\epsilon_m - m(m-2)/4 = 4\epsilon_{m+2}\delta_m - m(m+2)/4, \end{aligned} \tag{1.3c}$$

where $z = x + iy$. As explained in the introduction, our main object of study is an automorphic form f on H satisfying an inhomogeneous equation $(L_m - \lambda)f = p$ with $\lambda \in \mathbf{C}$ and a function p on H. In this section, we prove some basic formulas on such an f when p is a function of Dirac's type, with no reference to L-functions. The connection with L-functions will be made in Section 5.

We fix a discrete subgroup Γ of $SL_2(\mathbf{R})$ containing -1 such that $\Gamma\backslash H$ has a finite measure. For a point w that is either a cusp of Γ or a point of H we put

$$\Gamma_w = \{\gamma \in \Gamma \mid \gamma(w) = w\}, \qquad e_w = e(w) = [\Gamma_w : \{\pm 1\}], \tag{1.4}$$

e_w being defined only when $w \in H$. We take a subset Ξ of $SL_2(\mathbf{R})$ so that

(1.5a) $\{\xi(\infty) \mid \xi \in \Xi\}$ *is a complete set of* Γ*-inequivalent cusps*;

(1.5b) $$\xi\begin{pmatrix}1 & 1\\ 0 & 1\end{pmatrix}\xi^{-1} \textit{ generates } \Gamma_{\xi(\infty)}/\{\pm 1\}.$$

Though Ξ can be empty, we are primarily interested in the case of non-empty Ξ.

We now consider a $\mathbf{C}$-valued measurable function f on H such that

(1.6) $$f\|_m\gamma = \theta(\gamma)f \textit{ for every } \gamma \in \Gamma,$$

where θ is a character of Γ with values in $\{z \in \mathbf{C} \mid |z| = 1\}$ such that $\theta(-1) = (-1)^m$. If g is another such function, we put

(1.7a) $$\langle f, g\rangle_\Gamma = \int_{\Gamma\backslash H} \overline{f(z)}g(z)d_Hz \qquad (d_Hz = y^{-2}dxdy),$$

(1.7b) $$\|f\|_\Gamma = (\langle f, f\rangle_\Gamma)^{1/2},$$

whenever the integral is convergent. The subscript Γ will often be suppressed.

We now specify a set of singularities $\mathcal{Z}$ such that $\mathcal{Z} = \bigcup_{\nu=1}^{p} \Gamma\zeta_\nu$ with Γ-inequivalent points $\zeta_1, \ldots, \zeta_p$ of H, and impose the following conditions (1.8a,b,c) on f:

(1.8a) *f is C^∞ outside of $\mathcal{Z}$ and $L_mf = \lambda f$ with $\lambda \in \mathbf{C}$ there.*

(1.8b) *For every $\zeta \in \mathcal{Z}$ there is a neighborhood U of ζ, a complex number c, and a continuous function r on U such that*: (i) $f(z) = c \log|z - \zeta| + r(z)$ *for* $\zeta \neq z \in U$, (ii) *r is C^∞ on $U - \{\zeta\}$, and* (iii) $\lim_{z\to\zeta}(z - \zeta)\log|z - \zeta|(|\partial r/\partial z| + |\partial r/\partial \bar{z}|) = 0$.

(1.8c) *For every $\xi \in \Xi$, $(f\|_m\xi)(x + iy) = 0(y^\sigma)$ as $y \to \infty$ uniformly in x with a real number $\sigma < 1/2$.*

If f and g satisfy (1.6) and (1.8a,b,c), then $\langle f, g\rangle$ is meaningful; in particular $|f|$ belongs to $L^2(\Gamma\backslash H)$. We also consider a stronger condition:

(1.9) *For every $\xi \in \Xi$, $(f\|_m\xi)(x + iy) = O(e^{-ay})$ as $y \to \infty$ uniformly in x with $a > 0$.*

We say that f is *rapidly decreasing* (at every cusp of Γ) if (1.9) is satisfied.

Proposition 1.1. *Let f and g be two functions satisfying* (1.6) *and* (1.8a,b,c) *with eigenvalues λ and μ respectively and with the same $\mathcal{Z}$ and m; put*

$$f(z) = c_\nu \log|z - \zeta_\nu| + r_\nu(z), \qquad g(z) = b_\nu \log|z - \zeta_\nu| + q_\nu(z)$$

with constants c_ν, b_ν and functions r_ν, q_ν as in (1.8b). *Then*

$$(\bar{\lambda} - \mu)\langle f, g\rangle_\Gamma = 2\pi \sum_{\nu=1}^{p} [\bar{c}_\nu q_\nu(\zeta_\nu) - b_\nu \overline{r_\nu(\zeta_\nu)}]/e(\zeta_\nu),$$

where e is defined by (1.4). *In particular, we have*

$$-\mathrm{Im}(\lambda)\langle f, f\rangle_\Gamma = 2\pi \sum_{\nu=1}^{p} \mathrm{Im}[\bar{c}_\nu r_\nu(\zeta_\nu)]/e(\zeta_\nu).$$

Proof. Changing Γ for its suitable subgroup, we may assume that Γ has no elliptic elements. Let $L'_m = 4\delta_{m-2}\epsilon_m$. A simple calculation shows that

$$d(\bar{f}\epsilon_m g \cdot y^{-1} d\bar{z}) = 2i(\overline{\epsilon_m f} \cdot \epsilon_m g - (1/4)\bar{f}L'_m g)d_H z.$$

Take a sufficiently small positive number ρ so that the sets

$$\xi(\{x + iy \mid 0 \leqq x < 1, y > 1/\rho\}) \qquad (\xi \in \Xi),$$

$$K_\nu = \{z \in H \mid |z - \zeta_\nu| < \rho\} \qquad (1 \leqq \nu \leqq p)$$

are mapped into $\Gamma\backslash H$ with no overlapping. Let D be the complement of the union of these sets on the Riemann surface $\Gamma\backslash H$. Then the above equality shows that

$$2i \int_D \overline{\epsilon_m f} \cdot \epsilon_m g d_H z - (i/2) \int_D \bar{f} \cdot L'_m g d_H z = \int_{\partial D} \bar{f}\epsilon_m g \cdot y^{-1} d\bar{z}.$$

Exchange f and g, and take the complex conjugate of the resultant equality; add it to the original one. Since $L_m - L'_m = m(2 - m)/4$, we thus obtain

(1.10)

$$(i/2)\int_D (\overline{L_m f}\cdot g - \bar{f} L_m g) d_H z = \int_{\partial D} (\bar{f}\epsilon_m g\cdot y^{-1} d\bar{z} + g\overline{\epsilon_m f}\cdot y^{-1} dz).$$

Let S denote the line segment connecting $i\rho$ to $1+i\rho$. Then $-\partial D = \Sigma_{\nu=1}^{p} \partial K_\nu + \Sigma_{\xi\in\Xi} \xi(S)$. Let σ be a real number as in (1.8c), which we can assume to be common to both f and g and to all $\xi \in \Xi$. Now we have

$$(1.11) \qquad (\epsilon_m f)\|_{m-2}\xi = O(y^\sigma) \textit{ as } y \to \infty \textit{ uniformly in } x.$$

This will be shown at the end of the proof. Therefore, since $\sigma < 1/2$, the integral over $\xi(S)$ tends to 0 as ρ tends to 0. To compute

$$(*) \qquad \int_{\partial K_\nu} g\overline{\epsilon_m f}\cdot y^{-1} dz$$

we observe that

$$\overline{\epsilon_m f} = -(\bar{c}_\nu y/2)(z-\zeta_\nu)^{-1} - (\bar{c}_\nu mi/4)\log\rho - y\partial\bar{r}_\nu/\partial z - (mi/4)\overline{r_\nu(z)}$$

on ∂K_ν. Then it is easy to see that the integral of (*) is of the form

$$-b_\nu \bar{c}_\nu \pi i\cdot\log\rho - \bar{c}_\nu \pi i q_\nu(\zeta_\nu) + o(1) \qquad (\rho\to 0).$$

The integral of $\bar{f}\epsilon_m g\cdot y^{-1} d\bar{z}$ over ∂K_ν can be computed in a similar way. Since the left-hand side of (1.10) tends to $(i/2)(\bar{\lambda}-\mu)\langle f, g\rangle$, we obtain the desired formula.

It remains to prove (1.11). Fix an element ξ of Ξ; let $\theta\left(\xi\begin{pmatrix}1&1\\0&1\end{pmatrix}\xi^{-1}\right) = e^{2\pi i p}$ with $p\in\mathbf{R}$. Then $f\|_m\xi$ is C^∞ on the half plane

$$H_M = \{z\in H \mid \operatorname{Im}(z) > M\}$$

for some $M > 0$. A well known argument combining the equation $L_m(f\|_m\xi) = \lambda f\|_m\xi$ with (1.8c) shows that $(f\|_m\xi)(z) = \Sigma_{r-p\in\mathbf{Z}}\, \alpha_r(y) e^{2\pi(irx-|r|y)}$ on H_M with functions α_r of the forms

$$(1.12a) \qquad \alpha_0(y) = \begin{cases} by^s + b'y^{1-s} & \text{if } \lambda \neq 1/4, \\ by^{1/2} + b'y^{1/2}\log y & \text{if } \lambda = 1/4, \end{cases}$$

(1.12b) $\alpha_r(y) = b_r y^s \Psi(s - (m/2)\operatorname{sgn}(r), 2s; 4\pi|r|y)$ if $r \neq 0$,

where b, b', and b_r are constants, s is a complex number such that $s(1-s) = \lambda$ and $\operatorname{Re}(s) \leqq 1/2$, and $\Psi(a, c; z)$ is a holomorphic function on $\mathbf{C}^2 \times (-i)H$ such that

$$\Psi(a, c; z) = \Gamma(a)^{-1} \int_0^\infty e^{-tz} t^{a-1}(1+t)^{c-a-1} dt \tag{1.13}$$

if $ia \in H$ and $iz \in H$.

We now need

LEMMA 1.2. *Let V be a compact subset of $\mathbf{C}$ and $h(z, s)$ a function on $H_M \times V$. Suppose that h is C^∞ in z and each derivative $\partial^{a+b}h/\partial x^a \partial y^b$ is continuous on $H_M \times V$, and h has an expansion of the form*

$$h(z, s) = \sum_{0 \neq r \equiv p (\mathrm{mod}\ \mathbf{Z})} b_r(s) y^s e^{2\pi(irx - |r|y)} \Psi(s_r, 2s; 4\pi|r|y),$$

where $p \in \mathbf{R}$, $s_r = s - (m/2)\operatorname{sgn}(r)$, and the b_r are continuous functions on V. Then $h(z, s) = O(e^{-cy})$ as $y \to \infty$ uniformly in (x, s) with any $c > 0$ such that $c < 2\pi|r|$ for all r appearing in the sum.

Proof. It is well known that $\lim_{t\to\infty} t^a \Psi(a, c; t) = 1$ uniformly for (a, c) in any compact subset of $\mathbf{C}^2$ (see [2, p. 278], [14, (10.8)], for example). Therefore we can find $\eta > M$ such that

$$1/2 < |t^{s_r} \Psi(s_r, 2s; t)| < 2 \quad \text{if} \quad t > \eta \quad \text{and} \quad s \in V.$$

It is also well known that the Fourier expansion of a C^∞ function with continuous parameters is absolutely and locally uniformly convergent. Hence we see that

$$\sum_r e^{-(2\pi|r| - c)y} |b_r(s) y^s (4\pi|r|y)^{-s_r}| \tag{1.14}$$

is convergent for $y > \eta$ uniformly in s with any c as in our lemma. Now we can find $\tau > \eta$ such that each term of (1.14) is decreasing as y is increasing and $y \geqq \tau$. Therefore, if $y \geqq \tau$, we have

$$\sum_r e^{-2\pi|r|y}|b_r(s)y^s\Psi(s_r, 2s; 4\pi|r|y)|$$
$$\leqq 2e^{-cy}\sum_r e^{-(2\pi|r|-c)\tau}|b_r(s)\tau^s(4\pi|r|\tau)^{-s_r}|.$$

Since the last sum is bounded for $s \in V$, we obtain our assertion.

Applying this lemma to $f\|_m\xi$, we see from condition (1.8c) that $\alpha_0(y) = by^s$ with $\mathrm{Re}(s) < 1/2$. This combined with the relation $L_{m-2}\epsilon_m f = \lambda\epsilon_m f$ shows that

$$(\epsilon_m f)\|_{m-2}\xi = cy^s + \sum_{r\neq 0}\beta_r(y)e^{2\pi(irx-|r|y)}$$

on H_M with β_r of type (1.12b) with $m - 2$ instead of m. Therefore (1.11) follows from the above lemma.

PROPOSITION 1.3. *In the setting of Proposition* 1.1, *suppose* $\lambda = \bar{\mu} = s(1-s)$ *with* $s \in \mathbf{C}$; *suppose* f *and* g *satisfy* (1.6), (1.8a,b), *and, instead of* (1.8c), *the following condition*:

(1.15) *For every* $\xi \in \Xi$, *there exist constants* b_ξ, c_ξ, *and a positive number* a *such that*

$$(f\|_m\xi)(z) = c_\xi y^{1-s} + O(e^{-ay})$$
$$(g\|_m\xi)(z) = b_\xi y^{\bar{s}} + O(e^{-ay}) \qquad (y \to \infty).$$

Then we have

$$(1-2\bar{s})\sum_{\xi\in\Xi} b_\xi\bar{c}_\xi = 2\pi\sum_{\nu=1}^{p}[\bar{c}_\nu q_\nu(\zeta_\nu) - b_\nu\overline{r_\nu(\zeta_\nu)}]/e(\zeta_\nu).$$

Proof. We repeat the proof of Proposition 1.1. This time the left-hand side of (1.10) is 0. The limits of the integrals over $\xi(S)$, however, may be nonzero. In fact,

$$(\epsilon_m f)\|_{m-2}\xi = (ic_\xi/2)(s - 1 + m/2)y^{1-s} + O(e^{-cy})$$

for some $c > 0$, and hence we find that

$$\int_{\xi(S)}\overline{\epsilon_m f}\cdot gy^{-1}dz \to (ib_\xi\bar{c}_\xi/2)(1 - \bar{s} - m/2) \qquad (\rho \to 0).$$

Computing the other integral over $\xi(S)$ in the same way, we obtain the desired formula.

Let us conclude this section by recalling the definition of a cusp form. If f satisfies (1.6) and (1.8a,c) with $\mathfrak{Z} = \varnothing$ and $\sigma < 0$, then f is called a *cusp form.* The following facts are well known: (i) every cusp form is rapidly decreasing; (ii) a function satisfying (1.6) and (1.8a) with $\mathfrak{Z} = \varnothing$ belongs to $L^2(\Gamma\backslash H)$ if and only if it is either a cusp form or a constant; the constant is precluded if $m \neq 0$ or $\theta \neq 1$; (iii) if f is a nonzero cusp form and $L_m f = \lambda f$, then $|m|(2 - |m|)/4 \leqq \lambda \in \mathbf{R}$.

Given (Γ, θ, m), we take a maximal orthonormal set $\Phi = \Phi(\Gamma, \theta, m) = \{\varphi\}$ of cusp forms of type (Γ, θ, m), adding a constant function if $m = 0$ and $\theta = 1$. For each $\varphi \in \Phi$, we put $L_m\varphi = \lambda_\varphi\varphi$. Let $\Lambda(\Gamma, \theta, m)$ denote the set of such eigenvalues λ_φ for all $\varphi \in \Phi$. Then we let $S(\Gamma, \theta, m)$ denote the set of all complex numbers s such that $s(1 - s) \in \Lambda(\Gamma, \theta, m)$. Thus $S(\Gamma, \theta, m)$ contains 0 and 1 if $m = 0$ and $\theta = 1$.

2. The kernel function $G_s(z, w)$. Let Γ, θ, and m be the same as in Section 1. We define functions $k_s(z, w)$ and $G_s(z, w; \theta, m)$ for $s \in \mathbf{C}$ and $(z, w) \in H^2$ by

(2.1a) $$G_s(z, w; \theta, m) = (1/2) \sum_{\gamma\in\Gamma} \theta(\gamma) J_\gamma(w)^m k_s(z, \gamma w),$$

(2.1b) $k_s(z, w) =$

$$-(4\pi)^{-1} i^m |z - \bar{w}|^m (z - \bar{w})^{-m} (1 - t)^s \tilde{F}(s + m/2, s - m/2, 2s; 1 - t),$$

where $t = |z - w|^2/|z - \bar{w}|^2$ and

(2.1c) $$\tilde{F}(a, b, c; z) = \Gamma(a)\Gamma(b)\Gamma(c)^{-1} F(a, b, c; z)$$

with the standard hypergeometric function F. As mentioned in the introduction, the function G_s was introduced by Roelcke [9], and later investigated by several authors [1, 3, 4, 7, 8]. Let us now recall some of its basic properties given in these papers, in particular in Hejhal [7], following the formulation of that article (with a few notational changes). First of all, we have

(2.2) $$k_s(\alpha z, \alpha w) = J_\alpha(z)^m J_\alpha(w)^{-m} k_s(z, w) \textit{ for every } \alpha \in \mathrm{SL}_2(\mathbf{R}),$$

(2.3a) $$G_s(z, w) = (1/2) \sum_{\gamma\in\Gamma} \theta(\gamma)^{-1}J_\gamma(z)^{-m}k_s(\gamma z, w),$$

(2.3b) $G_s(z, w) = \theta(\gamma)^{-1}J_\gamma(z)^{-m}G_s(\gamma z, w) = \theta(\gamma)J_\gamma(w)^m G_s(z, \gamma w)$ *for every* $\gamma \in \Gamma$.

Here and henceforth we simply write $G_s(z, w)$ for our function if there is no fear of confusion.

The series for G_s in (2.1a) is meaningful and convergent for $\mathrm{Re}(s) > 1$, $\Gamma(s \pm m/2) \neq \infty$, and for $z \notin \Gamma w$. Now $G_s(z, w)$ when $z \notin \Gamma w$ can be continued to the whole plane as a meromorphic function of s, with poles only in the following set:

(2.4) $$P(\Gamma, \theta, m) = S(\Gamma, \theta, m) \cup \{1/2\} \cup P'(\Gamma, \theta, m),$$

where P' is the set of poles of certain Eisenstein series which we will describe below. More precisely, for every fixed $s_0 \in \mathbf{C}$, there exists a nonnegative integer α and a neighborhood U of s_0 such that $(s - s_0)^\alpha G_s(z, w)$ is C^∞ on $U \times \{(z, w) \in H^2 \mid z \notin \Gamma w\}$ and holomorphic in s; we can take $\alpha = 0$ if $s_0 \notin P(\Gamma, \theta, m)$. When G_s is finite, it satisfies

(2.5) $$L_m G_s(z, w) = s(1 - s)G_s(z, w),$$

where z is the variable of differentiation.

Fix an element ξ of Ξ: put $\theta\left(\xi\begin{pmatrix}1 & 1\\ 0 & 1\end{pmatrix}\xi^{-1}\right) = e^{2\pi i p}$ with $p \in \mathbf{R}$. Then G_s has the following Fourier expansion at $\xi(\infty)$:

(2.6) $$J_\xi(z)^{-m}G_s(\xi z, w) = (1 - 2s)^{-1}y^{1-s}E_\xi(w, s; \theta^{-1}, -m)$$

$$\sum_{0\neq r\equiv p \pmod{\mathbf{Z}}} e^{2\pi(irx-|r|y)}F_{\xi r}(w, s)y^s\Psi(s_r, 2s; 4\pi|r|y).$$

Here the symbols are as follows: $z = x + iy$, s_r and Ψ are as in Lemma 1.2 and (1.13): $F_{\xi r}$ is a certain explicitly defined function; E_ξ is given by

(2.7) $$E_\xi(w, s; \theta, m) = \begin{cases} \sum_{\alpha\in\Gamma_t\backslash\Gamma} \theta(\alpha)^{-1}\,\mathrm{Im}(w)^s\|_m\xi^{-1}\alpha & \text{if } p \in \mathbf{Z}, \\ 0 & \text{if } p \notin \mathbf{Z}, \end{cases}$$

where $t = \xi(\infty)$. The expansion of (2.6) is valid for sufficiently large y

when w stays in a compact set and $s \notin P(\Gamma, \theta, m)$. It can be shown that E_ξ of (2.7) can be continued as a meromorphic function of s to the whole $\mathbf{C}$. Now $P'(\Gamma, \theta, m)$ of (2.4) is the set of poles of the E_ξ for all $\xi \in \Xi$. It is known that $s = 1$ or $\mathrm{Re}(s) < 1/2$ for every $s \in P'(\Gamma, \theta, m)$. We also note here

$$(2.8) \quad G_s(z, w) - G_{1-s}(z, w) = (1 - 2s)^{-1} \sum_{\xi \in \Xi} E_\xi(z, s: \theta, m) E_\xi(w, 1 - s; \bar{\theta}, -m)$$

(see [7, p. 319, p. 414]). Fix a real number $a > \mathrm{Max}(1, |m|/2)$. Then

$$(2.9) \quad G_s(z, w) - G_a(z, w) = Q_s(z, w) + \sum_{\varphi \in \Phi} \{[s(1 - s) - \lambda_\varphi]^{-1} - [a(1 - a) - \lambda_\varphi]^{-1}\} \varphi(z)\overline{\varphi(w)}$$

with a function $Q_s(z, w)$ which is meromorphic on the whole s-plane with poles only in $\{1/2\} \cup P'(\Gamma, \theta, m)$ (see [7, p. 250, p. 318, p. 414]).

Let us now study the nature of $G_s(z, w)$ as a function of z when w is fixed, by modifying it as follows: for each $s \in \mathbf{C}$ not belonging to $\{1/2\} \cup P'(\Gamma, \theta, m)$, we put

$$(2.10) \quad f_s(z, w) = f_{s,w}(z) = \lim_{S \to s} \{G_S(z, w) - \sum [S(1 - S) - \lambda_\varphi]^{-1} \varphi(z)\overline{\varphi(w)}\},$$

where the sum is taken over all φ in Φ such that $s(1 - s) = \lambda_\varphi$. Obviously $f_s = G_s$ if $s \notin S(\Gamma, \theta, m)$. We easily see that $f_{s,w}$ is well defined for $z \notin \Gamma w$.

PROPOSITION 2.1. *For a fixed w and $s \notin \{1/2\} \cup P'(\Gamma, \theta, m)$, $f_{s,w}$ as a function of z satisfies*

$$(2.11) \qquad [L_m - s(1 - s)] f_{s,w} = \sum_{s(1-s)=\lambda_\varphi} \overline{\varphi(w)}\varphi;$$

$f_{s,w}$ is not identically equal to 0 if and only if $\theta(\alpha)J_\alpha(w)^m = 1$ for every $\alpha \in \Gamma_w$. If it is not 0, we have

$$(2.12) \qquad f_{s,w}(z) = (2\pi)^{-1} e_w \log|z - w| + r_{s,w}(z)$$

for z in a sufficiently small neighborhood U of w with a continuous function $r_{s,w}$ on U. Moreover

$$(2.13) \qquad r_{s,w}(z) = p_{s,w}(z) + q_{s,w}(z)\log|z - w|$$

with C^∞ functions $p_{s,w}$ and $q_{s,w}$ on U such that $q_{s,w}(w) = 0$. Furthermore, for every $\zeta \in \Xi$, we have

$$(f_{s,w}\|_m \xi)(z) - (1 - 2s)^{-1}y^{1-s}E_\xi(w, s; \bar{\theta}, -m) = O(e^{-cy})$$

with some $c > 0$ uniformly in x.

Proof. Put $P_S(z, w) = G_S(z, w) - \Sigma' [S(1 - S) - \lambda_\varphi]^{-1}\varphi(z)\overline{\varphi(w)}$ with φ running over Φ under the condition $\lambda_\varphi = s(1 - s)$. Then P is holomorphic in S on a small disc $|S - s| < \epsilon$ and $L_m P_S - S(1 - S)P_S = \Sigma' \varphi(z)\overline{\varphi(w)}$ from which we obtain (2.11). Now we see from (2.6) and Lemma 1.2 that for every $\xi \in \Xi$ and for $|S - s| = \epsilon/2$,

$$J_\xi(z)^{-m}P_S(\xi z, w) - (1 - 2S)^{-1}y^{1-S}E_\xi(w, S; \bar{\theta}, -m) = O(e^{-cy})$$

with some $c > 0$. Applying the maximum principle to the left-hand side, we obtain the last assertion. (The result also follows from (2.15) below.) Next we observe that f_s satisfies formula (2.3b), from which the "only-if"-part of the second assertion follows immediately. To prove the remaining part, we employ a formula (see [2, p. 74, (2)])

$$\tilde{F}(a, b, a + b; 1 - t) = A(t) - B(t)\log(t) \qquad (0 < t < 1)$$

with convergent power series A and B on $|t| < 1$ such that $B(0) = 1$, provided Γ is finite at a, b, and $a + b$. From this we can easily derive that

(2.14)

$$k_s(z, w) = (2\pi)^{-1} \log|z - w| + A_s(z, w) + B_s(z, w)\log|z - w|$$

with real analytic A_s and B_s on H^2 such that $B_s(z, z) = 0$. Take a neighborhood U of w so that $\gamma U \cap U = \varnothing$ for every $\gamma \in \Gamma$, $\notin \Gamma_w$. Suppose $\theta(\alpha)J_\alpha(w)^m = 1$ for every $\alpha \in \Gamma_w$. If $\mathrm{Re}(s) > 1$ and $\Gamma(s \pm m/2) \neq \infty$, the existence of $r_{s,w}$, $p_{s,w}$, and $q_{s,w}$ immediately follows from (2.14) and (2.1a).

If $\mathrm{Re}(s) \leqq 1$ or $\Gamma(s \pm m/2) = \infty$, it follows from (2.9) and (2.10). Once we have (2.12), it shows that $f_{s,w}$ is a nonzero function.

Now for a fixed (s, w) as above, $f_{s,w}$ satisfies (1.6) and (1.8b) with $\mathcal{Z} = \Gamma w$. Moreover the above proposition shows that $f_{s,w}\|_m \xi = O(y^{1-s})$ for every $\xi \in \Xi$. Therefore $f_{s,w}$ satisfies (1.8c) if $\mathrm{Re}(s) > 1/2$. In any case, $r_{s,w}(w)$ is meaningful; for simplicity, we put $r_s(w) = r_{s,w}(w)$. As for (1.8a), however, it is satisfied by $f_{s,w}$ only if $s \notin S(\Gamma, \theta, m)$ or if $\varphi(w) = 0$ for every φ such that $\lambda_\varphi = s(1-s)$. In such a case, the Fourier expansion of $f_{s,w}$ is a mere specialization of (2.6). If $s \in S(\Gamma, \theta, m)$, however, it can be shown that $f_{s,w}$ for a fixed (s, w) has an expansion of the form

$$(2.15)\quad f_{s,w}\|_m \xi = (1-2s)^{-1} y^{1-s} E_\xi(\tau, s; \bar{\theta}, -m) + \sum_{0 \neq r \equiv p(\mathbf{Z})} \{p_{\xi r} V_r(y, s) + q_{\xi r} \partial V_r/\partial s(y, s)\} e^{2\pi(irx - |r|y)}$$

for every $\xi \in \Xi$ and sufficiently large y, where $V_r(y, s) = y^s \Psi(s_r, 2s; 4\pi|r|y)$ and $p_{\xi r}$ and $q_{\xi r}$ are constants.

LEMMA 2.2. *If $s \notin \{1/2\} \cup P'(\Gamma, \theta, m)$, we have $\overline{f_s(z, w)} = f_{\bar{s}}(w, z)$; moreover $\overline{r_s(w)} = r_{\bar{s}}(w)$ if $\theta(\alpha) J_\alpha(w)^m = 1$ for every $\alpha \in \Gamma_w$.*

Proof. We first observe that $\overline{G_s(z, w)} = G_{\bar{s}}(w, z)$ and $\overline{Q_s(z, w)} = Q_{\bar{s}}(w, z)$. Therefore the first equality follows immediately from (2.10). Suppose $\mathrm{Re}(s) > 1$ and $\Gamma(s \pm m/2) \neq \infty$; put

$$C_s(z, w) = (1/2) \sum_{\alpha \notin \Gamma_w} \theta(\alpha) J_\alpha(w)^m k_s(z, \alpha w).$$

Then $r_{s,w}(z) = C_s(z, w) + e_w A_s(z, w) + e_w B_s(z, w) \log|z - w|$ with A_s and B_s of (2.14). Since $\overline{k_s(z, w)} = k_{\bar{s}}(w, z)$, we see that $A_{\bar{s}}(z, w) = \overline{A_s(w, z)}$. By (2.2), we have $\overline{C_s(z, w)} = (1/2) \sum_{\beta \notin \Gamma_w} \theta(\beta) J_\beta(z)^m k_{\bar{s}}(w, \beta z)$, and hence $\overline{C_s(w, w)} = C_{\bar{s}}(w, w)$. Thus we obtain $\overline{r_s(w)} = r_{\bar{s}}(w)$ when $\mathrm{Re}(s) > 1$ and $\Gamma(s \pm m/2) \neq \infty$. In the general case, we have

$$r_s(w) = Q_s(w, w) + r_a(w) + \sum{}' \{[s(1-s) - \lambda_\varphi]^{-1} - [a(1-a) - \lambda_\varphi]^{-1}\} |\varphi(w)|^2 - \sum{}'' [a(1-a) - \lambda_\varphi]^{-1} |\varphi(w)|^2,$$

where a is as in (2.9), Σ' resp. Σ'' is the sum over all $\varphi \in \Phi$ such that $\lambda_\varphi \neq s(1-s)$ resp. $\lambda_\varphi = s(1-s)$. Therefore we obtain $\overline{r_s(w)} = r_{\bar{s}}(w)$.

We now fix a finite set of points T of Γ-inequivalent points of H. Fixing an s not contained in $\{1/2\} \cup P'(\Gamma, \theta, m)$, put

$$f = \sum_{\tau \in T} c(\tau) f_{s,\tau}, \qquad g = \sum_{\tau \in T} b(\tau) f_{s,\tau}, \tag{2.16a}$$

$$\psi_c = \sum_{\tau \in T} \sum_{\varphi} c(\tau)\overline{\varphi(\tau)}\varphi, \qquad \psi_b = \sum_{\tau \in T} \sum_{\varphi} b(\tau)\overline{\varphi(\tau)}\varphi \tag{2.16b}$$

with complex numbers $c(\tau)$ and $b(\tau)$, and with φ running over Φ under the condition $\lambda_\varphi = s(1-s)$. Then for each $\tau \in T$ we have

$$f(z) = c(\tau)(2\pi)^{-1} e_\tau \log|z - \tau| + r_{f,\tau}(z), \tag{2.17a}$$

$$g(z) = b(\tau)(2\pi)^{-1} e_\tau \log|z - \tau| + r_{g,\tau}(z) \tag{2.17b}$$

in a neighborhood of τ, where

$$r_{f,\tau} = c(\tau) r_{s,\tau} + \sum_{\tau' \neq \tau} c(\tau') f_{s,\tau'}, \tag{2.18a}$$

$$r_{g,\tau} = b(\tau) r_{s,\tau} + \sum_{\tau' \neq \tau} b(\tau') f_{s,\tau'}. \tag{2.18b}$$

PROPOSITION 2.3. *The notation being as above, the following assertions hold*:

(1) *If* $\mathrm{Re}(s) > 1/2$ *or both* f *and* g *are rapidly decreasing, then*

$$2i \cdot \mathrm{Im}[s(1-s)]\langle f, g\rangle + \langle f, \psi_b\rangle - \langle \psi_c, g\rangle$$

$$= \sum_{\tau \in T} [b(\tau)\overline{r_{f,\tau}(\tau)} - \overline{c(\tau)} r_{g,\tau}(\tau)]$$

$$= \sum_{\tau \in T} b(\tau)\overline{c(\tau)}[\overline{r_s(\tau)} - r_s(\tau)] + \sum_{\tau \neq \tau'} b(\tau')\overline{c(\tau)}[\overline{f_s(\tau', \tau)} - f_s(\tau, \tau')].$$

(2) *If* $s = (1/2) + it$ *with* $0 \neq t \in \mathbf{R}$, *we have* $\langle \psi_c, g\rangle = \langle f, \psi_b\rangle$ *and*

$$\sum_{\xi \in \Xi} \Big\{\sum_{\tau \in T} b(\tau) E_\xi(\tau, \bar{s}; \bar{\theta}, -m)\Big\}\Big\{\sum_{\tau \in T} \overline{c(\tau) E_\xi(\tau, s; \theta, m)}\Big\}$$

$$= 2it \sum_{\tau \in T} [b(\tau)\overline{r_{f,\tau}(\tau)} - \overline{c(\tau)}r_{g,\tau}(\tau)]$$

$$= 4t \sum_{\tau \in T} b(\tau)\overline{c(\tau)}\operatorname{Im}[r_s(\tau)] + 2it \sum_{\tau \neq \tau'} b(\tau')\overline{c(\tau)}[\overline{f_s(\tau', \tau)} - f_s(\tau, \tau')].$$

We note here special cases of these formulas:

(2.19a) $$\operatorname{Im}[r_s(\tau)] = \operatorname{Im}[\langle \Sigma\, \overline{\varphi(\tau)}\varphi, f_{s,\tau}\rangle - s(1-s)\|f_{s,\tau}\|^2]$$

if $\operatorname{Re}(s) > 1/2$ *or* $f_{s,\tau}$ *is rapidly decreasing.*

(2.19b) $$\langle \Sigma\, \overline{\varphi(\tau)}\varphi, f_{s,\tau}\rangle \in \mathbf{R} \quad \text{if} \quad \operatorname{Re}(s) = 1/2.$$

In these two formulas, φ runs over Φ under the condition $\lambda_\varphi = s(1-s)$.

(2.19c) $$4t \cdot \operatorname{Im}[r_s(\tau)] = \sum_{\xi \in \Xi} |E_\xi(\tau, s; \theta, m)|^2$$

$$\text{if} \quad s = (1/2) + it \quad \text{with} \quad 0 \neq t \in \mathbf{R}.$$

(2.19d) $$2it[\overline{f_s(\tau', \tau)} - f_s(\tau, \tau')] = \sum_{\xi \in \Xi} E_\xi(\tau, s; \theta, m)E(\tau', \bar{s}; \bar{\theta}, -m)$$

if $\tau \notin \Gamma\tau'$ *and* $s = (1/2) + it$ *with* $0 \neq t \in \mathbf{R}$.

Proof. If $s \notin S(\Gamma, \theta, m)$, the formula of (1) is an immediate consequence of Proposition 1.1. Suppose $s \in S(\Gamma, \theta, m)$; let P_S be as in the proof of Proposition 2.1. For every $\xi \in \Xi$ we see from Lemma 1.2 and its proof that $\epsilon_m(P_S(z, \tau)\|_m \xi)$ is of the form $Ay^{1-S} + O(e^{-by})$ with $b > 0$ uniformly for $|S - s| = \rho$ with a sufficiently small $\rho > 0$. Then the maximum principle tells that $\epsilon_m(f_{s,\tau}\|_m \xi)$ is of the form $ay^{1-s} + O(e^{-by})$. Repeating the proof of Proposition 1.1, we find that the left-hand side of (1.10) tends to $\operatorname{Im}[s(1-s)]\langle f, g\rangle + (i/2)\{\langle \psi_c, g\rangle - \langle f, \psi_b\rangle\}$ and therefore we obtain (1). To prove (2), we first observe that

$$(1 - 2s)[f_s(z, w) - f_{1-s}(z, w)] = \sum_{\xi \in \Xi} E_\xi(z, s; \theta, m)E_\xi(w, 1 - s; \bar{\theta}, -m).$$

If $s = (1/2) + it \neq 1/2$, this together with Lemma 2.2 yields (2.19c,d). Taking a linear combination, we obtain the second part of (2). Define c_ξ

and b_ξ as in (1.15) for the present f and g. Then we have

$$-2itc_\xi = \sum_{\tau\in T} c(\tau)E_\xi(\tau, s; \bar{\theta}, -m),$$

$$-2itb_\xi = \sum_{\tau\in T} b(\tau)E_\xi(\tau, s; \bar{\theta}, -m),$$

and hence

$$4t^2 \sum_\xi b_\xi\bar{c}_\xi = \sum_\xi \Big\{\sum_{\tau\in T} b(\tau)E_\xi(\tau, s; \bar{\theta}, -m)\Big\}\Big\{\sum_{\tau\in T} \overline{c(\tau)}E_\xi(\tau, \bar{s}; \theta, m)\Big\}$$

$$= \sum_{\tau,\tau'} b(\tau')\overline{c(\tau)} \sum_\xi E_\xi(\tau', s; \bar{\theta}, -m)E_\xi(\tau, 1-s; \theta, m).$$

By (2.8), we see that the last sum over ξ is invariant under $s \mapsto 1-s$. Hence $4t^2 \sum_\xi b_\xi\bar{c}_\xi$ is equal to the last sum of (2). Repeat the proofs of Propositions 1.1 and 1.3 with the present f and g. Then the left-hand side of (1.10) tends to $(i/2)\{\langle\psi_c, g\rangle - \langle f, \psi_b\rangle\}$ and the right-hand side to

$$(2\bar{s}-1)(i/2)\sum_\xi b_\xi\bar{c}_\xi + (i/2)\sum_{\tau\in T}[\overline{c(\tau)}r_{g,\tau}(\tau) - b(\tau)\overline{r_{f,\tau}(\tau)}]$$

as shown in the proof of Proposition 1.3. This sum must be 0, because of the last formula of (2). Thus we obtain $\langle\psi_c, g\rangle = \langle f, \psi_b\rangle$, which completes the proof.

It should be noted that a formula for $\|f_{s,\tau}\|$ of the same type as (1) of Proposition 2.3 was given by Elstrodt [1, II, Satz 7.2] when $\mathrm{Re}(s) > 1$ and $\Gamma(s \pm m/2) \neq \infty$.

We now specialize our group Γ to the congruence subgroup

$$\Gamma_0(W) = \{\gamma \in SL_2(\mathbf{Z}) \mid c_\gamma \equiv 0 \pmod{W}\}. \tag{2.20}$$

Here W is a positive integer. We take a Dirichlet character θ modulo W such that $\theta(-1) = (-1)^m$ and put $\theta(\gamma) = \theta(d_\gamma)$ for $\gamma \in \Gamma_0(W)$; we understand that $\theta(d_\gamma) = 1$ for every γ if $W = 1$. Recall that every cusp of $\Gamma_0(W)$ is $\Gamma_0(W)$-equivalent to $t = u/v$ with a positive divisor v of W and an integer u prime to v such that $0 < u \leqq (v, W/v)$. Now we have

LEMMA 2.4. *Suppose W is the conductor of θ; let $t = u/v$ with u and*

v as above. Then Γ_t contains an element β such that $\mathrm{Tr}(\beta) = 2$ *and* $\theta(\beta) \neq 1$ *if and only if v is not prime to W/v.*

Proof. Take integers x and y so that $uy - vx = 1$; let $\epsilon = \begin{pmatrix} u & x \\ v & y \end{pmatrix}$, $\alpha = \begin{pmatrix} 1 & q \\ 0 & 1 \end{pmatrix}$ with $q \in \mathbf{Q}$, and $\beta = \epsilon\alpha\epsilon^{-1}$. Then $\epsilon(\infty) = t$ and

$$(*) \qquad \beta = \begin{pmatrix} 1 - quv & qu^2 \\ -qv^2 & 1 + quv \end{pmatrix}.$$

This is contained in Γ_t if and only if $\beta \in \mathrm{SL}_2(\mathbf{Z})$ and W divides qv^2; moreover Γ_t consists of the elements $\pm\beta$ with all such q's. Suppose $(v, W/v) = 1$ and $\beta \in \Gamma_t$. Then $q = kW/v$ with $k \in \mathbf{Z}$, so that $quv \in W\mathbf{Z}$. Thus $\theta(d_\beta) = 1$, which proves the "only-if"-part. Suppose $(v, W/v) \neq 1$. Fix a prime p dividing $(v, W/v)$, and put $W = p^e M$ and $v = p^f N$ with integers M and N prime to p. Then $0 < f < e$. Put $r = \mathrm{Max}(0, e - 2f)$ and take $p^r M/N$ as q in (*). Then $\beta \in \Gamma_t$ and $1 + quv = 1 + p^{r+f}Mu$. Considering $(\mathbf{Z}_p/W\mathbf{Z}_p)^\times$ as a direct product factor of $(\mathbf{Z}/W\mathbf{Z})^\times$ in a natural way, denote by θ_p the restriction of θ to $(\mathbf{Z}_p/W\mathbf{Z}_p)^\times$. Then $\theta(1 + quv) = \theta_p(1 + p^{r+f}Mu)$. Let $G_i = \{a \in \mathbf{Z}_p^\times \mid a - 1 \in p^i\mathbf{Z}_p\}$. Since $p \nmid Mu$ and $e \leqq 2r + 2f < 2e$, we see that $1 + p^{r+f}Mu$ generates G_{r+f}/G_e. Therefore $\theta_p(1 + p^{r+f}Mu) \neq 1$, and hence $\theta(d_\beta) \neq 1$, which completes the proof.

Choose Ξ for $\Gamma_0(W)$ and define G_s and E_ξ with respect to $\Gamma_0(W)$ and a primitive θ. Then the above lemma shows that $E_\xi = 0$ unless $\xi(\infty)$ is equivalent to $1/v$ with a positive divisor v of W such that v is prime to W/v. In particular, if W is a prime power >1, then E_ξ is a nonzero function only if $\xi(\infty)$ is equivalent to 0 or ∞.

3. Main theorems in the first case. In this section we confine ourselves to the case $\Gamma = \mathrm{SL}_2(\mathbf{Z})$, though many, if not all, of our results can be generalized to the case of congruence subgroups. We consider G_s and E_ξ of Section 2 with $\Gamma = \mathrm{SL}_2(\mathbf{Z})$ by taking θ to be the trivial character; thus we write $G_s(z, w; m)$ or $G_s(z, w)$; naturally m must be even and Ξ consists of the identity element. Thus the function of (2.7) simply becomes

$$(3.1a) \qquad E(z, s; m) = \sum_{\alpha \in \Gamma_\infty \backslash \Gamma} \mathrm{Im}(z)^s \|_m \alpha.$$

We put then

$$(3.1b) \qquad E^*(z, s; m) = \pi^{-s}\Gamma(s + |m|/2)\zeta(2s)E(z, s; m),$$

where ζ is Riemann's zeta function. It is well known that $s(1-s)E^*$ can be continued to a real analytic function on $H \times \mathbf{C}$ that is entire in s and invariant under $s \mapsto 1-s$ (cf. Section 5).

We are going to consider the inhomogeneous equation $(L_m - \lambda)f = p$ with a function p on H satisfying the following condition:

(3.2a) *p is C^∞ and satisfies $p\|_m\gamma = p$ for every $\gamma \in \Gamma$.*

We then have a Fourier expansion

$$p(z) = \sum_{n\in\mathbf{Z}} e^{2\pi inx}\psi_n(y) \tag{3.2b}$$

with C^∞ functions ψ_n. We impose the following condition on them:

(3.2c) *For every $\eta > 0$, there exist positive constants A, B, and c such that $|\psi_n(y)| \leqq A(|n|+1)^B e^{-(2\pi|n|+c)y}$ for every n and $y > \eta$.*

We easily see that $e^{cy}p(z)$ as well as $e^{cy}\psi_n(y)$ is bounded; thus p is rapidly decreasing. We now put

$$R(s) = \int_{\Gamma\backslash H} p(z)E(z, s; -m)d_H z, \tag{3.3a}$$

$$R^*(s) = \pi^{-s}\Gamma(s + |m/2|)\zeta(2s)R(s). \tag{3.3b}$$

The integral of (3.3a) is convergent if E is finite at s. Moreover we see that $s(1-s)R^*(s)$ is an entire function and $R^*(s) = R^*(1-s)$. The purpose of this section is to study the function

$$f(z, s) = -\int_{\Gamma\backslash H} G_s(z, w; m)p(w)d_H w \tag{3.4}$$

in connection with R. The properties of G_s recalled in Section 2 imply that for every $s_0 \in \mathbf{C}$, there is a nonnegative integer k such that $(s-s_0)^k\Gamma(s + |m/2|)\zeta(2s)G_s(z, w)$ as a function of w is slowly increasing uniformly on a neighborhood of s_0; we can take $k = 0$ if $s \notin \{1/2\} \cup S(\Gamma, 1, m)$ as will be seen in Lemma 5.1 below. Therefore $f(z, s)$ is well-defined as a meromorphic function of s as stated in (i) of the following

THEOREM 3.1. *The function $f(z, s)$ has the following properties*:

(i) $\Gamma(s + |m/2|)\zeta(2s)f(z, s)$ *is meromorphic on the whole s-plane with poles only in* $\{1/2\} \cup S(\Gamma, 1, m)$.

(ii) $f(z, s)$ *is a* C^∞ *function in* (z, s) *if* $s \notin P(\Gamma, 1, m)$; *moreover, for every* $s_0 \in \mathbf{C}$, *there exists a neighborhood U of* s_0 *and a nonnegative integer n such that* $(s - s_0)^n f(z, s)$ *is* C^∞ *on* $H \times U$.

(iii) $f(\gamma z, s) = J_\gamma(z)^m f(z, s)$ *for every* $\gamma \in \Gamma$.

(iv) $[L_m - s(1 - s)]f(z, s) = p(z)$ *if* $s \notin P(\Gamma, 1, m)$.

(v) $(2s - 1)[f(z, s) - f(z, 1 - s)] = R(s)E(z, 1 - s; m)$
$= R(1 - s)E(z, s; m)$.

(vi) *If K is a compact subset of* $\mathbf{C}$ *disjoint with* $P(\Gamma, 1, m)$, *then* $f(z, s) - (2s - 1)^{-1}y^{1-s}R(s) = O(e^{-by})$ *as* $y \to \infty$ *uniformly for* $(x, s) \in \mathbf{R} \times K$, *where* $b = \mathrm{Min}(c, \pi)$.

Assertion (iii) follows immediately from (2.3b). If $\mathrm{Re}(s) > 1$ and $\Gamma(s \pm m/2) \neq \infty$, (ii) and (iv) follow from a well known principle. In fact, if f is defined by (3.4) with any bounded C^∞ function p and such an s, then f is C^∞ and satisfies (iv) (see [9], [3], and [7, p. 645]). To prove (ii) for an arbitrary s, take any $a > \mathrm{Max}(1, |m/2|)$, and put $Z_s = G_a - G_s$. Then

$$f(z, s) - f(z, a) = \int_{\Gamma\backslash H} Z_s(z, w)p(w)d_H w.$$

Thus our problem is the differentiability of the right-hand side. Take a point $(z_0, s_0) \in H \times \mathbf{C}$; let U denote a neighborhood of (z_0, s_0) which will be made smaller according to our requirements. Since $G_s(z, w; m) = G_s(w, z; -m)$, (2.6) implies that for a suitable nonnegative integer α, G_s as a function of w has an expansion

$$(3.5)\quad (s - s_0)^\alpha G_s(z, u + iv) = r_0(z, s)v^{1-s} + \sum_{0 \neq n \in \mathbf{Z}} e^{2\pi(inu - |n|v)} r_n(z, s)v^s \Psi(s_n', 2s; 4\pi|n|v)$$

for $v > M_U$ and $(z, s) \in U$, where $s_n' = s + (m/2)\mathrm{sgn}(n)$, r_0 and r_n are C^∞ on U, and M_U is a constant depending on U. Since the left-hand side is C^∞ in (z, w, s), we see that

$$(3.6)\quad (\partial^{a+b+c}/\partial x^a \partial y^b \partial s^c)[(s - s_0)^\alpha G_s(z, u + iv)]$$

for every nonnegative a, b, c has a similar expansion. By Lemma 1.2, we see that (3.6) is $O(v^{1-s})$ uniformly on a suitable U. The same is true for G_a with a in place of s. Hence if a is sufficiently large, the partial derivatives of $(s - s_0)^\alpha Z_s$ with respect to (x, y, s) are all $O(v^{1-s})$ uniformly on U. This proves (ii). By analytic continuation, we then obtain (iv) for any $s \notin P(\Gamma, 1, m)$. Now (2.8) specialized to the present case shows

$$(1 - 2s)[G_s(z, w) - G_{1-s}(z, w)] = E(z, 1 - s; m)E(w, s; -m),$$

which together with (3.3a) proves (v).

The principal feature of our theorem is in (vi), whose proof is not so simple. We first observe that the automorphy property of f together with its differentiability shows that

$$f(z, s) = \sum_{n\in\mathbf{Z}} q_n(y, s)e^{2\pi inx} \tag{3.7}$$

with functions q_n which are holomorphic or meromorphic in s according as f itself is holomorphic or meromorphic. To compute q_n, put

$$-G_s^*(z, w) = \sum_{n\in\mathbf{Z}} k_s(z + n, w) = \sum_{n\in\mathbf{Z}} k_s(z, w + n). \tag{3.8}$$

This is meaningful for $z - w \notin \mathbf{Z}$, $\mathrm{Re}(s) > 1$, and $\Gamma(s \pm m/2) \neq \infty$. Obviously

$$f(z, s) = \int_T G_s^*(z, w)p(w)d_H w \qquad (T = \Gamma_\infty\backslash H). \tag{3.9}$$

Now G_s^* has a Fourier expansion of the form

$$G_s^*(z, w) = \sum_{n\in\mathbf{Z}} e^{2\pi in(x-u)-2\pi|n|(y+v)}g_n(y, v, s) \tag{3.10}$$

with functions g_n given as follows (see [7, p. 351], [8, Lemma 4.3]):

$$g_0(y, v; s) = (2s - 1)^{-1}\cdot\begin{cases} y^s v^{1-s} & (y < v), \\ y^{1-s}v^s & (y > v), \end{cases} \tag{3.11a}$$

(3.11b) $g_n(y, v; s) = |4\pi n|^{2s-1}\Gamma(s_n)y^s v^s$

$$\cdot\begin{cases}\Omega(s_n, 2s; 4\pi|n|y)\Psi(s_n, 2s; 4\pi|n|v) & (n \neq 0, y < v),\\ \Psi(s_n, 2s; 4\pi|n|y)\Omega(s_n, 2s; 4\pi|n|v) & (n \neq 0, y > v).\end{cases}$$

Here $s_n = s - (m/2)\mathrm{sgn}(n)$, $z = x + iy$, $w = u + iv$, Ψ is the function of (1.13), and $\Omega(a, c; z)$ is a holomorphic function on $\mathbf{C}^3$ with an integral representation

(3.12) $$\Omega(a, c; z) = \Gamma(a)^{-1}\Gamma(c-a)^{-1}\int_0^1 e^{zu}u^{a-1}(1-u)^{c-a-1}du,$$

valid for $\mathrm{Re}(c) > \mathrm{Re}(a) > 0$. The expansion of (3.10) is valid if $y \neq v$, $\mathrm{Re}(s) > 1$, and $\Gamma(s \pm m/2) \neq \infty$.

LEMMA 3.2. *Given any compact subset X of $\mathbf{C}^2$ and a positive number η, there exist positive constants A and B depending only on X and η such that if $(a, c) \in X$, one has*

$$|\Psi(a, c; y)| < Ay^B \quad \text{for} \quad y \geqq \eta,$$

$$|\Psi(a, c; y)| < Ay^{-B} \quad \text{for} \quad 0 < y \leqq \eta,$$

$$|\Omega(a, c\, y)| < Ay^B e^y \quad \text{for} \quad y \geqq \eta,$$

$$|\Omega(a, c; y)| < A \quad \text{for} \quad 0 < y \leqq \eta.$$

Moreover $\Psi(a, c; y)y^a \to 1$ as $y \to \infty$ uniformly for $(a, c) \in X$.

These are well known; see [2, p. 262, p. 278]. For the function Ψ, see also [12, Theorem 3.1] and [14, Section 10]. The method in [14] is also applicable to Ω. We need an estimate of the incomplete gamma function

$$\gamma(a, b; y) = \int_y^\infty e^{-bt}t^{a-1}dt \qquad (a \in \mathbf{C}, b > 0, y > 0).$$

We easily find that $\gamma(a, b; y) = y^a e^{-by}\Psi(1, a+1; by)$. Hence Lemma 3.2 shows that

(3.13) $$|\gamma(a, b; y)| \leqq Ae^{-by}(by)^B|y^a| \quad \text{for} \quad by > \eta > 0$$

when a stays in a compact subset L of $\mathbf{C}$, where A and B are positive constants depending only on L and η.

From (3.3a), (3.1a), and (3.2b), we obtain

$$R(s) = \int_T p(u+iv)v^s d_H w \qquad (T = \Gamma_\infty \backslash H) \tag{3.14}$$

$$= \int_0^\infty \psi_0(v) v^{s-2} dv$$

for $\operatorname{Re}(s) > 1$. Now (3.9) shows that

$$f(z, s) = \int_0^\infty \int_0^1 G_s^*(z, u+iv) p(u+iv) du \cdot v^{-2} dv.$$

From (3.10) and (3.2b) we obtain

$$\int_0^1 G_s^*(z, u+iv) p(u+iv) du = \sum_{n \in \mathbf{Z}} e^{2\pi(inx - |n|y - n|v)} g_n(y, v, s) \psi_n(v).$$

Therefore we have formally

$$f(z, s) = \sum_{n \in \mathbf{Z}} e^{2\pi(inx - |n|y)} f_n(y, s), \tag{3.15a}$$

$$f_n(y, s) = \int_0^\infty e^{-2\pi|n|v} \psi_n(v) g_n(y, v, s) v^{-2} dv. \tag{3.15b}$$

This formal calculation can be justified if

$$\sum_{n \neq 0} e^{-2\pi|n|y} \int_0^\infty e^{-2\pi|n|v} |\psi_n(v) g_n(y, v, s)| v^{-2} dv \tag{3.16}$$

is convergent. To see this convergence, we take a positive number η and a compact subset K of $\{s \in \mathbf{C} \mid \Gamma(s \pm m/2) \neq \infty\}$ and make estimates of integrals when $s \in K$ and $y > \eta$. Let us first consider f_0. By (3.11a) and (3.14) we have

$$(2s-1)f_0(y,s) = y^{1-s}\int_0^y \psi_0(v)v^{s-2}dv + y^s\int_y^\infty \psi_0(v)v^{-s-1}dv$$

$$= R(s)y^{1-s} - y^{1-s}\int_y^\infty \psi_0(v)v^{s-2}dv + y^s\int_y^\infty \psi_0(v)v^{-1-s}dv$$

if $\mathrm{Re}(s) > 1$. From (3.13) and (3.2c), we obtain

$$|(2s-1)f_0(y,s) - R(s)y^{1-s}| \leqq A_1 e^{-cy}y^{B_1}$$

for $y > \eta$ and $s \in K$. Here and in the following $A_i, B_i, C_i, \ldots$ are positive constants which depend only on K, η, and φ and are independent of n.

Next suppose $n \neq 0$. Then

$$f_n(y,s) = |4\pi n|^{2s-1}\Gamma(s_n)y^s[U(y,s) + V(y,s)],$$

$$U(y,s) = \Psi(s_n, 2s; 4\pi|n|y)\int_0^y e^{-2\pi|n|v}\psi_n(v)\Omega(s_n, 2s; 4\pi|n|v)v^{s-2}dv,$$

$$V(y,s) = \Omega(s_n, 2s; 4\pi|n|y)\int_y^\infty e^{-2\pi|n|v}\psi_n(v)\Psi(s_n, 2s; 4\pi|n|v)v^{s-2}dv.$$

Let $U^*(y,s)$ and $V^*(y,s)$ denote the functions obtained from the above expressions for U and V by replacing Ψ, Ω, and $\psi_n(v)v^{s-2}$ by their absolute values. By (3.2c), Lemma 3.2, and (3.13) we have

$$V^*(y,s) \leqq A_2|n|^{B_2}y^{C_2}e^{4\pi|n|y}\int_y^\infty e^{-cv-4\pi|n|v}v^{D_2}dv$$

$$\leqq A_3|n|^{B_3}y^{C_3}e^{-cy}$$

if $s \in K$ and $y > \eta$. As for $U(y,s)$, we first put

(3.17a) $$b_n(s) = |4\pi n|^{2s-1}\Gamma(s_n)c_n(s),$$

(3.17b) $$c_n(s) = \int_0^\infty e^{-2\pi|n|v}\psi_n(v)\Omega(s_n, 2s, 4\pi|n|v)v^{s-2}dv,$$

and let $c_n^*(s)$ denote the integral whose integrand is the absolute value of that of c_n. By (3.2c) and Lemma 3.2, we see that c_n is holomorphic in s for $\mathrm{Re}(s) > M$ with a sufficiently large M and moreover $c_n^*(s) \leqq A_4|n|^{B_4}$ if $s \in K$ and $\mathrm{Re}(s) > M$. Now

$$\begin{aligned} c_n^*(s)|\Psi(s_n, 2s; 4\pi|n|y)| &- U^*(y, s) \\ &= |\Psi(s_n, 2s; 4\pi|n|y)| \int_y^\infty e^{-2\pi|n|v}|\psi_n(v)\Omega(s_n, 2s; 4\pi|n|v)v^{s-2}|dv \\ &\leqq A_5|n|^{B_5}y^{C_5}e^{-cy} \end{aligned}$$

by (3.2c), Lemma 3.2, and (3.13), if $s \in K$ and $y > \eta$. Combining these estimates, we obtain

$$f_0(y, s) = (2s-1)^{-1}y^{1-s}R(s) + h_0(y, s), \tag{3.18a}$$

$$f_n(y, s) = b_n(s)y^s\Psi(s_n, 2s; 4\pi|n|y) + h_n(y, s) \qquad (n \neq 0) \tag{3.18b}$$

with functions $h_n(y, s)$ such that

$$|h_n(y, s)| \leqq A_6(|n|+1)^{B_6}y^{C_6}e^{-cy} \quad \text{for} \quad s \in K \quad \text{and} \quad y > \eta. \tag{3.19}$$

At the same time we see that (3.16) is majorized by

$$A_7 \sum_{n \neq 0} e^{-2\pi|n|y}|n|^{B_7}y^{C_7}$$

if $s \in K$, $\mathrm{Re}(s) > M$, and $y > \eta$. Since the last series is convergent, we have thus established (3.15b) for $\mathrm{Re}(s) > M$ with some M. Since h_n is essentially a sum of several integrals over $[y, \infty)$ which are absolutely and locally uniformly convergent on $\mathbf{C}$, we easily see that $\Gamma(s_n)^{-1}h_n$ is an entire function of s. Moreover it can be easily seen that $\Gamma(s_n)^{-1}h_n$ is C^∞ in (y, s) and in particular $\Gamma(s_n)^{-1}\partial h_n/\partial y$ is holomorphic in s. Since f_n is well-defined by (3.15a) as a function with the differentiability as stated in (ii), (3.18b) shows that b_n can be continued to a meromorphic function on the whole $\mathbf{C}$ independent of y. Therefore we have

$$\begin{aligned} f(z, s) - (2s-1)^{-1}y^{1-s}R(s) &- \sum_{n\in\mathbf{Z}} e^{2\pi(inx-|n|y)}h_n(y, s) \\ &= \sum_{0\neq n\in\mathbf{Z}} e^{2\pi(inx-|n|y)}b_n(s)y^s\Psi(s_n, 2s; 4\pi|n|y). \end{aligned} \tag{3.20}$$

Now the left-hand side of (3.20) is holomorphic on the domain

$$\{s \in \mathbf{C} \mid s \notin P(\Gamma, 1, m), \Gamma(s \pm m/2) \neq \infty\}. \tag{3.21}$$

Since $\Psi(a, b; y) \neq 0$ for sufficiently large y, b_n must be holomorphic on this domain. Applying Lemma 1.2 to (3.20), we find that it is $O(e^{-\pi y})$ locally uniformly on (3.21). Combining this with (3.19), we obtain (vi) of our theorem for K contained in (3.21). Suppose now s_0 is a pole of $\Gamma(s + m/2)$ or $\Gamma(s - m/2)$, but $s_0 \notin P(\Gamma, 1, m)$. Put

$$A(z, s) = e^{by}\{f(z, s) - (2s - 1)^{-1}y^{1-s}R(s)\}.$$

Then the above argument shows that $(s - s_0)A(z, s)$ is bounded when $y > \eta$ and $|s - s_0| \leqq \epsilon$ for some η and ϵ. Now $A(z, s)$ is holomorphic at s_0 and therefore the maximum principle shows that $A(z, s)$ itself is bounded for $y > \eta$ and $|s - s_0| \leqq \epsilon$. This completes the proof.

THEOREM 3.3. *Define b_n and c_n by* (3.17a,b) *with ψ_n of* (3.2b), Ω *of* (3.12), *and $s_n = s - (m/2)\operatorname{sgn}(n)$. Then the integral for c_n is convergent for sufficiently large* $\operatorname{Re}(s)$, *and c_n can be continued to a meromorphic function on the whole plane such that $\Gamma(s + |m/2|)\zeta(2s)c_n(s)$ has poles only in $\{1/2\} \cup S(\Gamma, 1, m)$. Moreover, put*

$$g(z, s) = (2s - 1)^{-1}y^{1-s}R(s) + \sum_{0 \neq n \in \mathbf{Z}} e^{2\pi(inx - |n|y)}b_n(s)y^s\Psi(s_n, 2s; 4\pi|n|y).$$

Then g is a C^∞ function for $z \in H$ and s in the set of (3.21), *meromorphic on the whole s-plane; moreover we have $L_m g(z, s) = s(1 - s)g(z, s)$,*

$$(2s - 1)[g(z, s) - g(z, 1 - s)] = R(s)E(z, 1 - s; m) = R(1 - s)E(z, s; m),$$

$$(2s - 1)[|4\pi n|^{-s}b_n(s) - |4\pi n|^{s-1}b_n(1 - s)] = R(s)\omega_n(1 - s) = R(1 - s)\omega_n(s),$$

where ω_n is defined by (3.23b) *below.*

Proof. We have already proved these assertions except the last few equalities. That g is an eigenfunction follows immediately from its expression. Put $W_n(y, s) = |4\pi ny|^s \Psi(s_n, 2s, 4\pi|n|y)$ for $n \neq 0$ and $y > 0$. A well known functional equation for Ψ (see [2, p. 257, (6)], [14, (10.10)]) implies that $W_n(y, s) = W_n(y, 1 - s)$. Now we have

$$g(z, s) = (2s - 1)^{-1} y^{1-s} R(s) + \sum_{n \neq 0} e^{2\pi(inx - |n|y)} |4\pi n|^{-s} b_n(s) W_n(y, s).$$

It is well known that $E(z, s; m)$ has the following expansion (see [14, p. 17], for example):

$$(3.22)\quad E(z, s; m) = y^s + \omega_0(s) y^{1-s} + \sum_{0 \neq n \in \mathbf{Z}} e^{2\pi(inx - |n|y)} \omega_n(s) W_n(y, s),$$

$$(3.23\text{a})\quad \omega_0(s) =$$

$$\pi i^{-m} 2^{2-2s} \Gamma(2s - 1) \zeta(2s - 1) / [\Gamma(s + m/2) \Gamma(s - m/2) \zeta(2s)],$$

$$(3.23\text{b})\quad \omega_n(s) = i^{-m} \pi^s |n|^{-s} [\Gamma(s + (m/2)\operatorname{sgn}(n)) \zeta(2s)]^{-1} \sum_{0 < d | n} d^{2s-1}.$$

Comparing the Fourier coefficients of both sides of (v) of Theorem 3.1 and those of (3.20), we find for $n \neq 0$ that

$$(3.24)\quad h_n(y, s) - h_n(y, 1 - s) = W_n(y, s)\{(2s - 1)^{-1} R(s) \omega_n(1 - s)$$

$$- |4\pi n|^{-s} b_n(s) + |4\pi n|^{s-1} b_n(1 - s)\}.$$

For a generic s, $|4\pi ny|^k W_n(y, s)$ tends to 1 as $y \to \infty$, where $k = -\operatorname{sgn}(n) m/2$. Therefore (3.19) shows that (3.24) must vanish. A similar vanishing holds also for $n = 0$. This proves the desired equalities.

It should be noted that $g(z, s)$ is not an automorphic form. If it were (for a generic s), it must be an Eisenstein series, which is a contradiction, as it has no term for y^s.

THEOREM 3.4. *The notation being the same as in Theorem* 3.1, *put* $f_s(z) = f(z, s)$ *for* $s \notin P(\Gamma, 1, m)$. *Then the following assertions hold*:

(1) f_s *is not identically equal to* 0 *on* H;
(2) f_s *is rapidly decreasing if and only if* $R(s) = 0$;

(3) $\mathrm{Im}[\langle p, f_s\rangle] = \mathrm{Im}[s(1-s)]\|f_s\|^2$ *if* $R(s) = 0$ *or* $\mathrm{Re}(s) > 1/2$;
(4) $\mathrm{Im}[\langle f_s, p\rangle] = (4t)^{-1}|R(s)|^2$ *if* $s = (1/2) + it$ *with* $0 \neq t \in \mathbf{R}$;
(5) $(1-2s)\langle E(z, \bar{s}; m), f_s\rangle = dR(s)/ds$ *if* $R(s) = 0$.

Proof. The first two assertions follow directly from Theorem 3.1, (iv) and (vi). To prove the remaining ones, let us write s_0 for a fixed point outside $P(\Gamma, 1, m)$ and put $f_* = f(z, s_0)$ and $\lambda_0 = s_0(1-s_0)$. We observe that

$$(3.25)\quad \epsilon_m f_* =$$

$$(i/2)(s_0 - 1 + m/2)(2s_0 - 1)^{-1}y^{1-s_0}R(s_0) + O(e^{-by}) \qquad (y \to \infty).$$

In fact, Lemma 3.2 combined with the formulas

$$\partial\Omega(a, c; y)/\partial y = a\Omega(a+1, c+1; y),$$

$$\partial\Psi(a, c; y)/\partial y = -a\Psi(a+1, c+1; y)$$

gives an estimate of $\partial h_n/\partial y$ similar to (3.19), and also that of the effect of ϵ_m on the right-hand side of (3.20) in view of Lemma 1.2. If $R(s_0) = 0$, $\langle f_*, f_*\rangle$ is meaningful, and $\epsilon_m f_*$ is rapidly decreasing. Since $(L_m - \lambda_0)f_* = p$, we have $\langle L_m f_*, f_*\rangle - \bar{\lambda}_0\langle f_*, f_*\rangle = \langle p, f_*\rangle$. Moreover $\langle L_m f_*, f_*\rangle = \langle f_*, L_m f_*\rangle$ as can be seen from the proof of Proposition 1.1. Therefore we obtain (3). We can easily verify that the same conclusion holds if $\mathrm{Re}(s) > 1/2$. To prove (5), put $\lambda = s(1-s)$ and $E(z, s) = E(z, s; m)$. That f_* and $\epsilon_m f_*$ are rapidly decreasing guarantees the equality

$$\langle E(z, \bar{s}), L_m f_*\rangle = \langle L_m E(z, \bar{s}), f_*\rangle,$$

from which we can easily derive

$$(\lambda - \lambda_0)\langle E(z, \bar{s}), f_*\rangle = \langle E(z, \bar{s}), \varphi\rangle = R(s).$$

Taking the limit when s tends to s_0, we obtain (5). Finally as for (4), a modification of Proposition 1.2 as in the proof of (2) of Proposition 2.3 gives the desired formula. (See also Remark (C) below.)

So far we have treated $f(z, s)$ with p only under conditions (3.2a,c). To make it more specific, take

$$p(z) = y^{(k+l)/2}\overline{g(z)}h(z) \tag{3.26}$$

with holomorphic cusp forms g and h of weight k and l, respectively, with respect to Γ, with expansions

$$g(z) = \sum_{n=1}^{\infty} a_n e^{2\pi inz} \quad \text{and} \quad h(z) = \sum_{n=1}^{\infty} b_n e^{2\pi inz}. \tag{3.27}$$

For simplicity, put

$$m = l - k, \qquad \kappa = (k + l)/2. \tag{3.28}$$

Then p satisfies (3.2a) with this m. Moreover, if we define ψ_n by (3.2b), then

$$\psi_n(y) = y^{\kappa} e^{-2\pi|n|y} \sum_{r=1}^{\infty} e^{-4\pi ry} \cdot \begin{cases} \bar{a}_r b_{n+r} & \text{if} \quad n \geqq 0, \\ \bar{a}_{r-n} b_r & \text{if} \quad n < 0. \end{cases} \tag{3.29}$$

Since $a_n = O(n^{\alpha})$ and $b_n = O(n^{\beta})$ with some α and β, we can easily verify (3.2c) with any $c < 4\pi$. Put

$$D(s) = \sum_{n=1}^{\infty} \bar{a}_n b_n n^{-s}, \tag{3.30a}$$

$$R_0(s) = (4\pi)^{-s}\Gamma(s)D(s). \tag{3.30b}$$

Define $R(s)$ and $f(z, s)$ for the present p. Then (3.14) shows that $R(s) = R_0(s + \kappa - 1)$. Thus the assertions of the above theorems for p of (3.26) concern the series of (3.30a). In this case $c_n(s)$ can be expressed as an explicit infinite series. In fact we have

COROLLARY 3.5. *For cusp forms g and h as in* (3.27) *and* $0 \neq n \in \mathbf{Z}$, *put*

$$A_n(s) = |n|^{s-1}\Gamma(2s)^{-1}\Gamma(s + \kappa - 1)\Gamma(s_n)$$
$$\cdot \begin{cases} \displaystyle\sum_{r=1}^{\infty} \bar{a}_r b_{r+n}(r + n)^{1-\kappa-s} F(s_n, s + \kappa - 1, 2s; n/(r + n)) & \textit{if} \quad n > 0, \\ \displaystyle\sum_{r=1}^{\infty} \bar{a}_{r-n} b_r (r - n)^{1-\kappa-s} F(s_n, s + \kappa - 1, 2s; n/(r - n)) & \textit{if} \quad n < 0, \end{cases}$$

where $s_n = s - (m/2)\operatorname{sgn}(n)$. *Then these series are absolutely convergent for sufficiently large* $\operatorname{Re}(s)$. *Moreover* A_n *can be continued to a meromorphic function on the whole plane that satisfies*

$$(2s-1)(4\pi)^{-\kappa}[A_n(s) - A_n(1-s)]$$
$$= \omega_n(s)R_0(\kappa - s) = \omega_n(1-s)R_0(s+\kappa-1)$$

with ω_n *of* (3.23b), R_0 *of* (3.30b), *and* $\kappa = (k+l)/2$. *Moreover* $\Gamma(s_n)^{-1}\Gamma(s+|m/2|)\zeta(2s)A_n(s)$ *has poles only in* $\{1/2\} \cup S(\Gamma, 1, m)$.

A result of the same type was obtained by Goldfeld in [5] in the case where g and h are of the same weight.

Proof. Employing a well known formula

$$F(a, b, c; z) = \frac{\Gamma(c)}{\Gamma(a)\Gamma(c-a)} \int_0^1 t^{a-1}(1-t)^{c-a-1}(1-tz)^{-b}dt,$$

valid for $\operatorname{Re}(c) > \operatorname{Re}(a) > 0$ and $|\arg(1-z)| < \pi$, we see from (3.12) that

$$\int_0^\infty e^{-\beta y}\Omega(a, c; \gamma y)y^{b-1}dy = \Gamma(b)\Gamma(c)^{-1}\beta^{-b}F(a, b, c; \gamma/\beta)$$

if $\beta > \gamma > 0$. Therefore b_n of (3.17a) is equal to $(4\pi)^{s-\kappa}|n|^s A_n(s)$. Thus our assertions immediately follow from Theorem 3.3.

Let us conclude this section by a few remarks.

(A) If we take p of (3.26) assuming that $g = h$ and that g is a common eigenfunction of all Hecke operators, then

$$\pi^{-s}\Gamma(s)\zeta(2s)R(s)/[\pi^{-s/2}\Gamma(s/2)\zeta(s)]$$

is an entire function as shown in [11]. Therefore, if $\zeta(s) = 0$, $\operatorname{Re}(s) > 0$, and $\zeta(2s) \neq 0$ (which is the case if $\operatorname{Re}(s) \geqq 1/2$), then $R(s) = 0$. Thus Theorem 3.4 is applicable to such zeros of the Riemann zeta function.

(B) Coming back to the general case in which p is a function satisfying (3.2a,b,c), we see from (vi) of Theorem 3.1 that if $\operatorname{Re}(s) > 1/2$, $s \notin P(\Gamma, 1, m)$, and $R(s) \neq 0$, then f_s is L^2, but not rapidly decreasing, a fact which makes a notable contrast to the case of cusp forms.

(C) As to the quantity $\langle p, f_s \rangle$ appearing in Theorem 3.4, it can also be written $[p, p]_s$, if we put

$$(3.31) \qquad [p, q]_s = -\int_{\Gamma\backslash H}\int_{\Gamma\backslash H} G_s(z, w)\overline{p(z)}q(w)d_Hzd_Hw$$

for two rapidly decreasing functions p and q satisfying (3.2a). Therefore the Riemann hypothesis for R concerns whether $[p, p]_s \in \mathbf{R}$ when $R(s) = 0$. It should also be noted that (4) of the theorem can be derived by combining (2.8) with (3.3a); also (4) can be proved under milder conditions than (3.2c).

(D) Next we consider the function $F(z, s) = q(s)f(z, s)$ with $q(s) = \pi^{-s}\Gamma(s + |m/2|)\zeta(2s)$. Then $F(z, s)$ is finite if $s \notin \{1/2\} \cup S(\Gamma, 1, m)$, and

$$[L_m - s(1 - s)]F(z, s) = q(s)p(z),$$

$$F(z, s) - (2s - 1)^{-1}y^{1-s}R^*(s) = O(e^{-by}) \qquad (y \to \infty)$$

with R^* of (3.3b). Now suppose $q(s_0) = 0$, but still $s_0 \notin \{1/2\} \cup S(\Gamma, 1, m)$. Then $F(z, s_0)$ is an eigenform of L_m. If $R^*(s_0) = 0$, $F(z, s_0)$ must be a cusp form. However, the Riemann hypothesis for ζ and R^* would imply the impossibility of the simultaneous vanishing of q and R^* for $0 < \mathrm{Re}(s_0) < 1$. (If $F(z, s_0)$ is not identically equal to 0, which is the case if s_0 is a pole of $f(z, s)$ and s_0 is a simple zero of q, then that it is a cusp form contradicts the assumption $s_0 \notin S(\Gamma, 1, m)$.) Thus it is natural to assume $R^*(s_0) \neq 0$ when $q(s_0) = 0$. Then $F(z, s_0)$ is a cyclopean form in the sense of [14, Section 9] and it must be an Eisenstein series. In other words, if s_0 is a simple zero of q, the residue of $f(z, s)$ at $s = s_0$ is such an Eisenstein series.

(E) Finally we note that (3.2c) is a rather strong condition. For instance, a nonzero holomorphic cusp form does not satisfy it. Even when it is satisfied, the corresponding f may not necessarily be a natural object. As such an example, we can take $p(z) = y^{1/2}|\eta(z)|^2$ with $\eta(z) = q^{1/24}\prod_{n=1}^{\infty}(1 - q^n)$, $q = e^{2\pi iz}$. In this case R^* of (3.3b) is essentially $\zeta(2s)\zeta(2s - 1)$ times gamma factors and therefore $\mathrm{Re}(s) = 1/2$ is not the line of its zeros. If $p(z) = y^2|\eta(z)|^8$, however, R^* is essentially $\zeta(s)^2L(s, \chi)$ times gamma factors, where χ is the Hecke (ideal) character of $\mathbf{Q}(e^{2\pi i/3})$ such that $\chi((\alpha)) = \alpha^2|\alpha|^{-2}$ if $\alpha \equiv 1 \pmod 2$ (cf. [11, (5.6)]). Therefore $\mathrm{Re}(s) = 1/2$ is the critical line in this case.

4. L-functions of a CM-field as special values of Eisenstein series. This and the following sections concern the second case in which a

Hecke L-function of an imaginary quadratic field gives the dominant term of a certain automorphic form. Though we shall eventually consider only quadratic fields, in this section we treat more generally CM-fields of arbitrary degree. Our aim is to find a *canonical expression* for a Hecke L-function of such a field as a special value of an Eisenstein series. This necessitates the introduction of nonmaximal orders and their ideals.

For an algebraic number field K of finite degree, we denote by $K_{\mathbf{A}}^{\times}$, $K_{\mathbf{a}}^{\times}$, $K_{\mathbf{f}}^{\times}$, and O_K the idele group of K, its archimedean and nonarchimedean factors, and the maximal order of K. Throughout this section, we fix a subfield F of K such that $[K : F] = 2$, and denote by ρ the nontrivial automorphism of K over F. A finite prime of F resp. K will be denoted by p resp. v; F_p and K_v then denote their completions; also we put $K_p = K \otimes_F F_p$. For $x \in K_{\mathbf{A}}^{\times}$, we denote by x_v and x_p its v-component and p-component. By an O_F-*lattice* in K, we understand a finitely generated O_F-submodule of K that spans K over F. If A is an O_F-lattice, we denote by A_p its p-closure in K_p. For $x \in K_{\mathbf{A}}^{\times}$, we denote by xA the O_F-lattice such that $(xA)_p = x_pA_p$ for every finite prime p of F. For simplicity, we put $O_p = (O_F)_p$ and similarly denote by O_v the closure of O_K in K_v. By an O_F-*order* in K, we understand an O_F-lattice in K which is a subring of K containing O_F. (Every O_F-order R in K is of the form $R = O_F + BO_K$ with an integral ideal B in F; moreover B is unique for R. We shall not need these facts, however.) If R is an O_F-order, we understand by an R-*ideal* an O_F-lattice of the form xR with $x \in K_{\mathbf{f}}^{\times}$. We call an R-ideal A *integral* if $A \subset R$. Given an integral ideal C in K, we denote by I_R^C the group of all the R-ideals of the form xR with an element x of $K_{\mathbf{f}}^{\times}$ such that $x_v = 1$ for every v dividing C. If $C = BO_K$ with an ideal B in F, we write I_R^C also I_R^B. Notice that if $\alpha \in K^{\times}$ and $\alpha R \in I_R^C$, then α is a v-unit for every $v \mid C$, but the converse is not necessarily true.

Let us now take a *Hecke character* χ of K, by which we understand a continuous homomorphism of $K_{\mathbf{A}}^{\times}$ into $\{z \in \mathbf{C} \mid |z| = 1\}$ that is trivial on $K^{\times}$. We denote by χ_v and $\chi_{\mathbf{a}}$ the restrictions of χ to $K_v^{\times}$ and $K_{\mathbf{a}}^{\times}$; if Y is an integral ideal in F or K, we put $\chi_Y = \prod_{v \mid Y} \chi_v$; we view χ_v and $\chi_{\mathbf{a}}$ as characters of $K_{\mathbf{A}}^{\times}$. Further we denote by χ^* the ideal character attached to χ. Let ψ be the Hecke character of F obtained by composing χ with the injection of $F_{\mathbf{A}}^{\times}$ into $K_{\mathbf{A}}^{\times}$. Let Z resp. W be the integral ideal in K resp. F which represents the finite part of the conductor of χ resp. ψ. We can decompose Z into the product $Z = Z_0 Z_1$ with two integral ideals Z_0 and Z_1 in K such that Z_0 and Z_0^{ρ} are divisible by the same prime ideals in K and that $Z^{\rho} + Z_1 = O_K$. Such a decomposition is obviously unique. Let $S = F \cap Z$, $S_0 =$

$F \cap Z_0$, and $Q = F \cap Z_1$. Then $QO_K = Z_1 Z_1^\rho$ and $S = S_0 Q$. Now $Q_p = W_p$ for every $p|Q$. In fact, let v be the prime of K dividing p and Z_1. Then the other prime of K dividing p does not divide Z, and hence $\psi_p(a) = \psi_v(a)$ for every p-unit a. This shows that $W_p = Z_v \cap F_p = Q_p$. Thus $W = W_0 Q$ with an integral ideal W_0 in F dividing S_0. We then put $C = W_0 Z_1$ and consider an O_F-order R of the form $R = O_F + BO_K$ with an ideal B in F satisfying condition (4.1e) below. To facilitate our later reference, we list here the properties of these ideals and order:

(4.1a) *Z is the conductor of χ; $Z = Z_0 Z_1$; Z_0 and Z_0^ρ have the same prime factors in K; $Z_1 + Z^\rho = O_K$.*

(4.1b) $$S = Z \cap F = S_0 Q, \qquad S_0 = Z_0 \cap F, \qquad Q = Z_1 \cap F.$$

(4.1c) *W is the conductor of ψ; $W = W_0 Q$, $W_0 \supset S_0$.*

(4.1d) $$C = W_0 Z_1.$$

(4.1e) $$S_0 \subset B \subset W_0^{-1} S_0.$$

(4.1f) $$R = O_F + BO_K.$$

Now we denote by $\mathfrak{N}_R^C$ the subgroup of $K^\times$ consisting of the elements α satisfying the following set of conditions:

(4.2) *$\alpha R \in I_R^C$; there is an element r of F prime to W_0 such that $\alpha - r \in W_0 R_p$ for every $p|W_0$.*

We understand that $\mathfrak{N}_R^C = K^\times$ if $C = O_K$.

LEMMA 4.1. *If $xR = R$, $x \in K_{\mathfrak{f}}^\times$, and $v \nmid C$, then $\chi_v(x) = 1$.*

Proof. This is obvious if $v \nmid Z$ since $xR = R$ implies that x_v is a v-unit. Suppose $v|Z$ and $v \nmid C$. Then v divides a prime p of F such that $p|S_0$ and $p \nmid W_0$. Since $x_p \in R_p^\times$ and $R_p \subset O_p + Z_p$, we have $x_p - s \in Z_p$ with $s \in O_p^\times$. Then $\chi_v(x) = \chi_v(s) = \psi_p(s) = 1$ since $p \nmid W$, which proves our lemma.

We define a character χ^R of I_R^C by

(4.3) $$\chi^R(xR) = \prod_{v \nmid C} \chi_v(x) \quad \textit{for} \quad xR \in I_R^C, \qquad x \in K_{\mathfrak{f}}^\times.$$

This is well-defined in view of the above lemma. We have obviously

(4.4a) $$\chi^R(A) = \chi^*(AO_K) \quad \textit{if} \quad A \in I_R^Z,$$

(4.4b) $$\chi^R(\alpha R) = \chi_C(\alpha)^{-1}\chi_{\mathbf{a}}(\alpha)^{-1} \quad \textit{if} \quad \alpha \in K \quad \textit{and} \quad \alpha R \in I_R^C.$$

Let us now assume that F is totally real and K totally imaginary. For $x \in F_{\mathbf{A}}^{\times}$ (in particular for $x \in F^{\times}$) we write $x \gg 0$ if all the archimedean components of x are positive. We define groups G, G^1, and G_+ by

(4.5)

$$G = \mathrm{GL}_2(F), \qquad G^1 = \mathrm{SL}_2(F), \qquad G_+ = \{\alpha \in G \mid \det(\alpha) \gg 0\}.$$

We consider their parabolic subgroups by putting

(4.6) $$P = \{\alpha \in G \mid c_\alpha = 0\}, \qquad P^1 = P \cap G^1, \qquad P_+ = P \cap G_+.$$

We define the adelizations $G_{\mathbf{A}}$, $G_{\mathbf{A}}^1$, $P_{\mathbf{A}}$ etc, as well as the symbols $G_{\mathbf{a}}$, $G_{\mathbf{f}}$, G_p, etc. as usual. Hereafter we use the symbols **a** and **f** to denote the sets of archimedean and nonarchimedean primes of F. Given two fractional ideals X and X' in F such that XX' is integral, we put

(4.7a) $$S[X', X] = G_{\mathbf{a}} \prod_{p \in \mathbf{f}} S_p[X', X],$$

(4.7b) $$S_p[X', X] = \{y \in G_p \mid \det(y) \in O_p^{\times},\ a_y \in O_p,$$

$$b_y \in X_p',\ c_y \in X_p,\ d_y \in O_p\},$$

(4.7c) $$S^1[X', X] = G_{\mathbf{A}}^1 \cap S[X', X],$$

(4.7d) $$\Gamma[X', X] = G_+ \cap S[X', X], \qquad \Gamma^1[X', X] = G^1 \cap S[X', X].$$

Let $H^{\mathbf{a}}$ denote the product of **a** copies of the upper half plane H, which consists of the indexed elements $(z_u)_{u \in \mathbf{a}}$ with $z_u \in H$. We let G_+ act on $H^{\mathbf{a}}$ as usual by componentwise action.

We fix a *CM-type* Φ of K, by which we mean a set of embeddings of K into $\mathbf{C}$ such that for a generic element α of K, the elements α^φ and $\alpha^{\rho\varphi}$ for $\varphi \in \Phi$ are exactly the conjugates of α over $\mathbf{Q}$. We then denote by K_Φ the subset of K consisting of the elements τ of K such that $\mathrm{Im}(\tau^{\varphi}) > 0$ for every $\varphi \in \Phi$. Identifying Φ with $\mathbf{a}$ in a natural way, we can consider K_{Φ} a subset of $H^{\mathbf{a}}$ through the map $\tau \mapsto (\tau^{\varphi})_{\varphi \in \Phi}$. Obviously K_{Φ} is stable under G_{+}.

Take an element ζ of K_{Φ} such that $\zeta^{\rho} = -\zeta$. For an O_F-lattice A in K, denote by $\lambda_{\zeta}(A)$ the O_F-linear span of $\mathrm{Tr}_{K/F}(\zeta\alpha\beta^{\rho})$ for all α and β in A, and by $\lambda(A)$ the ideal class of $\lambda_{\zeta}(A)$ modulo $\mathbf{a}$ in F. This is independent of the choice of ζ. Given a fractional ideal X in F, we put

$$I_R^C(X) = \{A \in I_R^C \mid X \in \lambda(A)\}, \tag{4.8a}$$

$$H_X^C = \{\tau \in K_{\Phi} \mid XQ\tau + O_F \in I_R^C,\ X\tau \subset O_v \text{ for every } v|Z_1\}. \tag{4.8b}$$

LEMMA 4.2.

(i) *H_X^C is stable under $\Gamma[X^{-1}, WX]$.*

(ii) *For every fractional ideal T in F prime to W, the map $\tau \mapsto TXQ\tau + T$ gives a bijection of $\Gamma[X^{-1}, WX]\backslash H_X^C$ onto $I_R^C(T^2XQ)/M_R^C$, where $M_R^C = \{\alpha R \mid \alpha \in \mathfrak{N}_R^C\}$.*

The proof will be given in Section 7.

LEMMA 4.3. *$G_{\mathbf{A}} = P_{\mathbf{A}}S[X', X]$ if $X'X = O_F$. If $XX' \neq O_F$, $P_{\mathbf{A}}S[X', X]$ consists of the elements y of $G_{\mathbf{A}}$ such that $(d_y)_p \neq 0$ and $(d_y)_p^{-1}(c_y)_p \in X_p$ for every $p|XX'$.*

This is an easy exercise. At least the first assertion is well known.

Thus for any fixed X, we have $G_{\mathbf{A}} = P_{\mathbf{A}}S[X^{-1}, X]$ and $G_{\mathbf{A}}^1 = P_{\mathbf{A}}^1S^1[X^{-1}, X]$. To every $y \in G_{\mathbf{A}}$, we can assign a fractional ideal $\mathrm{il}_X(y)$ as follows: let $y = gh$ with $g \in P_{\mathbf{A}}$ and $h \in S[X^{-1}, X]$; then we put $\mathrm{il}_X(y) = d_gO_F$. It is also given by

$$\mathrm{il}_X(y)_p = (c_y)_pX_p^{-1} + (d_y)_pO_p \text{ for every } p \in \mathbf{f}. \tag{4.9}$$

We can easily verify that

$$\mathrm{il}_{stX}(y) = t^{-1}\,\mathrm{il}_X(y \cdot \mathrm{diag}[s^{-1}, t]) \text{ for every } s \text{ and } t \in F_{\mathbf{A}}^{\times}. \tag{4.10}$$

The ideal class of $\mathrm{il}_X(y)$ depends only on the coset $PyS[X^{-1}, X]$. In addition to X, we fix an integral ideal W in F.

LEMMA 4.4.

(i) $G \cap P_{\mathbf{A}}S[X^{-1}, XW] = P(G^1 \cap P_{\mathbf{A}}^1S^1[X^{-1}, XW])$.

(ii) *There is a finite subset* $\mathfrak{B}$ *of* G^1 *such that* $G^1 \cap P^1_{\mathbf{A}}S^1[X^{-1}, XW]$ *is a disjoint union of* $P^1\beta\Gamma^1[X^{-1}, XW]$ *for* $\beta \in \mathfrak{B}$.

(iii) $P\alpha\Gamma[X^{-1}, XW] = P\alpha\Gamma^1[X^{-1}, XW]$ *for every* $\alpha \in G$.

(iv) *If* $\mathfrak{B}$ *is a set as in* (ii), *then the ideals* $\mathrm{il}_X(\beta)$ *for* β *in* $\mathfrak{B}$ *form a complete set of representatives for the ideal classes in F.*

(v) *If* $\mathfrak{B}$ *is a set as in* (ii), *then* $G \cap P_{\mathbf{A}}S[X^{-1}, XW]$ *is a disjoint union of* $P\beta\Gamma[X^{-1}, XW]$ *for all* $\beta \in \mathfrak{B}$.

Assertions (ii) and (iv) were given in [13, Lemma 1.6] when $X = O_F$. The general case can be proved in a similar way. The remaining assertions either follow from these or can be verified in a straightforward way.

Before proceeding further, we introduce a notational convention: for two elements $x = (x_v)_{v \in \mathbf{a}}$ and $y = (y_v)_{v \in \mathbf{a}}$ in $\mathbf{C}^{\mathbf{a}}$, we put

$$x^y = \prod_{v \in \mathbf{a}} (x_v)^{y_v} \tag{4.11a}$$

whenever the factors are meaningful. Denoting by u the identity element of the ring $\mathbf{C}^{\mathbf{a}}$, we have the following special case of (4.11a):

$$x^{su} = \Bigl(\prod_{v \in \mathbf{a}} x_v\Bigr)^s \qquad (s \in \mathbf{C}). \tag{4.11b}$$

Now, for $\alpha \in G_+$ and $z \in H^{\mathbf{a}}$ we define a factor of automorphy $j(\alpha, z)$ to be the element of $\mathbf{C}^{\mathbf{a}}$ given by

$$j(\alpha, z) = (j_v(\alpha, z))_{v \in \mathbf{a}}, \qquad j_v(\alpha, z) = \det(\alpha_v)^{-1/2}(c_v z_v + d_v), \tag{4.12}$$

where α_v is the image of α under v and (c_v, d_v) is the lower half of α_v.

Given an integral ideal W in F as above, we now consider a Hecke character φ of F such that

$$\varphi(x) = x^m |x|^{i\mu - m} \quad \textit{for} \quad x \in F^\times_{\mathbf{a}} = (\mathbf{R}^\times)^{\mathbf{a}}, \tag{4.13a}$$

$$\varphi(O_p^\times) = 1 \quad \textit{for every} \quad p|W. \tag{4.13b}$$

Here $m \in \mathbf{Z}^{\mathbf{a}}$, $\mu \in \mathbf{R}^{\mathbf{a}}$, and the right-hand side of (4.13a) should be understood in the sense of (4.11a). We assume that $\Sigma_{v \in \mathbf{a}}\, \mu_v = 0$. For $\alpha \in G_+$ and a function f on $H^{\mathbf{a}}$ we define a function $f\|_{m,\mu}\alpha$ by

$$(f\|_{m,\mu}\alpha)(z) = |j(\alpha, z)|^{m - i\mu} j(\alpha, z)^{-m} f(\alpha z). \tag{4.14}$$

Taking a complete set of representatives $\mathfrak{A}$ for $P_+\backslash(G_+ \cap P_{\mathbf{A}}S[X^{-1}, XW])$, we define an Eisenstein series $E(z, s)$, depending on X, W, φ, and m, by

$$(4.15)\quad E(z, s) = E(z, s; X, W; \varphi, m)$$

$$= \sum_{\alpha \in \mathbf{A}} N(\det(\alpha)^{-1}\,\mathrm{il}_X(\alpha)^2)^s \varphi_{\mathbf{a}}(\det(\alpha)^{-1/2} d_\alpha)$$

$$\cdot\, \varphi^*(d_\alpha\,\mathrm{il}_X(\alpha)^{-1})\,\mathrm{Im}(z)^{su}\,\|_{m,\mu}\alpha.$$

Here $z \in H^{\mathbf{a}}$, $s \in \mathbf{C}$, and we understand that the factor $\varphi_{\mathbf{a}}(\det(\alpha)^{-1/2}d_\alpha)\varphi^*(d_\alpha\,\mathrm{il}_X(\alpha)^{-1})$ means $\varphi_{\mathbf{a}}(\det(\alpha)^{-1/2})\varphi^*(\mathrm{il}_X(\alpha)^{-1})$ if $W = O_F$. Notice that if $W \neq O_F$ and $\alpha \in G \cap P_{\mathbf{A}}S[X^{-1}, XW]$, then $a_\alpha d_\alpha \neq 0$ and $d_\alpha\,\mathrm{il}_X(\alpha)^{-1}$ is prime to W and that E is independent of the choice of $\mathfrak{A}$. It can easily be seen that

$$(4.16)\qquad E\|_{m,\mu}\gamma = \det(\gamma)^{i\mu/2}\varphi_W(d_\gamma)E \textit{ for every } \gamma \in \Gamma[X^{-1}, XW].$$

For simplicity, let us put $\Gamma^1 = \Gamma^1[X^{-1}, XW]$. Take a set $\mathfrak{B}$ as in Lemma 4.4; let A_β be a complete set of representatives for $(P^1 \cap \beta\Gamma^1\beta^{-1})\backslash\beta\Gamma^1$. Then that lemma shows that $\bigcup_{\beta\in\mathfrak{B}} A_\beta$ can be taken as $\mathfrak{A}$, and therefore

$$(4.17\mathrm{a})\qquad E(z, s) = \sum_{\beta \in \mathfrak{B}} E_\beta(z, s),$$

$$(4.17\mathrm{b})\quad E_\beta(z, s) = N(\mathrm{il}_X(\beta))^{2s} \sum_{\alpha \in A_\beta} \varphi_{\mathbf{a}}(d_\alpha)\varphi^*(d_\alpha\,\mathrm{il}_X(\beta)^{-1})\,\mathrm{Im}(z)^{su}\,\|_{m,\mu}\alpha.$$

If $X = O_F$, this is essentially the same as the series of [13] and [14]. The series with an arbitrary X is not much different from that special case in its nature. We need to introduce another type of series

$$(4.18)\quad E_Y(z, s) = E_Y(z, s; X, W; \varphi, m)$$

$$= N(Y)^{2s}\,\mathrm{Im}(z)^{su} \sum_{(c,d)} \varphi_{\mathbf{a}}(d)\varphi^*(dY^{-1})(cz + d)^{-m}|cz + d|^{m - i\mu - 2su}.$$

Here Y is a fractional ideal in F and (c, d) runs over $T[X, Y; W]/O_F^\times$, where

$$T[X, Y; W] = \{(c, d) \in WXY \times Y \mid (c, d) \neq (0, 0),\ dY^{-1} + W = O_F\}.$$

If $W = O_F$, we understand that $\varphi_{\mathbf{a}}(d)\varphi^*(dY^{-1}) = \varphi^*(Y^{-1})$ even when $d = 0$. We can easily verify that E_Y depends only on the ideal class of Y.

PROPOSITION 4.5. *Let $L_W(s, \varphi) = \Sigma\, \varphi^*(Z)N(Z)^{-s}$, where Z runs over all the integral ideals in F prime to W, and let $L_W(s, \varphi; Y) = \Sigma_{Z\sim Y}\, \varphi^*(Z)N(Z)^{-s}$, where $Z \sim Y$ means the additional condition that Z belongs to the ideal class of Y. Then, for a fixed X, we have*

$$E_Y(z, s) = \sum_{\beta\in\mathfrak{B}} L_W(2s, \varphi; \mathfrak{il}_X(\beta)Y^{-1})E_\beta(z, s),$$

$$\sum_Y E_Y(z, s) = L_W(2s, \varphi)E(z, s),$$

where Y runs over a complete set of representatives for the ideal classes in F.

Proof. A similar result was given in [15, Lemma 6.5]. The present case is somewhat different, but the proof can be given in the same way.

We now take a Hecke character χ of K and define ψ, Z, W, Q, C, and R as before with a choice of B as in (4.1e). Define the L-function $L(s, \chi)$ of χ as usual; define also χ^R by (4.3). Then the main result of this section can be given as follows:

THEOREM 4.6. *Let h be the class number of F and let $\mathfrak{X}$ be a complete set of representatives for the ideal classes modulo $\mathbf{a}$ in F. Suppose*

$$\chi(t) = t^m|t|^{i\mu-m} \quad \textit{for} \quad t \in K_{\mathbf{a}}^\times \tag{4.19}$$

with $m \in \mathbf{Z}^{\mathbf{a}}$ and $\mu \in \mathbf{R}^{\mathbf{a}}$, $\Sigma_{v\in\mathbf{a}}\,\mu_v = 0$, where $|t| = ((x_v{}^2 + y_v{}^2)^{1/2})_{v\in\mathbf{a}}$ if $t_v = x_v + iy_v$ with real x_v and y_v. Then

$$h[U : O_F^\times]L(s, \chi) = N(d(K/F))^{-s/2}N(2B^{-1}Q)^sL(2s, \psi)$$
$$\cdot \sum_{X\in\mathfrak{X}} N(X)^s\, \Sigma\, \chi^R(XQ\tau + O_F)^{-1}E(\tau, s; X, W; \psi, m),$$

where $U = R^\times \cap \mathfrak{N}_R^C$, $d(K/F)$ is the different of K relative to F, and τ runs over $\Gamma[X^{-1}, XW]\backslash H_X^C$ with H_X^C defined by (4.8b).

The proof will be given in Section 7.

The reader may have wondered if the nonmaximal order R is really necessary. Indeed, if we merely seek *some* expression for $L(s, \chi)$, then that can be obtained in a rather straightforward fashion. For instance, with S as in (4.1b), define $L_S(s, \chi)$ in the same manner as for $L_W(s, \varphi)$ in Proposition 4.5. Then $L_S(s, \chi)$ has an expression which is similar to that of the above theorem and which involves only ordinary ideals and χ^* instead of χ^R, but $E(\ldots; X, S; \ldots)$ instead of $E(\ldots; X, W; \ldots)$. The main trouble with this expression is that the functional equations for L_S and $E(\ldots; X, S; \ldots)$ are not clear-cut, which presents serious technical difficulties in our later application. That is why nonmaximal orders and χ^R have been considered.

5. Main theorems in the second case. Let us first specialize the material of Section 4 to the case $F = \mathbf{Q}$. We naturally take $X = \mathbf{Z}$ and eliminate the symbol X. We also understand that the letters W, B, Q, etc. which previously meant ideals in F now denote the positive rational numbers that generate them. For instance, $\Gamma[X^{-1}, XW]$ is nothing else but the group $\Gamma_0(W)$ of (2.21). The Eisenstein series of (4.15) takes the form

$$E(z, s; W, \varphi, m) = \sum_{\alpha \in V} \varphi_W(d_\alpha)^{-1} \mathrm{Im}(z)^s \|_m \alpha, \tag{5.1}$$

where φ is a Hecke character of $\mathbf{Q}$ whose conductor divides W, m is a fixed integer, and $V = [P \cap \Gamma_0(W)] \backslash \Gamma_0(W)$. Condition (4.13a) becomes simply

$$\varphi_{\mathbf{a}}(-1) = (-1)^m. \tag{5.2}$$

Let us now put

(5.3)

$$E^*(z, s; W, \varphi, m) = \pi^{-s}\Gamma(s + |m/2|)L_W(2s, \varphi)E(z, s; W, \varphi, m).$$

Then $s(1 - s)E^*$ can be continued to a real analytic function on $H \times \mathbf{C}$ that is holomorphic in s; the factor $s(1 - s)$ is unnecessary if $\varphi \neq 1$ or $m \neq 0$. These facts are well known. A more general result holds for the function of (4.15) (see [14, Theorem 4.1]). It should be remembered that $L_W(s, \varphi) = \sum_{(n, W)=1} \varphi^*(n\mathbf{Z})n^{-s} = \sum_{(n, W)=1} \varphi_W(n)^{-1}n^{-s}$. If W is the conductor of φ, then

(5.4)

$$E^*(z, 1 - s; W, \varphi, m) = \mathrm{sgn}(m)^m q(\overline{\varphi}) W^{3s-2} z^{-m} |z|^m E^*(\delta z, s; W, \overline{\varphi}, m),$$

where $q(\overline{\varphi}) = \Sigma_t\, \varphi_W(t)^{-1} e^{2\pi i t/W}$ with t running over $(\mathbf{Z}/W\mathbf{Z})^\times$, and

$$\delta = W^{-1/2}\begin{pmatrix} 0 & -1 \\ W & 0 \end{pmatrix}. \tag{5.5}$$

Naturally $q(\overline{\varphi}) = 1$ if $W = 1$. Formula (5.4) is also well known and easily follows from [10, Lemma 3.3], for example.

We now consider the function G_s of Section 2 with $\Gamma = \Gamma_0(W)$ and $\theta = \varphi_W$. We denote $S(\Gamma, \theta, m)$, $P(\Gamma, \theta, m)$, and $P'(\Gamma, \theta, m)$ by $S(W, \varphi, m)$, $P(W, \varphi, m)$, and $P'(W, \varphi, m)$. Transforming G_s, E_ξ, and cusp forms by δ, we easily see that these sets S, P, and P' are invariant under the change of φ for $\overline{\varphi}$. The symbols being as in (2.7), we have

$$E_1(z, s; \theta, m) = E(z, s; W, \varphi, m). \tag{5.6}$$

LEMMA 5.1. *Suppose that W is 1 or a power of a prime and that W is the conductor of φ. Then the following assertions hold*:

(i) $(1 - 2s)[G_s(z, w; \varphi_W, m) - G_{1-s}(z, w; \varphi_W, m)]$

$$= \begin{cases} E(z, s; 1, 1, m)E(w, 1 - s; 1, 1, -m) & \text{if } W = 1, \\ E(z, s; W, \varphi, m)E(w, 1 - s; W, \overline{\varphi}, -m) & \\ \qquad + E(z, 1 - s; W, \varphi, m)E(w, s; W, \overline{\varphi}, -m) & \text{if } W > 1. \end{cases}$$

(ii) *Let $\beta_m(s, \varphi) = \pi^{-s}\Gamma(s + |m/2|)L(2s, \varphi)$. Suppose $w \notin \Gamma_0(W)z$. If $W > 1$ or $m \neq 0$, $\beta_m(s, \varphi)\beta_m(s, \overline{\varphi})G_s(z, w; \varphi_W, m)$ has poles only in $\{1/2\} \cup S(W, \varphi, m)$; if $\varphi = \overline{\varphi}$, the factor $\beta_m(s, \overline{\varphi})$ is unnecessary. If $W = 1$ and $m = 0$, $\beta_m(s, 1)G_s(z, w; 1, 0)$ has poles only in $\{1/2\} \cup S(W, 1, 0)$.*

Proof. If $W = 1$, (i) follows immediately from (2.8). If $W > 1$, the right-hand side of (2.8) has only two nonvanishing terms corresponding to the cusps 0 and ∞ for the reason explained at the end of Section 2. The term corresponding to ∞ presents no problem. The other term has the form

$$J_\delta(z)^{-m}E(\delta z, s; W, \overline{\varphi}, m)J_\delta(w)^m E(\delta w, 1 - s; W, \varphi, -m) \tag{5.7}$$

with δ of (5.5). By (5.3) and (5.4) we have

$$J_\delta(z)^{-m}E(\delta z, s; W, \varphi, m) = \eta_m x(s, \varphi)E(z, 1-s; W, \bar{\varphi}, m)$$

with $\eta_m = \operatorname{sgn}(m)^m$ and a function $x(s, \varphi)$ depending only on s, φ, and $|m|$. Substituting δz for z, we easily see that $x(s, \varphi)x(1-s, \bar{\varphi}) = (-1)^m$, and hence (5.7) is equal to the last term of the formula of (i). To prove (ii), assume $w \notin \Gamma_0(W)z$. Recall that G_s for $\mathrm{Re}(s) \geqq 1/2$ has poles only in $\{1/2\} \cup S(W, \varphi, m)$. Therefore the formula of (i) implies that if $\mathrm{Re}(s) < 1/2$ and $s \notin S(W, \varphi, m)$, then the poles of G_s are those of the right-hand side of (i). Suppose $m \neq 0$ or $W > 1$. As already mentioned, $\beta_m(s, \varphi)E(z, s; W, \varphi, m)$ is finite everywhere, and therefore we obtain assertion (ii) in this case. Suppose $m = 0$ and $W = 1$. Then $E^*(z, s; 1, 1, 0)$ is holomorphic in s except at simple poles at 0 and 1, from which we obtain the desired result.

Now take an imaginary quadratic field K embedded in $\mathbf{C}$ and a Hecke character χ of K such that

$$\chi(t) = t^m|t|^{-m} \quad \text{for} \quad t \in K_{\mathbf{a}}^\times. \tag{5.8}$$

We then consider its L-function $L(s, \chi)$ and put

$$\mathcal{R}(s, \chi) = |D_K N(Z)|^{s/2}(2\pi)^{-s}\Gamma(s + |m/2|)L(s, \chi), \tag{5.9}$$

where D_K denotes the discriminant of K and Z the conductor of χ. The formula of Theorem 4.6 in the present case yields

$$\mathcal{R}(s, \chi) = 2[U:1]^{-1}(QB^{-1}N(Z)^{1/2})^s \cdot \sum_{\tau \in T} \chi^R(\mathbf{Z}Q\tau + \mathbf{Z})^{-1}E^*(\tau, s; W, \psi, m), \tag{5.10}$$

where ψ is the restriction of χ to $\mathbf{Q}$ (see Section 4), W is the conductor of ψ, Q is defined by (4.1b), and $T = \Gamma_0(W)\backslash H_Z^C$ with H_Z^C defined by

$$H_Z^C = \{\tau \in K \cap H \mid \mathbf{Z}Q\tau + \mathbf{Z} \in I_R^C, \tau \in O_v \text{ for every } v|Z_1\}. \tag{5.11}$$

We now state the main result in the present case:

Theorem 5.2. *The notation being as above, suppose that W is either 1 or a power of a prime; if $W > 1$, suppose further that $\bar{\chi} = \chi$ or $\psi \circ N_{K/\mathbf{Q}} = 1$ or B of (4.1e) can be chosen so that $N(Z_0) \neq W_0^3B^2$. Put $S =$*

$S(W, \psi, m)$ for simplicity. Then there exists a function $f(z, s)$ on $H \times \mathbf{C}$, with singularities, satisfying the following conditions (i–vi):

(i) *$\beta_m(s, \bar{\psi})f$ is C^∞ in (z, s) if $s \notin S \cup \{1/2\}$ and z does not belong to the set $\mathcal{Z} = \delta(H_{\mathbf{Z}}^C) \cup \omega(H_{\mathbf{Z}}^C)$, where β_m is as in Lemma* 5.1 *and $\omega(z) = -\bar{z}$.*

(ii) *$f(\gamma z, s) = \psi_W(d_\gamma)^{-1}J_\gamma(z)^m f(z, s)$ for every $\gamma \in \Gamma_0(W)$.*

(iii) *If $z \notin \mathcal{Z}$, $\beta_m(s, \bar{\psi})f(z, s)$ is meromorphic in s with poles only in $S \cup \{1/2\}$.*

(iv) *If $s \notin S \cup \{1/2\}$, then $\beta_m(s, \bar{\psi})f(z, s)$ as a function of z satisfies* (1.8a,b) *with $\lambda = s(1 - s)$ and $\mathcal{Z}$ of* (i).

(v) *Suppose $s_0 \notin S \cup \{1/2\}$. Then $(1 - N^{1-2s})\beta_m(s, \bar{\psi})\mathcal{R}(s, \chi)$ vanishes at s_0 if and only if $[\beta_m(s, \bar{\psi})f(z, s)]_{s=s_0}$ is rapidly decreasing, where $N = W_0^3B^2/N(Z_0)$.*

(vi) *If $\psi = \bar{\psi}$, the factor $\beta_m(s, \bar{\psi})$ is unnecessary in the above assertions. If $\chi = \bar{\chi}$ or $\psi \circ N_{K/\mathbf{Q}} = 1$, then the factor $(1 - N^{1-2s})\beta_m(s, \bar{\psi})$ in* (v) *is unnecessary and we can take $\mathcal{Z}$ to be any of the four sets $H_{\mathbf{Z}}^C$, $\omega(H_{\mathbf{Z}}^C)$, $\delta(H_{\mathbf{Z}}^C)$, and $\delta\omega(H_{\mathbf{Z}}^C)$.*

Proof. We first observe that

$$E^*(z, s; W, \varphi, m) = E^*(-\bar{z}, s; W, \varphi, -m), \tag{5.12}$$

and therefore (5.10) can be transformed into

$$\mathcal{R}(s, \chi) = \mu A^s \sum_{\tau \in T} \chi^R(\mathbf{Z}Q\tau + \mathbf{Z})^{-1}E^*(-\bar{\tau}, s; W, \psi, -m), \tag{5.13}$$

where $\mu = 2/[U : 1]$ and $A = QB^{-1}N(Z)^{1/2}$. Put

$$g(z, s) = \mu A^s \pi^{-s}\Gamma(s + |m/2|)L(2s, \psi) \cdot \sum_{\tau \in T} \chi^R(\mathbf{Z}Q\tau + \mathbf{Z})^{-1}G_s(z, -\bar{\tau}; \bar{\psi}_W, m). \tag{5.14}$$

For a fixed s, g is a function of type (2.16a), except that here it is not subject to "modification" of (2.10). Now (5.13) combined with (2.6) and (5.6) shows that the constant term of the Fourier expansion of g is exactly $y^{1-s}(1 - 2s)^{-1}\mathcal{R}(s, \chi)$. If $W = 1$, we can take this g as the desired f. In fact, the required properties (i–vi) (with $\mathcal{Z} = \omega(H_{\mathbf{Z}}^C)$) follow from the corresponding properties of G_s, or rather, from Proposition 2.1, in view of Lemma 5.1. If $W > 1$, g does not necessarily satisfy (v), since $\Gamma_0(W)$ has more than one cusp. Therefore we need to study $g\|_m\delta$ with δ of (5.5). We observe that

$$J_\delta(z)^{-m}G_s(\delta z, w; \theta, m) = (-J_\delta(w))^m G_s(z, \delta w; \bar{\theta}, m) \tag{5.15}$$

and recall that

$$\mathcal{R}(1-s, \chi) = r(\chi)\mathcal{R}(s, \bar{\chi}) \tag{5.16}$$

with a constant $r(\chi)$ such that $\overline{r(\chi)} = r(\bar{\chi}) = r(\chi)^{-1}$. By (2.6), (5.4), (5.6), and (5.15), the constant term of the expansion of $J_\delta(z)^{-m}g(\delta z, s)$ is

$$\begin{aligned} y^{1-s}(1-2s)^{-1}\mu\epsilon q(\psi)W^{1-3s}A^s M(s, \psi) \\ \cdot \sum_{\tau\in T} \chi^R(\mathbf{Z}Q_\tau + \mathbf{Z})^{-1}E^*(-\bar{\tau}, 1-s; W, \psi, -m), \end{aligned} \tag{5.17}$$

where $\epsilon = \operatorname{sgn}(-m)^m$ and $M(s, \psi) = L(2s, \psi)/L(2s, \bar{\psi})$. By (5.13) and (5.16), we see that (5.17) is equal to $y^{1-s}(1-2s)^{-1}A_1A_2^sM(s, \psi)\mathcal{R}(s, \bar{\chi})$ with $A_1 = \epsilon r(\chi)q(\psi)WA^{-1}$ and $A_2 = W^{-3}A^2$. We need to consider another function

$$\begin{aligned} h(z, s) = \mu A^s\pi^{-s}\Gamma(s + |m/2|)L(2s, \bar{\psi}) \\ \cdot \sum_{\tau\in T} \chi^R(\mathbf{Z}Q_\tau + \mathbf{Z})G_s(z, \tau; \psi_W, m). \end{aligned} \tag{5.18}$$

This satisfies (i–vi) with the same m, but with ψ instead of $\bar{\psi}$ and $\mathcal{Z} = H_{\mathbf{Z}}^C$. Moreover, by (5.10), its constant term is $y^{1-s}(1-2s)^{-1}\mathcal{R}(s, \bar{\chi})$; similarly the constant term of $J_\delta(z)^{-m}h(\delta z, s)$ is

$$(-1)^m y^{1-s}(1-2s)^{-1}\bar{A}_1A_2^sM(s, \psi)^{-1}\mathcal{R}(s, \chi).$$

As observed at the end of Section 2, G_s as a function of z vanishes at the cusps inequivalent to 0 and ∞; hence the same is true for g and h. Suppose now $\chi = \bar{\chi}$ or $\psi \circ N_{K/\mathbf{Q}} = 1$. Then we have $\mathcal{R}(s, \chi) = \mathcal{R}(s, \bar{\chi})$ and $\psi = \bar{\psi}$ in both cases and observe that any of the four functions g, h, $g\|_m\delta$ and $h\|_m\delta$ has the required properties (i–vi); $\mathcal{Z}$ is $\omega(H_{\mathbf{Z}}^C)$, $H_{\mathbf{Z}}^C$, $\delta\omega(H_{\mathbf{Z}}^C)$, and $\delta(H_{\mathbf{Z}}^C)$, respectively.

In a more general case, we consider f defined by

(5.19) $$f(z, s) = g(z, s) - (-1)^m A_1 A_2^s M(s, \psi) J_\delta(z)^{-m} h(\delta z, s).$$

Its constant term is

(5.20) $$y^{1-s}(1 - 2s)^{-1}[1 - (W^3 A^{-2})^{1-2s}]\mathfrak{R}(s, \chi),$$

while the constant term of $f\|_m \delta$ is 0. In view of (4.1a–e), we have $W^3 A^{-2} = W_0^3 B^2 N(Z_0)^{-1}$. This is different from 1 because of our assumption, and hence (5.20) is nonvanishing. Thus the only remaining point is the location of poles of f. Lemma 5.1 shows that $\beta_m(s, \bar{\psi})g$ and $\beta_m(s, \psi)h$ are finite if $z \notin \omega(H_{\mathbf{Z}}^C) \cup \delta(H_{\mathbf{Z}}^C)$ and $s \notin S \cup \{1/2\}$. If $\psi = \bar{\psi}$, we see that g and h without the factor β_m have the same property. This completes the proof.

Proposition 5.3. *The assumptions being the same as in Theorem* 5.2, *define* g, h, *and* f *by* (5.14), (5.18), *and* (5.19). *Let* $s_0 \notin S \cup \{1/2\}$; *suppose* $\beta_m(s_0, \psi)\beta_m(s_0, \bar{\psi}) \neq 0$ *with* β_m *of Lemma* 5.1. *Then* $g(z, s_0)$ *and* $h(z, s_0)$ *are nonzero functions. Furthermore, if* $\mathrm{Re}(s_0) \neq 1/2$ *or* $W_0 \neq 1$, $f(z, s_0)$ *is a nonzero function.*

Proof. For each $\tau \in \omega(H_{\mathbf{Z}}^C)$ and $s \notin P(W, \psi, m)$ we have

(5.21a)

$$g(z, s) = \mu A^s \beta_m(s, \psi)\chi^R(\mathbf{Z}Q\bar{\tau} + \mathbf{Z})^{-1} \cdot [(2\pi)^{-1} e_\tau \log|z - \tau| + r_{\tau,s}^g(z)]$$

in a neighborhood of τ with a function $r_{\tau,s}^g$ continuous at τ. This is a special case of (2.17a). We have to verify that this is so even when $e_\tau > 1$, which will be seen from Lemma 5.4 below combined with Proposition 2.1. Similarly, for each $\tau \in H_{\mathbf{Z}}^C$ we have

(5.21b)

$$h(z, s) = \mu A^s \beta_m(s, \bar{\psi})\chi^R(\mathbf{Z}Q\tau + \mathbf{Z}) \cdot [(2\pi)^{-1} e_\tau \log|z - \tau| + r_{\tau,s}^h(z)]$$

with a function $r_{\tau,s}^h$ continuous at τ. Therefore $g(z, s_0)$ and $h(z, s_0)$ are nonzero functions. As for $f(z, s_0)$, it is sufficient to show that the singularities of g cannot be cancelled by those of $h\|_m \delta$. For that purpose, let $\tau \in H_{\mathbf{Z}}^C$ and $\sigma \in H_{\mathbf{Z}}^C$; suppose $-\bar{\tau} = \delta(\sigma)$. For every $p | W_0$, we have $\mathbf{Z}_p + (BO_K)_p = R_p = \mathbf{Z}_p\sigma + \mathbf{Z}_p = \mathbf{Z}_p\tau + \mathbf{Z}_p$, and hence $W\sigma R_p = W\sigma(\mathbf{Z}_p\bar{\tau} + \mathbf{Z}_p) = \mathbf{Z}_p W\sigma + \mathbf{Z}_p = \mathbf{Z}_p + WR_p = \mathbf{Z}_p + (WBO_K)_p$. This is a contradiction, since the last module is not an R_p-ideal. Thus we cannot have $-\bar{\tau} = \delta(\sigma)$ if

$W_0 \neq 1$. Suppose $-\bar{\tau} = \delta(\sigma)$ and $W_0 = 1$. Then, in a neighborhood of $-\bar{\tau}$, we have

$$f(z, s) = \mu A^s \beta_m(s, \psi)(2\pi)^{-1} e(-\bar{\tau}) \log|z + \bar{\tau}| \cdot [u_1 + u_2(W^{-3}A^2)^{s-1/2}] + \text{(finite part)}$$

with $|u_1| = |u_2| = 1$. Since $W^{-3}A^2 \neq 1$, the singularity vanishes only when $\mathrm{Re}(s) = 1/2$, which completes the proof.

LEMMA 5.4

(i) *$H_{\mathbf{Z}}^C$ contains an elliptic point of $\Gamma_0(W)$ only if $C = Z = Z_1$ and $W = Q = N(Z)$.*

(ii) *Let σ be a point of $K \cap H$ fixed by an elliptic element γ of $\Gamma_0(W)$. Then $J_\gamma(\sigma)^m = \psi_W(d_\gamma)$ if $-\bar{\sigma} \in H_{\mathbf{Z}}^C$ and $J_\gamma(\sigma)^m = \psi_W(d_\gamma)^{-1}$ if $\sigma \in H_{\mathbf{Z}}^C$.*

Proof. Let σ and γ be as in (ii); put $\zeta = c_\gamma \sigma + d_\gamma$. Then ζ is a root of unity other than ± 1, and $\sigma = \alpha(\zeta)$ with $\alpha \in GL_2(\mathbf{Z})$. Hence $(c_\alpha \zeta + d_\alpha)(\mathbf{Z}\sigma + \mathbf{Z}) = \mathbf{Z}\zeta + \mathbf{Z} = O_K$. Suppose $\sigma \in H_{\mathbf{Z}}^C$. Then $R = O_K$, so that $B = 1$. By (4.1e), this is possible only when $W_0 = S_0$. For every p dividing W_0, we have $\mathbf{Z}_p \sigma + \mathbf{Z}_p = R_p$, and hence $R_p = \mathbf{Z}_p \zeta + \mathbf{Z}_p = \mathbf{Z}_p c_\gamma \sigma + \mathbf{Z}_p = c_\gamma R_p + \mathbf{Z}_p$. Since $p | c_\gamma$, this happens only when $W_0 = 1$, in which case we have $W = Q$ and $C = Z = Z_1$, in view of (4.1a–d). By (5.11), we have $\zeta - d_\gamma = c_\gamma \sigma \in Z_v$ for every $v | Z$. Thus $\psi_W(d_\gamma) = \chi_Z(d_\gamma) = \chi_Z(\zeta) = \chi_{\mathbf{a}}(\zeta)^{-1} = \zeta^{-m} = J_\gamma(\sigma)^{-m}$, which proves the last fact of (ii). The case $-\bar{\sigma} \in H_{\mathbf{Z}}^C$ can be proved by observing that $J_\gamma(\sigma)^{-1} = J_{\omega\gamma\omega}(\omega\sigma)$.

The main point of Theorem 5.2 is in (v), which is parallel to Theorem 3.4, (2). As explained in the introduction, we have the analogues of Theorem 3.4, (3), (4), and (5) in the present case. For simplicity, we state them only for the function g of (5.14) as follows:

THEOREM 5.5. *Let the notation be the same as in Theorem* 5.2, (5.10), (5.13), (5.14), *and* (5.21a); *let β_m be as in Lemma* 5.1.

(i) *If $\mathcal{R}(s, \chi) = \mathcal{R}(s, \bar{\chi}) = 0$, $\beta_m(s, \psi) \neq 0$, and $s \notin \{1/2\} \cup S$, we have $-\mathrm{Im}[s(1 - s)]\|g(z, s)\|^2 = |\mu A^s \beta_m(s, \psi)|^2 \,\mathrm{Im}[\Sigma_{\tau \in \omega(T)} r_{\tau,s}^g(\tau)]$.*

(ii) *For $s = (1/2) + it$ with $0 \neq t \in \mathbf{R}$, we have*

$$4t|\mu A^s \beta_m(s, \psi)|^2 \,\mathrm{Im}[\sum_{\tau \in \omega(T)} r_{\tau,s}^g(\tau)] = \begin{cases} |\mathcal{R}(s, \chi)|^2 & \text{if } W = 1, \\ |\mathcal{R}(s, \chi)|^2 + |\mathcal{R}(s, \bar{\chi})|^2 & \text{if } W > 1. \end{cases}$$

(iii) *Put* $E(z, s) = E(z, s; W, \bar{\psi}, m)$. *For s as in* (i), *we have*

$$d\mathcal{R}(s, \chi)/ds = (2s - 1)\langle E(z, \bar{s}), g(z, s)\rangle.$$

Proof. The first two assertions are special cases of Proposition 2.3. In (ii), we need Lemma 5.1 and (5.14). To prove (iii), let s_0 denote a particular s as in (i), and use s for the variable. Put $\lambda_0 = s_0(1 - s_0)$, $\lambda = s(1 - s)$, $F_s(z) = E(z, \bar{s})$, $g_0(z) = g(z, s_0)$, and $\alpha(s) = \kappa A^s \beta_m(s, \psi)$. Observe that the conclusion of Proposition 1.1 is applicable to F_s and g_0, even if F_s may not satisfy (1.8c), since g_0 is rapidly decreasing. Thus the formula of the proposition with $c_\nu = 0$ gives

$$(\lambda - \lambda_0)\langle F_s, g_0\rangle = -2\pi \sum_{\nu=1}^{p} b_\nu \overline{F_s(\zeta_\nu)}/e(\zeta_\nu).$$

By (5.21a) and (5.13), the right-hand side is $-\mathcal{R}(s, \chi)\alpha(s_0)/\alpha(s)$. Taking the limit when s tends to s_0, we obtain (iii).

Remark 5.6. (1) If $Z_1 = O_K$, then $Q = 1$, $C = W_0 O_K$, and $H_{\mathbf{Z}}^C$ is stable under the map $\tau \mapsto -\bar{\tau}$. Therefore, replacing $-\bar{\tau}$ by τ in (5.14), we obtain

$$(5.22)\quad g(z, s)$$
$$= \mu A^s \pi^{-s}\Gamma(s + |m/2|)L(2s, \psi) \cdot \sum_{\tau \in T} \chi^R(\mathbf{Z}\bar{\tau} + \mathbf{Z})^{-1} G_s(z, \tau; \bar{\psi}_W, m).$$

In particular, if $\psi \circ N_{K/\mathbf{Q}} = 1$, then $\psi = \bar{\psi}$ and $g = h$.

(2) In Theorem 5.2, we have associated an eigenform of weight m to a character χ satisfying (5.8). We can also associate to χ a form of weight $-m$ satisfying (i–vi) with the change of m for $-m$. In fact, let g^* and h^* denote the functions defined by (5.14) and (5.18) with $\bar{\chi}$, $\bar{\psi}$, and $-m$ instead of χ, ψ, and m. If $\psi \circ N_{K/\mathbf{Q}} = 1$, we can take both g^* and h^* as the desired function. In the general case, we take

$$h^*(z, s) - A_1 A_2^s M(s, \psi) J_\delta(z)^m g^*(\delta z, s)$$

with A_1, A_2, and $M(s, \psi)$ as in the proof of Theorem 5.2.

6. A few more comments. Though we have treated only two types of zeta functions, many more cases are likely to be comprehended in the same fashion. Let us first examine possible analogues of $\mathcal{R}(s, \chi)$ when we take $\mathrm{SL}_2(\mathbf{C})$ in place of $\mathrm{SL}_2(\mathbf{R})$. Naturally we are led to the special values $\mathcal{E}(\sigma, s)$ of an Eisenstein series $\mathcal{E}$ of $\mathrm{SL}_2(M)$ with an imaginary quadratic field M, where σ is a nontrivial fixed point of an element of $\mathrm{SL}_2(M)$. It can easily be seen that σ determines a definite quaternion algebra A over $\mathbf{Q}$. In the simplest case where $\mathcal{E}$ is of level 1 and M has class number 1, $\zeta_M(2s)\mathcal{E}(\sigma, s)$ is of the form $\Sigma_{0 \neq x \in X} N_{A/\mathbf{Q}}(x)^{-2s}$ with a lattice X in A. In a more general case, we have to deal with linear combinations of such sums under some congruence conditions on x. Now, by virtue of the results of Eichler, Shimizu, Jacquet, and Langlands, we know that there are many elliptic cusp forms $g(z) = \Sigma_{n=1}^{\infty} a_n e^{2\pi i n z}$ such that the series $L(s, g) = \Sigma_n a_n n^{-s}$ can be obtained as such a linear combination. Therefore it is plausible that $L(s, g)$ takes the place of $L(s, \chi)$ when we consider our problem for $\mathrm{SL}_2(M)$ instead of $\mathrm{SL}_2(\mathbf{Q})$.

Coming back to $\mathrm{SL}_2(\mathbf{Q})$, there is another possibility that concerns a real quadratic field, say F. It was shown by Siegel in [16] that the zeta function of F or more generally an L-function of a Hecke character of F can be obtained as a linear combination of integrals of the type

$$\int_1^{\lambda} E(\gamma_t(i), s; 0)t^{-1}dt,$$

where $t \mapsto \gamma_t$ is a homomorphism of $\mathbf{R}^{\times}$ into $\mathrm{SL}_2(\mathbf{R})$ and λ is a unit of F. This suggests that we may be able to obtain a function similar to $f(z, s)$ in Theorem 5.2 for the L-functions of F.

More generally, let $E(z, s)$ denote an Eisenstein series of a reductive algebraic group G over a number field, where z is the variable on $G_{\mathbf{R}}$ and s is a complex parameter. Suppose that there is a homomorphic embedding h of another such group G' into G. Then, with a suitable function p on $G'_{\mathbf{R}}$ and a suitable quotient Φ of $G'_{\mathbf{R}}$, one may obtain a zeta function $R(s)$ by an integral

$$R(s) = \int_{\Phi} p(w)E(h(w), s)dw.$$

This is a general principle of producing zeta functions, and those mentioned in the present paper belong to this type. On the other hand, no general theory of the resolvent kernel has been developed in the higher-dimensional case, and it may be too simple-minded to expect the existence of a function which generalizes G_s and whose constant term is $E(z, s)$. Still, there is no reason to believe that the functions investigated in this paper are phenomena found only in the low-dimensional cases. It may fairly be said that further investigations of the inhomogeneous equations similar to $(L_m - \lambda)f = p$ in the higher-dimensional case are very much worth undertaking.

7. Proofs of the results of Section 4. Define K, F, χ, ψ, R, and other symbols as in Section 4, in particular as in (4.1a–f); define also $\mathfrak{N}_R^C$ by (4.2).

LEMMA 7.1. *An element α of $K^\times$ belongs to $\mathfrak{N}_R^C$ if and only if there is an element t of F prime to C such that $\alpha - t \in Z_v$ for $v|Z_1$ and $\alpha - t \in WR_p$ for $p|W_0$.*

Proof. Suppose $\alpha \in \mathfrak{N}_R^C$. Then $\alpha R \in I_R^C$, so that α is a v-unit for $v|Z_1$. Since K_v is isomorphic to F_q if such a v divides q, there is an element s of F such that $\alpha - s \in Z_v$ for $v|Z_1$. With r as in (4.2), we can take $t \in F$ so that $t - r \in WR_p$ for $p|W_0$ and $t - s \in Q_p$ for $p|Q$. This proves the "only if"-part. The "if"-part can be seen by observing that $R_p = (O_K)_p$ if $p|Q$ and that the existence of t implies that $\alpha \in R_p^\times$ for $p|W_0$ and $\alpha \in O_v^\times$ for $v|Z_1$.

LEMMA 7.2. *Let E be an ideal of F divisible by C. Then every R-ideal in I_R^C can be written in the form αA with $\alpha \in \mathfrak{N}_R^C$ and $A \in I_R^E$.*

Proof. Take an element x of $K_\mathfrak{f}^\times$ such that $x_v = 1$ for $v|C$. We can find an element α of K such that $\alpha - x_p \in x_p ER_p$ for every $p|E$. Then $\alpha x_p^{-1} R_p = R_p$ for such p. Therefore $\alpha^{-1} xR \in I_R^E$ and hence $\alpha R \in I_R^C$. Since $\alpha - 1 \in W_0 R_p$ for $p|W_0$, we have $\alpha \in \mathfrak{N}_R^C$, which proves our assertion.

For $\alpha \in \mathfrak{N}_R^C$, denote by t_α any element t as in Lemma 7.1, or rather its class modulo W. Then $\alpha \mapsto t_\alpha$ defines a homomorphism of $\mathfrak{N}_R^C$ into $(O_F/W)^\times$.

LEMMA 7.3.

(i) $\chi_C(\alpha) = \psi_W(t_\alpha)$ *if* $\alpha \in \mathfrak{N}_R^C$.

(ii) $\chi_Z(\alpha) = \chi_C(\alpha)$ *if* $\alpha \in K$ *and* $\alpha R \in I_R^Z$.

Proof. Let $\alpha \in \mathfrak{N}_R^C$. Since $\alpha - t_\alpha \in (W_0 + S_0 O_K)_p$ for $p|W_0$, there is an element r of W such that $\alpha - t_\alpha - r \in (S_0 O_K)_p$ for such p. Then $\alpha - t_\alpha - r \in Z_v$ for $v|C$. Hence $\chi_C(\alpha) = \chi_C(t_\alpha + r) = \psi_W(t_\alpha)$, which proves (i). If $\alpha R \in I_R^Z$, we see that $\alpha \in R_p^\times$ for $p|S_0$, and hence $\chi_v(\alpha) = 1$ for $v|Z_0$, $v \nmid W_0$ by Lemma 4.1. Therefore we obtain (ii).

Define χ^R by (4.3); let $M_R^C = \{\alpha R \mid \alpha \in \mathfrak{M}_R^C\}$. Since $I_R^C = M_R^C I_R^Z$ by Lemma 7.2, we see from (4.4a,b) that χ^R can also be defined by

$$\chi^R(\alpha A) = \chi_C(\alpha)^{-1}\chi_{\mathbf{a}}(\alpha)^{-1}\chi^*(AO_K) \text{ for } \alpha \in \mathfrak{M}_R^C \text{ and } A \in I_R^Z. \tag{7.1}$$

For an R-ideal $A = xR$ with $x \in K_{\mathbf{f}}^{\times}$, we define a fractional ideal $N_{K/F}(A)$ in F and a positive rational number $N(A)$ by $N_{K/F}(A) = N_{K/F}(x)O_F$ and $N(A) = \prod_{v\in\mathbf{f}} |x_v|_v^{-1}$. Obviously $N_{K/F}(A) = N_{K/F}(AO_K)$ and $N(A) = N(AO_K)$.

Define the L-function $L(s, \chi)$ of χ as usual and another function $L(s, \chi^R)$ by

$$L(s, \chi^R) = \sum_A \chi^R(A)N(A)^{-s}, \tag{7.2}$$

where A runs over all the integral R-ideals in I_R^C. Since $(O_K)_p = R_p$ if $p \nmid S_0$, we can easily verify that $A \mapsto AO_K$ gives an isomorphism of I_R^Z onto I^Z and that $A \subset R$ if and only if $AO_K \subset O_K$.

PROPOSITION 7.4. $L(s, \chi^R) = L(s, \chi)$.

Proof. The nontrivial point of our assertion is that $L(s, \chi^R)$ involves R-ideals which are not necessarily prime to Z. For every finite prime p not dividing W, put $L_p(s) = \sum \chi_p(x)|x|^s$, where x runs over the nonzero elements of R_p modulo $R_p^{\times}$. Then we easily see that

$$L(s, \chi^R) = L(s, \chi) \prod_{p|S,\, p\nmid W} L_p(s).$$

Thus our task is to show that $L_p(s) = 1$ if $p|S$ and $p \nmid W$. For such a p, fix a prime element π of F_p and put $P = \pi O_p$. Then $S_p = P^e$ with $0 < e \in \mathbf{Z}$. We first treat the case where p splits in K. Then $(O_K)_p$ can be identified with $O_p \times O_p$ and R_p with $\{(a, b) \in O_p \times O_p \mid a - b \in P^e\}$. Let $x = (\pi^m c, \pi^n d) \in R_p$ with p-units c and d. Then $x \in R_p$ only in the following two cases: (i) $m \geqq e$ and $n \geqq e$; (ii) $m = n < e$ and $c - d \in P^{e-m}$. Now we have

$$L_p(s) = \sum_{(m,n),(c,d)} \chi_p(\pi^m c, \pi^n d)|\pi|^{(m+n)s} \tag{7.3}$$

with all possible $(\pi^m c, \pi^n d)$ modulo $R_p^{\times}$. Since $p \nmid W$, we have $\chi_p(c, c) = 1$ for every p-unit c. This shows that $Z_p = P^e(O_K)_p$. Fix (m, n) in Case (i).

The terms of L_p with this (m, n) produce

$$\chi_p(\pi^m, \pi^n)|\pi|^{(m+n)s} \Sigma\, \chi_p(1, d), \tag{7.4}$$

where d runs over U_0/U_e, $U_k = \{y \in O_p^\times \mid y - 1 \in P^k\}$. Since both prime divisors of p in K divide Z, the sum must be 0. If $m = n < e$, the terms of (7.3) with this (m, n) produce (7.4) with d running over U_{e-m}/U_e. This sum is not 0 only if $m = 0$, in which case the sum gives 1. Next suppose p is ramified; let γ be a prime element of K_p; let $Y = (\gamma O_K)_p$, $Z_p = Y^e$, and $f = [(e + 1)/2]$. Then $S_p = P^f$, $R_p = O_p + Y^{2f} = O_p + \gamma P^f$, and

$$L_p(s) = \sum_{m,u} \chi_p(\gamma^m u)|\gamma|^{ms},$$

where (m, u) runs over $\mathbf{Z} \times (O_K)_p^\times / R_p^\times$ under the condition $\gamma^m u \in R_p$. Put

$$U_k = O_p^\times \cdot \{x \in (O_K)_p^\times \mid x - 1 \in Y^k\}.$$

If $m \geqq 2f$, the terms with this fixed m produce $\chi_p(\gamma^m)|\gamma|^{ms} \Sigma\, \chi_p(u)$ with u running over U_0/U_{2f}. Obviously the sum is 0. Suppose $m = 2n < 2f$ with $n \in \mathbf{Z}$. Then $\pi^n u \in R_p$ with $u \in (O_K)_p^\times$ if and only if $u \in U_{2f-2n}$. Thus the terms of L_p with this m can be written in the form $\chi_p(\pi^n)|\gamma|^{2ns} \Sigma\, \chi_p(u)$ with u running over U_{2f-2n}/U_{2f}. The sum is nonzero only if $f = n$, in which case it is 1. Suppose $m = 2n + 1 < 2f$ and $\gamma\pi^n u \in R_p$ with $u \in (O_K)_p^\times$. Then $\gamma u \in F_p + \gamma P^{f-n}$, so that $\gamma u = a + \gamma\pi b$ with $a \in F_p$ and $b \in O_p$. Then a must be divisible by π, which is impossible, since $\pi \nmid \gamma u$. Thus (7.3) has no terms with $m = 2n + 1 < 2f$. It remains to consider the case where p remains prime in K. This case can be handled in a similar fashion and in fact it is simpler than the ramified case, and so we omit the details.

Let $\mathcal{J}$ be a complete set of representatives for I_R^C/M_R^C and let $U = R^\times \cap \mathfrak{N}_R^C$. By Lemma 7.4, we have

$$L(s, \chi) = \sum_{A \in \mathcal{J}} \sum_{\alpha} \chi^R(\alpha A^{-1}) N(\alpha A^{-1})^{-s}, \tag{7.5}$$

where α runs over $(A \cap \mathfrak{N}_R^C)/U$. By (4.4b) and Lemma 7.3, this can also be written

$$L(s, \chi) = \sum_{A \in \mathcal{J}} \chi^R(A)^{-1} N(A)^s \sum_{\alpha} \psi_W(t_\alpha)^{-1} \chi_{\mathbf{a}}(\alpha)^{-1} N_{K/\mathbf{Q}}(\alpha)^{-s}. \tag{7.6}$$

By Lemma 7.2, we may take $\mathcal{I}$ from $\mathcal{I}_R^Z$. If we do so, $\chi^R(A)^{-1}$ in (7.6) can be written $\chi^*(AO_K)^{-1}$.

LEMMA 7.5. *Let A be an O_F-lattice in K and let $X = A \cap F$. Then $A = X + Y\xi$ with a fractional ideal Y in F and $\xi \in K$.*

Proof. Since A/X is a torsion-free O_F-module, it is O_F-isomorphic to a fractional ideal Y in F. Then $Y^{-1}A/Y^{-1}X$ is isomorphic to O_F. Let ξ be an element of $Y^{-1}A$ which corresponds to 1 by this isomorphism. Then $Y^{-1}A = Y^{-1}X + O_F\xi$, which proves our lemma.

Fix an element ζ of $K^\times$ such that $\zeta^\rho = -\zeta$. For an O_F-lattice A in K, denote by $\lambda_\zeta(A)$ the O_F-linear span of $\mathrm{Tr}_{K/F}(\zeta\alpha\beta^\rho)$ for all α and β in A.

LEMMA 7.6. *Let A be an O_F-lattice in K of the form $A = X\xi + Y\eta$ with $\xi, \eta \in K$ and fractional ideals X and Y in F. Then $\lambda_\zeta(A) = \zeta(\xi\eta^\rho - \xi^\rho\eta)XY$.*

This is an easy exercise.

Let $d(K/F)$ denote the different of K relative to F. Given ζ as above, let D_ζ denote the fractional ideal in F such that $D_\zeta O_K = \zeta d(K/F)$.

LEMMA 7.7. *Let $R = O_F + BO_K$ with an integral ideal B in F. Then $\lambda_\zeta(A) = N_{K/F}(A)BD_\zeta$ for every R-ideal A. Moreover, if $A = X\xi + Y\eta$ as in Lemma 7.6, then $N_{K/F}(A) = \zeta(\alpha - \alpha^\rho)XYB^{-1}D_\zeta^{-1}$ and $N(A) = N(B^{-1}XY)N((\alpha - \alpha^\rho)d(K/F)^{-1})^{1/2}$, where $\alpha = \xi\eta^\rho$.*

Proof. We can easily verify that $\lambda_\zeta(A) = N_{K/F}(A)\lambda_\zeta(R)$, $\lambda_\zeta(R) = B\lambda_\zeta(O_K)$, and $\lambda_\zeta(O_K) = D_\zeta$. Combining these and Lemma 7.6, we obtain our formulas.

We now assume that F is totally real and K totally imaginary, take ζ from K_Φ, and define $\lambda(A)$ as in Section 4. The notation being as in Lemma 7.6, we see that $XY \in \lambda(A)$ if $\xi/\eta \in K_\Phi$.

LEMMA 7.8. *Let $A \in I_R^C(X)$ (see (4.8a)) with a fractional ideal X in F. Then there exists an element τ of K_Φ and μ in $\mathfrak{N}_R^C$ such that $\mu A = X\tau + O_F$.*

Proof. By Lemma 7.2, we may assume that $A \in I_R^W$. Let $Y = A \cap F$. We easily see that Y is prime to W. There is an element g of F such that gY is integral and prime to W. Replacing A by gA, we may assume that Y is integral. By Lemma 7.5, we have $A = Y + E\xi$ with a fractional ideal E and $\xi \in K$. Replacing E and ξ by hE and $h^{-1}\xi$ with a suitable h in F, we may assume that $E + Y = O_F$. Take $r \in Y$ and $s \in EW$ so that $r - s = 1$. Let $\alpha = r + s\xi$ and $\beta = 1 + \xi$. Then we easily see that $Y + E\xi = O_F\alpha +$

$YE\beta$ and $\alpha - 1 = s(1 + \xi) \in WR_p$ for $p|W$, and hence $\alpha \in \mathfrak{N}_R^C$. Thus $\alpha^{-1}A = O_F + YE\alpha^{-1}\beta$. Take $\gamma \in F$ so that $\gamma\alpha^{-1}\beta \in K_\Phi$. By Lemma 7.6 we have $\lambda(A) = \lambda(\alpha^{-1}A) \ni \gamma YE$. Therefore $\gamma YE = \delta X$ with a totally positive δ in F. Put $\tau = \delta\gamma^{-1}\alpha^{-1}\beta$. Then $\tau \in K_\Phi$ and $\alpha^{-1}A = O_F + X\tau$ as desired.

LEMMA 7.9. *The element τ of Lemma 7.8 can be chosen so that $X\tau \subset Z_v$ for every v dividing Z_1.*

Proof. If $v|Z_1$, we see that $X\tau \subset O_v$. Take a totally positive a in F so that aX is integral and prime to Z_1. Then $a^{-1}\tau \in O_v$. Since $O_K = O_F + Z_1$, we can find an element b of O_F such that $a^{-1}\tau - b \in Z_v$ for $v|Z_1$. Putting $\tau' = \tau - ab$, we obtain $\mu A = O_F + X\tau'$ and $X\tau' \subset Z_v$ for every such v.

Proof of Lemma 4.2. Let $\gamma = \begin{pmatrix} a & b \\ c & d \end{pmatrix} \in \Gamma[X^{-1}, WX]$ and $A_\tau = TXQ\tau + T$. Then $A_\tau = (c\tau + d)A_{\gamma(\tau)}$. Suppose $\tau \in H_X^C$. Observe that $X(a\tau + b) \subset O_v$ and $c\tau \in Z_v$ for $v|Z_1$, and $c\tau \in W_0R_p$ for $p|W_0$. Since d is prime to W, we see that $c\tau + d \in \mathfrak{N}_R^C$ by Lemma 7.1 and $X\gamma(\tau) \subset O_v$ for $v|Z_1$, and hence $\gamma(\tau) \in H_X^C$. This proves (i) and that the map of (ii) is well-defined. If $B \in I_R^C(T^2XQ)$, then $T^{-1}B \in I_R^C(XQ)$. Then Lemma 7.9 proves the surjectivity of the map of (ii). Suppose $A_\tau = \alpha A_\sigma$ for $\sigma, \tau \in H_X^C$ and $\alpha \in \mathfrak{N}_R^C$. We can put $\alpha\sigma = a\tau + b$ and $\alpha = c\tau + d$ with a, b, c, d in F. Put $\gamma = \begin{pmatrix} a & b \\ c & d \end{pmatrix}$. Then $\sigma = \gamma\tau$, and we easily see that $\gamma \in \Gamma[Q^{-1}X^{-1}, QX]$. Since $\alpha \in \mathfrak{N}_R^C$, we have $c\tau + d - t_\alpha \in W_0R_p = W_0(X_p\tau + O_p)$ for $p|W_0$. Thus $c \in W_0X_p$ for every such p. Since $c \in QX$, this shows that $c \in WX$. Now $X\tau \subset O_v$ and $(a\tau + b)X = \alpha\sigma X \subset O_v$ for $v|Z_1$, so that $Xb \subset O_p$ for $p|Q$. Since $b \in Q^{-1}X^{-1}$, we have $b \in X^{-1}$, so that $\gamma \in \Gamma[X^{-1}, WX]$. This completes the proof.

LEMMA 7.10. *Let $A = XTQ\tau + T$ with X and T as in Lemma 4.2 and $\tau \in H_X^C$; let $\alpha = a\tau + b$ with $a \in XTQ$ and $b \in T$. Then α belongs to $\mathfrak{N}_R^C$ if and only if $\alpha \neq 0$, $a \in XTW$, and b is prime to W, in which case we have $\chi_C(\alpha) = \psi_W(t_\alpha) = \psi_W(b)$.*

Proof. If $\tau \in H_X^C$, $a \in XTW$, and b is prime to W, then we easily see from Lemma 7.1 that $a\tau + b \in \mathfrak{N}_R^C$. Conversely, suppose $a\tau + b - t \in W_0R_p$ for $p|W_0$ and $a\tau + b - t \in Z_v$ for $v|Z_1$, with an element t of F prime to W. Then $a\tau + b - t \in W_0(XTQ\tau + T)_p$, and hence $a \in W_0X_p$ and $b - t \in W_{0p}$ for every such p. Since $a \in XTQ$, we have $a \in XTW$, and so $a\tau \in Z_v$ for $v|Z_1$. Therefore $b - t \in Z_v$ for every such v, and hence b is prime to W. Since we can take b as t_α, we obtain the last assertion from Lemma 7.3.

Proof of Theorem 4.6. Let χ and $\mathfrak{X}$ be as in the theorem. For any fixed ideal Y in F prime to W, we can take $\bigcup_{X\in\mathfrak{X}} I_R^C(QXY^2)/M_R^C$ as $\mathfrak{I}$ of (7.6). By Lemma 4.2, (ii), we can take the R-ideals $YXQ\tau + Y$ with $\tau \in \Gamma[X^{-1}, WX]\backslash H_X^C$ as a complete set of representatives for $I_R^C(QXY^2)/M_R^C$. Therefore by (7.6) we have

$$(7.7)\quad L(s, \chi) = \sum_{X\in\mathfrak{X}} \sum_{\tau} \chi^R(YXQ\tau + Y)^{-1} \cdot N(YXQ\tau + Y)^s \sum_{\alpha} \psi_W(t_\alpha)^{-1}\chi_{\mathbf{a}}(\alpha)^{-1}N_{K/\mathbf{Q}}(\alpha)^{-s},$$

where τ runs over $\Gamma[X^{-1}, WX]\backslash H_X^C$ and α over $[\mathfrak{N}_R^C \cap (YXQ\tau + Y)]/U$, $U = \mathfrak{N}_R^C \cap R^\times$. By Lemma 7.10, the last sum over α can be written in the form

$$[U : O_F^\times]^{-1} \sum_{a,b} \psi_W(b)^{-1}\chi_{\mathbf{a}}(a\tau + b)^{-1}N_{K/\mathbf{Q}}(a\tau + b)^{-s},$$

where (a, b) runs over all the nonzero elements of $(YXW \times Y)/O_F^\times$ under the condition that $bY^{-1} + W = O_F$. Applying Lemma 7.7 to $YXQ\tau + Y$ and using E_Y of (4.18), we obtain

$$[U : O_F^\times]L(s, \chi) = N(d(K/F))^{-s/2}N(2B^{-1}Q)^s \cdot \sum_{X\in\mathfrak{X}} N(X)^s \sum_{\tau} \chi^R(XQ\tau + O_F)^{-1}E_Y(\tau, s; X, W; \psi, m).$$

Taking the sum over all Y belonging to a complete set of representatives for the ideal classes in F, we obtain the formula of Theorem 4.6 in view of Proposition 4.5.

PRINCETON UNIVERSITY

REFERENCES

[1] J. Elstrodt, Die Resolvente zum Eigenwertproblem der automorphen Formen in der hyperbolischen Ebene I, II, III, *Math. Ann.* **203** (1973), 295-330, *Math. Z.* **132** (1973), 99-134, *Math. Ann.* **208** (1974), 99-132.

[2] A. Erdélyi et al., *Higher Transcendental Functions*, Vol. 1, McGraw-Hill, 1953.

[3] L. D. Fadeev, Expansion in eigenfunctions of the Laplace operator on the fundamental domain of a discrete group on the Lobacevskii plane, *Trans. Moscow Math. Soc.* **17** (1967), 357–386.

[4] J. D. Fay, Fourier coefficients of the resolvent for a Fuchsian group, *J. Reine Angew. Math.* **294** (1977), 143–203.

[5] D. Goldfeld, Analytic and arithmetic theory of Poincaré series, *Astérisque* **61** (1979), 95–107.

[6] D. Hejhal, Some observations concerning eigenvalues of the Laplacian and Dirichlet L-series, in *Recent Progress in Analytic Number Theory*, Vol. 2, Academic Press, 1981, 95–110.

[7] ———, The Selberg trace formula for *PSL*(2, **R**), Vol. 2, *Springer Lecture Notes in Math.* No. 1001, 1983.

[8] H. Neunhöffer, Über die analytische Fortsetzung von Poincaréreihen, *Sitz.-Ber.Heidelberg Akad. Wiss. Math.* 1973, 33–90.

[9] W. Roelcke, Das Eigenwertproblem der automorphen Formen in der hyperbolischen Ebene I, II, *Math. Ann.* **167** (1966), 292–337, **168** (1967), 261–324.

[10] G. Shimura, On modular forms of half integral weight, *Ann. of Math.* **97** (1973), 440–481.

[11] ———, On the holomorphy of certain Dirichlet series, *Proc. London Math. Soc.* **31** (1975), 79–98.

[12] ———, Confluent hypergeometric functions on tube domains, *Math. Ann.* **260** (1982), 269–302.

[13] ———, On Eisenstein series, *Duke Math. J.* **50** (1983), 417–476.

[14] ———, On the Eisenstein series of Hilbert modular groups, *Revista Mat. Iberoamer.* 1, No. 3 (1985), 1–42.

[15] ———, On Hilbert modular forms of half-integral weight, *Duke Math. J.* **55** (1987), 765–838.

[16] C. L. Siegel, *Lectures on Advanced Analytic Number Theory*, Tata Institute of Fundamental Research, 1961.

THE CHEEGER-SIMONS INVARIANT AS A CHERN CLASS

By STEVEN ZUCKER

What follows is a report on work done jointly with R. Hain at Max-Planck-Institut für Mathematik (Bonn) during the fall of 1987, when there was a special year in Hodge theory and algebraic geometry. The focus during that period was the generalized Hodge conjectures of Beilinson [Be: Section 6], relating algebraic K-theory and Hodge theory on algebraic varieties. It was in this setting that interest in the Cheeger-Simons invariant (in the case of flat vector bundles) was aroused.

The Cheeger-Simons invariant is defined for vector bundles with connection on smooth manifolds. For holomorphic vector bundles on nonsingular complex varieties, one can define Chern classes in Deligne cohomology. Both are "refinements," or "liftings," of the usual Chern classes in deRham cohomology. One sees that it makes sense to compare the two for bundles with F^1-connections on varieties. Our main result (5.2.1) asserts that in this case, the Cheeger-Simons invariant maps to the Chern class in Deligne cohomology. The sense in which the comparison is made becomes clear by the end of Section 3.

In Section 4, we present some discussion of the problem of defining Chern classes in Deligne cohomology in the nonalgebraic case. Otherwise, the specific contents of this article is adequately summarized by the following:

Contents

Manuscript received 8 November 1988.
Supported in part by The National Science Foundation through Grant DMS-8800355.

6. Proof of the main theorem (with extra hypotheses) via universal connections.
7. Proof of the main theorem by simplicial methods.
References.

I want to thank R. Hain, M. Kashiwara, and Y. Shimizu for helpful discussions during the preparation of this article, and H. Esnault for several conversations in Bonn.

1. Chern classes. We recall some basic facts about Chern classes of complex vector bundles in general, beginning with those in ordinary cohomology.

(1.1). The original construction by Chern [Ch] was for bundles E on a smooth manifold M. For any C^∞ connection ∇ on E, one can use its curvature Θ, an $\mathrm{End}(E)$-valued 2-form, to produce closed $2p$-forms (Chern forms)

$$c_p(E, \nabla) = (-1)^p \operatorname{tr}(\Lambda^p \Theta) \tag{1.1.1}$$

for all positive integers p. Moreover, there is a formula that shows that for any two connections, the difference between the corresponding Chern forms is (canonically) exact. Thus, (1.1.1) defines, via the deRham theorem, a class

$$c_p(E) \in H^{2p}(M, \mathbf{C}) \tag{1.1.2}$$

(1.2). It is not hard to see that $c_p(E)$ lies in the image of

$$H^{2p}(M, \mathbf{Z}(p)) \to H^{2p}(M, \mathbf{C}), \tag{1.2.1}$$

where $\mathbf{Z}(p)$ denotes $(2\pi i)^p\mathbf{Z}$ (note the normalization in (1.1.1)); the kernel of (1.2.1) is precisely the torsion subgroup. To spare ourselves notational complications, *we will permit ourselves to identify* $\mathbf{Z}(p) \xrightarrow{\sim} \mathbf{Z}$ (with $(2\pi i)^p$ corresponding to 1), *which is compatible with products*.

The lifting of $c_p(E)$ to $H^{2p}(M, \mathbf{Z})$ was originally described in terms of obstruction classes [Ch: Chapter III] (see also [St: Section 41.4]). It can also be accomplished via a form of the "splitting principle":

a) The definition of $c_1(E) \in H^2(M, \mathbf{Z})$, when E is a line bundle, is easily made (it, in fact, represents the isomorphism class of E), and one sets, of course, $c_p(E) = 0$ for $p > 1$.

b) For E of rank r, its pullback to the associated projective space bundle $\mathbf{P} = \mathbf{P}(E)$ splits off a tautological line bundle, whose Chern class

$$\xi \in H^2(\mathbf{P}, \mathbf{Z})$$

has been defined by (a).

c) One then has

$$H^\bullet(\mathbf{P}, \mathbf{Z}) \simeq H^\bullet(M, \mathbf{Z})[\xi]/(f(\xi)); \tag{1.2.2}$$

here, f is a monic polynomial of degree r:

$$f(\xi) = \xi^r + \sum_{1 \leq p \leq r} (-1)^p c_p \xi^{r-p}, \tag{1.2.3}$$

with $c_p \in H^{2p}(M, \mathbf{Z})$; then, in fact, c_p is the p-th Chern class of E.

(1.2.4) *Remark.* Note that the above construction is quite general, and in particular does not require that M be a manifold.

(1.3). The procedure in (1.2) allows for the construction of Chern classes with values in other rings. Given a suitable category of spaces, and a contravariant functor from it into graded rings with unit,

$$M \mapsto R^\bullet(M),$$

and a functorial definition of

$$c_1(E) \in R^2(M)$$

whenever E is a line bundle over M, one can define the Chern classes of a bundle E of any rank,

$$c_p(E) \in R^{2p}(M),$$

to be the coefficients in (1.2.3), provided the analogue of (1.2.2) holds, i.e., if

$$\{\xi^k : 0 \le k \le r-1\}$$

form a system of generators of $R^\bullet(\mathbf{P}(E))$ as a *free* $R^\bullet(M)$-module. Examples of this include the case where $R^\bullet$ is the Chow ring of a smooth projective algebraic variety [Gr], and Deligne-Beilinson cohomology [Be] (see also [Bl], [EZ], and (3.3) below).

2. The Cheeger-Simons invariant.

(2.1). A ($\mathbf{C}$-valued) C^∞ i-form on M, $\phi \in A^i(M)$, defines a smooth singular cochain $T(\phi) \in S^i(M, \mathbf{C})$ by the following natural and inevitable formula: if

$$\sigma : \Delta^i \to M$$

is a C^∞ mapping (smooth to the boundary),

$$T(\phi)\sigma = \int_{\Delta^i} \sigma^*\phi. \tag{2.1.1}$$

This gives a morphism of complexes

$$T : A^\bullet(M) \to S^\bullet(M, \mathbf{C}), \tag{2.1.2}$$

though not one of differential graded algebras (i.e., $T \circ d = \delta \circ T$, but T is not compatible with the usual products).

(2.2). In the above, take ϕ to be a Chern form $c_p(E, \nabla)$ for some bundle with connection on M (recall (1.1)). The reduction of (2.1.1) modulo $\mathbf{Z}$:

$$\overline{T}(\phi) \in S^{2p}(M, \mathbf{C}/\mathbf{Z}) \tag{2.2.1}$$

is necessarily cohomologous to zero. Therefore, one can write

$$\overline{T}(\phi) = \delta y \qquad (y \in S^{2p-1}(M, \mathbf{C}/\mathbf{Z})), \tag{2.2.2}$$

with y determined only up to cocycles. One wants to make a functorial choice of y, at least up to coboundaries. In the case when $\phi = 0$, where y is itself a cocycle, this will select a unique element of $H^{2p-1}(M, \mathbf{C}/\mathbf{Z})$; as such, one is constructing a sort of secondary characteristic class for (E, ∇).

(2.3). The construction of [CS] can be reformulated as

(2.3.1) THEOREM. *Let (E, ∇) be a vector bundle with connection on M. There exists a unique choice of*

$$\bar{y} \in S^{2p-1}(M, \mathbf{C}/\mathbf{Z})/\delta S^{2p-2}(M, \mathbf{C}/\mathbf{Z}),$$

such that $\delta\bar{y} = \overline{T}(c_p(E, \nabla))$, and which is compatible with pullbacks.

(2.3.2) *Remark.* Note that there is only one possibility for $\bar{y}$ whenever $H^{2p-1}(M, \mathbf{C}/\mathbf{Z}) = 0$, i.e., if $H^{2p-1}(M, \mathbf{C}) = 0$ and $H^{2p}(M, \mathbf{Z})$ is torsion-free.

We give a quick sketch of the proof of (2.3.1). Uniqueness follows from the existence of universal connections on the universal bundle over a Grassmannian [NR]: (E, ∇) can be realized as a pullback of the universal connection. Since the Grassmannians satisfy the condition of (2.3.2), uniqueness follows from functoriality.

While use of classifying spaces can also be made to define $\bar{y}$, it is more useful to have a "direct" construction. One way of doing this is to pull (E, ∇) back to the Stiefel bundle $V^p(E)$ of partial frames ($p - 1$ vectors short of a full frame)—an idea akin to the realization of Chern classes as obstruction classes—where the Chern forms become exact, and almost canonically so. One obtains $\bar{y}$ therefrom by elementary homological nonsense.

(2.3.3) *Definition.* The pair $(c_p(E, \nabla), -\bar{y})$ is the *Cheeger-Simons invariant* of (E, ∇).

3. Cones. It is necessary, at this point, to bring in the notion of the mapping cone of a morphism of complexes, for it is the common algebraic thread in the definitions of the characteristic classes we wish to compare.

(3.1). Let $f : K^\bullet \to L^\bullet$ be a morphism of (cochain) complexes. The *cone* $C^\bullet = C^\bullet(f)$ is, by definition,[1] the complex with terms

[1] There is another convention, under which the right-hand side of (3.1.1) gives C^{i-1}.

(3.1.1) $$C^i = K^i \oplus L^{i-1},$$

and differential

(3.1.2) $$D(k, \ell) = (-dk, d\ell + f(k)).$$

One has a short exact sequence of complexes:

(3.1.3) $$0 \to L^\bullet[-1] \to C^\bullet \to \tilde{K}^\bullet \to 0,$$

where $\tilde{K}^\bullet$ is essentially $K^\bullet$, and therefore the long exact sequence of cohomology:

(3.1.4) $$\cdots \to H^{2p-1}(L^\bullet) \to H^{2p}(C^\bullet) \to H^{2p}(K^\bullet) \xrightarrow{f_*} H^{2p}(L^\bullet) \to \cdots.$$

(3.1.5) *Remark.* Note that $(k, -\ell)$ is a cocycle in the cone if and only if $dk = 0$ and $f(k) = d\ell$.

(3.2). The following description of the Cheeger-Simons invariant is essentially in [Es: Section 4]. From (2.1), there is a morphism of complexes:

(3.2.1) $$A^{\geq k}(M) \to S^\bullet(M, \mathbf{C}/\mathbf{Z})$$

(on the left, the notation indicates truncation from below), and the resulting cone will be denoted $C^\bullet(M, k)$. Then

(3.2.2) $$\hat{H}^\bullet(M, \mathbf{Z}) \simeq \bigoplus_k H^k(C^\bullet(M, k)),$$

with a suitable multiplication, is the *ring of differential characters* from [CS]. It is here that the Cheeger-Simons invariant lives, and where its "secondary" nature disappears, as a Chern class with values in the ring of differential characters.

(3.3). Let X be a compact Kähler (complex) manifold, D the union of a finite number of smooth complex hypersurfaces with transversal intersections (normal crossings) and put $M = X - D$. Denote by $D^\bullet_M(p)$ the mapping cone of

(3.3.1) $$F^pA^\bullet(X, \log D) \to S^\bullet(M, \mathbf{C}/\mathbf{Z}(p))$$

(for the notion of forms with logarithmic poles, appearing above, see [De2:II, Section 3]), and put

(3.3.2) $$H^\bullet_{\mathfrak{D}}(M, \mathbf{Z}(p)) = H^\bullet(D^\bullet_M(p)).$$

This is the *Deligne cohomology* of M with $\mathbf{Z}(p)$-coefficients (one checks that it is actually independent of the choice of compactification X of M). One sees in (3.3.1) that it can be written as the hypercohomology of the cone $\mathfrak{D}^\bullet_M(p)$ of the morphism of complexes of sheaves

(3.3.3) $$F^p\Omega^\bullet_X(\log D) \to Rj_*(\mathbf{C}/\mathbf{Z}(p)),$$

where j denotes the inclusion of M in X. Underlying this is the central quasi-isomorphism

(3.3.4) $$\Omega^\bullet_X(\log D) \sim Rj_*\mathbf{C}.$$

For instance, one has for $\mathfrak{D}^\bullet_M(1)$ when M is compact (i.e., $D = \phi$)

(3.3.5) $$[F^1\Omega^\bullet_M \to \mathbf{C}/\mathbf{Z}(1)] \sim [\mathbf{Z}(1) \to \mathcal{O}_M] \sim \mathcal{O}^*_M[-1],$$

so then

(3.3.6) $$H^2_{\mathfrak{D}}(M, \mathbf{Z}(1)) \simeq H^1(M, \mathcal{O}^*_M).$$

From pairings

$$D^\bullet_M(p) \otimes D^\bullet_M(q) \to D^\bullet_M(p+q),$$

there are pairings defined in (3.3.2), under which

(3.3.7) $$\bigoplus_p H^{2p}_{\mathfrak{D}}(M, \mathbf{Z}(p))$$

becomes a ring.

A useful variant of (3.3.2) is obtained by replacing (3.3.1) by something we shall denote $\overline{D}^\bullet_M(p)$:

(3.3.8) $$(\mathrm{Dec}\, W)_{2p} F^p A^\bullet(X, \log D) \to S^\bullet(M, \mathbf{C}/\mathbf{Z}(p)).$$

Here, the left-hand side is the largest subcomplex of $F^p A^\bullet(X, \log D)$ consisting of forms whose weight and degree add up to at most $2p$ (see [De2: II(1.3.3)] or [Zu: Section 3(12)]). As this is the construction used in [Be: Section 5], we will write here

(3.3.9) $$H^\bullet_{\mathfrak{D}\mathfrak{B}}(M, \mathbf{Z}(p)) = H^\bullet(\overline{D}^\bullet(p)),$$

the *Deligne-Beilinson cohomology* of M with coefficients in $\mathbf{Z}(p)$.

(3.4). We display in a single diagram the consequences of (3.1.4) in the instances from (3.2) and (3.3), and the relations among them:

(3.4.1)

$$\begin{array}{ccccccccc}
0 \longrightarrow & H^{2p-1}(M, \mathbf{C}/\mathbf{Z}(p)) & \longrightarrow & \hat{H}^{2p}(M, \mathbf{Z}(p)) & \longrightarrow & \{\phi \in A^{2p}(M) : d\phi = 0,\ [\phi] \in I^{2p}(M, \mathbf{Z}(p))\} & \longrightarrow 0 \\
 & \updownarrow & & & & & \\
0 \longrightarrow & W\tilde{J}^p(M) & \longrightarrow & H^{2p}_{\mathfrak{D}\mathfrak{B}}(M, \mathbf{Z}(p)) & \longrightarrow & W_{2p} F^p I^{2p}(M, \mathbf{Z}(p)) & \longrightarrow 0 \\
 & \updownarrow & & \downarrow & & \cap\downarrow & \\
0 \longrightarrow & \tilde{J}^p(M) & \longrightarrow & H^{2p}_{\mathfrak{D}}(M, \mathbf{Z}(p)) & \longrightarrow & F^p I^{2p}(M, \mathbf{Z}(p)) & \longrightarrow 0;
\end{array}$$

in the above,

(3.4.2) $$I^k(M, \mathbf{Z}(p)) = \mathrm{im}\{H^k(M, \mathbf{Z}(p)) \to H^k(M, \mathbf{C})\},$$

and $W\tilde{J}^p(M)$ and $\tilde{J}^p(M)$ are extensions of $H^{2p}(M, \mathbf{Z}(p))_{\mathrm{tors}}$ by

(3.4.3)
$$WJ^p(M) = I^{2p-1}(M, \mathbf{Z}(p))\backslash H^{2p-1}(M, \mathbf{C})/W_{2p} F^p H^{2p-1}(M, \mathbf{C}),$$
$$J^p(M) = I^{2p-1}(M, \mathbf{Z}(p))\backslash H^{2p-1}(M, \mathbf{C})/F^p H^{2p-1}(M, \mathbf{C})$$

respectively.

(3.4.4) *Remarks.* i) When M is compact, $WJ^p(M) = J^p(M)$ is the p-th intermediate Jacobian of M. Moreover,

$$F^p I^{2p}(M, \mathbf{Z}(p)) = W_{2p} F^p I^{2p}(M, \mathbf{Z}(p))$$

is the space of Hodge classes (type (p, p)) in $I^{2p}(M, \mathbf{Z}(p))$.

ii) In general, $W_{2p}F^pI^{2p}(M, \mathbf{Z}(p))$ is the image of the space of Hodge classes on X.

iii) (Cf. (2.3.2)). When $H^{2p-1}(M, \mathbf{C}/\mathbf{Z}(p)) = 0$, one has

$$\hat{H}^{2p}(M, \mathbf{Z}(p)) = \{\text{closed } (2p)\text{-forms with } \mathbf{Z}(p)\text{-valued periods}\},$$

$$H^{2p}_{\mathfrak{D}\mathfrak{B}}(M, \mathbf{Z}(p)) = W_{2p}F^pI^{2p}(M, \mathbf{Z}(p)),$$

$$H^{2p}_{\mathfrak{D}}(M, \mathbf{Z}(p)) = F^pI^{2p}(M, \mathbf{Z}(p));$$

in other words, there is "nothing new" in differential characters and Deligne cohomology for such M.

(3.5). We can introduce the amalgamation of the constructions in (3.2) and (3.3), namely the cone of the mapping

$$F^pA^{\geq 2p}(X, \log D) \to S^\bullet(M, \mathbf{C}/\mathbf{Z}(p)). \tag{3.5.1}$$

The resulting cohomology groups will be denoted

$$H^\bullet_{\mathfrak{G}}(M, \mathbf{Z}(p)), \tag{3.5.2}$$

for it is the complex form of the "G-cohomology" of [Es: (4.2)]. From the definition, there are mappings

$$\begin{array}{ccc} H^\bullet_{\mathfrak{G}}(M, \mathbf{Z}(p)) & \longrightarrow & \hat{H}^\bullet(M, \mathbf{Z}(p)) \\ \downarrow & & \\ H^\bullet_{\mathfrak{D}}(M, \mathbf{Z}(p)); & & \end{array} \tag{3.5.3}$$

in fact, it gives rise to an intermediary row in (3.4.1):

$$0 \to H^{2p-1}(M, \mathbf{C}/\mathbf{Z}(p)) \to H^{2p}_{\mathfrak{G}}(M, \mathbf{Z}(p)) \tag{3.5.4}$$

$$\to \{\phi \in F^pA^{2p}(X, \log D) : d\phi = 0, [\phi] \in I^{2p}(M, \mathbf{Z}(p))\} \to 0.$$

Thus, the mapping

$$H^{2p}_{\mathfrak{G}}(M, \mathbf{Z}(p)) \hookrightarrow \hat{H}^{2p}(M, \mathbf{Z}(p)) \tag{3.5.5}$$

is injective, as indicated.

4. Chern classes in Deligne cohomology. From (3.4.1), one sees that Deligne (also Deligne-Beilinson) cohomology satisfies the formula (1.2.2) that underlies the "splitting principle." Because of this, it is sometimes possible to define Chern classes in the ring (3.3.7) by that method, for instance when M is compact, or for algebraic vector bundles on algebraic manifolds (see [Es: Section 8]). Though these cases are sufficient for most purposes, I wish to present here some discussion of the noncompact analytic case. We seek to determine a natural setting for defining the Chern class.

(4.1). Let $M = X - D$, as before, though we are free to modify X by blowing up submanifolds of D. The fundamental notion for the purposes at hand is that of a holomorphic vector bundle on M *with meromorphic structure along* D. These are given by elements of

(4.1.1) $$H^1(X, \mathcal{G}\ell_r(\mathcal{O}_X)(*D)),$$

where $\mathcal{G}\ell_r(\mathcal{O}_X)(*D)$ is the sheaf of invertible $r \times r$ matrix valued functions on M with at worst poles or zeroes along D. When $r = 1$, one writes

$$\mathcal{O}_X(*D)^* \quad \text{for} \quad \mathcal{G}\ell_1(\mathcal{O}_X)(*D);$$

there are mappings

(4.1.2) $$H^1(X, \mathcal{O}_X^*) \xrightarrow{\beta} H^1(X, \mathcal{O}_X(*D)^*) \xrightarrow{\alpha} H^1(M, \mathcal{O}_M^*).$$

From the exact sequence of sheaves on X

(4.1.3) $$1 \to \mathcal{O}_X^* \to \mathcal{O}_X(*D)^* \xrightarrow{\text{ord}} a_*\mathbf{Z}_{\tilde{D}} \to 0,$$

where $\tilde{D}$ is the normalization of D (i.e., the disjoint union of its irreducible components), and $a : \tilde{D} \to X$ is the obvious finite mapping, one can describe the nature of the mapping β. The latter fits into a diagram

(4.1.4)
$$\begin{array}{ccccccccc} & & & & & & & & H^2(X, \mathcal{O}_X) \\ & & & & & & & & \downarrow \\ H^0(\tilde{D}, \mathbf{Z}) & \longrightarrow & H^1(X, \mathcal{O}_X^*) & \xrightarrow{\beta} & H^1(X, \mathcal{O}_X(*D)^*) & \longrightarrow & H^1(\tilde{D}, \mathbf{Z}) & \xrightarrow{\delta} & H^2(X, \mathcal{O}_X^*) \\ & & & & & & & \searrow^{\gamma} & \downarrow \\ & & & & & & & & H^3(X, \mathbf{Z}(1)). \end{array}$$

From this, one sees easily:

(4.1.5) PROPOSITION.

i) $\ker \beta$ *is generated by the line bundles associated to the components of D*;

ii) *A necessary and sufficient condition that every line bundle on M with meromorphic structure along D extend across D is that* δ *be injective*;

iii) *If* $H^2(X, \mathcal{O}_X) = 0$, *and the Gysin mapping* γ *has a nontrivial kernel, then* β *is not surjective.*

(4.1.6) *Remark.* For the *Zariski* topology in the algebraic case, β is always a surjection.

(4.2). The following is probably well-known:

(4.2.1) PROPOSITION. *The cohomology sheaves of* $\mathfrak{D}^\bullet_M(1)$ *are*:

$$\mathcal{H}^i(\mathfrak{D}^\bullet_M(1)) \simeq \begin{cases} 0 & \text{if } i = 0, \\ \mathcal{O}_X(*D)^* & \text{if } i = 1, \\ R^i j_* \mathbf{Z}_M(1) & \text{if } i \geq 2. \end{cases}$$

Proof. The interesting case is $i = 1$, which is all we shall consider here. The exact sequence of the cone gives

$$0 \to \mathcal{H}^1(\mathfrak{D}^\bullet_M(1)) \to \Omega^1_X(\log D) \cap \ker d$$
$$\to \mathcal{H}^1(j_*(\Omega^\bullet_M/\mathbf{Z}_M(1)) \simeq R^1 j_*(\mathbf{C}/\mathbf{Z}(1)).$$

Our assertion then follows from the fact that

$$\begin{array}{ccccc} \mathcal{O}_X(*D)^* & \hookrightarrow & j_*(\mathcal{O}^*_M) & \xleftarrow{\sim} & j_*(\mathcal{O}_M/\mathbf{Z}_M(1)) \\ & \searrow & & \searrow^{d\log} & \downarrow d \\ & & \Omega^1_X(\log D) & \hookrightarrow & j_*\Omega^1_M \end{array}$$

is Cartesian.

(4.2.2) COROLLARY. *There is a natural injection*

$$H^1(X, \mathcal{O}_X(*D)^*) \to H^2_{\mathfrak{D}}(M, \mathbf{Z}(1)).$$

Thus, we again see that the definition of c_1 is tautological. However, the attempt to use the splitting argument leads to the annoying question of whether one can compactify the projective bundles that arise. To circumvent this difficulty, we restrict ourselves from now on to bundles on M that admit coherent extensions to X, hence admit bundle extensions on some modification of X. For this, we need:

(4.2.3) PROPOSITION. *Let E and E' be holomorphic vector bundles on X, with total Chern classes $c(E)$, $c(E')$ in Deligne cohomology* ((3.3.7)). *Assume that E and E' have meromorphically isomorphic restrictions to M. Then*

$$c_p(E) - c_p(E') \in \ker\{H^{2p}_{\mathfrak{D}}(X, \mathbf{Z}(p)) \xrightarrow{\rho} H^{2p}_{\mathfrak{D}}(M, \mathbf{Z}(p))\}.$$

Proof. The above is clear for line bundles, by (4.1.5,i). That this, together with the properties of the Chern class on X, implies the same for bundles of higher rank is a consequence of the following argument (shown to me by Kashiwara). Let $\mathbf{F}(E)$ denote the (projective) flag bundle of E, with projection π_E onto X. Then $\pi_E^*(E)$ has a tautological filtration, whose successive quotients are line bundles. The analogous statement holds for E'. Now, the meromorphic isomorphism of E and E' induces a bimeromorphic mapping between $\mathbf{F}(E)$ and $\mathbf{F}(E')$. The closure of its graph is then an analytic variety, so let Z be a desingularization of it, and $\pi : Z \to X$ the obvious mapping. By construction, π^*E and π^*E' possess meromorphically isomorphic filtrations on M, with one-dimensional successive quotients. Since the total Chern class is multiplicative for exact sequences, it follows that for any p,

$$\pi^*c_p(E) - \pi^*c_p(E') \in \ker\{H^{2p}_{\mathfrak{D}}(Z, \mathbf{Z}(p)) \to H^{2p}_{\mathfrak{D}}(Z - \pi^{-1}(D), \mathbf{Z}(p))\}.$$

From this, one gets the desired assertion from the injectivity of

$$\pi^* : H^{2p}_{\mathfrak{D}}(M, \mathbf{Z}(p)) \to H^{2p}_{\mathfrak{D}}(Z - \pi^{-1}(D), \mathbf{Z}(p)).$$

(4.2.4) *Definition.* Let E be a vector bundle on M with meromorphic structure along D, and assume that E is given as the restriction of a bundle $\overline{E}$ (determined up to meromorphic equivalence) on X. Then

$$c_p(E) \in H^{2p}_{\mathfrak{D}}(M, \mathbf{Z}(p))$$

is defined to be $\rho(c_p(\overline{E}))$.

(4.2.5) *Remark.* The same argument applies to $H^{2p}_{\mathfrak{D}\mathfrak{B}}(M, \mathbf{Z}(p))$, for $\mathcal{O}_X(*D)^*$ gives the first cohomology sheaf there as well (cf. (4.2.1)).

5. The main theorem: comparison of the Chern classes.

(5.1). We assume again that $M = X - D$, with X compact Kähler and D a union of smooth hypersurfaces with normal crossings. From (3.4.1), it makes sense to compare the Cheeger-Simons Chern class in the ring of differential characters with the one in Deligne-Beilinson cohomology for a holomorphic vector bundle E on M, extendable to $\overline{E}$ on X, equipped with a connection whose Chern forms are zero. This condition is fulfilled if $\overline{E}$ has a meromorphic connection with only logarithmic poles along D, for then its curvature form

$$\Theta \in H^0(X, \Omega^2_X(\log D) \otimes \mathrm{End}(\overline{E})) \tag{5.1.1}$$

gives

$$c_p(E) \in F^{2p}H^{2p}(M, \mathbf{C}), \tag{5.1.2}$$

which is otherwise incompatible with fact that $c_p(E)$ is of Hodge type (p, p) in $W_{2p}H^{2p}(M, \mathbf{C})$. In particular, flat vector bundles on M fall into this class, for they have canonical extensions in the sense of [Del: p. 95].

With the aid of (3.5), we can make the above comparison for any bundle with F^1-connection, i.e. a logarithmic connection compatible with the complex structure, whose curvature Θ then lies in $F^1A^2(X, \log D)$.

(5.2). We now can state the principal result:

(5.2.1) THEOREM. *In the setting of* (5.1), *the Cheeger-Simons Chern class of* (E, ∇) *in* $H^{2p}_{\mathfrak{g}}(M, \mathbf{Z}(p))$ *maps to the Chern class of* E *in Deligne*[*-Beilinson*] *cohomology.*

(5.2.2) COROLLARY. *If the Chern forms of* (E, ∇) *vanish, the Cheeger-Simons Chern class of* E *in* $H^{2p-1}(M, \mathbf{C}/\mathbf{Z}(p))$, *when projected in* $[W]\tilde{J}^p(M)$, *gives the Chern class of* E *in Deligne*[*-Beilinson*] *cohomology.*

One motivation for proving this theorem is the following conjecture of Bloch:

(5.2.3) CONJECTURE [Bl: p. 104]. *Let X be a projective algebraic manifold, E a flat vector bundle on X. Then the Chern classes in Deligne cohomology*

$$c_p(E) \in \tilde{J}^p(X)$$

are torsion elements if $p > 1$.

(The restriction $p > 1$ is needed, since for a flat line bundle $c_1(E)$ can be *anything* in $\tilde{J}^1(X)$. On the other hand, the product of two elements of $\tilde{J}^1(X)$ is a torsion element.) The hope had been to throw this conjecture back to investigating the same for the more topological Cheeger-Simons invariant. That is, however, a nontrivial question, one that has been considered before (cf. [CS: Section 8]).

(5.3). The underlying theme behind the proof of (5.2.1) is that all bundles, even with connection, can be realized as the pullback of a universal bundle on a Grassmannian. There, the condition of (3.4.4, iii) is satisfied, so the two classes correspond for trivial reasons. One then uses functoriality to finish the argument; here one must be a little careful, since there is no mapping from differential characters to Deligne cohomology in general (recall (3.4.1)); there is, however, the intermediary from (3.5).

6. Proof of the main theorem (with extra hypotheses) via universal connections.

(6.1). Instead of using the construction from [NR], we consider the problem of classifying bundles with connection as a lifting problem. Let E be a vector bundle on M, and assume that it is classified by a mapping

$$f : M \to B, \tag{6.1.1}$$

i.e., $E \simeq f^*U$, where U is the universal bundle on a suitable Grassmannian B. We may assume without loss that f is an immersion.

(6.1.2) PROPOSITION.

i) *The set of all connections on the universal bundle U is in natural one-to-one correspondence with the space of cross-sections of an affine space bundle* $\psi : \tilde{B} \to B$,

ii) $\tilde{U} = \psi^*U$ *has a tautological connection* $\check{\nabla}$.

iii) *Given any immersion* $f : M \to B$, *and connection* ∇ *on* $E = f^*U$, *there is a lifting* $\tilde{f} : M \to \tilde{B}$ $(\psi \circ \tilde{f} = f)$ *such that* $\tilde{f}^*\check{\nabla} = \nabla$.

In fact, if P denotes the frame bundle of U, with structure group $G = \mathrm{GL}(r, \mathbf{C})$,

(6.1.3)

$$\tilde{B} = \{\phi \in \mathrm{Hom}(TB, TP/G) : \phi \text{ projects to the identity mapping of } TB\}.$$

(6.2). Let E now be a holomorphic vector bundle on $M = X - D$ (as in (3.3)), having extension $\bar{E}$ to X, and F^1-connection on M. We first make the following:

(6.2.1) *Assumption.* $\bar{E}$ is generated by its global holomorphic sections.

Under (6.2.1), E has a classifying mapping (6.1.1) f that is holomorphic, and which extends to X. We further assume:

(6.2.2) *Assumption.* The mapping f is an immersion. Let f be a lifting of f, as in (6.1.2, iii), that classifies the connection.

Because

$$H^{2p-1}(\tilde{B}, \mathbf{C}/\mathbf{Z}(p))' \simeq H^{2p-1}(B, \mathbf{C}/\mathbf{Z}(p)) = 0, \tag{6.2.3}$$

the Chern form $c_p(\tilde{U}, \check{\nabla})$ has a unique lifting to $\hat{H}^{2p}(\tilde{B}, \mathbf{C}/\mathbf{Z}(p))$, the universal Cheeger-Simons class, represented by

$$\tilde{c}_p = (c_p(\tilde{U}, \check{\nabla}), -\tilde{y}) \tag{6.2.4}$$

in the cone. Then

$$\hat{c}_p = \tilde{f}^*\tilde{c}_p = (c_p(E, \nabla), -\tilde{f}^*\tilde{y})$$

represents the Cheeger-Simons invariant of E, which lies in the subgroup $H^{2p}_{\mathcal{G}}(M, \mathbf{Z}(p))$ (see (3.5)). We wish to show that $\hat{c}_p$ induces the Chern class of E in Deligne cohomology. The difficulty is, of course, that $\tilde{c}_p$ is not itself in $H^{2p}_{\mathcal{G}}(\tilde{B}, \mathbf{Z}(p))$.

Let ∇_0 be a metric connection on U. We have $\nabla_0 = \tau^*\check{\nabla}$ for some cross-section τ of ψ (by (6.1.2)). The Cheeger-Simons class of (U, ∇_0) does lie in $H^{2p}_{\mathcal{G}}(B, \mathbf{Z}(p))$, —it is represented by $(c_p(U, \nabla_0), -\tau^*\tilde{y}))$—and for trivial reasons again, induces its Chern class in $H^{2p}_{\mathfrak{D}}(B, \mathbf{Z}(p))$. Therefore, the same holds for $(E, f^*\nabla_0)$ on M. Put $\tilde{g} = \tau \circ f$; then $\tilde{g}^*\tilde{c}_p$ represents the Chern class in Deligne cohomology.

(6.3). We must compare $\tilde{g}^*\tilde{c}_p$ and $\hat{c}_p$. Since $\tilde{g}$ and $\tilde{f}$ both lift f, there is the linear homotopy $\tilde{h}$ from $\tilde{f}$ to $\tilde{g}$, which gives the linear interpolation between the connections $\tilde{f}^*\check{\nabla}$ and $\tilde{g}^*\check{\nabla}$. Let H be the corresponding homotopy operator from $S^\bullet(\tilde{B}, \mathbf{C})$ to $S^\bullet(M, \mathbf{C})$:

$$(H(s))\sigma = s(\tilde{h}_*(\sigma \times I)) \tag{6.3.1}$$

for $s \in S^\bullet(\tilde{B}, \mathbf{C})$. It satisfies

$$\tilde{g}^*s - \tilde{f}^*s = \delta H(s) + H(\delta s), \tag{6.3.2}$$

and preserves both $A^\bullet(\ )$ (recall (2.1.2)) and $S^\bullet(\ , \mathbf{Z}(p))$.

In the cone $D^\bullet_M(p)$, we have (we put $c_p = c_p(\tilde{U}, \check{\nabla})$):

$$\begin{aligned}
\tilde{g}^*\tilde{c}_p - \hat{c}_p &= (\tilde{g}^*c_p - \tilde{f}^*c_p, -\tilde{g}^*\tilde{y} + \tilde{f}^*\tilde{y}) \\
&= (dHc_p, -\delta H\tilde{y} - H\delta\tilde{y}) \\
&= (dHc_p, -\delta H\tilde{y} - Hc_p) \\
&= D(-Hc_p, -H\tilde{y}),
\end{aligned}$$

for $Hc_p \in F^pA^{2p-1}(X, \log D)$, as one readily checks. This shows that $\hat{c}_p$ gives the Chern class in Deligne cohomology, as desired, when assumptions (6.2.1) and (6.2.2) are satisfied.

(6.4). So, how strong are the assumptions from (6.2)? When M is an algebraic variety, and E is algebraic, one can arrange that the conditions are satisfied by $E \otimes L$, where L is a sufficiently ample line bundle. Since they are also satisfied by L, it is possible to adapt the preceding argument to the general algebraic vector bundle. One gets stuck here in the nonalgebraic case.

7. Proof of the main theorem by simplicial methods.

(7.1). Let Δ^n denote the standard n-simplex, with vertices labeled by

$$[\mathbf{n}] = \{0, 1, \ldots, n\}. \tag{7.1.1}$$

The *face mapping* $f_j : \Delta^{n-1} \to \Delta^n$ $(0 \leq j \leq n)$ is induced by the strictly increasing function from $[\mathbf{n} - \mathbf{1}]$ to $[\mathbf{n}]$ that omits the value j. (For some purposes, one also considers the *degeneracies* $d_j : \Delta^n \to \Delta^{n-1}$ $(0 \leq j < n)$ induced by the otherwise strictly increasing mapping $[\mathbf{n}] \to [\mathbf{n} - \mathbf{1}]$ that repeats the value j.)

A *strict simplicial C^∞ manifold $M_\bullet$* is a sequence of C^∞ manifolds $\{M_n\}_{n=0}^\infty$, together with a system of mappings

$$M_n \xrightarrow{\hat{f}_j} M_{n-1}$$

(subject to obvious compatibilities). This gives the data for forming a topological space $|M_\bullet|$, its *geometric realization*, as a quotient of $\coprod_n (M_n \times \Delta^n)$ via the identifications

$$M_n \times \Delta^n \ni (x, f(t)) \sim (\hat{f}(x), t) \in M_{n-1} \times \Delta^{n-1} \tag{7.1.2}$$

for all face mappings f.

From the above description, there is a natural definition of smooth differential forms (likewise, smooth singular cochains) on $|M_\bullet|$:

(7.1.3) *Definition.* $A^\bullet(|M_\bullet|) \subset \Pi_{n\geq 0} A^\bullet(M_n \times \Delta^n)$ is the subcomplex consisting of sequences of forms $\{\phi_n\}$ for which the two pullbacks via

$$\begin{array}{ccc} & \xrightarrow{1\times f} & M_n \times \Delta^n \\ M_n \times \Delta^{n-1} & & \\ & \xrightarrow{\hat{f}\times 1} & M_{n-1} \times \Delta^{n-1} \end{array}$$

coincide. One defines $S^\bullet(|M_\bullet|, \mathbf{C})$ analogously.

As in (2.1), there is a morphism of complexes,

$$T : A^\bullet(|M_\bullet|) \to S^\bullet(|M_\bullet|, \mathbf{C}). \tag{7.1.4}$$

The cohomology of both sides gives $H^\bullet(|M_\bullet|, \mathbf{C})$.

If one wishes to "forget the glue" in (7.1.2), one can form the double complex

(7.1.5) $$\bigoplus_{n \geq 0} A^{\bullet}(M_n)[-n],$$

with differentials d and

$$\delta = \sum_j (-1)^j \hat{f}_j^*$$

Then, integration over the simplex factor defines a morphism from $A^{\bullet}(|M_{\bullet}|)$ to (7.1.5), which induces an isomorphism on cohomology. One uses the complex (7.1.5) for doing Hodge theory on simplicial varieties.

In the same spirit as (7.1.3), one defines a bundle with connection on a simplicial manifold $M_{\bullet}$ to be a system of such, (E_n, ∇^n) on $M_n \times \Delta^n$, together with isomorphisms

$$(1 \times f)^*(E_n, \nabla^n) \rightrightarrows (\hat{f} \times 1)^*(E_{n-1}, \nabla^{n-1}).$$

We can then define its Chern forms, Cheeger-Simons invariant, etc. in the inevitable manner.

(7.2). We present two fundamental examples of simplicial manifolds (actually, of simplicial anythings).

(7.2.1) *Example* 1. Let M be a manifold. Define $M_{\bullet}$ to be:

$$M_n = \begin{cases} M & \text{if } n = 0 \\ \phi & \text{if } n > 0. \end{cases}$$

Then $|M_{\bullet}| = M$, and likewise $A^{\bullet}(|M_{\bullet}|) = A^{\bullet}(M)$. [When one imposes degeneracies, one takes $M_n = M$ in all degrees, and all face and degeneracy mappings are the identity mapping.]

(7.2.2) *Example* 2. Given M, define $\Pi_{\bullet}M$ by

$$\Pi_n M = M^{[n]} \qquad ((n + 1)\text{-fold product}),$$

with canonically determined face mappings [and degeneracies], given by the various projections [and partial diagonal mappings respectively]. Then $|\Pi_\bullet M|$ is always a contractible space.

(7.3). Given the bundle with connection (E, ∇) on M, one makes an auxilliary construction from its frame bundle P (structure group $G = \mathrm{GL}(r, \mathbf{C})$) and the associated connection on it. The simplicial manifold $\Pi_\bullet P$ has a diagonal action of G; specifically, each $P^{[n]}$ does, providing a collection of principal bundles

$$P^{[n]} \to P^{[n]}/G, \tag{7.3.1}$$

which, via any projection π_j, is the pullback of

$$P = P^{[0]} \to P^{[0]}/G = M.$$

On $P^{[n]} \times \Delta^n$, one takes as connection

$$\nabla^n = \sum_{j=0}^{n} t_j \pi_j^* \nabla, \tag{7.3.2}$$

where $(t_0, \ldots, t_n)$ are the barycentric coordinates of Δ^n. This is clearly compatible with all face [and degeneracy] mappings. One can pass back to the vector bundle $E_n = P^{[n]} \times^G \mathbf{C}^r$, if one wishes. From (7.2.2), one gets (see [De2: III(6.1)]):

(7.3.3) PROPOSITION. *$|\Pi_\bullet P| \to |\Pi_\bullet P|/G$ is a model for the universal G-bundle*

$$E_G \to B_G.$$

(7.3.4) *Observation.* (P, ∇) coincides with the pullback of $(\Pi_\bullet P, \nabla^\bullet)$ via the embedding of the simplicial manifold $P_\bullet$ in $\Pi_\bullet P$ (modulo the action of G).

(7.4). Given (7.3.4), the proof of Theorem (5.2.1) is easy. Suppose that $M = X - D$, and E is a bundle on M with extension $\overline{E}$ to X. Since $\mathbf{P}(\mathfrak{gl}(\overline{E}) \oplus \mathcal{O}_X)$ is Kähler ([SS: (6.37)]), the principal bundle P of E is a manifold of the same sort as M, as are the spaces $P^{[n]}/G$. The definitions of all of our cones from Section 3 admit extensions to the simplicial setting, so we get (cf. (3.5.3)):

$$\begin{array}{ccc} H^{2p}_{\mathcal{G}}(\Pi_{\bullet}P/G, \mathbf{Z}(p)) & \to & \hat{H}^{2p}(\Pi_{\bullet}P/G, \mathbf{Z}(p)) \\ \downarrow & & \\ H^{2p}_{\mathcal{D}}(\Pi_{\bullet}P/G, \mathbf{Z}(p)). & & \end{array} \tag{7.4.1}$$

By (7.3.3) and (3.4.4, iii)—recall that B_G is (homotopically) a direct limit of Grassmannians—the p-th Chern form of the bundle $E_{\bullet}$ determines unique elements in each of the groups in (7.4.1), which are automatically compatible via the mappings. Pulling back to $P_{\bullet}/G \simeq M$, we get compatible elements, there, which gives the desired assertion.

THE JOHNS HOPKINS UNIVERSITY

REFERENCES

[Be] A. Beilinson, Notes on absolute Hodge cohomology. In: *Applications of Algebraic K-Theory to Algebraic Geometry and Number Theory*, Contemp. Math. **55**, Part I, 35–68, AMS, 1986.

[Bl] S. Bloch, Applications of the dilogarithm function in algebraic K-theory and algebraic geometry. In: *Int'l Symp. on Algebraic Geometry, Kyoto, 1977*, 103–114.

[CS] J. Cheeger and J. Simons, Differential characters and geometric invariants. In: *Geometry and Topology.* Lect. Notes in Math., 1167, 50–80, Springer, 1985.

[Ch] S.-S. Chern, Characteristic classes of Hermitian manifolds, *Ann. of Math.*, **47** (1946), 85–121.

[De1] P. Deligne, *Equations Différentielles à Points Singuliers Réguliers*, Lect. Notes in Math., **163**, Springer, 1970.

[De2] ———, Théorie de Hodge, II, III. Publ. Math. IHES **40** (1971), 5–57; **44** (1974), 5–77.

[EZ] F. El Zein and S. Zucker, Extendability of normal functions associated to algebraic cycles. In: *Topics in Transcendental Algebraic Geometry*, Ann. of Math. Studies **106** (1984), Princeton Univ. Press, 269–288.

[Es] H. Esnault, Characteristic classes of flat bundles, 1987.

[EV] ——— and E. Viehweg, Deligne-Beilinson cohomology. In: Beilinson's Conjectures on Special Values of L-Functions, Perspectives in Math., **4**, 43–91, Academic Press, 1988.

[Gr] A. Grothendieck, La théorie des classes de Chern. *Bull. Soc. Math. France*, **86** (1958), 137–154.

[NR] M. S. Narasimhan and S. Ramanan, Existence of universal connections, I, II, *Amer. J. Math.*, **83** (1961), 563–572; **85** (1963), 223–231.

[SS] B. Shiffman and A. Sommese, *Vanishing Theorems on Complex Manifolds*, Progress in Math., **56**, Birkhäuser, 1985.
[St] N. Steenrod, *The Topology of Fibre Bundles*, Princeton Univ. Press, 1951.
[Zu] S. Zucker, Degeneration of mixed Hodge structures, *Proc. Symp. Pure Math.*, **46**(2) (1987), AMS, 283-293.